Communications
in Computer and Information Science 2438

Series Editors

Gang Li , *School of Information Technology, Deakin University, Burwood, VIC, Australia*

Joaquim Filipe , *Polytechnic Institute of Setúbal, Setúbal, Portugal*

Zhiwei Xu, *Chinese Academy of Sciences, Beijing, China*

Rationale

The CCIS series is devoted to the publication of proceedings of computer science conferences. Its aim is to efficiently disseminate original research results in informatics in printed and electronic form. While the focus is on publication of peer-reviewed full papers presenting mature work, inclusion of reviewed short papers reporting on work in progress is welcome, too. Besides globally relevant meetings with internationally representative program committees guaranteeing a strict peer-reviewing and paper selection process, conferences run by societies or of high regional or national relevance are also considered for publication.

Topics

The topical scope of CCIS spans the entire spectrum of informatics ranging from foundational topics in the theory of computing to information and communications science and technology and a broad variety of interdisciplinary application fields.

Information for Volume Editors and Authors

Publication in CCIS is free of charge. No royalties are paid, however, we offer registered conference participants temporary free access to the online version of the conference proceedings on SpringerLink (http://link.springer.com) by means of an http referrer from the conference website and/or a number of complimentary printed copies, as specified in the official acceptance email of the event.

CCIS proceedings can be published in time for distribution at conferences or as post-proceedings, and delivered in the form of printed books and/or electronically as USBs and/or e-content licenses for accessing proceedings at SpringerLink. Furthermore, CCIS proceedings are included in the CCIS electronic book series hosted in the SpringerLink digital library at http://link.springer.com/bookseries/7899. Conferences publishing in CCIS are allowed to use Online Conference Service (OCS) for managing the whole proceedings lifecycle (from submission and reviewing to preparing for publication) free of charge.

Publication process

The language of publication is exclusively English. Authors publishing in CCIS have to sign the Springer CCIS copyright transfer form, however, they are free to use their material published in CCIS for substantially changed, more elaborate subsequent publications elsewhere. For the preparation of the camera-ready papers/files, authors have to strictly adhere to the Springer CCIS Authors' Instructions and are strongly encouraged to use the CCIS LaTeX style files or templates.

Abstracting/Indexing

CCIS is abstracted/indexed in DBLP, Google Scholar, EI-Compendex, Mathematical Reviews, SCImago, Scopus. CCIS volumes are also submitted for the inclusion in ISI Proceedings.

How to start

To start the evaluation of your proposal for inclusion in the CCIS series, please send an e-mail to ccis@springer.com

Quanying Liu · Youzhi Qu · Haiyan Wu · Yu Qi ·
An Zeng · Dan Pan
Editors

Human Brain and Artificial Intelligence

4th International Workshop, HBAI 2024
Jeju Island, South Korea, August 3, 2024
Proceedings

Editors
Quanying Liu
Southern University of Science
and Technology
Guangdong, China

Haiyan Wu
University of Macau
Taipa, China

An Zeng
Guangdong University of Technology
Guangzhou, Guangdong, China

Youzhi Qu
Qingdao University
Qingdao, Shandong, China

Yu Qi
Zhejiang University
Hangzhou, Zhejiang, China

Dan Pan
Guangdong Polytechnic Normal University
Guangzhou, Guangdong, China

ISSN 1865-0929   ISSN 1865-0937 (electronic)
Communications in Computer and Information Science
ISBN 978-981-96-4000-3   ISBN 978-981-96-4001-0 (eBook)
https://doi.org/10.1007/978-981-96-4001-0

© The Editor(s) (if applicable) and The Author(s), under exclusive license
to Springer Nature Singapore Pte Ltd. 2025

This work is subject to copyright. All rights are solely and exclusively licensed by the Publisher, whether the whole or part of the material is concerned, specifically the rights of translation, reprinting, reuse of illustrations, recitation, broadcasting, reproduction on microfilms or in any other physical way, and transmission or information storage and retrieval, electronic adaptation, computer software, or by similar or dissimilar methodology now known or hereafter developed.
The use of general descriptive names, registered names, trademarks, service marks, etc. in this publication does not imply, even in the absence of a specific statement, that such names are exempt from the relevant protective laws and regulations and therefore free for general use.
The publisher, the authors and the editors are safe to assume that the advice and information in this book are believed to be true and accurate at the date of publication. Neither the publisher nor the authors or the editors give a warranty, expressed or implied, with respect to the material contained herein or for any errors or omissions that may have been made. The publisher remains neutral with regard to jurisdictional claims in published maps and institutional affiliations.

This Springer imprint is published by the registered company Springer Nature Singapore Pte Ltd.
The registered company address is: 152 Beach Road, #21-01/04 Gateway East, Singapore 189721, Singapore

If disposing of this product, please recycle the paper.

Preface

The Fourth International Workshop on Human Brain and Artificial Intelligence (HBAI 2024), a joint workshop of the 33rd International Joint Conference on Artificial Intelligence, was held in Jeju Island, South Korea. As the trend of interdisciplinary convergence between brain science and artificial intelligence (AI) continues to accelerate, HBAI 2024 brought together researchers at the forefront of human brain research and AI, showcasing the latest discoveries in this interdisciplinary field. The workshop focused on exploring how AI techniques can contribute to human brain research and how brain science can inspire advancements in AI.

We received 74 submissions, most of which presented intriguing and valuable new insights related to human brain research and artificial intelligence. Through a double-blind peer review process, with three reviewers evaluating each paper, HBAI 2024 selected 29 high-quality papers. These papers were shared via oral presentations during the one-day workshop. To encourage deeper discussions of these fascinating insights, the workshop was divided into oral presentations, poster sessions, and panel discussions, covering key topics such as AI for Brain Science, AI for Brain Technology, and Brain-Inspired AI.

We extend our deepest gratitude to the authors of the accepted papers, as well as the members of the Organizing Committee and the Program Committee, whose contributions were instrumental to the success of the workshop. In addition to those who contributed to the content of HBAI 2024, we would like to thank Wenliang (Kevin) Li, who participated in the AI for Brain Science panel discussion, as well as our dedicated members of NCC Lab, Xinke Shen, Zhichao Liang, Chen Wei, Dongyang Li, and Menglong Zhang, who worked tirelessly to support this event and deserve special recognition.

Despite the inevitable challenges and uncertainties encountered during the organization of the conference, the participants' enthusiasm and engagement turned these challenges into achievements. We sincerely thank every participant for their support and contribution, as it was your involvement that made HBAI 2024 a truly inspiring and intellectually stimulating academic event.

October 2024

Quanying Liu
Youzhi Qu
Haiyan Wu
Yu Qi
An Zeng
Dan Pan

Organization

Organizing Committee

Quanying Liu	Southern University of Science and Technology, China
Haiyan Wu	University of Macau, China
An Zeng	Guangdong University of Technology, China
Dan Pan	Guangdong Polytechnic Normal University, China
Yu Qi	Zhejiang University, China
Youzhi Qu	Southern University of Science and Technology, China

Program Committee

Amy Wenxuan Ding	EMLYON Business School, France
Aniruddha Sinha	Indian Institute of Technology Bombay, India
Benjamin Becker	University of Hong Kong, China
Bin Hu	Beijing Institute of Technology, China
Bo Hong	Tsinghua University, China
Camillo Porcaro	University of Padua, Italy
Changsong Zhou	Hong Kong Baptist University, China
Dante Mantini	KU Leuven, Belgium
Daqing Guo	University of Electronic Science and Technology of China, China
Dezhong Yao	University of Electronic Science and Technology of China, China
Dong Ming	Tianjin University, China
Gang Pan	Zhejiang University, China
Huiguang He	Chinese Academy of Science, China
Jia Liu	Tsinghua University, China
Jian Liu	University of Birmingham, UK
Juyang Weng	Michigan State University, USA
Kai Du	Peking University, China
Nicole Wenderoth	ETH Zurich, Switzerland
Qunxi Dong	Beijing Institute of Technology, China
Si Wu	Peking University, China
Tianzi Jiang	Chinese Academy of Science, China

Wanzeng Kong	Hangzhou Dianzi University, China
Xiaolin Hu	Tsinghua University, China
Xiaowei Song	Simon Fraser University, Canada
Xiaowei Xu	Guangdong Provincial People's Hospital, China
Yalin Wang	Arizona State University, USA
Yiyu Shi	University of Notre Dame, USA
Yong He	Beijing Normal University, China
Yu Hu	Hong Kong University of Science and Technology, China
Yuanyuan Mi	Tsinghua University, China
Yueming Wang	Zhejiang University, China
Zhiguo Zhang	Harbin Institute of Technology, China
Zhuliang Yu	South China University of Technology, China

Contents

AI for Brain Science

Comparison of the Brain Visual Cortex and CNN Under Continuous Object Property Space 3
Qiande Zhao, Deying Li, Congying Chu, and Lingzhong Fan

CoCoG-2: Controllable Generation of Visual Stimuli for Understanding Human Concept Representation 18
Chen Wei, Jiachen Zou, Dietmar Heinke, and Quanying Liu

Uncovering Cognitive Taskonomy Through Transfer Learning in Masked Autoencoder-Based fMRI Reconstruction 35
Youzhi Qu, Junfeng Xia, Xinyao Jian, Wendu Li, Kaining Peng, Zhichao Liang, Haiyan Wu, and Quanying Liu

Interpersonal Relationship Analysis with Dyadic EEG Signals via Learning Spatial-Temporal Patterns 51
Wenqi Ji, Fang Liu, Xinxin Du, Niqi Liu, Chao Zhou, Minjing Yu, Xinyue Guo, Guozhen Zhao, and Yong-Jin Liu

Effect of Music Training in Neural Responses to Emotional Speech Prosody: Insights from EEG and Brain Network Analysis 67
Ci-Jun Gao, Xucheng Liu, Gufeng Jia, Hongzhe Yang, Wei Tao, Haiyan Wu, Ruey-Song Huang, and Feng Wan

Potential Indicator for Continuous Emotion Arousal by Dynamic Neural Synchrony 89
Guandong Pan, Zhaobang Wu, Yaqian Yang, Xin Wang, Longzhao Liu, Zhiming Zheng, and Shaoting Tang

Exploring EEG-Based Neural Correlates of Multivariate Ordinal Emotion Representations 105
Xuyang Chen, Xin Xu, Dan Zhang, Quanying Liu, and Xinke Shen

The Co-varying Multimodal Pattern in Treatment-Resistant and Non-treatment-Resistant Schizophrenia 120
Siyuan Cao, Shuzhan Gao, Chuang Liang, Vince D. Calhoun, Xuyun Wen, Zening Fu, Lei Wu, Rongtao Jiang, Daoqiang Zhang, Shile Qi, and Xijia Xu

Investigating the Dynamics of Seizure Neuroactivities Using Hidden
Markov Model 130
   Tao Feng, Huihang Ke, Hui Yao, and Chao Wu

Suppressing Seizure via Optimal Electrical Stimulation to the Hub
of Epileptic Brain Network 144
   *Zhichao Liang, Guanyi Zhao, Yinuo Zhang, Weiting Sun, Jingzhe Lin,
Jialin Wang, and Quanying Liu*

AI for Brain Technology

SVFormer: A Direct Training Spiking Transformer for Efficient Video
Action Recognition 161
   *Liutao Yu, Liwei Huang, Chenlin Zhou, Han Zhang, Zhengyu Ma,
Huihui Zhou, and Yonghong Tian*

BL-BERT: Extracting Body Language from Behavior Sequences in Freely
Moving Mice 181
   *Yaning Han, Zhiwei Jiang, Furong Ju, Liping Wang, Quanying Liu,
and Pengfei Wei*

Benchmarking Neural Decoding Backbones Towards Enhanced On-Edge
iBCI Applications 192
   *Zhou Zhou, Guohang He, Zheng Zhang, Luziwei Leng, Qinghai Guo,
Jianxing Liao, Xuan Song, and Ran Cheng*

Enhanced Local Attention with Deep Neural Networks for EEG Decoding 207
   Wei Tao, Jiawo Ye, Chio-In Ieong, Haiyan Wu, and Feng Wan

Mirror Contrastive Loss Based Sliding Window Transformer
for Subject-Independent Motor Imagery Based EEG Signal Recognition 222
   *Jing Luo, Qi Mao, Weiwei Shi, Zhenghao Shi, Xiaofan Wang,
Xiaofeng Lu, and Xinhong Hei*

Active Urination Detection Using EEG Based on FBCNet 240
   *Anan Gan, Banghua Yang, Yonghuai Zhang, Xingye He, Fenqi Rong,
Liang Chang, and Aolei Yang*

D2CAN: Domain-Guided Contrastive Adversarial Network for EEG-Based
Cross-Subject Cognitive Workload Decoding 250
   Ruichao Zhan, Dongyang Li, Song Wang, and Quanying Liu

Group-Specific Fusion Model and Its Application in Identifying
Multimodal Co-varying Diagnostic Patterns for Psychiatric Disorders 265
   Siyuan Cao, Chuang Liang, Qi Zhu, Rongtao Jiang, Daoqiang Zhang,
   Vince D. Calhoun, and Shile Qi

Multi-category Brain Tumor Segmentation via Multi-scale
and Cross-category Relation Modeling 275
   Dongzhe Li, Baoyao Yang, Yuebin Xie, Weide Zhan, and Jingsong Lin

Consistent Brain Age Difference in Childhood Autism Spectrum Disorder
and its Subtypes 287
   Fangling Sun, Chuang Liang, Wei Shao, Zening Fu, Daoqiang Zhang,
   Rongtao Jiang, Shile Qi, and Vince D. Calhoun

Brain-Aware Readout Layers in GNNs: Advancing Alzheimer's Early
Detection and Neuroimaging 297
   Jiwon Youn, Dong Woo Kang, Hyun Kook Lim, and Mansu Kim

TSICNet: Importance of Connectome Information for Epilepsy
Classification 312
   Zonghan Du and Zhongyuan Lai

Brain-Inspired AI

A Brain-Inspired Distributed Long-Term Memory Guided Online
Continual Learning Method 331
   Yuyang Han, Xiuxing Li, Qixin Wang, Tianyuan Jia, and Xia Wu

Memory Sequence Length of Data Sampling Impacts the Adaptation
of Meta-reinforcement Learning Agents 344
   Menglong Zhang, Fuyuan Qian, and Quanying Liu

Parameter-Efficient Fine-Tuning of ChatGLM to Mitigate Hallucinations
in Chinese Abstractive Summarization 359
   Yongjian Huang and Simin Wu

TUN-GCA: A Novel Approach for Organ Segmentation in Nasopharyngeal
Carcinoma CT Images 368
   Wenxin Che, Penghui Du, Rihan Huang, Quanying Liu, Youzhi Qu,
   and Ziyuan Ye

Convolutional Neural Networks Based on Axial Counting Attention
for Deburring Cross-Sectional Images of Aluminum Profiles with Burrs 382
   Weidong Huang, Dan Pan, An Zeng, and Baijing Liu

Assessing the Feasibility of Using AI Models to Simplify Brain Imaging
Reports for Patients: A Comparative Analysis of Four Large Language
Models 396
   Min Xu and Yiwen Wang

How Do Transformers Integrate Meanings? An Investigation Using
Interpretable Brain-Based Componential Semantics in Two-Word Phrases 407
   Shaonan Wang

Author Index 421

AI for Brain Science

Comparison of the Brain Visual Cortex and CNN Under Continuous Object Property Space

Qiande Zhao[1,2], Deying Li[1], Congying Chu[1], and Lingzhong Fan[1,2,3]($\boxtimes$)

[1] Brainnetome Center, Institute of Automation, Chinese Academy of Sciences, Beijing 100190, China
{zhaoqiande2020,congying.chu,lingzhong.fan}@ia.ac.cn
[2] School of Future Technology, University of Chinese Academy of Sciences, Beijing 100049, China
[3] School of Health and Life Sciences, University of Health and Rehabilitation Sciences, Qingdao 266000, Shandong, China

Abstract. One important area of artificial intelligence (AI) is the alignment of deep learning models and the human brain. It's been proved that Convolutional Neural networks (CNN) are candidate computational models of the primate brain visual stream. Recent advances in neuroscience showed that object properties, including animacy and size, played an important role in visual object representations. But the property representations in CNN and whether these representations are similar with those in the visual cortex are unclear. Using the THINGS dataset, we created a continuous object property space and applied voxel-wise encoding models to map properties to human brain fMRI data and CNN layer features. Dimension reduction of human object property ratings identified three key dimensions: grasp, animacy, and feeling, which organize an object property space. Then, model weight analysis produced property representation maps, highlighting the role of the higher-level visual cortex in property representations. Cluster analysis using these maps functionally parcellated the visual cortex into three regions, each associated with specific preferred objects and properties. Finally, property analysis in CNN indicated that property could predicted responses across all layers. In conclusion, representations of object continuous properties in CNN and the human brain are different, which enlightens direction of the future alignment study between deep learning models and the brain.

Keywords: Continuous Object Properties · Convolutional Neural Network · Human Visual Cortex · Alignment

1 Introduction

Developments of artificial intelligence are often accompanied by related neuroscience achievements, such as the convolution and attention mechanism. An important driving force is to develop the computational models that could match human performance on multi-tasks, behave like human and have similar representations like human brain, which is the alignment of computational models and the brain[1]. A lot of studies have been done in this field. Yamins et al. claimed that convolution neural network (CNN) was good candidate model of human ventral visual cortex [2]. CNN followed a similar hierarchical processing structure with visual cortex [3]. Alignment of human object space and deep neural network (DNN) could improve the task performance of DNN [4]. These studies proved that the alignment of DNN and brain could enlighten the new computational models.

Object recognition is a core ability of the human beings. Recent studies focused on the object representations of the visual cortex in large-scale [5]. These studies showed that object property played an important role in object representation, such as object shape, texture, animacy and so on. Specifically, property helped to identify and classify objects [6]. More importantly, property helped to organize objects in the brain space [7]. For example, human ventral temporal cortex (VTC) represented objects according to animacy and real-world size principles, with category selective regions like FFA (fusiform face area), PPA (para-hippocampal place area) organized along these two dimensions [7]. But these results were based mainly on the discrete property views and a very limited number of images, leaving representations of the visual cortex under the continuous property dimensions unknown. Property representations in CNN and whether these representations are similar with those in the visual cortex are also unclear.

In this study, we proposed a continuous property space and identified the representations of visual cortex under this space using the THINGS dataset. The THINGS dataset provided human cognitive ratings for 12 continuous properties of a large number of natural object images [8] and brain functional Magnetic Resonances Imaging(fMRI) signals of three human subjects when viewing these natural object images [9]. We used voxel-wise linear encoding methods to build mappings from property space to brain responses. Object property map could be obtained in the human brain through this model and then the functional patterns of the visual cortex. Property representations in CNN were analyzed finally. We found that:

(i) Object property was represented mainly on higher visual cortex but rarely on early visual cortex.
(ii) Property maps unveiled detailed functional patterns, which organized category-selective regions in the visual cortex.
(iii) Not like the visual cortex, object property could predict responses across all CNN layers.

This study provided a continuous properties space, which underlie the functional organization of the visual cortex, and CNN showed a different representation patterns compared to the visual cortex. These results enlighten that CNN performance might be improved if aligned with the visual cortex under object property space.

2 Related Work

2.1 Large-Scale Representations of the Visual Cortex

In recent years, lots of studies focused on the functional organization of the visual cortex in large-scale, providing a more comprehensive understanding of the visual cortex in primates including humans and macaques [5]. Object properties played an important role in visual representations, which were key to decide whether object was safe or not and how we could interact with it. Many potential object properties which were important to recognize object were not been studied systematically. Konkle et al. found that the object representation pattern of human visual cortex was arranged according to animacy and real-world size based on contrast analysis [7]. And further study showed that visual cortex shared a unifying space with CNN [10]. In addition, Bao et al. used a pretrained CNN model to construct a low dimensional object space, and found that the distribution of object representations in the inferior temporal (IT) lobe of macaques was similar to the patterns of the model object space, which could be described as two properties: animate to inanimate and spiky to stubby (or curved to not curved) [11]. Other important properties in object recognition included manipulability (manipulability, grasp/hold), mobility (movability, move/be moved), and naturalness (naturalness, natural/manmade) [12–14]. In addition, objects also changed in terms of subjective value (subjective value, precious/pleasant), weight (weight, heavy), emotional valence (emotional valence) and arousal (arousal) [12, 15]. However, only a few properties were included in individual work and these works were analyzed on very small datasets. The role of object properties in object representation have not yet been systematically studied.

2.2 CNNs as Visual Computational Models

With the development of AI, more and more aspects related to alignment of DNN and brain in functions and structures have been studied, especially on field of the visual cortex. Yamins et al. designed brain-score to quantify the shared variance between activations of brain visual area and CNN layers [2]. Similar results were also found using decoding methods, which was that early visual area represented objects similar to CNN shallow layers and higher visual cortex to CNN deeper layers [3]. Further study used task-related variance to analyze the relations between visual cortex and CNN and found that all visual areas are more related to deep CNN layers, which indicated that early and higher visual areas contained category-related information [16]. Constrained by brain behavioral or functional data, DNN model was trained to improve the classification accuracy, or learn better latent representations and features [4, 17]. 2.1 illustrates the importance of property in visual object representation, but whether CNN shares an object space with the visual cortex under property views is not clear. If not, then the alignment of CNN with the brain property space would probably contribute to better brain-like CNN models.

3 Methods

This study used voxel-wise linear encoding methods to build mappings from object property space to human brain functional and CNN responses separately, shown in Fig. 1. Firstly, we analyzed the relations of the 12 object properties provided by the THINGS dataset using dimension reduction and found a low dimensional manifold. Then, encoding model was trained to map object properties to brain fMRI signals. Through the model, property representation maps in the cortex were obtained. Finally, linear mapping of object properties to the CNN feature space was built to find the property representations in each layer, and results were compared with the human brain.

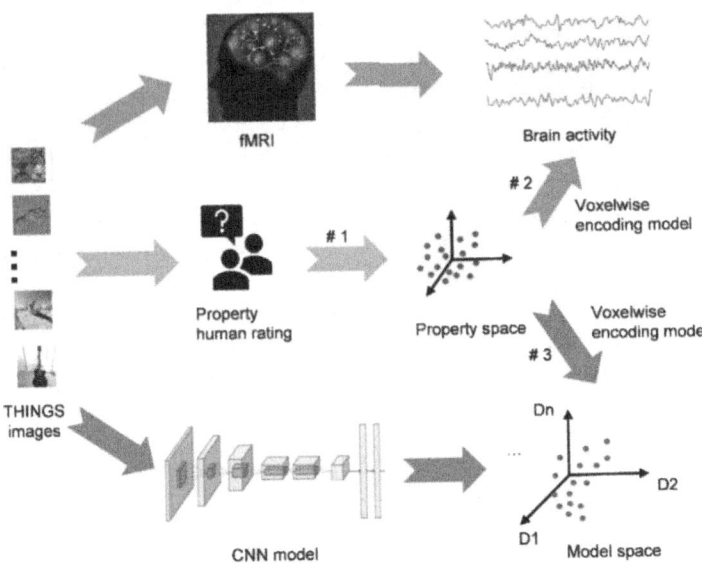

Fig. 1. Pipeline of property representation analysis. Including 3 steps, built object property space, analyzed property map in the brain and CNN models by training voxel-wise encoding models.

3.1 THINGS Dataset

The THINGS dataset was a multimodal neuroimage dataset including fMRI and EEG data when 3 subjects viewing images, human ratings of 12 object properties, and triplet odd-one-out behavior task data [9].

Neuroimaging Data and Preprocessing. fMRI experiment scanned 3 subjects when viewing THINGS images. All magnetic resonance images (MRI) were collected using a 3 T Siemens Magnetom Prisma scanner and a 32-channel head coil. fMRI was 2 mm isotropic resolution (60 axial slices, 2 mm slice thickness, no slice gap, matrix size 96 × 96, FOV = 192 × 192 mm, TR = 1.5 s, TE = 33 ms, flip angle = 75°). And additional high resolution T1-weighted and T2-weighted images were collected.

The MRI data was preprocessed with fMRIPrep[20]. The steps included slice time correction, head motion correction, susceptibility distortion correction, co-registration between functional and anatomical images, brain tissue segmentation and cortical surface reconstruction. The retinotopic mapping and functional localizer data were used to define retinotopic regions in early visual cortex and category-selective regions in higher visual cortex. GLM was used to estimate image-wise responses across session. All analysis used in this study was based on preprocessed data for sub-01 provided by this dataset.

Human Property Ratings. THINGS dataset used crowdsourcing to generate a broad set of object property ratings, including manmadeness, preciousness, animacy, pleasantness, and arousal level and real-world size, 12 properties in total [18]. For example, Animacy ratings for each object concept were collected by presenting raters with the respective noun and asking them to respond to the property 'something that lives' on a Likert scale. Real-world size ratings for each object concept were obtained in two steps. First, raters were instructed to indicate the size of a given object noun on a continuous scale, defined by nine reference objects spanning the size range of all objects. Others follows the same procedure.

3.2 Voxel-Wise Encoding Model

We divided the whole fMRI data and related property ratings into training set and testing set as 9:1 ratio randomly. We mapped the object property space to the cortex through voxel-wise linear encoding models based on training set [19]. For each voxel, we modeled its response to an image as a linear combination of the property features in the object space.

$$x_i = a_i + b_i y + \varepsilon_i \tag{1}$$

where x_i is the beta value at the i-th voxel obtained from GLM analysis, y is the object property represented as a 3-dimensional column vector with each element corresponding to one axis (or feature) in the property space, b_i is a row vector of regression coefficients, a_i is the bias term, and ε_i is the error or noise.

Training the Encoding Model. We had 720 object concepts beta values for each voxel in human brain, which can be viewed as voxel responses across time t, with each time point relating to an image object. It follows that the response of the i-th voxel at time t was expressed as Eq. (2)

$$x_i(t) = a_i + b_i y(t) + \varepsilon_i(t) \tag{2}$$

We estimated the coefficients $(a_i; b_i)$ given time samples of $(x_i; y)$ by using least-squares estimation with L2-norm regularization. That is, to minimize the following loss function defined separately for each voxel.

$$L_I = \frac{1}{T} \sum_{t=1}^{T} (x_i(t) - a_i - b_i(t))^2 + \lambda_i \|b_i\|_2^2 \tag{3}$$

where T is the number of objects, and λ_i is the regularization parameter for the i-th voxel. We applied 10-fold generalized cross-validation in order to determine the regularization parameter. Statistical significance of every voxel based on a block-wise permutation test was applied. Specifically, we divided the training data into blocks and kept the property features intact within each block but randomly shuffled the block sequence for each of 100,000 trials of permutation, resulting in a null distribution. Against this null distribution, we compared the cross-validation without permutation and calculated the one-sided p value while testing the significance with FDR q < 0.05. For more details of statistical significance, please refer to [19].

Testing the encoding Model. We also tested how well the encoding model could be generalized to test set. For this purpose, the trained encoding model was applied to the test set, generating a voxel-wise model prediction of the fMRI response to the test image set.

$$\hat{x}_i(t) = \hat{a}_i + \hat{b}_i y(t) \tag{4}$$

where y(t) is the vector of object property across test set images. Then, predicted voxel responses were correlated with actual responses to calculate Pearson correlation as model performance.

3.3 Property Representations in CNN

Because AlexNet was used in many related studies, here we used it as example model [10]. AlexNet included 8 layers, with 5 convolutional layers and 3 fully-connected layers. We extracted THINGS images features across all layers of a pretrained AlexNet sourced from the Torchvision (PyTorch) model zoo, from layer 1 to layer 7. The image feature of each layer was defined as the output vector of this layer, such as the output of maxpooling operation of layer1, the output of ReLu opreation of layer 6 and 7. Before extract the model feature, images in THINGS dataset was normalized just as the ImageNet dose.

Images features and their properties were divided as did in 3.1. Then a voxel-wise encoding model was trained to map from property space to each model layer separately, just as did to the human brain fMRI. Through the model, we could obtain the significantly predicted units and the prediction accuracy (Pearson correlation) of these units in each layer.

Brain-score is a paradigm which identify the similarity of model and the brain region [1]. It trained partial least square regression (PLSR) model to predict one feature with another feature. Here, we calculated brain-score of each CNN layer and brain responses to identify the relations of them. During the calculation, the brain voxels and layer units only include property significantly predicted ones.

4 Results

In this section, we used human object property ratings to predict human fMRI and CNN layer responses and got representations of each property. Then we tried to answer whether these property representations follow the same patterns in the brain and CNN. The following parts would show these results in details.

4.1 Object Property Space

Because object properties correlated with each other, the property relations were first analyzed. THINGS dataset provided the human cognitive ratings of 12 important properties for 1854 object concepts, including animacy, natural, manmade and so on. Manmade objects and animals were gathered at both ends of the animacy dimension, while small particles, such as sand, and huge man-made objects are gathered at both endings of the size axis, such as trains, etc.

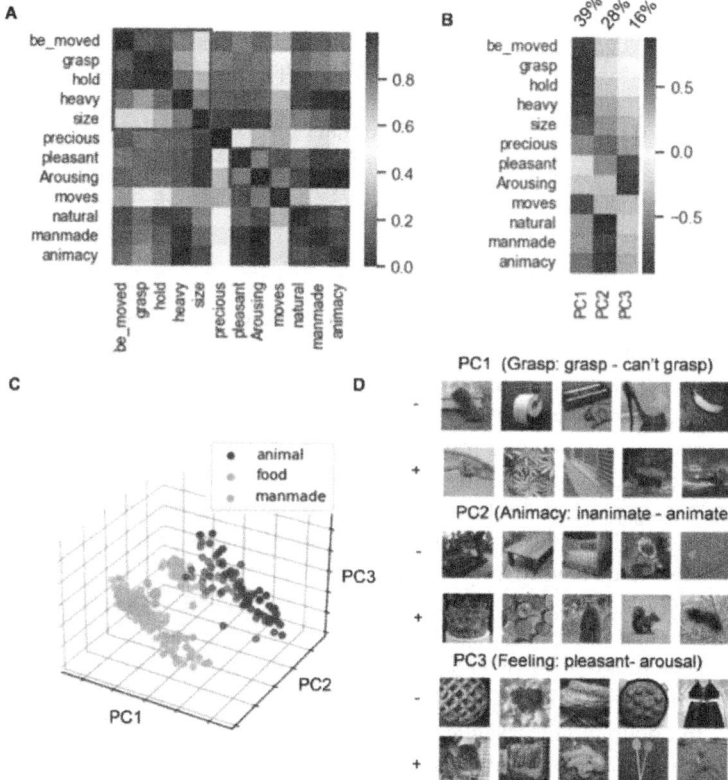

Fig. 2. Object property space. Figure A. Representational dissimilarity matrix across 12 object properties. Three blocks of property were strengthened by red box. Figure B. The first three PCs correlated with each property. Figure C. Distribution of objects in the 3 PCs space. The colors represented different object categories. Figure D. Object images preferred by different properties. + and − was two endings of each PC. Combing the relations with property and preferred images, 3 PCs were referred to grasp, animacy and feeling. (Color figure online)

These object properties are not independent. Obviously, manmade and natural are two mutually exclusive properties. Heavy objects usually have big size. Thereafter, we calculated the representational dissimilarity matrix (RDM) across 12 properties and 3 blocks were observed clearly, with properties within each block are highly related, as shown in Fig. 2A. The properties were roughly divided into three categories, among which heavy, grasp, hold, be_moved and size were one block, natural, manmade and animacy were the second block, pleasant, arousal are the third block. Then, principal component analysis (PCA) was performed on the properties across all objects, and keep the first three principal components (PCs) with more than 80% explained variance. Three PCs and all 12 properties were correlated to show that these PCs were matched to three blocks separately (see Fig. 2B). In Fig. 2C, objects were plotted into this three-dimensional PC space and showed that objects distributed along property dimensions but not category. In Fig. 2D, we showed the objects preferred by positive and negative PC ends across three PCs. And found that PC1 was related to the object utility, PC2 was related to object animacy and PC3 was related to human's feeling to objects. In the end, we referred to three PCs as grasp to can't grasp, inanimate to animate and pleasant to arousal.

To conclude, dimension reduction of human object property ratings identified three key dimensions: grasp, animacy, and feeling. These three properties constructed a low dimensional manifold of object property space.

4.2 Property Representation Maps

After constructing the object property space, we studied how properties were represented in the human brain. The fMRI data was preprocessed using fMRIPrep [20] and GLM was analyzed to get betas value to each image for all voxels of the brain. In order to establish the mappings between object properties and brain functional signals, a voxel-wise linear encoding model was trained. Figure 3A showed the prediction performance of the encoding model. The highlighted areas were the Pearson correlation r value ($q < 0.05$, FDR corrected) between the predicted brain responses pattern and the actual brain responses for the given object property in the testing set. Following analysis were all based on the property significantly predicted brain areas. Brain areas which object properties could significantly predict were mainly distributed in the lateral occipital lobe and VTC, both of which were related to the higher level visual cortex. This result indicated that the higher visual cortex played an important role in object property representation.

The model weights corresponding to the object properties were mapped to the cerebral cortex to obtain the representations of each property. Figure 3B showed the representational maps of grasp, animacy and feeling. For the grasp map, the areas that prefer 'can't grasp' (red) coincides with the scene, face, and body-selective areas, which can be explained as animate and scene-related objects can't be grasped. Areas that prefer 'grasp' are distributed mainly on manmade object-selective areas. This was because manmade objects usually have certain kind of utility and are used by hand. For the animacy map, the areas that prefer 'animate' (red) are consistent with the face and body-selective areas. This result was basically consistent with previous research findings [7]. For the feeling map, the areas that prefer 'pleasant' only are located on the scene-selective areas. While areas that prefer 'arousal' includes face, body and manmade object-selective areas, which can be interpreted as human needs more attention to interact with animals and manmade tools.

In general, these results provided more refined property maps. These property maps could unite the organization of category-selective regions properly.

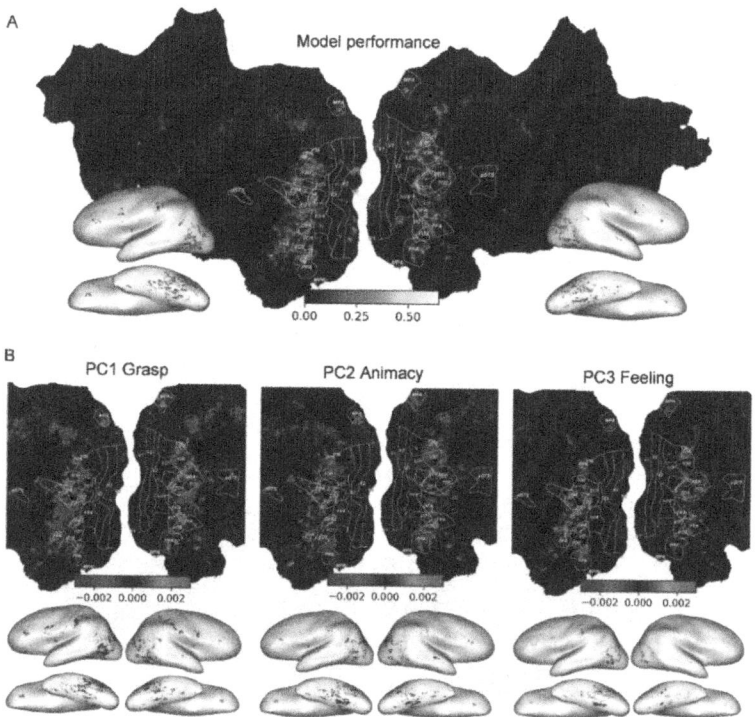

Fig. 3. Property representational map. Figure A. Pearson correlation r between property-predicted brain activation and real brain activation, FDR q < 0.05. Figure B. Representation map of each property in brain cortex according to the weight matrix. Color bar shows the betas range in each property map. (Color figure online)

4.3 Property-Driven Functional Parcellations

In order to study the functional organization of visual cortex under the property views, clustering analysis was used to three property maps. Voxels that could be significantly predicted by object property were clustered using K-Means algorithm. According to metric, the best clustering number was three (see Fig. 4A). Each voxel was assigned a cluster label and mapped into three-dimensional property space, in which 3 clusters could be observed clearly (see Fig. 4B). Next, property patterns, representations in visual cortex and preferred images of each cluster were analyzed. Each cluster has specific property pattern, with cluster 1 characterized by 'inanimate' (PC2 -), cluster 2 characterized by 'can't grasp' (PC1 +), and cluster 3 characterized by 'grasp' (PC1 -), as seen in Fig. 4C. Voxels in each cluster was shown in the brain cortex. And we observed that cluster 1 (red) was mainly located on scene-selective areas, cluster 2 (green) face and body-selective areas and cluster 3 (blue) manmade object-selective areas (see Fig. 4D). Most interestingly, we found that each cluster had specific represented objects. Cluster 1 preferred scene objects, such as indoor objects, cluster 2 preferred animals and human body parts, and cluster 3 preferred manmade objects and foods (see Fig. 4E).

In general, we found that each cluster driven by property maps had specific functions, with each functional parcellation associated with specific category-selective areas, preferred objects and properties.

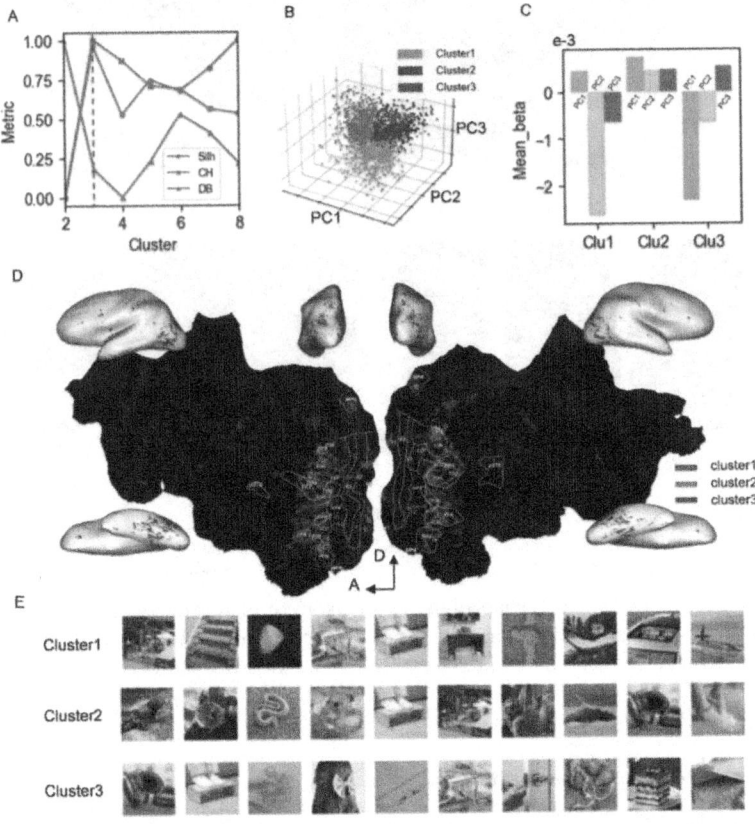

Fig. 4. Functional clusters of the visual cortex driven by object property. Figure A. Clustering metric across different number of clusters. Figure B. Distribution of significantly predicted voxels in the three-dimensional property space, and colors indicate clusters. Figure C. Average properties of the voxels in each cluster. Figure D. Distribution of each cluster in the cortex. Figure E. Object images preferred by each cluster. (Color figure online)

4.4 Property Representations in CNN

Previous studies showed that similarity of the responses between the brain and CNN captured the shared variance. The relations of the brain and CNN under the property-related variance are unclear. Therefore, property representations in CNN were analyzed and compared with result of the brain. Encoding model was trained to map from object property space to each CNN layers separately, just the same procedures as the brain analysis. And results showed that all layers could be predicted by object property and more than 60% of units could be significantly predicted by property (see Fig. 5A). Then, representations of property-driven functional clusters in CNN were analyzed using brain-score. First, we calculated brain-score between responses of each model layer and three functional parcellations separately, shown in Fig. 5C top. Next, the differences of the original responses and property predicted responses in each model layer were calculated firstly. Then, the brain-scores between the difference and the functional clusters were calculated separately. And property deprived brain score of difference activation between

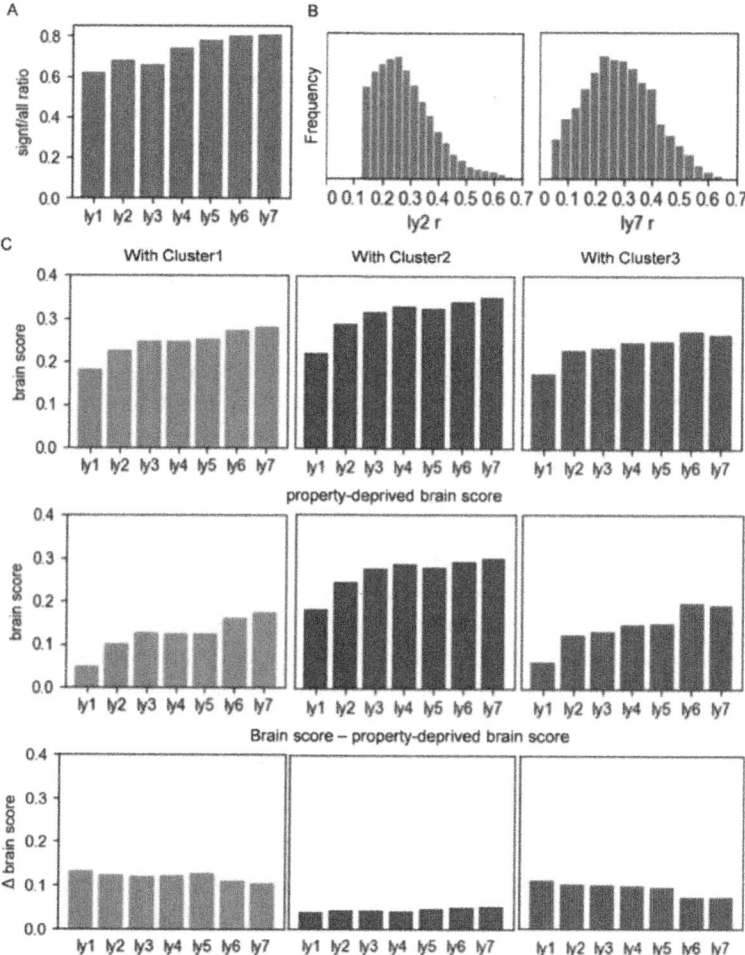

Fig. 5. Representation of properties and functional clusters in CNN. Figure A. All model layers can be significantly predicted by object property (FDR q < 0.05). Figure B. Prediction performance of AlexNet layer 2 and 7. Figure C. Top, brain-score between responses of each functional cluster and model layer. Middle, calculate the difference between actual responses and property predicted responses for each brain cluster and CNN layer separately. Then calculate property-deprived brain-scores between the difference responses between each brain cluster and CNN layer. Bottom, difference of the original brain-score and property-deprived brain-score.

brain and model was calculated, shown in Fig. 5C middle. Finally, we calculated the difference between the original brain-score and property deprived brain-score, and the results indicated that functional clusters were represented across all CNN layers (see in Fig. 5C bottom). In general, object property and related functional clusters were represented across all model layers.

5 Discussion

In this study, we firstly built a three-dimensional property space and these dimensions were grasp, animacy and feeling. Then encoding model showed property maps in the visual cortex and 3 functional clusters with each cluster corresponding to specific functions. Finally, property representation in CNN was analyzed and found that all model layers represented property and related functional clusters.

5.1 Object Property Space

Object have different level of properties, ranging from orientation, color, texture, shape to category [21], more abstract semantic property, such as animacy, arousal [22]. And these properties were represented mainly on hierarchical regions of the visual cortex, and some regions in the temporal, parietal and frontal cortex [23].

Many studies have proved that properties played important part in object representation and object recognition [6, 7]. The THINGS dataset provided 12 object properties, which were related to object visual appearance, such as animacy, heaviness and size, related to object functional utility, such as grasp and hold, and related to emotion triggered by objects such as precious, pleasant and arousal. These object properties reflected how we interacted with objects and object relations which could be used to organize objects by our brain. Although these properties were important, our results in Sect. 4.4 showed that these properties were not enough for a complete object space obviously, which should contain more properties than provided here. Data-driven methods would play a critical role in constructing a more complete object space [24]. But our low-dimensional object property space here proceeded for this endeavor.

5.2 Property Representation Maps

How property represented in the brain was usually based on a small number of images and the contrast methods [25], which were less precise than large number of images and property based encoding methods [26]. In this study, we used later approach to build property representation maps of object property space. According to Fig. 3B, animacy map mainly represents 'animate' on face and body-selective regions and leave rest areas 'inanimate', which is very similar to previous result [7], especially on the VTC. As for the grasp map, it mainly represents 'can't grasp' on face, body and scene-selective regions, because animals who have face and body usually can't grasp or don't have functional utility, and scene is usually the background of environment. Interestingly, grasp map is very similar to size map [27], which has proved to be important in visual object representation. But our grasp map shows more detailed patterns than size map in VTC and dorsal occipital lobe. Finally, feeling map represents 'pleasant' only on the scene-selective regions and 'arousal' on face, body and object-selective regions, including LOC. This is perhaps because animals who have face and body are alive and unpredictable, and manmade objects are of utility, which decide human needs attentions and be careful when interacting with them. On the contrary, scene is low risk for surviving so make it feel pleasant.

After we got property maps, functional parcellation of visual cortex under property maps was analyzed using clustering algorithm, as was shown in Fig. 4D. And a tripartite functional parcellation was obtained. Compared with previous study [7, 28], our result was more refined. For example, it showed new functional regions which is related to manmade object representation between FFA and PPA in VTC. Similar findings also existed in Dorsal Occipital cortex. Most importantly, this tripartite functional parcellation driven by object property unite the distribution of face, body and scene-selective regions and LOC in the higher visual cortex. Besides that, cluster 3 also represent on lateral parietal lobe, whose functions is probably related to tool use [29]. This study mainly showed a single subject result, but other subjects also showed similar patterns. Lastly, our result indicates that the higher visual cortex plays an important role in property representations.

5.3 Property Representations in CNN

Previous studies have shown that CNN is a good simulation of the visual cortex [2, 3]. First two PCs of layer 7 in AlexNet were high correlated with shape (aspect ratio) and animacy, which proved a good model of macaque Inferior temporal (IT) cortex [11]. But patterns of other properties in CNN, such as those used in this study, are not clear. And our results showed that activations of all layers could be predicted by object property well (see Fig. 5A). Differently, property was mainly represented on the higher visual cortex in the human brain. This result was conflicted with previous findings, that shallow model layers were similar to early visual cortex and deeper model layers were similar to higher visual cortex when it comes to response patterns [3, 30]. Therefore, we calculated brain-score of property related responses between the brain functional clusters and model layers (see in Fig. 5E). And results supported that property contributed to the shared variance between brain visual clusters and all model layers. But no obvious difference across all CNN layers was observed for every functional clusters in our results. It suggested that how patterns of property representations changed from shallow and deep layers was unclear, which needed further studies. The above analysis was based only on the AlexNet, more CNN models would be included in the future. Besides that, the contrastive-loss based CNN model proved a better task-general representation than supervised models and these models should also be involved in the future studies to get more evidences. But our results finally showed that property representations in CNN is not similar with those in the visual cortex.

These results also enlighten a new direction for the alignment of CNN and brain visual cortex. For example, CNN could be trained by adding additional loss function to make sure that CNN layer features could achieve better performance on object property decoding. This operation constraints CNN a better large-scale organization and perhaps could lower the effective dimensions of CNN representation space.

6 Conclusion

Overall, representations of object continuous properties in CNN and the human brain are different. The higher visual cortex plays an important part in property representations, while all CNN layers contribute to the property representations. And these results enlighten further deep learning models and brain alignment study.

Acknowledgments. This work was supported by the STI2030-Major Projects (grant no. 2021ZD0200203 to L.F.) and the Natural Science Foundation of China (grant nos. 82072099 to L.F. and 62250058 to C.C.). Data was supported by the Intramural Research Program of the National Institutes of Mental Health (ZIA-MH-002909) and a Feodor-Lynen fellowship of the Humboldt Foundation awarded to M.N.H.

Disclosure of Interests. This study declared no conflict of interests.

References

1. Schrimpf, M., Kubilius, J., Hong, H., et al.: Brain-score: which artificial neural network for object recognition is most brain-like?. BioRxiv 407007 (2018)
2. Yamins, D.L., Hong, H., Cadieu, C.F., et al.: Performance optimized hierarchical models predict neural responses in higher visual cortex. Proc. Natl. Acad. Sci. **111**(23), 8619–24 (2014)
3. Horikawa, T., Kamitani, Y.: Generic decoding of seen and imagined objects using hierarchical visual features. Nat. Commun. **8**(1), 15037 (2017)
4. Fu, K., Du, C., Wang, S., et al.: Improved video emotion recognition with alignment of CNN and human brain representations. IEEE Trans. Affect. Comput. (2023)
5. Grill Spector, K., Weiner, K.S.: The functional architecture of the ventral temporal cortex and its role in categorization. Nat. Rev. Neurosci. **15**(8), 536–48 (2014)
6. Hebart, M.N., Zheng, C.Y., Pereira, F., et al.: Revealing the multidimensional mental representations of natural objects underlying human similarity judgements. Nat. Hum. Behav. **4**(11), 1173–1185 (2020)
7. Konkle, T., Caramazza, A.: Tripartite organization of the ventral stream by animacy and object size. J. Neurosci. **33**(25), 10235–10242 (2013)
8. Hebart, M.N., Dickter, A.H., Kidder, A., et al.: THINGS: a database of 1,854 object concepts and more than 26,000 naturalistic object images. PLoS ONE **14**(10), e0223792 (2019)
9. Hebart, M.N., Contier, O., Teichmann, L., et al.: THINGS data, a multimodal collection of large scale da tasets for investigating object representations in human brain and behavior. Elife **12**, e82580 (2023)
10. Doshi, F.R., Konkle, T.: Cortical topographic motifs emerge in a self organized map of object space. Sci. Adv. **9**(25), eade8187 (2023)
11. Bao, P., She, L., McGill, M., Tsao, D.Y.: A map of object space in primate inferotemporal cortex. Nature **583**(7814), 103–108 (2020)
12. Sudre, G., Pomerleau, D., Palatucci, M., et al.: Tracking neural coding of perceptual and semantic features of concrete nouns. Neuroimage **62**(1), 451–463 (2012)
13. Huth, A.G., Nishimoto, S., Vu, A.T., et al.: A continuous semantic space describes the representation of thousands of object and action categories across the human brain. Neuron **76**(6), 1210–1224 (2012)

14. Magri, C., Konkle, T., Caramazza, A.: The contribution of object size manipulability, and stability on neural responses to inanimate objects. Neuroimage **237**, 118098 (2020)
15. Bradley, M.M., Lang, P.J.. Affective norms for English words (ANEW): instruction manual and affective ratings. Technical report C-1, The Center for Research in Psychophysiology, University of Florida (1999)
16. Sexton, N.J., Love, B.C.: Reassessing hierarchical correspondences between brain and deep networks through direct interface. Sci. Adv. **8**(28), eabm2219 (2022)
17. Takahashi, S., Sasaki, M., Takeda, K., et al.: Self-supervised learning facilitates neural representation structures that can be unsupervisedly aligned to human behaviors. In: ICLR 2024 Workshop on Representational Alignment (2024)
18. Stoinski, L.M., Perkuhn, J., Hebart, M.N.: THINGSplus: new norms and metadata for the THINGS database of 1854 object concepts and 26,107 natural object images. Behav. Res. Methods 1–21 (2023)
19. Zhang, Y., Han, K., Worth, R., Liu, Z.: Connecting concepts in the brain by mapping cortical representations of semantic relations. Nat. Commun. **2011**(1), 2020 (1877)
20. Esteban, O., Markiewicz, C.J., Blair, R.W., et al.: FMRIPrep: a robust preprocessing pipeline for functional MRI. Nat. Methods **16**, 111–116 (2019)
21. Peelen, M.V., Downing, P.E.: Category selectivity in human visual cortex: beyond visual object recognition. Neuropsychologia **105**, 177–183 (2017)
22. Lang, P.J., Bradley, M.M., Fitzsimmons, J.R., et al.: Emotional arousal and activation of the visual cortex: an fMRI analysis. Psychophysiology **35**(2), 199–210 (1998)
23. Landi, S.M., Freiwald, W.A.: Two areas for familiar face recognition in the primate brain. Science **357**(6351), 591–595 (2017)
24. Yao, M., Wen, B., Yang, M., et al.: High dimensional topographic organization of visual features in the primate temporal lobe. Nat. Commun. **14**(1), 5931 (2023)
25. Liu, J., Harris, A., Kanwisher, N.: Perception of face parts and face configurations: an fMRI study. J. Cogn. Neurosci. **22**(1), 203–211 (2010)
26. Naselaris, T., Olman, C.A., Stansbury, D.E., et al.: A voxel-wise encoding model for early visual areas decodes mental images of remembered scenes. Neuroimage **105**, 215–228 (2015)
27. Gabay, S., Kalanthroff, E., Henik, A., et al.: Conceptual size representation in ventral visual cortex. Neuropsychologia **81**, 198–206 (2016)
28. Coggan, D.D., Tong, F.: Spikiness and animacy as potential organizing principles of human ventral visual cortex. Cereb. Cortex **33**(13), 8194–8217 (2023)
29. Johnson-Frey, S.H.: The neural bases of complex tool use in humans. Trends Cogn. Sci. **8**(2), 71–78 (2004)
30. Xu, Y., Vaziri-Pashkam, M.: Limits to visual representational correspondence between convolutional neural networks and the human brain. Nat. Commun. **12**(1), 2065 (2021)

CoCoG-2: Controllable Generation of Visual Stimuli for Understanding Human Concept Representation

Chen Wei[1], Jiachen Zou[1], Dietmar Heinke[2], and Quanying Liu[1](✉)

[1] Southern University of Science and Technology, Shenzhen, China
liuqy@sustech.edu.cn
[2] University of Birmingham, Birmingham, UK

Abstract. Humans interpret complex visual stimuli using abstract concepts that facilitate decision-making tasks such as food selection and risk avoidance. Similarity judgment tasks are effective for exploring these concepts. However, methods for controllable image generation in concept space are underdeveloped. In this study, we present a novel framework called CoCoG-2, which integrates generated visual stimuli into similarity judgment tasks. CoCoG-2 utilizes a training-free guidance algorithm to enhance generation flexibility. CoCoG-2 framework is versatile for creating experimental stimuli based on human concepts, supporting various strategies for guiding visual stimuli generation, and demonstrating how these stimuli can validate various experimental hypotheses. CoCoG-2 will advance our understanding of the causal relationship between concept representations and behaviors by generating visual stimuli. The code is available at https://github.com/ncclab-sustech/CoCoG-2.

Keywords: Concept representation · Conditional generative model · Experimental design

1 Introduction

Humans receive abundant visual stimuli in their daily lives. By understanding visual input's abstract concepts such as function, edibility, and biological properties, humans can perform high-level cognitive decision-making tasks such as food selection and risk avoidance. A large number of human studies have shown that the similarity judgment task is an effective experimental paradigm for revealing human concept representation [7,13,15,21]. In the similarity judgment task, participants are required to evaluate various visual stimuli and make judgments based on the similarity of these images. Such similarity is measured in the concept space, rather than the original image space. So their distance in the concept space reflects the psychological space in the human mind [13,21]. The distance in main dimensions of concept representation can predict and explain human

---

C. Wei and J. Zou—Equal contribution.

decision-making behaviors; in turn, similarity judgments are expected to reflect key dimensions of concept representation. Previous studies have demonstrated that by controlling specific experimental inputs, we can directly observe the core factors that influence decision-making, a process referred to as counterfactual reasoning [4,18]. Similarly, by manipulating the concept representation within experimental stimuli, we can better uncover and understand the causal relationship between concept representations and behaviors. However, The concepts of complex natural visual stimuli often remain obscured and are quite challenging to manipulate. How to reveal the underlying concept space and generate the image by manipulating specific concepts remains largely unexplored.

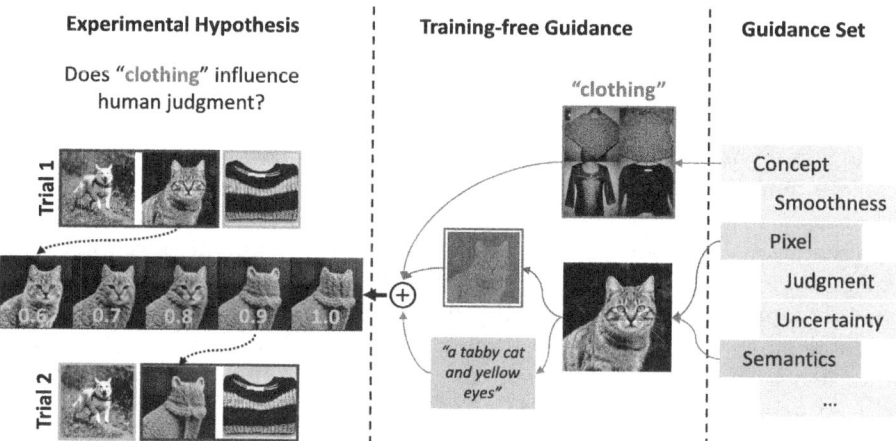

Fig. 1. Guided by the CoCoG-2, we can validate a specific hypothesis by generating visual stimuli in the controllable concept space using our framework. For instance, researchers propose an experimental hypothesis (*Does "clothing" influence human judgment?*), and then construct a loss function using a guidance set (aimed at preserving pixel and semantic features while modifying the concept of "clothing"), and finally generate alternative visual stimuli through training-free guidance (visual stimuli where the concept of "clothing" exceeds 0.9 lead to changes in human judgment). These synthetic images can be used to test the hypothesis with behavioral experiments.

Progress in controllable generation models has significantly shaped the AI field, especially through conditional generation models utilizing GANs and diffusion techniques [5,8,9,27,29]. Applications of these models span various domains such as text-to-image/video synthesis, image restoration, and biomedical imaging [2,11,14,17,20,22,26]. The generation conditions derive from diverse sources including text, sketches, and maps of edges, segmentation, or depth [1,14,20,22,34,35]. Despite these advances, the current models lack integration of human subjective inputs such as feelings or feedback, often resulting in outputs that do not align with human preferences. To better meet human expectations, recent studies have incorporated human feedback into these models, including visual preference scores and decisions from human-in-the-loop comparisons [3,12,23,28,31]. Furthermore, it is important to integrate human cognitive mechanisms, such as decision-making based on concepts, into generative models. This integration can make the models more human-like and uncover human concept representations by manipulating concepts in generated images.

To generate images by controlling the concept space, a pioneer study, Wei et al. [30], has proposed a concept-based controllable generation (CoCoG) framework. CoCoG utilizes concept embeddings as conditions for the image generation model, thereby indirectly manipulating behaviors through concept representations. CoCoG comprises two parts: a concept encoder that learns concept embeddings by predicting human behaviors, and a concept decoder that maps concept embeddings to visual stimuli using a conditional diffusion model with a two-stage generation strategy. The concept encoder in CoCoG can accurately predict human visual similarity decision-making behaviors, revealing a reliable and interpretable space of human concept representations. The concept decoder can generate visual stimuli controlled by concept embeddings. The generated visual stimuli are highly consistent with the target concept embeddings and can regulate human similarity decision-making behaviors by manipulating concept dimensions. Notably, CoCoG preliminarily proposes efficiently exploring concept representations through generated visual stimuli, thus providing counterfactual explanations for human behavior. However, CoCoG has its limitations. CoCoG can only input complete concept embeddings as guiding conditions, leading to a lack of flexibility in guided generation. Firstly, CoCoG cannot edit the concepts of an image while keeping other image features unchanged, and conflicts may arise between concepts. Secondly, CoCoG cannot directly generate images based on expected similarity judgments and requires manually selecting "key concept" to guide image generation. These limitations restrict the method's further application in cognitive research based on concept representations.

In this study, we upgrade the CoCoG framework to CoCoG-2. CoCoG-2 allows us to design experimental stimuli based on concepts and effectively validate hypotheses about concept representation. Moreover, we employ a training-free guidance algorithm for the controllable generation of diffusion models with high flexibility. Our work has three main contributions.

- We present a general framework for designing experimental stimuli based on human concept representations and integrating experimental conditions through training-free guidance.
- We have verified a variety of potential guidance strategies for guiding the generation of visual stimuli, controlling concepts, behaviors, and other image features.
- Our experimental results demonstrate that visual stimuli generated by combining different guidance strategies can validate a variety of experimental hypotheses and enrich our tools for exploring concept representation.

2 Preliminaries

2.1 Training-Free Guidance Diffusion Models

Diffusion Models. Diffusion models [10,27] operate through two distinct phases: the forward and the reverse processes. During the forward phase, which spans from time 0 to T, the model progressively converts an image into Gaussian noise. Conversely, during the reverse phase, it reconstructs the image starting from noise, reversing the timeline from T back to 0. Denote x_t as the state of the data at time t. In the forward phase, noise is methodically added according to a specific noise schedule:

$$x_t = a_t x_0 + b_t \epsilon_t, \qquad (1)$$

where, $a_t = \sqrt{\alpha_t}, b_t = \sqrt{1 - \alpha_t}$, α_t varies monotonically from 0 to 1 as t increases, and ϵ_t is sampled from a standard normal distribution $\mathcal{N}(0, I)$. The neural network in diffusion models is trained to approximate the noise at each timestep:

$$\min_\theta \mathbb{E}_{x_t,\epsilon_t} \left[\|\epsilon_\theta(x_t, t) - \epsilon_t\|_2^2 \right] = \min_\theta \mathbb{E}_{x_t,\epsilon_t} \left[\|\epsilon_\theta(x_t, t) + b_t \nabla_x \log p_t(x_t)\|_2^2 \right], \qquad (2)$$

where $p_t(x_t)$ represents the probability distribution of x_t. The dynamics of the reverse phase are governed by the following ordinary differential equation (ODE):

$$\frac{dx_t}{dt} = f(t) x_t - \frac{g^2(t)}{2} \nabla_x \log p_t(x_t) \qquad (3)$$

with $f(t) = -\frac{d \log a_t}{dt}$ and $g^2(t) = \frac{db_t^2}{dt} - 2 \frac{d \log \sqrt{\alpha_t}}{dt} b_t^2$. Through these equations, the reverse process manages to transform Gaussian noise back into coherent images.

In practical applications, the update rules for both the forward and reverse processes depend on the choice of the sampling algorithm. In this work, we employ the DDPM algorithm [8] and we represent the forward and reverse steps using the following notations respectively:

$$x_t = DDPM^+(x_{t-1}) \quad \text{and} \quad x_{t-1} = DDPM^-(x_t).$$

Training-Free Guidance. In diffusion models, conditional generation introduces a specific condition y into the generative process [9,27]. By applying Bayes' rule $p(x|y) = \frac{p(y|x)p(x)}{p(y)}$, it incorporates the condition through an additional likelihood term $p(x_t|y)$:

$$\nabla_{x_t} \log p_t(x_t|y) = \nabla_{x_t} \log p_t(x_t) + \nabla_{x_t} \log p_t(y|x_t). \qquad (4)$$

Recent approaches, known as training-free methods [1,2,25,32,34], utilize pre-trained diffusion models for conditional generation without the need for retraining or finetuning. These methods employ a pre-trained diffusion model and a differentiable loss function $\ell(f_\phi(x_0), y)$, defined using a neural network f_ϕ. Tweedie's formula is used to compute $\nabla_{x_t} \log p_t(y|x_t)$ by estimating $\hat{x}_0$ from x_t:

$$\hat{x}_0(x_t) \approx \mathbb{E}[x_0|x_t] = \frac{x_t - b_t \epsilon_t(x_t, t)}{a_t}. \tag{5}$$

$$\nabla_{x_t} \log p_t(y|x_t) = \nabla_{x_t} \ell(f_\phi(\hat{x}_0(x_t)), y), \tag{6}$$

Then, we can add an additional correction step in the reverse sampling process:

$$x_{t-1} = DDPM^-(x_t) - \eta \nabla_{x_t} \ell(f_\phi(\hat{x}_0(x_t)), \, y)$$

This approximation allows the use of an existing network, which is initially trained for processing clean data (x_0). The gradient of the final term in the loss function is determined through backpropagation across the guidance network and the diffusion framework. In our example, where the condition is concept dimension, f_ϕ is modeled using a concept encoder, and ℓ is the loss designed for the experiment.

2.2 Concept Based Controllable Generation

The CoCoG framework, introduced by Wei et al. [30], utilizes concept embeddings as conditional inputs for image generation, bridging cognitive science with artificial intelligence. CoCoG aims to enhance the production of natural visual stimuli in a controllable manner, closely aligning with human cognitive processes. It comprises two key components: a concept encoder for learning concept embeddings through human behavior prediction, and a concept decoder for transforming these embeddings into visual stimuli via a conditional diffusion model.

Concept Encoder. The concept encoder extracts concept embeddings from visual objects. Each image x from a dataset is processed by the CLIP image encoder f to produce a CLIP embedding h. This embedding is then transformed by a learnable projector g into a concept embedding c:

$$\begin{aligned} \text{CLIP embedding:} \quad & h = f(x), \\ \text{concept embedding:} \quad & c = g(h), \end{aligned} \tag{7}$$

Each dimension of c corresponds to an interpretable concept, with its activation value reflecting the concept's prominence in the visual object. Training is conducted using the *triplet odd-one-out* task from the THINGS dataset [6], employing dot product similarity and cross-entropy for decision prediction:

$$p(y) = Softmax(S_{jk}, S_{ik}, S_{ij}), \tag{8}$$

Here $S_{ij} = <c_i, c_j>$, and predicted behavior y are matched against human judgments to train the projector g, with L_1 regularization ensuring sparsity in the concept embedding c.

Concept Decoder. Following encoder training, triplets (x_i, h_i, c_i) are used by the concept decoder to control the generation of visual objects based on human concept representation. The process is formalized as:

$$p(x, h, c, y) = p(y)p(c|y)p(h|c)p(x|h). \tag{9}$$

The decoder operates in two phases. In stage I (prior diffusion), conditioned on c, a diffusion model learns the distribution of CLIP embeddings $p(h|c)$ using a lightweight U-Net $\epsilon_{prior}(h_t, t, c)$. The model trains on ImageNet pairs (h_i, c_i) using classifier-free guidance. In Stage II (CLIP-guided generation), with the CLIP embedding h from Stage I, a generator models $p(x|h)$ using pre-trained SDXL and IP-Adapter models [19,24,33], which facilitate the integration of h, to guide the U-Net's denoising process.

CLIP Embedding as an Intermediate Variable. They use CLIP embeddings as an intermediate variable for two reasons: 1) CLIP embeddings are low-level and retain key image information, effectively predicting human similarity judgments with linear probing [16]. 2) They enable the use of pre-trained generative models, adopting a two-stage generation strategy that requires only training a Prior Diffusion model, significantly reducing computational costs.

Despite its efficacy in generating controllable visual stimuli, CoCoG's reliance on full concept embeddings for input restricts flexibility. It cannot modify specific concepts without altering other image attributes, leading to potential conflicts. Furthermore, it cannot generate images directly from predicted similarity judgments (i.e., $p(x|y)$), requiring manual selection of key concepts for image generation (i.e., $p(x|c = c(y))$), which limits its utility in detailed cognitive research. To address these limitations, further work should explore methods that allow for more flexible manipulation of specific concepts while maintaining control over other image attributes. For instance, generating multiple trials with consistent visual stimuli or varying only specific concepts without altering other features could be beneficial for more complex experimental designs. This improvement, which will be discussed below, would enhance CoCoG's applicability in diverse research contexts.

3 Method

Compared to CoCoG, CoCoG-2 introduces improvements in the controllable generation strategy of the concept decoder. We simplify the distribution model from Eq. 9 to $p(x, h, e) = p(e)p(h|e)p(x|h)$. Here, e represents the conditions that the visual stimuli need to be controlled, including concepts c, similarity judgments y, and others.

3.1 Training-Free Guidance for Controlling Visual Stimuli

In the concept decoder of CoCoG-2, we continue to employ a two-stage strategy, first modeling $p(h|e)$ and then $p(x|h)$. In this work, we attempt to enhance the flexibility and applicability of CoCoG by introducing training-free guidance in the modeling of $p(h|e)$. As discussed in the Sect. 2.2, learning distribution $p(h|c)$ directly lacks flexibility. For training-free guidance, we only learn the prior distribution $p(h)$ and decompose $p(h|e)$ into the prior $p(h)$ and the likelihood $p(e|h)$ using Eq. 4. As shown in Eq. 6, for the likelihood term $p(e|h)$, we have $\nabla_{h_t} \log p_t(e|h_t) = \nabla_{h_t} \ell(f_\phi(\hat{h}_0(h_t)), e)$. By simply choosing the appropriate condition e and a differentiable loss function $\ell(f_\phi(\cdot), \cdot)$ based on the experimental hypothesis, we can effectively guide the generation of visual stimuli as demonstrated in Fig. 1.

3.2 Guidance Set and Corresponding Loss Functions

To construct the loss function, we often wish for the visual stimuli to satisfy multiple conditions via guiding the generative process. We can combine our various requirements to construct an overall loss function $\ell = \sum \lambda_k \ell_k$, where λ_k controls the weight of each type of loss. Specifically, in our experiments, we will define the guidance set and their corresponding loss functions as follows:

Concept Guidance: If we aim for the generated images to achieve predetermined concept values, we can define the loss function as:

$$\ell = \sum \|g(h)_i - c_i\| \tag{10}$$

where i indexes the concepts we wish to control, and c_i are their target values. We can control individual concepts or all concepts simultaneously (Like CoCoG).

Smoothness Guidance: If we desire a batch of images to maintain semantic smoothness, we can define the loss function as:

$$\ell = \sum_{i,j \in S} \|h_i - h_j\| \tag{11}$$

Here, $S = \{i, j\}$ includes the indices of images in the batch that we want to remain smooth. For example, we can generate a batch of pairwise similar images (as shown in Fig. 4).

Semantics Guidance: To make the generated images similar to a given image semantically, we can define the loss function as:

$$\ell = \|h - \bar{h}\| \tag{12}$$

where $\bar{h}$ is the CLIP embedding of the given image.

Judgment Guidance: If we want human similarity judgments to follow a target judgment distribution, we can define the loss function as:

$$\ell = -\sum p(y=i)\log(\bar{p}(y=i)) \qquad (13)$$

where $\bar{p}(y)$ is the target distribution. This loss function calculates the cross-entropy loss between the expected and target distributions. Given by Eq. 8, $E_{p(y)}[y] = Softmax \cdot S \cdot g(h)$, so ℓ is still a derivable function with respect to h.

Uncertainty Guidance: When we need to control the uncertainty in similarity judgments, we can define the loss function using the entropy of the expected judgment distribution:

$$\ell = H(p(y)) = -\sum p(y=i)\log(p(y=i)) \qquad (14)$$

Pixel Guidance: Since CLIP embedding h primarily controls the high-level features or semantic features of the generated images, we also try to introduce img2img [14] to control the low-level features of the images.

3.3 Improving Training-Free Guidance

Previous research has indicated that the generation process of training-free guidance is unstable and prone to guidance failures. To ensure the stability and effectiveness of guidance, we analyzed recent major advancements in this area and employed two techniques to enhance training-free guidance.

Adaptive Gradient Scheduling. Inspired by [25,32], the training guidance process can be viewed as optimizing the loss function of the guidance network. It stands to reason that adopting a more sophisticated optimizer could accelerate the convergence of the guidance. Following [32], we use a closed-form solution based on the spherical Gaussian constraint:

$$h_{t-1}^* = h_t - \eta\sqrt{n}\sigma_t \frac{\nabla_{h_t}\ell(f_\phi(\hat{h}_0(h_t)), e)}{\|\nabla_{h_t}\ell(f_\phi(\hat{h}_0(h_t)), e)\|_2}, \qquad (15)$$

where $\sigma_t = \sqrt{(1-\alpha_{t-1}/(1-\alpha_t))}\sqrt{1-\alpha_t/\alpha_{t-1}}$, n is the dimension of the h, and η is the guidance scale.

Resampling Trick. Following [25,34], we use resampling, or "time travel", to address sampling errors. This involves reintroducing random noise to reset the sampling state, reducing distributional divergence and aligning samples with the target distribution.

Algorithm 1. Improved training-free guidance for prior diffusion

1: **for** $t = T, \ldots, 0$ **do**
2:    **for** $i = 1, \ldots, s$ **do**
3:       $h_{t-1}^i = DDPM^-(h_t^i)$
4:       $\hat{h}_0^i = \frac{h_t^i - \sigma_t \epsilon_\theta(h_t^i, t)}{\sqrt{\alpha_t}}$
5:       $g_t = \nabla_{h_t} \ell(f_\phi(\hat{h}_0^i(h_t^i)), e)$
6:       $h_{t-1}^i = h_{t-1}^i - \eta\sqrt{n}\sigma_t \frac{g_t}{\|g_t\|_2}$
7:       **if** $i < s$ **then**
8:          $h_t^i = DDPM^+(h_{t-1}^i)$
9:       **end if**
10:      $h_{t-1}^0 \leftarrow h_{t-1}^s$
11:    **end for**
12: **end for**

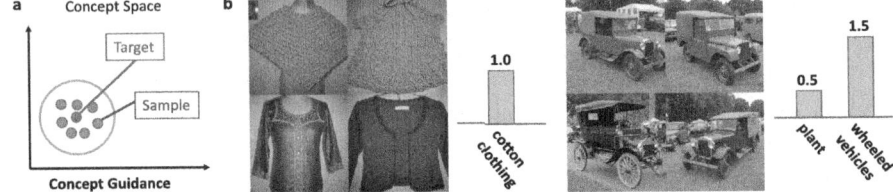

Fig. 2. Generate diverse visual stimuli for given target concepts. (a) Concept guidance used in this experiment. (b) Visual stimuli generated under the guidance of one or more target concepts.

4 CoCoG-2 for Versatile Experiment Design

4.1 Diverse Generation Based on Concept

In the first experiment, we tested the effectiveness of using CoCoG-2 to generate visual stimuli based on given target concepts. The objective of the experiment was to generate diverse samples specific activation values of single or multiple concept dimensions. Therefore we only used *Concept guidance* to guide the Prior Diffusion. As shown in Fig. 2b, whether guided by a single concept or multiple concepts, CoCoG-2 was able to generate images that were well-aligned with the target concepts, and these images exhibited good diversity in features unrelated to given concept.

4.2 Smooth Changes Based on Concepts

After confirming that target concepts can effectively guide the generation of visual stimuli, we modified the activation values of target concepts in the second experiment to generate a group of stimuli while maintaining smooth semantic features. We used *Concept guidance* to allow gradual changes in concept activation values and employed *Smoothness guidance* to maintain the similarity of semantic features between neighboring stimuli. We randomly generated 50 groups of

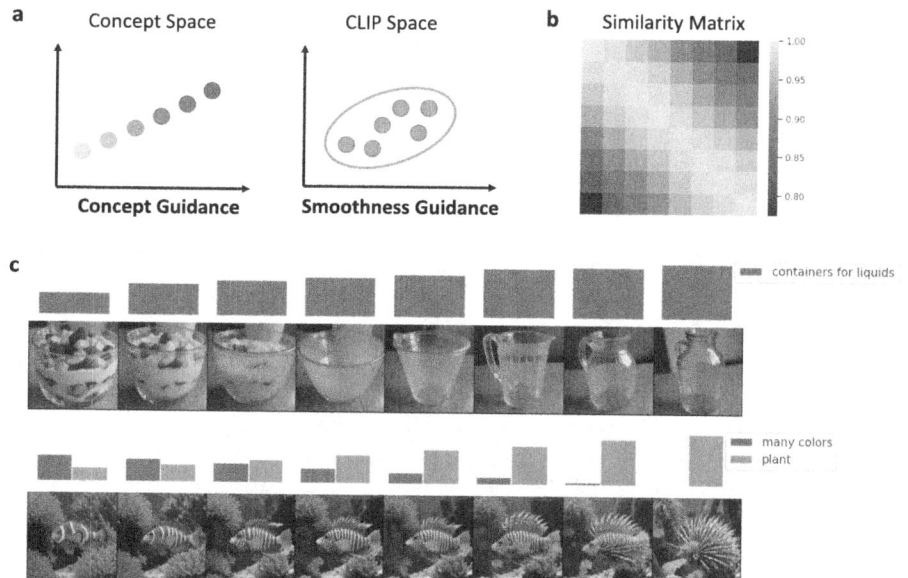

Fig. 3. The smooth change of visual stimuli generated based on concept. (a) Concept guidance and Smoothness guidance used in this experiment. (b) The average CLIP similarity matrix of 100 groups of stimuli. (c) Two trials of images generated under the guidance of one or more target Concepts.

visual stimuli and calculated the similarity matrix of CLIP embeddings for each group's stimuli, which was then averaged to obtain a mean similarity matrix. The similarity matrix shown in Fig. 3b indicates that the CLIP similarity of images within a trial ranged between 0.75 and 1.0, with similarity increasing as the distance of image position decreased, consistent with the expected effect of Smoothness guidance. Figure 3c displays visual stimuli generated under the guidance of one or more varying target concepts, showing clear and stable changes at the concept level, while other features were well maintained.

4.3 Image Editing in Concept

Next, we tested whether it was possible to start with an "original image" and edit target concepts to generate a group of stimuli that are similar to the original image and vary the concepts according to given values. For concept editing, we used *Concept guidance*; to maintain other features of images within a trial, we applied *Smoothness guidance*. Additionally, we utilized *Semantics guidance* in CLIP Space and *Pixel guidance* in Pixel Space to maximize the similarity between the generated images and the original image.

As shown in Fig. 4b, Visual stimuli generated by CoCoG-2 not only resembled the original image but also achieved clear and stable concept edits, with well-preserved low-level features. This experiment also highlighted a significant

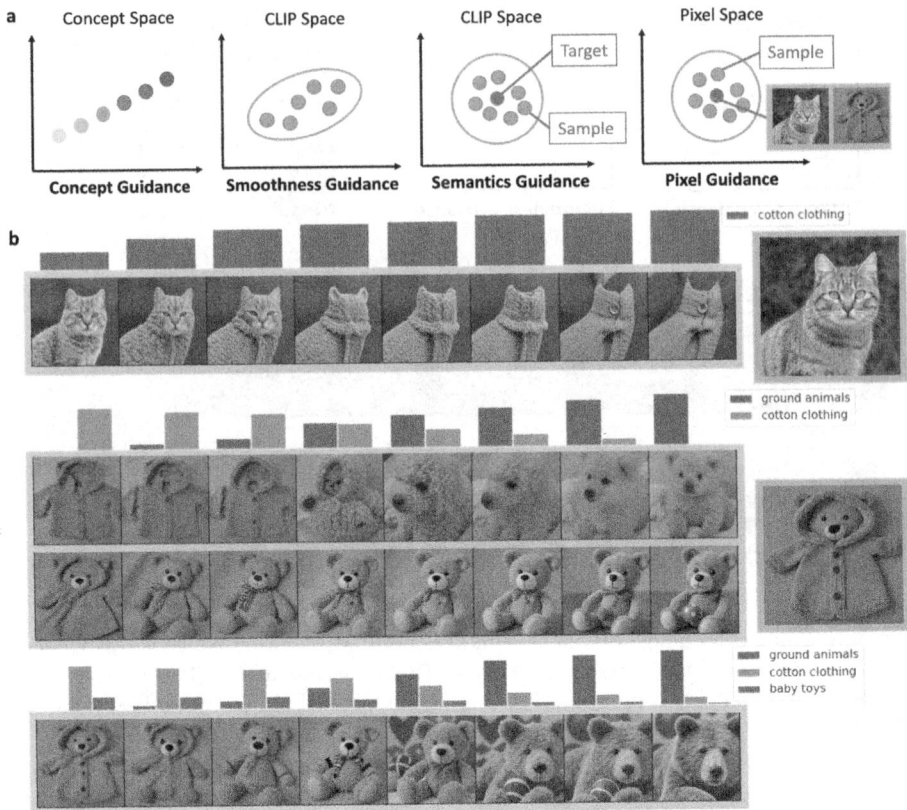

Fig. 4. Visual stimuli generated by image editing in concept. (a) *Concept, Smoothness, Semantics*, and *Pixel guidance* used in this experiment. (b) A trial of images generated under the guidance of a single concept (row 1); Visual stimuli generated by CoCoG-2 and CoCoG under the guidance of multiple concepts (row 2–4). When guided only by "ground animals" and "cotton clothing", the images generated by CoCoG-2 align well with the target concepts. However, the images generated by CoCoG cannot produce real animals due to conflicts between target concepts and the concept "baby toys" in the original image, necessitating manual adjustment of the "baby toys" value.

advantage of CoCoG-2 over CoCoG: it can automatically avoid conflicts between target concepts and the concepts in the original image, without the need for manually modifying conflicting concepts. As shown in rows 2–3 of Fig. 4b, both guided by "ground animals" and "cotton clothing", CoCoG-2 was able to generate real animals, whereas CoCoG could only generate fabric toy bears. This is because the concept "baby toys" in the original image conflicted with target concepts. Even if the activation value of "cotton clothing" was set to zero, the generated images still contained this concept. Therefore, as shown in row 4 of Fig. 4b, it was necessary to manually modify the activation value of "baby toys" to generate real animals.

CoCoG's manual modification process increased the complexity of visual stimuli generation, as conflicting concepts had to be identified and modified through multiple rounds of generation, and it is possible that contradictions could not be resolved by altering just a few key concepts. However, CoCoG-2's Training-Free Guidance method could automatically adjust conflicting concepts, greatly speeding up the generation process.

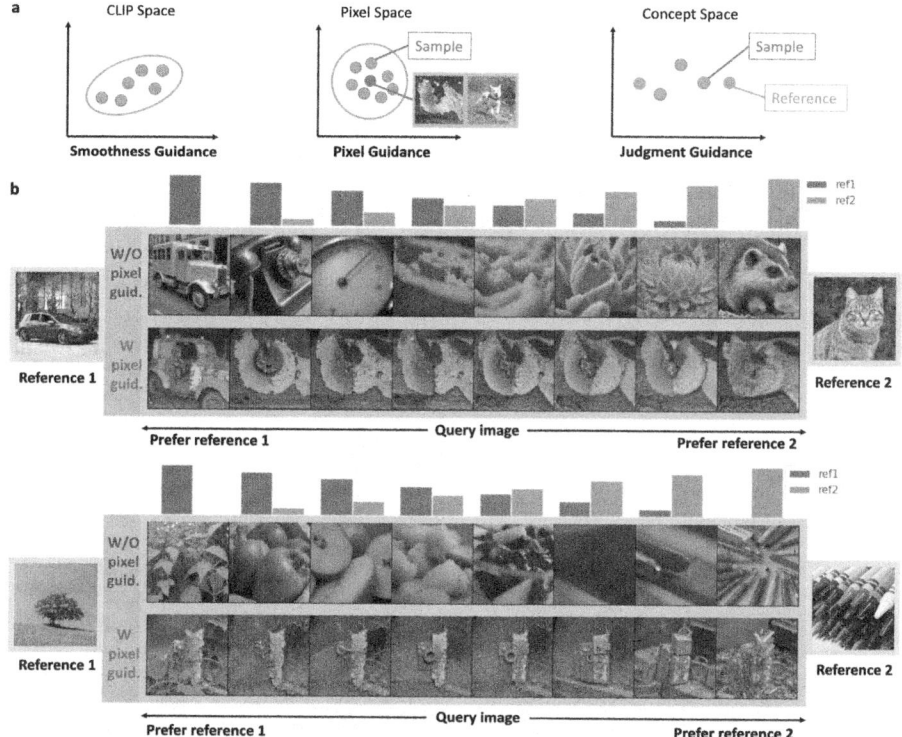

Fig. 5. Behavioral manipulation of similarity judgments with/without *Pixel guidance*. (a) *Smoothness, Pixel*, and *Judgment guidance* used in this experiment. (b) Images generated under the guidance of probability interpolation of two pairs of references. The top row of images for each set is generated without using the *Pixel guidance*, while the bottom row of images is generated with the *Pixel guidance*.

4.4 Behavioral Manipulation of Similarity Judgment

After verifying CoCoG-2's capability of concept editing based on the original image, we further tested whether CoCoG-2 could directly generate visual stimuli guided by experimental results. We used the *Two alternative forced choice* experiment as an example, where participants need to choose between reference

1 & 2, judging which is more similar to a query. In this scenario, the experimental results can be represented as the probability of choosing reference 1 or 2 as the more similar image. We used the probabilities of each reference image being chosen as the guidance, termed the *Judgment guidance*. Additionally, to study the causal relationship between visual cognitive behavior and concept representation, we needed to ensure that changes of behavioral results were solely caused by changes in concepts, which means maintaining other features of the images. Therefore, we used the *Smoothness guidance* to maintain other features within a group, and guided the query image's low-level features using an image unrelated to the reference images, termed the *Pixel guidance*.

In Fig. 5b, we show the generated query images for two references. The upper group of each set is without the *Pixel guidance*, and significant variations in low-level features can be observed among the generated images. The lower group is with the Pixel guidance, where the images in Fig. 5a labeled "ice" and "dog" were used to guide the low-level features respectively. Both groups of image align with the given results of similarity judgment. However, the lower group of images also maintains consistency with the guiding images in shape, color, and other low-level features. This greatly reduces the impact of low-level features on behavioral results, more reliably interpreting the causal relationship between behavior and concepts.

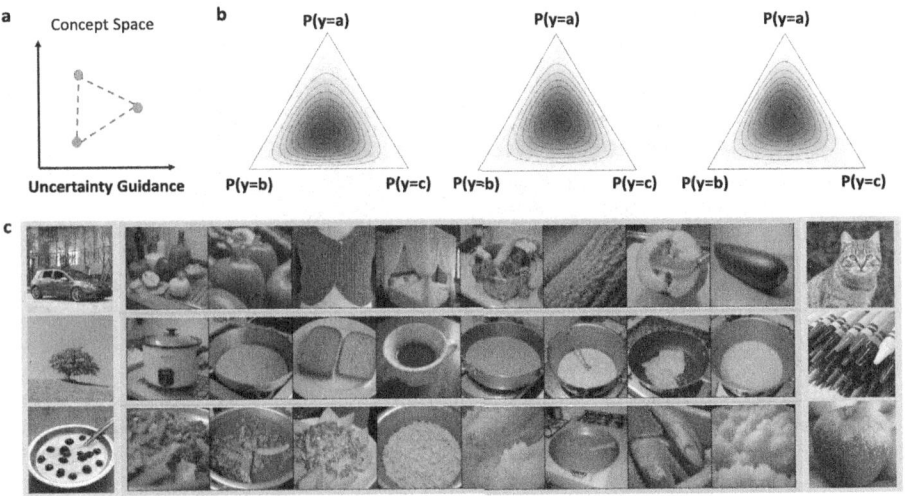

Fig. 6. Generated trials of similarity judgment leading to rich individual preference information. (a) *Uncertainty guidance* used in this experiment. (b) Probability distributions of three sets of generated images in c, with each side of the triangle representing the probability of choosing a particular image as the odd one out. (c) Images generated under the guidance of Uncertainty guidance for three pairs of references.

4.5 Optimal Design for Individual Preference

We demonstrated that CoCoG-2 can directly generate images guided by behavioral results, significantly facilitating the rapid production of desirable experimental outcomes. Beyond this, CoCoG-2 can also be used to design and generate visual stimuli that maximize information gain, thereby substantially reducing the number of experimental trails required in cognitive research. Taking individual preferences in the *Two alternative forced choice* experiment as an example, for the same set of inference images and query image, each participant may choose a different inference image as the more similar one. The greater the variance in participants' choices, and the more information each experiment can provide about individual preferences. Therefore, when the entropy of the probability distribution for choosing between the two reference images is maximized, each *Two alternative forced choice* experiment provides the maximum information gain.

Furthermore, we consider each experiment's three images from the perspective of the *Odd One Out* experiment. The Odd One Out experiment is a triplet choice task where participants select the image that is least similar to the others as the odd one out. This experiment can reveal the similarity relationships between the three images, and its results can be represented by the probability of each image being selected as the odd one out. Therefore, when the entropy of the probability distribution for the three images being chosen as the odd one out is maximized, each Odd One Out experiment provides the maximum information gain.

Consequently, our optimization goal is to maximize the entropy of the probability distribution for the three images being chosen as the odd one out, guided by *Uncertainty guidance*. The choice probability is determined by the similarity between the images, which in turn is dictated by their relative distances in Concept Space. Maximizing the entropy of the probability distribution implies that the three images form an equilateral triangle in Concept Space, as shown in Fig. 6a. Since Concept Space has 42 dimensions, there can be at most 43 possible query images that meet the equilateral triangle condition. Figure 6c displays generated images corresponding to three pairs of reference images. These generated images are positioned at the vertices of multiple equilateral triangles in the 42-dimensional space, exhibiting good diversity. Additionally, Fig. 6b shows the probability distribution for these three sets of images, with the distribution centers very close to the center of the equilateral triangle, indicating that the generated images well satisfy the target probability distribution. These images are diverse and informative, making them optimal experimental images for studying individual preferences.

5 Discussion

In conclusion, our proposed CoCoG-2 framework advances the field of controllable visual object generation by integrating concept representations with behavioral outcomes to guide the image generation process. By employing a versatile experiment designer and utilizing meticulously designed guidance, CoCoG-2 not only addresses the limitations of its predecessor, CoCoG, but also significantly enhances the flexibility and efficiency of generating visual stimuli for cognitive research. The experimental trials generated using CoCoG-2 are instrumental in probing the causal relationships between concept representations and human decision-making behaviors. This approach marks a substantial step towards understanding and manipulating concept-based representations in a controllable and meaningful way, offering a robust methodology for cognitive science research and practical applications in AI-driven visual content generation.

Ethical Statement. The human behavioral data in this study is from THINGS public dataset. No animal or human experiments are involved.

References

1. Bansal, A., et al.: Universal guidance for diffusion models. In: Proceedings of the IEEE/CVF Conference on Computer Vision and Pattern Recognition, pp. 843–852 (2023)
2. Chung, H., Sim, B., Ryu, D., Ye, J.C.: Improving diffusion models for inverse problems using manifold constraints. Adv. Neural. Inf. Process. Syst. **35**, 25683–25696 (2022)
3. Fan, Y., et al.: DPOK: reinforcement learning for fine-tuning text-to-image diffusion models. arXiv preprint arXiv:2305.16381 (2023)
4. Fernández-Loría, C., Provost, F., Han, X.: Explaining data-driven decisions made by AI systems: the counterfactual approach. arXiv preprint arXiv:2001.07417 (2020)
5. Goodfellow, I., et al.: Generative adversarial networks. Commun. ACM **63**(11), 139–144 (2020)
6. Hebart, M.N., et al.: Things-data, a multimodal collection of large-scale datasets for investigating object representations in human brain and behavior. Elife **12**, e82580 (2023)
7. Hebart, M.N., Zheng, C.Y., Pereira, F., Baker, C.I.: Revealing the multidimensional mental representations of natural objects underlying human similarity judgements. Nat. Hum. Behav. **4**(11), 1173–1185 (2020)
8. Ho, J., Jain, A., Abbeel, P.: Denoising diffusion probabilistic models. Adv. Neural. Inf. Process. Syst. **33**, 6840–6851 (2020)
9. Ho, J., Salimans, T.: Classifier-free diffusion guidance. arXiv preprint arXiv:2207.12598 (2022)
10. Karras, T., Aittala, M., Aila, T., Laine, S.: Elucidating the design space of diffusion-based generative models. Adv. Neural. Inf. Process. Syst. **35**, 26565–26577 (2022)
11. Kawar, B., Elad, M., Ermon, S., Song, J.: Denoising diffusion restoration models. Adv. Neural. Inf. Process. Syst. **35**, 23593–23606 (2022)
12. Kirstain, Y., Polyak, A., Singer, U., Matiana, S., Penna, J., Levy, O.: Pick-a-pic: an open dataset of user preferences for text-to-image generation. arXiv preprint arXiv:2305.01569 (2023)
13. Medin, D.L., Goldstone, R.L., Gentner, D.: Respects for similarity. Psychol. Rev. **100**(2), 254 (1993)

14. Meng, C., et al.: SDEdit: guided image synthesis and editing with stochastic differential equations. In: International Conference on Learning Representations (2021)
15. Murphy, G.L., Medin, D.L.: The role of theories in conceptual coherence. Psychol. Rev. **92**(3), 289 (1985)
16. Muttenthaler, L., Dippel, J., Linhardt, L., Vandermeulen, R.A., Kornblith, S.: Human alignment of neural network representations. In: The Eleventh International Conference on Learning Representations (2022)
17. Özbey, M., et al.: Unsupervised medical image translation with adversarial diffusion models. IEEE Trans. Med. Imaging (2023)
18. Peterson, J.C., Bourgin, D.D., Agrawal, M., Reichman, D., Griffiths, T.L.: Using large-scale experiments and machine learning to discover theories of human decision-making. Science **372**(6547), 1209–1214 (2021)
19. Podell, D., et al.: SDXL: improving latent diffusion models for high-resolution image synthesis. arXiv preprint arXiv:2307.01952 (2023)
20. Ramesh, A., Dhariwal, P., Nichol, A., Chu, C., Chen, M.: Hierarchical text-conditional image generation with clip latents. arXiv preprint arXiv:2204.06125, vol. 1, no. 2, p. 3 (2022)
21. Roads, B.D., Love, B.C.: Modeling similarity and psychological space. Annu. Rev. Psychol. **75** (2023)
22. Rombach, R., Blattmann, A., Lorenz, D., Esser, P., Ommer, B.: High-resolution image synthesis with latent diffusion models. In: Proceedings of the IEEE/CVF Conference on Computer Vision and Pattern Recognition, pp. 10684–10695 (2022)
23. von Rütte, D., Fedele, E., Thomm, J., Wolf, L.: Fabric: personalizing diffusion models with iterative feedback. arXiv preprint arXiv:2307.10159 (2023)
24. Sauer, A., Lorenz, D., Blattmann, A., Rombach, R.: Adversarial diffusion distillation. arXiv preprint arXiv:2311.17042 (2023)
25. Shen, Y., Jiang, X., Wang, Y., Yang, Y., Han, D., Li, D.: Understanding training-free diffusion guidance: mechanisms and limitations. arXiv preprint arXiv:2403.12404 (2024)
26. Song, Y., Shen, L., Xing, L., Ermon, S.: Solving inverse problems in medical imaging with score-based generative models. In: International Conference on Learning Representations (2021)
27. Song, Y., Sohl-Dickstein, J., Kingma, D.P., Kumar, A., Ermon, S., Poole, B.: Score-based generative modeling through stochastic differential equations. In: International Conference on Learning Representations (2020)
28. Tang, Z., Rybin, D., Chang, T.H.: Zeroth-order optimization meets human feedback: provable learning via ranking oracles. arXiv preprint arXiv:2303.03751 (2023)
29. Tao, M., Tang, H., Wu, F., Jing, X.Y., Bao, B.K., Xu, C.: DF-GAN: a simple and effective baseline for text-to-image synthesis. In: Proceedings of the IEEE/CVF Conference on Computer Vision and Pattern Recognition, pp. 16515–16525 (2022)
30. Wei, C., Zou, J., Heinke, D., Liu, Q.: CoCoG: controllable visual stimuli generation based on human concept representations. arXiv preprint arXiv:2404.16482 (2024)
31. Wu, X., Sun, K., Zhu, F., Zhao, R., Li, H.: Human preference score: better aligning text-to-image models with human preference. In: Proceedings of the IEEE/CVF International Conference on Computer Vision, pp. 2096–2105 (2023)
32. Yang, L., Ding, S., Cai, Y., Yu, J., Wang, J., Shi, Y.: Guidance with spherical gaussian constraint for conditional diffusion. arXiv preprint arXiv:2402.03201 (2024)
33. Ye, H., Zhang, J., Liu, S., Han, X., Yang, W.: IP-adapter: text compatible image prompt adapter for text-to-image diffusion models. arXiv preprint arXiv:2308.06721 (2023)

34. Yu, J., Wang, Y., Zhao, C., Ghanem, B., Zhang, J.: FreeDoM: training-free energy-guided conditional diffusion model. arXiv preprint arXiv:2303.09833 (2023)
35. Zhang, L., Rao, A., Agrawala, M.: Adding conditional control to text-to-image diffusion models. In: Proceedings of the IEEE/CVF International Conference on Computer Vision, pp. 3836–3847 (2023)

Uncovering Cognitive Taskonomy Through Transfer Learning in Masked Autoencoder-Based fMRI Reconstruction

Youzhi Qu[1], Junfeng Xia[1], Xinyao Jian[2], Wendu Li[1], Kaining Peng[1], Zhichao Liang[1], Haiyan Wu[3], and Quanying Liu[1(✉)]

[1] Department of Biomedical Engineering, Southern University of Science and Technology, Shenzhen, China
liuqy@sustech.edu.cn
[2] Department of Biostatistics, Epidemiology and Informatics, Perelman School of Medicine, The University of Pennsylvania, Philadelphia, USA
[3] Centre for Cognitive and Brain Sciences and Department of Psychology, University of Macau, Macau, China

Abstract. Data reconstruction is a widely used pre-training task to learn the generalized features for many downstream tasks. Although reconstruction tasks have been applied to neural signal completion and denoising, neural signal reconstruction is less studied. Here, we employ the masked autoencoder (MAE) model to reconstruct functional magnetic resonance imaging (fMRI) data, and utilize a transfer learning framework to obtain the *cognitive taskonomy*, a matrix to quantify the similarity between cognitive tasks. Our experimental results demonstrate that the MAE model effectively captures the temporal dynamics patterns and interactions within the brain regions, enabling robust cross-subject fMRI signal reconstruction. The cognitive taskonomy derived from the transfer learning framework reveals the relationships among cognitive tasks, highlighting subtask correlations within motor tasks and similarities between emotion, social, and gambling tasks. Our study suggests that the fMRI reconstruction with MAE model can uncover the latent representation and the obtained taskonomy offers guidance for selecting source tasks in neural decoding tasks for improving the decoding performance on target tasks.

Keywords: FMRI reconstruction · Transfer learning · Masked autoencoder · Cognitive tasks · Taskonomy

1 Introduction

In computer vision and natural language processing, reconstruction tasks serve as effective pre-training methods, guiding models to learn generalized features and achieving favorable performance on downstream tasks [7,16]. In computer

---

Y. Qu and J. Xia—These authors contributed equally to this work.

vision, reconstruction tasks are widely applied in image inpainting, denoising, and super-resolution [26,28,29]. In natural language processing, reconstruction tasks can improve the accuracy and fluency of translation, and enhance the completeness and creativity of content generation [3,4]. As pre-training tasks, reconstruction tasks enhance the models' ability to represent data by training them to reconstruct missing information. This capability not only enhances model performance on reconstruction tasks but also boosts performance on downstream tasks such as image classification, semantic understanding, and sentiment analysis. Classification tasks within neural decoding have demonstrated outstanding performance in neuroscience [18,33,35]. Nonetheless, the exploration of neural signal reconstruction remains nascent.

In the field of neuroscience, the reconstruction of neural signals is critically important, especially in signal completion and denoising [32]. Electroencephalogram (EEG) and functional magnetic resonance imaging (fMRI) studies are frequently disrupted by signal loss or noise, originating from equipment defects, poor electrode contact, and physiological activities [5,39]. Masked reconstruction pre-training methods have achieved remarkable achievements in computer vision and natural language processing, demonstrating potential in neural signal reconstruction [7,16]. In natural language processing, bidirectional encoder representations from transformers (BERT) employ a masking strategy during pre-training tasks such as language modeling and next sentence prediction, enabling the model to acquire more profound and comprehensive language features [7]. The masking strategy involves randomly masking words in the input sequence, forcing the model to focus on all potential word combinations within the context. The masked autoencoder (MAE) model masks random patches of input images and reconstructs them to learn the latent representation of the image [16,30], after the vision transformer (ViT) model addressed issues related to mask tokens and positional embeddings [9]. The MAE has not only achieved outstanding results in image reconstruction but has also been extended to multimedia applications such as audio and video [10,17]. This study proposes applying the MAE model to reconstruct resting-state and task-based fMRI, which is trained by randomly masking temporal and spatial dimensions within the fMRI data.

The brain demonstrates exceptional capabilities, notably the ability to generalize learned knowledge to new tasks and to handle multiple tasks efficiently [11,12]. Understanding the relationships among cognitive tasks is essential for comprehending how the brain processes and coordinates cognitive functions. Relationships exist not only among cognitive tasks but also among various tasks processed by deep learning models. The effectiveness of transfer learning is influenced by the relationships between tasks. Transfer learning can enhance performance on target tasks by transferring knowledge from source tasks. The closer the relationship between the tasks, the better the results of transfer learning tend to be [22,36]. Researchers quantify the relationships among computer vision tasks by analyzing the changes in performance of transfer learning [37]. The cognitive taskonomy not only deepens our understanding of the relationships among tasks but also guides the selection of more relevant source tasks

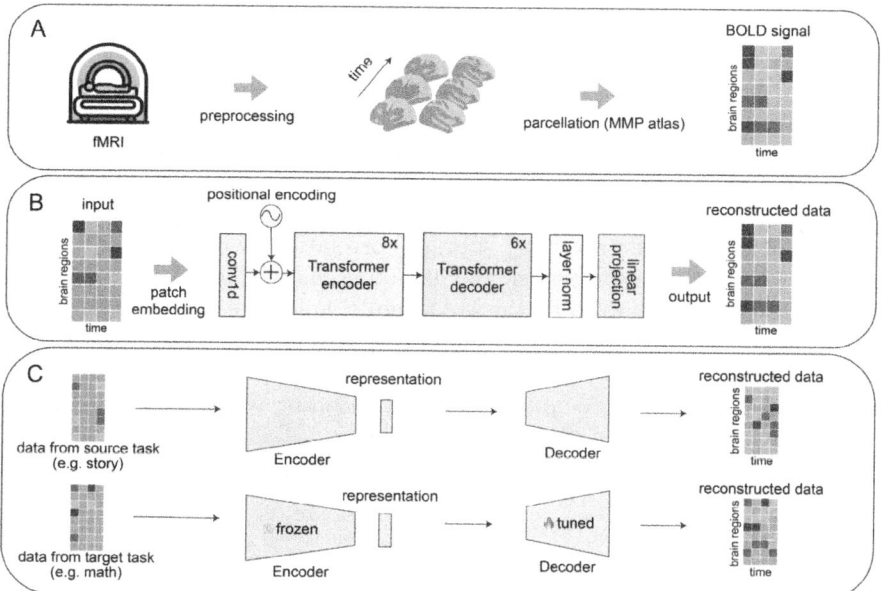

Fig. 1. Framework for fMRI reconstruction and transfer learning: (A) fMRI preprocessing, which involves preprocessing fMRI data and parcellating the brain using the MMP atlas. (B) The MAE model for fMRI reconstruction. Masked fMRI data are first processed through patch embedding, then passed through the transformer-based encoder and decoder, and finally linearly projected to reconstruct the fMRI signals. (C) Transfer learning is utilized to estimate the relationships among the source and target tasks. Initially, the MAE model is trained on the source task using masked fMRI data from the source task. In the second step, the MAE model, trained on the source task, is transferred to the target task. During this specific training phase, which utilizes masked fMRI data from the target task, the encoder is frozen to prevent updates, while only the decoder is tuned.

to improve the performance on target tasks [37]. Compared to neural decoding classification models [24], the MAE model provides a more profound approach to investigating the relationships among cognitive tasks. The MAE model effectively captures the temporal dynamic patterns and the interactions between brain regions, thereby enriching the study of cognitive task relationships. Investigating the performance changes in transfer learning via the MAE model helps to illuminate the relationships among cognitive tasks.

In this study, we employ the MAE model to reconstruct fMRI signals in both resting and task states, and utilize a transfer learning framework to quantify the relationships among cognitive tasks. The contributions of this research are as follows: (1) The MAE model demonstrates robust fMRI reconstruction capabilities across individuals in resting and task states, effectively learning the temporal dynamic patterns and interactions within the brain in fMRI data. (2) Mask testing reveals the difficulty of fMRI reconstruction of different brain networks, highlighting the diversity of these networks. Furthermore, the reconstruction difficulty varies among cognitive tasks, with working memory tasks being more challenging, while language and motor tasks are comparatively easier. (3) The application of the transfer learning framework further quantifies the relationships between cognitive tasks, revealing subtask correlations within motor tasks and similarities among emotion, social, and gambling tasks. The robust cross-subject fMRI reconstruction enabled by the MAE model holds the potential for addressing common issues of signal loss in neuroimaging data collection [32]. Moreover, the task relationship analysis derived from the transfer learning framework offers guidance for selecting source tasks in neural decoding research.

2 Method

2.1 HCP Dataset

In this study, the fMRI data utilized were obtained from the public large-scale Human Connectome Project (HCP) S1200 dataset [27]. The fMRI data utilized in this study comprises resting-state and task-based fMRI data from 1,000 healthy participants. These fMRI data were preprocessed using the standard HCP pipeline [14], and then parcellated into 360 brain regions based on the multi-modal parcellation (MMP) atlas [13]. The brain regions were divided into 7 networks according to the Yeo-7 network, including the visual network (VIS), somatomotor network (SOM), dorsal attention network (DAN), ventral attention network (VAN), limbic network (LIM), frontoparietal network (FPN), and default mode network (DMN) [34].

The task-based fMRI data from the HCP comprises seven categories of cognitive tasks: working memory, motor, emotion processing, gambling, language processing, relational processing, and social cognition. These categories are subdivided into 23 subtasks, utilized by the MAE model for task-based fMRI reconstruction. The working memory task requires participants to memorize and recognize a series of images, divided into eight subtasks based on target categories (body, faces, places, tools) and recall steps (0-back, 2-back): 0bk body, 0bk faces, 0bk places, 0bk tools, 2bk body, 2bk faces, 2bk places, and 2bk tools. The motor task elicits simple bodily responses through visual cues, divided into five subtasks based on the body parts involved: left foot, right foot, left hand, right hand, and tongue. The emotion processing task, which assesses the ability to recognize emotions, is divided into two subtasks: fear and neutral. The gambling task examines decision-making and risk assessment through a simulated card game and includes two subtasks: win and loss. The language processing task involves auditory comprehension and language reasoning, and is divided into two subtasks: math and story. The relational processing task tests logical reasoning by comparing the shapes and textures of objects and comprises two subtasks: relation and match. The social cognition task evaluates understanding of social interactions by judging the intentions of geometric shapes' actions and includes two subtasks: mental and random.

2.2 A Masked fMRI Modeling Framework

Inspired by the MAE and SIMM models [16,30], we propose a masked fMRI reconstruction model that incorporates masking strategies, patch embedding with position encoding, an encoder, and a decoder. The encoder of this model initially maps the masked fMRI signals to a latent representation space, followed by the decoder, which reconstructs fMRI signals based on latent representations. In this study, we utilized three masking strategies: brain masking, time masking, and a brain and time masking, as illustrated in Fig. 1. In brain masking, signals from brain regions are masked throughout the entire fMRI time series. In time masking, the signal at the masked time points across all brain regions is masked. The brain and time masking randomly mask both dimensions to learn the temporal dynamic patterns of brain activity and the interactions within the brain regions for the MAE model.

The preprocessed fMRI data were initially divided into patches and masked accordingly. Unlike MAE, which encodes only the unmasked patches, our approach, similar to simMIM [30], encodes all patches using 1D convolutional neural network. After the patch embedding process, the encoded patches were given sinusoidal positional encoding. The encoder is an 8-layer transformer encoder structure that processes the input data. The decoder, which is a 6-layer transformer decoder structure, applies layer normalization and performs a linear projection to reconstruct the fMRI data. The model utilizes mean squared error (MSE) as the loss function to evaluate the difference between the reconstructed data and the original data, with optimization of model parameters occurring through backpropagation.

2.3 Transfer Learning

The fMRI reconstruction model trained on each cognitive subtask serves as the gold standard for the source tasks. When transferring the source task to the target task, the encoder parameters of the gold standard model were fixed, and the decoder was fine-tuned using 1% of the target task data. To construct the cognitive task relationship matrix (cognitive taskonomy), we conducted transfer experiments among the 23 cognitive tasks. Each cognitive subtask was used as a source task and transferred to the other 22 tasks, forming a 23×23 matrix. The elements (i, j) represented the transfer from source task i to target task j. The value for each matrix element was defined as the MSE of the transferred model minus the MSE of the gold standard model (trained with 80% of the data). Finally, we applied z-score normalization to each column of the matrix, resulting in the cognitive taskonomy, as shown in Fig. 6.

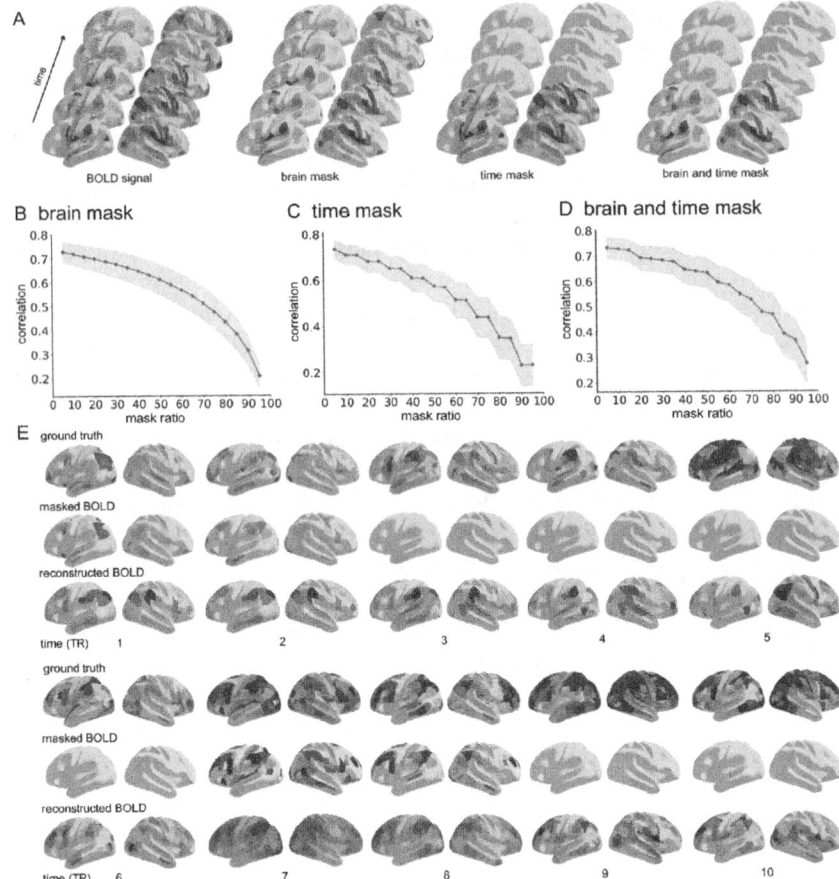

Fig. 2. Results of resting-state fMRI reconstruction using the MAE model. (A) Different masking strategies. Original data, brain-masked data, time-masked data, both brain and time masked data are shown from left to right. (B), (C), and (D) show the reconstruction results of the MAE model with varying mask ratios using the brain mask strategy, the time mask strategy, and the brain and time mask strategy, respectively. (E) Illustration of fMRI reconstruction by the MAE model of the brain and time mask strategy. The original signal, the masked BOLD signal, and the MAE reconstruction results are shown from top to bottom.

2.4 Experimental Setup

In this study, we comprehensively tested the performance of fMRI reconstruction under different configurations of MAE models. The preprocessed fMRI data were divided into a 2D matrix consisting of 360 brain regions and 20 time frames, ensuring consistent time lengths for both resting-state and task-based fMRI. Specifically, we explored masking ratios of 25%, 50%, and 75%, and evaluated the impact of patch sizes of 2, 5, and 10 on the fMRI reconstruction performance.

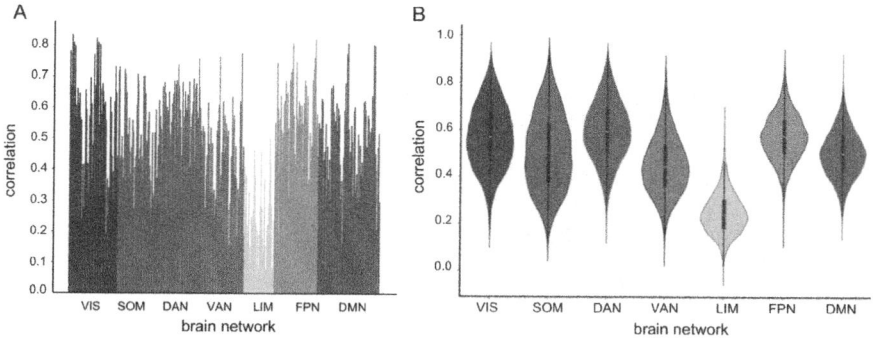

Fig. 3. Results of brain mask testing. (A) The MAE reconstruction results for 360 brain regions (sorted according to brain networks). (B) The MAE reconstruction results for each brain network.

Additionally, we evaluated the effect of varying the encoder and decoder with different hidden size of 512, 768, and 1024 on the reconstruction results. Based on the ablation study results (Table 1), we selected a random masking ratio of 50%, set the patch size in the temporal dimension to 2, and chose hidden size of 1024 for both the encoder and decoder. The fMRI data were divided into training and test sets at a 4:1 ratio based on different subjects. During training, the learning rate for the MAE model was set to 3e-5, using the AdamW optimizer, and a cosine annealing strategy was employed to adjust the learning rate.

3 Results

3.1 Reconstruction of Resting-State and Task-Based fMRI

To assess the impact of different masking strategies on fMRI reconstruction, we used the Pearson correlation coefficient to evaluate the reconstruction results, as shown in Fig. 2. The reconstruction results of the MAE model with varying mask ratios indicate that the model can effectively learn latent features in both spatial and temporal dimensions, achieving high quality fMRI reconstruction. Although increasing the masking ratio usually leads to worse reconstruction performance, the Pearson correlation coefficient of the model reconstruction results still exceeds 0.5 at high mask ratios (*e.g.*, 0.6), demonstrating the model's excellent reconstruction ability. To comprehensively evaluate the impact of different parameter settings on performance, we conducted ablation experiments involving various masking ratios, patch sizes, and hidden sizes, as shown in Table 1. A comparison of different patch sizes showed that smaller patch sizes resulted in lower reconstruction loss. Table 1 shows that encoders and decoders with larger hidden size achieve better reconstruction performance.

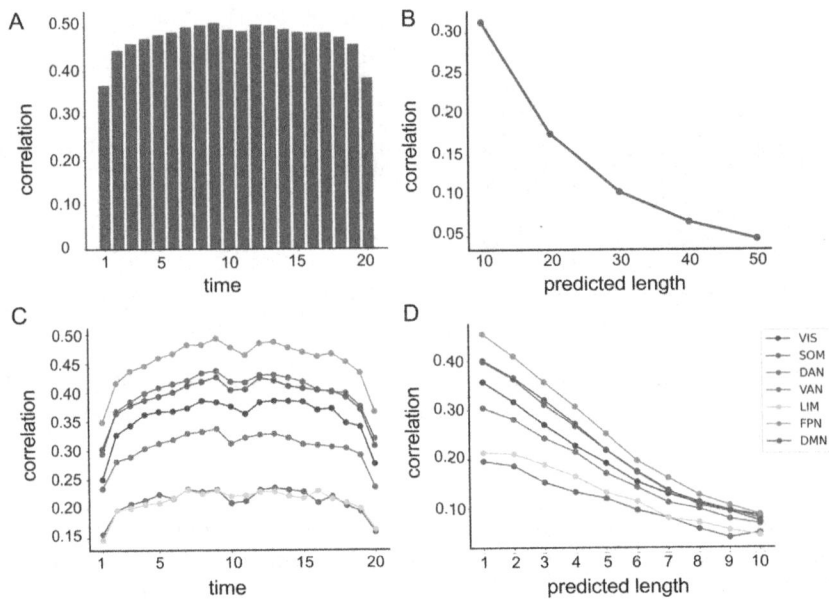

Fig. 4. Results of time mask testing. (A) The MAE reconstruction results for masked fMRI data, with each time frame individually masked. (B) The MAE prediction results for fMRI signals at various future time scales. (C) The MAE reconstruction results of different brain networks for masked fMRI data, with each time frame individually masked. (D) The MAE prediction results of different brain networks for fMRI signals at various future time scales.

To explore the task-based fMRI reconstruction performance, MAE models were trained using fMRI data from various cognitive tasks. Figure 5 depicts the reconstruction results for each task. The experimental results indicate that the Pearson correlation coefficient of fMRI reconstruction for most cognitive tasks are above 0.5, and relatively similar among subtasks within the same cognitive category. Among 23 cognitive subtasks, the reconstruction results for language tasks were the best, whereas those for working memory tasks were relatively lower. Figure 5 illustrates the differing difficulties encountered in reconstructing fMRI data under various cognitive tasks, with language and motor tasks being easier to reconstruct and working memory and gambling tasks being more challenging.

3.2 Brain Mask and Time Mask Testing

To comprehensively analyze the differences in reconstruction difficulty among brain regions, we masked each brain region individually and calculated the Pearson correlation coefficients for the reconstruction results. Figure 3 shows

Table 1. Model ablation experiment for resting-state fMRI reconstruction. We report masking ratio (mr), correlation (corr), encoder hidden size (ehs) and decoder hidden size (dhs). If not specified, the default settings are a 50% masking ratio, a patch size of 2, and encoder and decoder hidden sizes of 1024. Default settings are marked in gray.

(a) Masking ratio

mr	corr	mae	mse
0.25	0.576	0.228	0.092
0.5	**0.614**	**0.220**	**0.082**
0.75	0.593	0.221	**0.082**

(b) Patch size

ps	corr	mae	mse
2	**0.614**	**0.220**	**0.082**
5	0.527	0.241	0.098
10	0.438	0.256	0.112

(c) Encoder hidden size

ehs	corr	mae	mse
512	0.586	0.228	0.087
768	0.589	0.227	0.088
1024	**0.614**	**0.220**	**0.082**

(d) Decoder hidden size

dhs	corr	mae	mse
512	0.589	0.227	0.087
768	0.601	0.224	0.085
1024	**0.614**	**0.220**	**0.082**

the results of the reconstruction of each brain region and brain networks. The reconstruction correlations for brain regions within the LIM network have a high probability of falling below 0.4, indicating suboptimal reconstruction performance. Conversely, the median reconstruction correlations for the VIS, DAN, and FPN networks exceed 0.5, suggesting these regions more effectively utilize information from other regions for fMRI reconstruction.

In addition to analyzing the differences in reconstruction among various brain regions, the properties of MAE in fMRI reconstruction along the temporal dimension were further investigated. Figure 4A displays the Pearson correlation coefficients for the reconstructed results of fMRI data, with each time frame individually masked. The results indicate that signals in the middle of the time series are relatively easier to reconstruct, while those at the beginning and end of the series pose greater challenges. Figure 4B illustrates the correlation between the length of the prediction interval and the reconstruction performance. In this analysis, we use a fixed set of 20 frames of data to predict future signals of different lengths. The results demonstrate that as the prediction interval increases, the reconstruction performance gradually declines. Figure 4C and Fig. 4D show that brain networks with strong signal reconstruction capabilities at various time points also perform well in predicting signals of varying lengths. For example, LIM and SOM networks do not have a strong ability to reconstruct signals at different points in time, nor can they effectively predict future signals.

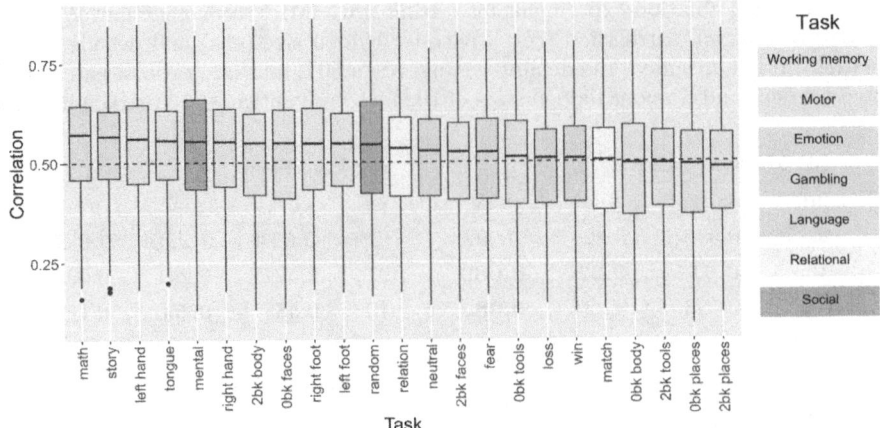

Fig. 5. Results of fMRI reconstruction across cognitive tasks using the MAE model. The tasks are sorted by the correlation coefficient of the reconstruction results, from highest to lowest. The colors correspond to the cognitive categories to which each subtask belongs.

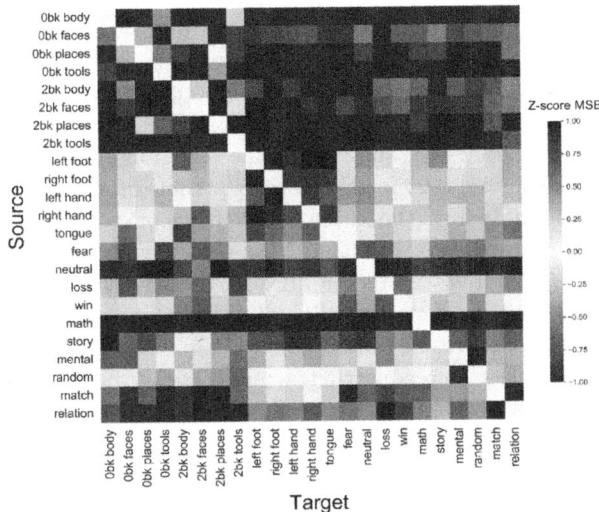

Fig. 6. Cognitive taskonomy derived from transfer learning, demonstrates the similarity among 23 cognitive tasks. These values represent changes in the reconstruction result when transferring from source tasks to target tasks. Positive values indicate that the source task improves the reconstruction results of the target task, and negative values indicate the opposite.

3.3 Relationships Among Cognitive Tasks

This study employs a masked fMRI modeling framework with transfer learning to quantify the relationships between cognitive tasks. As illustrated in Fig. 1C, the transfer learning process comprises two steps. Initially, the MAE model is trained on masked fMRI data from the source task. After training, the MAE model is transferred to the target task to evaluate the transfer performance. During fine-tuning on the target task, the parameters of the encoder are fixed, while only the parameters of the decoder are updated using 1% of the target task data. Finally, the relationship between tasks is determined by the transfer performance. Better transfer performance indicates a closer relationship between the source and target tasks, whereas poorer performance suggests a more distant relationship.

Figure 6 depicts the transfer results among the 23 cognitive tasks, reflecting the relationships between these tasks. Tasks within the same cognitive category exhibit closer relationships. For instance, the five subtasks within the motor category demonstrate strong transferability among each other, particularly between the left foot and right foot, and between the left hand and right hand. Furthermore, transferability varies across different cognitive tasks. For example, tasks such as math and neutral tasks exhibit effective transferability to other tasks, whereas working memory tasks face challenges in transferring to other tasks.

Figure 7 presents the hierarchical clustering results of cognitive tasks based on the Euclidean distance matrix calculated from reconstruction results between tasks. The results demonstrate that tasks within the same cognitive category are more closely clustered. For instance, the five subtasks in the motor category are closely clustered, and the random and mental tasks within the social category are in close proximity. Notably, within the working memory category, 0-back places are closer to 0-back tools, and 2-back faces are closer to 2-back bodies. This indicates a stronger relationship between places and tools, as well as between faces and bodies. Figure 7B, Fig. 7C, and Fig. 7D employ various hierarchical clustering strategies, which generally group tasks within the same cognitive category together, further illustrating the close relationships among similar cognitive tasks.

4 Discussion

In the brain masking experiments, it was observed that the reconstruction results for the VIS, DAN, and FPN networks were better. The redundancy and consistency of information within brain networks may account for these differences [19,20]. Information from other related regions that are more consistent with the masked brain region will assist in its reconstruction. Furthermore, some regions within these networks may demonstrate functional specialization in processing specific tasks, indicating a functional similarity among these regions [38]. This enables related regions to more effectively reconstruct missing information during reconstruction. Conversely, it was found that the LIM network presents more challenges in reconstruction compared to other brain networks. This is

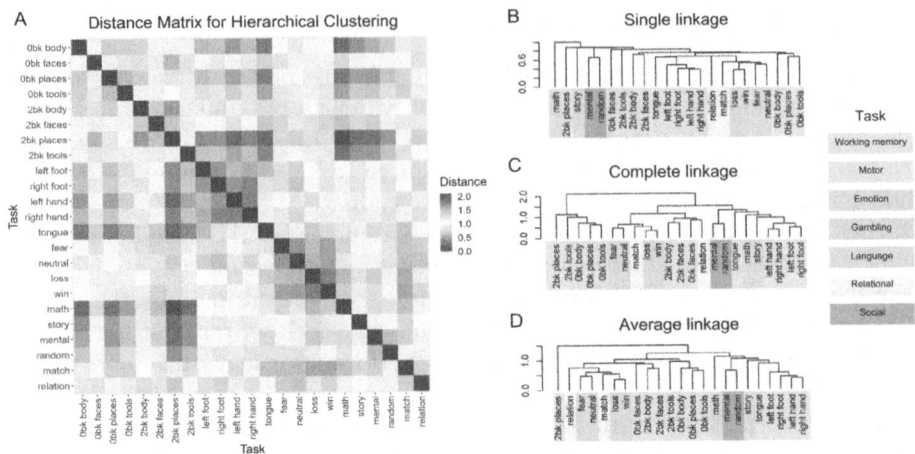

Fig. 7. Hierarchical clustering results of cognitive tasks. (A) The distance matrix among 23 cognitive tasks, calculated using Euclidean distances between pairs based on the reconstruction results of brain regions for each task. (B), (C), and (D) show the results of hierarchical clustering performed with single linkage, complete linkage, and average linkage methods, respectively. The colors on the right indicate the cognitive categories to which each subtask belongs.

potentially due to the regions within the limbic system performing multiple functions [25]. This multifunctionality amplifies the complexity of neural activity patterns, making it more difficult for the MAE model to capture the dynamic patterns of these brain regions compared to other brain networks. In the time masking experiments, we observed that reconstruction performance was relatively poor at the beginning and end of the sequence. In tests predicting fMRI signals across various time lengths, it was found that the longer the prediction time, the poorer the performance. This can be attributed to the inherent complexity and spontaneous activation of resting-state fMRI signals [23,31]. Brain activity is characterized by highly complex and dynamic features, influenced by a variety of internal and external factors. As the prediction length increases, this complexity and dynamism render accurate predictions increasingly challenging.

In this study, reconstruction experiments on task-based fMRI signals reveal that language and motor tasks were easier to reconstruct, while working memory tasks were more difficult. This difference is probably related to the inherent complexity of these tasks. Firstly, the brain regions activated by motor and language tasks tend to be localized [2]. In contrast, working memory tasks involve multiple cognitive processes that require coordination of the brain [6,8], which increases the complexity of reconstruction. Furthermore, individual variability across different cognitive tasks may also contribute to the observed differences in reconstruction difficulty [1,31]. Working memory tasks often exhibit greater variability between individuals, making it more difficult for the model to reconstruct fMRI signals across different subjects [8].

To explore the relationships between cognitive tasks, transfer learning experiments were conducted across cognitive tasks. Better transfer performance between cognitive tasks indicates a closer relationship among them. Figure 6 reveals that subtasks within the same category tend to have better transfer results. For instance, effective transfer performance is observed among the five motor subtasks. These subtasks primarily engage the primary motor cortex and premotor cortex, where the activation patterns and representations are more similar, thereby facilitating generalization between these tasks during transfer learning [15]. The distance matrix in Fig. 7 illustrates that the five motor subtasks cluster closely, reflecting the similarity among these subtasks. Additionally, differences exist in transfer ability between different cognitive tasks, with not all tasks within the same category demonstrating good transfer performance. The transfer performance of working memory tasks is poor, potentially due to the inconsistent brain regions and dynamic patterns, such as the different task difficulties with varying n-back [8]. Similar to findings in neural decoding classification models [24], fMRI reconstruction results also reveal a high similarity between emotion, social, and gambling tasks. This similarity may arise from the involvement of complex decision-making processes, which activate similar brain regions such as the prefrontal cortex [2,21]. Overall, the transfer performance between cognitive tasks can elucidate the relationships among them.

This study demonstrates that the MAE model can robustly reconstruct fMRI signals across subjects. This highlights the potential of using MAE models to address the common issue of signal loss in applications [32]. The analysis of reconstruction difficulty across brain regions provides a reliability assessment, enabling a more cautious interpretation of the results in brain regions that are challenging to reconstruct. Future research could extend to exploring the application of MAE models to other modalities of neural signals, such as EEG reconstruction. Furthermore, the cognitive taskonomy provides a more reliable guidance for selecting source tasks in neural decoding studies. This relationship matrix can assist researchers in more effectively utilizing source task data, thereby improving decoding performance in target tasks.

5 Conclusion

In this study, we employ the MAE model to reconstruct both resting-state and task-based fMRI signals and utilize a transfer learning framework to quantify the relationships between cognitive tasks. The MAE model effectively captures the temporal dynamics and interactions among brain regions, thereby enabling robust reconstruction of fMRI signals. By utilizing the transfer learning framework, we further quantify the relationships between cognitive tasks, revealing subtask correlations within motor tasks and identifying similarities among emotion, social, and gambling tasks. This study not only demonstrates the potential of the MAE model to reconstruct lost signals but also provides guidance for selecting source tasks in neural decoding studies.

Acknowledgments. This work was funded in part by the National Key R&D Program of China (2021YFF1200804), Shenzhen Science and Technology Innovation Committee (2022410129, KCXFZ20201221173400001).

References

1. Baldassarre, A., Lewis, C.M., Committeri, G., Snyder, A.Z., Romani, G.L., Corbetta, M.: Individual variability in functional connectivity predicts performance of a perceptual task. Proc. Natl. Acad. Sci. **109**(9), 3516–3521 (2012)
2. Barch, D.M., et al.: Function in the human connectome: task-fMRI and individual differences in behavior. Neuroimage **80**, 169–189 (2013)
3. Brown, T., et al.: Language models are few-shot learners. Adv. Neural. Inf. Process. Syst. **33**, 1877–1901 (2020)
4. Bubeck, S., et al.: Sparks of artificial general intelligence: early experiments with GPT-4. arXiv preprint arXiv:2303.12712 (2023)
5. Chen, J.E., Rubinov, M., Chang, C.: Methods and considerations for dynamic analysis of functional MR imaging data. Neuroimaging Clin. **27**(4), 547–560 (2017)
6. Collette, F., Van der Linden, M.: Brain imaging of the central executive component of working memory. Neurosci. Biobehav. Rev. **26**(2), 105–125 (2002)
7. Devlin, J., Chang, M.W., Lee, K., Toutanova, K.: BERT: pre-training of deep bidirectional transformers for language understanding. In: Proceedings of the 2019 Conference of the North American Chapter of the Association for Computational Linguistics: Human Language Technologies, Volume 1 (Long and Short Papers), pp. 4171–4186 (2019)
8. Dong, S., Reder, L.M., Yao, Y., Liu, Y., Chen, F.: Individual differences in working memory capacity are reflected in different ERP and EEG patterns to task difficulty. Brain Res. **1616**, 146–156 (2015)
9. Dosovitskiy, A., et al.: An image is worth 16×16 words: transformers for image recognition at scale. In: International Conference on Learning Representations (2021). https://openreview.net/forum?id=YicbFdNTTy
10. Feichtenhofer, C., Li, Y., He, K., et al.: Masked autoencoders as spatiotemporal learners. Adv. Neural. Inf. Process. Syst. **35**, 35946–35958 (2022)
11. Flesch, T., Saxe, A., Summerfield, C.: Continual task learning in natural and artificial agents. Trends Neurosci. **46**(3), 199–210 (2023)
12. Garner, K.G., Dux, P.E.: Knowledge generalization and the costs of multitasking. Nat. Rev. Neurosci. **24**(2), 98–112 (2023)
13. Glasser, M.F., et al.: A multi-modal parcellation of human cerebral cortex. Nature **536**(7615), 171–178 (2016)
14. Glasser, M.F., et al.: The minimal preprocessing pipelines for the human connectome project. Neuroimage **80**, 105–124 (2013)
15. Gordon, E.M., et al.: A somato-cognitive action network alternates with effector regions in motor cortex. Nature **617**(7960), 351–359 (2023)
16. He, K., Chen, X., Xie, S., Li, Y., Dollár, P., Girshick, R.: Masked autoencoders are scalable vision learners. In: Proceedings of the IEEE/CVF Conference on Computer Vision and Pattern Recognition, pp. 16000–16009 (2022)

17. Huang, P.Y., et al.: Masked autoencoders that listen. Adv. Neural. Inf. Process. Syst. **35**, 28708–28720 (2022)
18. Luo, W., Yin, W., Liu, Q., Qu, Y.: A hybrid brain-computer interface using motor imagery and SSVEP based on convolutional neural network. Brain-Apparatus Commun.: J. Bacomics **2**(1), 2258938 (2023)
19. Luppi, A.I., et al.: A synergistic core for human brain evolution and cognition. Nat. Neurosci. **25**(6), 771–782 (2022)
20. Luppi, A.I., Rosas, F.E., Mediano, P.A., Menon, D.K., Stamatakis, E.A.: Information decomposition and the informational architecture of the brain. Trends Cogn. Sci. (2024)
21. Olsson, A., Ochsner, K.N.: The role of social cognition in emotion. Trends Cogn. Sci. **12**(2), 65–71 (2008)
22. Pan, S.J., Yang, Q.: A survey on transfer learning. IEEE Trans. Knowl. Data Eng. **22**(10), 1345–1359 (2009)
23. Power, J.D., Schlaggar, B.L., Petersen, S.E.: Studying brain organization via spontaneous fMRI signal. Neuron **84**(4), 681–696 (2014)
24. Qu, Y., Jian, X., Che, W., Du, P., Fu, K., Liu, Q.: Transfer learning to decode brain states reflecting the relationship between cognitive tasks. In: International Workshop on Human Brain and Artificial Intelligence, pp. 110–122. Springer (2022)
25. Rolls, E.T.: The cingulate cortex and limbic systems for emotion, action, and memory. Brain Struct. Funct. **224**(9), 3001–3018 (2019). https://doi.org/10.1007/s00429-019-01945-2
26. Tian, C., Fei, L., Zheng, W., Xu, Y., Zuo, W., Lin, C.W.: Deep learning on image denoising: an overview. Neural Netw. **131**, 251–275 (2020)
27. Van Essen, D.C., et al.: The human connectome project: a data acquisition perspective. Neuroimage **62**(4), 2222–2231 (2012)
28. Wang, Z., Chen, J., Hoi, S.C.: Deep learning for image super-resolution: a survey. IEEE Trans. Pattern Anal. Mach. Intell. **43**(10), 3365–3387 (2020)
29. Xiang, H., Zou, Q., Nawaz, M.A., Huang, X., Zhang, F., Yu, H.: Deep learning for image inpainting: a survey. Pattern Recogn. **134**, 109046 (2023)
30. Xie, Z., et al.: SimMIM: a simple framework for masked image modeling. In: Proceedings of the IEEE/CVF Conference on Computer Vision and Pattern Recognition, pp. 9653–9663 (2022)
31. Xing, X.X., Gao, X., Jiang, C.: Individual variability of human cortical spontaneous activity by 3T/7T fMRI. Neuroscience **528**, 117–128 (2023)
32. Yan, Y., et al.: Reconstructing lost bold signal in individual participants using deep machine learning. Nat. Commun. **11**(1), 5046 (2020)
33. Ye, Z., Qu, Y., Liang, Z., Wang, M., Liu, Q.: Explainable fMRI-based brain decoding via spatial temporal-pyramid graph convolutional network. Hum. Brain Mapp. **44**(7), 2921–2935 (2023)
34. Yeo, B.T., et al.: The organization of the human cerebral cortex estimated by intrinsic functional connectivity. J. Neurophysiol. (2011)
35. Yin, W., Qu, Y., Ma, Z., Liu, Q.: HyperNTF: a hypergraph regularized nonnegative tensor factorization for dimensionality reduction. Neurocomputing **512**, 190–202 (2022)
36. Yosinski, J., Clune, J., Bengio, Y., Lipson, H.: How transferable are features in deep neural networks? Adv. Neural Inf. Process. Syst. **27** (2014)
37. Zamir, A.R., Sax, A., Shen, W., Guibas, L.J., Malik, J., Savarese, S.: Taskonomy: disentangling task transfer learning. In: Proceedings of the IEEE Conference on Computer Vision and Pattern Recognition, pp. 3712–3722 (2018)

38. Zeki, S., Watson, J., Lueck, C., Friston, K.J., Kennard, C., Frackowiak, R.: A direct demonstration of functional specialization in human visual cortex. J. Neurosci. **11**(3), 641–649 (1991)
39. Zhang, H., Zhao, M., Wei, C., Mantini, D., Li, Z., Liu, Q.: EEGdenoiseNet: a benchmark dataset for deep learning solutions of EEG denoising. J. Neural Eng. **18**(5), 056057 (2021)

Interpersonal Relationship Analysis with Dyadic EEG Signals via Learning Spatial-Temporal Patterns

Wenqi Ji[1], Fang Liu[2], Xinxin Du[1], Niqi Liu[1], Chao Zhou[3], Minjing Yu[4], Xinyue Guo[1], Guozhen Zhao[5], and Yong-Jin Liu[1(✉)]

[1] Department of Computer Science and Technology, Tsinghua University, Beijing 100084, People's Republic of China
{jwq21,lnq22,xy-guo21}@mails.tsinghua.edu.cn,
{duxx,liuyongjin}@tsinghua.edu.cn

[2] State Key Laboratory of Media Convergence and Communication, Communication University of China, Beijing 100024, People's Republic of China
fangliu@cuc.edu.cn

[3] Institute of Software, Chinese Academy of Sciences, Beijing 100190, People's Republic of China
zhouchao@iscas.ac.cn

[4] College of Intelligence and Computing, Tianjin University, Tianjin 300072, People's Republic of China
minjingyu@tju.edu.cn

[5] CAS Key Laboratory of Behavioral Science, Institute of Psychology, Beijing 100049, People's Republic of China
zhaogz@psych.ac.cn

Abstract. Interpersonal relationship quality is pivotal in social and occupational contexts. Existing analysis of interpersonal relationships mostly rely on subjective self-reports, whereas objective quantification remains challenging. In this paper, we propose a novel social relationship analysis framework using spatio-temporal patterns derived from dyadic EEG signals, which can be applied to quantitatively measure team cooperation in corporate team building, and evaluate interpersonal dynamics between therapists and patients in psychiatric therapy. First, we constructed a dyadic-EEG dataset from 72 pairs of participants with two relationships (stranger or friend) when watching emotional videos simultaneously. Then we proposed a deep neural network on dyadic-subject EEG signals, in which we combine the dynamic graph convolutional neural network for characterizing the interpersonal relationships among the EEG channels and 1-dimension convolution for extracting the information from the time sequence. To obtain the feature vectors from two EEG recordings that well represent the relationship of two subjects, we integrate deep canonical correlation analysis and triplet loss for training the network. Experimental results show that the social relationship type (stranger or friend) between two individuals can be effectively identified through their EEG data.

Keywords: Interpersonal relationship quality · Dyadic EEG signals · Spatio-temporal patterns · Deep canonical correlation analysis

1 Introduction

With the development of artificial intelligence technologies, EEG data has been extensively explored across various research areas and application scenarios, including emotion recognition [2], depression or epilepsy detection [24], and various implementations of brain-computer interface [21]. In deep-learning-based EEG analysis, the majority of existing studies focus on tasks for a single person [6,33]. However, we note that psychological investigations [1] have found that there exists mutual influence in EEG data amongst individuals in a group, and their EEG data shows synchrony under certain tasks and stimuli.

Inspired by this, in this paper we propose to study interpersonal relationships, which are categorized basically to communal sharing relationships (e.g., friends, family members or couples) or equality matching relationships (e.g., strangers) [13], by dyadic EEG data from two individuals. The findings of our study can be used to objectively measure the quality of interpersonal relationships, for example, to guide the selection of team members and to circumvent the problem of cognitive bias that can arise when employees assess the level of rapport between them on their own. Our work aims to expand and enlighten EEG-based research, shifting the focus from the neural variations of individuals to the interconnections in EEG patterns across individuals.

Unlike the analysis of single-subject EEG data, the exploration of dyadic-subject EEG data presents significant disparities. Primarily, from a psychological perspective, the EEG data of paired participants reflect distinct mental phenomenon when faced with specific tasks compared to data from individual participants. Technologically, the analysis of dyadic-subject EEG data differs from the widely-researched individual EEG tasks like emotion recognition [3], requiring a unique experimental design and procedure. Moreover, for the construction and subsequent processing of datasets, studies of interpersonal interaction require additional mechanisms to analyze the relationship between the two participants. This multifaceted analysis extends beyond the scope of traditional single-subject EEG research and presents the unique challenges and complexities inherent to dyadic-subject EEG investigation.

To address the aforementioned challenges, our paper takes a pioneering step in the exploration of interpersonal relationships using dyadic-subject EEG data and AI methodologies. We have established a comprehensive process for analyzing dyadic-subject EEG data, covering problem definition, experimental design, dataset creation for dyadic-subject EEG, and the final algorithmic approach. As illustrated in Fig. 1, we use hyperscanning, designed for simultaneously recording the brain activity of two or more individuals participating in the same cognitive activity [5], to construct our proposed dataset from the hyperscanned EEG recordings. Furthermore, we design a social relationship classification network tailored for dyadic-subject EEG data, which extracts spatial-temporal features from EEG sequence, inspired by [27], combine the classification with canonical correlation analysis (CCA). To enhance the classification process, we have also integrated a triplet framework for enhancing the representations of the social relationship between the two subjects.

Our contributions are summarized as follows:

- We propose a novel EEG data analysis task to explore the social relationship between two individuals, and establish a pipeline for collecting and analyzing dyadic EEG data.
- We propose a network to model the temporal and spatial information of EEG data, where the EEG channels are modeled as nodes in a graph and processed with graph convolution operations while the time sequence is analyzed with 1-D convolution. For determining the relationships, we integrate the attention mechanism to fuse two types of features from dyadic EEG data and propose a training scheme with triplet loss and CCA loss.
- We construct a dyadic-subject EEG dataset for the analysis of social relationship between two individuals.

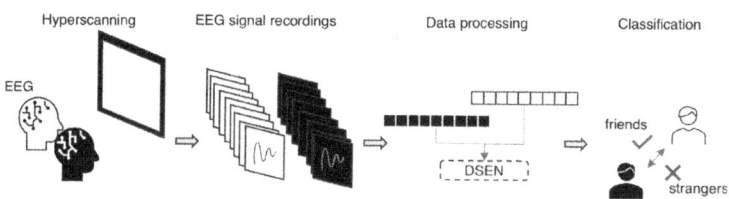

Fig. 1. The procedure of exploring the relationships between two individuals. In our proposed framework, the relationship between the two participants (stranger or friend) can be determined based on their EEG recordings while watching emotional video clips.

2 Related Works

In this section, we first briefly delve into methods for EEG analysis based on deep learning. As mentioned previously, the majority of these techniques primarily concentrate on analyzing individual EEG data. Contrastively, in this paper we shift our focus to studies that specifically center on interpersonal relations. It is noteworthy that some of these works on multi-subjects do not employ EEG data, while others refrain from leveraging AI-related technologies.

2.1 Deep-Learning-Based EEG Processing

The exploration of emotion types or psychological states using EEG can be traced back to the 1980s [22]. Machine-learning-based methodologies for EEG analysis, including signal prediction and classification, were introduced in the early 2000s [8,14].

In order to explore the relationship between two individuals, discriminative features are required. Although most focused on a single subject, existing studies on EEG-based emotion measurement have spawned effective methods for extracting effective EEG features and have performed well on tasks like emotion recognition. Regarding EEG-derived features, [3,9] indicated that features such as power spectral density (PSD) or differential entropy (DE) can be extracted. Further, many machine-learning-based algorithms have been proposed to tackle tasks related to emotions or brain activities using features from EEG. Among these studies, a subset based on traditional machine learning techniques, e.g., the support vector machine (SVM) and random forest (RF), being the most prevalently utilized [3]. Within deep learning approaches, convolutional neural network (CNN) stands out given its capability to automatically learn spatial and temporal patterns from data. In the context of EEG-based CNN algorithms, for instance, EEG-Net [17] effectively learns information both temporally and inter-electrodes in EEG data. As a representative of recursive network architectures, [4] applied the LSTM to EEG and achieved good results in emotion recognition. Furthermore, based on graph that can effectively model the electrodes in EEG, [26] introduced a dynamic graph convolutional neural network (DGCNN) to simultaneously optimize network parameters and a weighted graph characterizing the strength of functional relation between each pair of two electrodes. SparseDGCNN [30] modified DGCNN by imposing a sparseness constraint on the graph. Besides, other mainstream EEG analysis studies integrate various CNN and LSTM architectures to process EEG data, including deep belief network (DBN) [32], variational pathway reasoning (VPR) [31], etc. Moreover, there is a knowledge distillation framework [20] and the SSLAPP method [19] that employs attention modules for encoding the EEG signal and GANs for limited labeled data.

2.2 Multi-subjects Studies

Hyperscanning [12] is a technique that involves the simultaneous recording of brain activity from two or more individuals engaged in the same task or social interaction. It is a widely used experimental paradigm for multi-subject studies.

To analyze synchronization and mutual psychological influence on individuals within a group, existing studies often involve physiological signals other than EEG or do not employ AI-based methods. [10] investigated paired subjects performing tasks together using functional near-infrared spectroscopy (fNIRS), and their study revealed that certain brain regions, specifically the inferior frontal gyrus (IFG), played a pivotal role during the task. Similarly, [23] also cope up with fNIRS to investigate dyads consisting of lovers, friends, and strangers. With EEG data, [7] conducted a comparative analysis to identify EEG features suitable for inter-brain exploration. These studies validated the existence of brain synchronization during multi-subject activities, thereby enabling the exploration of relationships between two subjects through their neural signal data. In this paper, we investigate a novel task to determine the relationship type of the two participants using dyadic-subject EEG data.

3 Methods

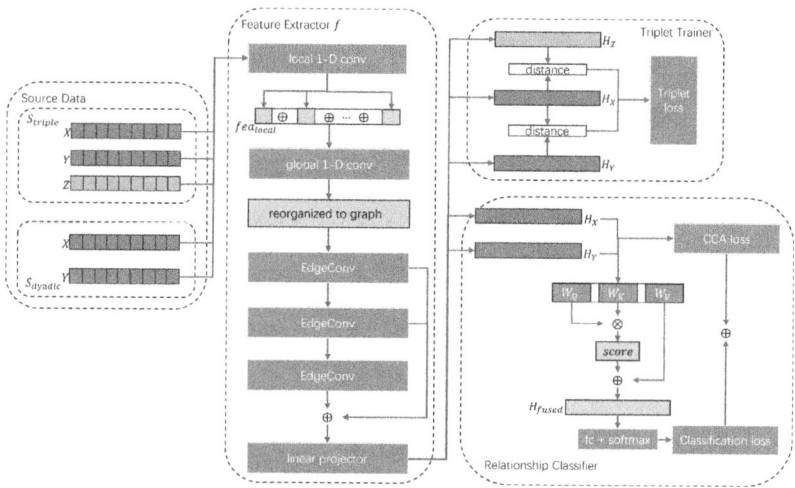

Fig. 2. The framework of our proposed Dyadic-Subject EEG Network (DSEN). DSEN takes dyadic-subject EEG data as input to extract temporal and spatial features with the feature extractor f. The features are fused by an attention-based module to form a cohesive representation for classification. The feature extractor f is trained individually via a triplet loss. The fused features are utilized by a classifier, which is trained using a combined CCA loss and classification loss.

3.1 Dyadic-Subject EEG Dataset

EEG Data Collection. In our experiment, EEG signals were recorded using the NuAmps™ 40-channel unipolar system (Compumedics Neuroscan, USA) with a sampling frequency of 1000 Hz. Two reference electrodes were attached to the mastoid areas on both sides of the subjects. The impedance of all electrodes was maintained below 5 kΩ. A low-pass filter was set to 60 Hz.

Seventy-two pairs of healthy participants, aged between 19 and 30, were recruited from universities and participated in the experiment. Among all these pairs of participants, 18 pairs were strangers to each other, while the remaining 54 pairs were friends or couples. During the collection of EEG data, the subjects were seated half a meter apart and were not allowed to communicate with each other.

For main experimental tasks, the two subjects who participated in the hyperscanning experiments were asked to watch nine video clips. Given that positive emotion exhibits more synchronization than negative emotions in interpersonal relationships [15], video clips labeled with positive emotions were selected from the Chinese emotional film database proposed by [11] for the experiments. Moreover, to provide a richer emotional perspective, nine video clips representing the positive emotions of friendship, awe, kinship, respect, gratitude, humor, longing, pride, and love were chosen. The duration of the video clips ranged from 160 s to 296 s, with an average length of 216 s. After watching each emotional video clip, subjects were required to perform a distractive task to neutralize the emotional effect of the previously watched video clip, and were given a mandatory half-minute break before the next video clip was presented. The above process is repeated until all 9 video materials have been played.

All procedures of the study were reviewed and approved by the Ethics Committee of Tsinghua University. Before the experiment, all the participants were explained well about the details of the experiment and signed the informed consent.

Preprocessing. Through the aforementioned experiment, we collected our dyadic-subject EEG data that includes hyperscanned EEG signal recordings of 144 participants as 72 pairs[1], who were co-watching nine emotional video clips. For subsequent processing and analysis, we downsampled the data to 200 Hz, applied a digital filter from 1–45 Hz, and then re-referenced using the average of the left and right mastoids. Additionally, on this basis, ocular artifacts were removed using independent component analysis (ICA). In order to evaluate our proposed DSEN framework (Sect. 3.4), we segmented the EEG signal recordings of subjects watching each video clip into 2-s windows. Then, we concatenated the 2-s segments from all nine video clips without overlaps to form a data sample, thereby establishing our dyadic-subject EEG dataset.

3.2 Preliminary Study

Currently widely-used EEG datasets, such as DEAP [16], SEED [32], and MPED [25], only contain data of individual subjects and are not suitable for analyzing the relationship between two subjects. We set up an experiment to collect dyadic-subject EEG data for exploring the social relationship between two individuals. Section 3.1 provides the detailed construction process of our dyadic-subject EEG dataset. Then we conduct a specific statistical test on the dyadic-subject EEG data to validate the correlation between the participants relationships and the EEG signals.

Notations. In our dataset, each data sample represents EEG data of a pair of two subjects, denoted by (X, Y). As input for the subsequent classification and analyses tasks, we derive the amplitude sequences denoted by (A_X, A_Y) from

---

[1] All EEG data comprise a total of 5,688 data samples after window-based segmentation and concatenation.

(X, Y). Let n represent the number of emotional video clips that the two subjects co-watched, each input sample $(a_x, a_y) = \{(a_x, a_y)^1, (a_x, a_y)^2, \cdots, (a_x, a_y)^n\} \in (A_X, A_Y)$ contains n EEG segments from the two subjects co-watching all videos clips with different labels of emotions.

Statistic Analysis. In our proposed innovative analysis process, we first verify the data distribution, and then conduct preliminary feature selection and processing.

We employ a Shapiro-Wilk (S-W) test to verify that the EEG data we collected conforms to a normal distribution. Consequently, we can use the t-test method on the band-pass filtered EEG data to verify the influence of the relationship and gender pairing of the dyadic subjects. This initial analysis serves to explore the correlation between the EEG data of the two subjects with respect to their relationships. Additionally, these procedures also effectively alleviate the potential influence of gender on synchrony differences. For S-W test and t-tests, the instantaneous amplitude and phase samples are generated from the analytic signals $X(t) + iX_H(t)$ and $Y(t) + iY_H(t)$, obtained from Hilbert transform [18]. The amplitude and phase samples are derived as:

$$A_X(t) = \sqrt{X(t)^2 + X_H(t)^2}, A_Y(t) = \sqrt{Y(t)^2 + Y_H(t)^2}$$
$$\phi_X(t) = tan^{-1}\frac{X_H(t)}{X(t)}, \phi_Y(t) = tan^{-1}\frac{Y_H(t)}{Y(t)} \quad (1)$$

Subsequently, the inter-subject correlation ISC_{XY} and the phase locking value PLV_{XY} are derived by:

$$ISC_{XY} = \frac{cov(A_X(t), A_Y(t))}{\sigma A_X(t)\sigma A_Y(t)}$$
$$PLV_{XY} = |\frac{1}{N}\sum_{n=1}^{N} e^{i(\phi_X(tn) - \phi_Y(tn))}| \quad (2)$$

With the ISC_{XY} and PLV_{XY}, we applied a t-test for both the relationship status of the pair (whether they know each other or not). We also compared the significance of dyadic relationship differences across various frequency bands, namely θ (4–7 Hz), α (8–12 Hz), β (13–29 Hz), and γ (30–45 Hz). This analyses enabled us to pre-select the optimal frequency band combination, which are β and γ bands, for subsequent processing.

T-Test Results. We used the ISC and PLV features in different frequency bands as dependent variables. The type of relationship between them was used as an independent variable for the t-test. Specifically, the types of relationships were divided into freinds and strangers. These categorizations facilitated the exploration of potential influences on the dependent variables.

The significance level was set to 0.05. Table 1 demonstrates that the EEG synchrony features ISC for dyadic subjects show significant differences across

Table 1. t-test results with ISC and PLV from dyadic-subject EEG data as dependent variables and relationships of them as independent variables.

IV	feature	statistic	p-value
relation	PLV_θ	−0.400	0.691
	PLV_α	0.587	0.559
	PLV_β	0.282	0.779
	PLV_γ	−1.080	0.284
	ISC_θ	0.735	0.465
	ISC_α	−0.250	0.803
	ISC_β	**1.787**	**0.078**
	ISC_γ	**2.287**	**0.025**

different relationship types in β frequency band and marginally significant in γ band. Our experimental results show that synchronized dyadic EEG activities in the β and γ frequency bands are effective in distinguishing the relationships between two individuals. This is also supported by some existing findings that the β or γ frequency bands play a more significant role in regression analysis for inter-brain emotion prediction [7]. These discoveries in the preliminary study provided theoretical foundations for our algorithmic design of predicting relationships based on EEG data and our pre-selected EEG features.

3.3 Feature Extraction

In order to extract efficient features from the raw EEG data i.e., (A_X, A_Y), we propose a sequential processing framework that mainly contains two modules (see the feature extractor block in Fig. 2): 1) a CNN module to extract the temporal information from the EEG amplitude sequences, and 2) a DGCNN module to identify the relationships among EEG electrodes.

Considering that our input includes EEG data segments of dyadic subjects co-watching nine different video clips (see Sect. 3.1 for details), a grouped 1-D convolutional layer is used to extract the local temporal features $\{fea^1_{local}, \cdots, fea^9_{local}\}$ from each segment of a_x and a_y, and the outputs are concatenated into $F = fea^1_{local} \oplus \cdots \oplus fea^9_{local}$ as local temporal feature outputs and then are used as the input of following global 1-D convolutional layer. The second 1-D convolutional layer is applied to obtain the overall temporal features T_x, T_y from F_x, F_y. Within the CNN module described above, we apply convolutions separately for each channel of EEG.

Although the CNN-based layers ensure that our temporal information extraction module can effectively learn the time-series features in EEG data, they do not account for inter-electrode relationship information. Consequently, DGCNN, as a method that can effectively extract information between electrode channels, is employed for feature extraction from the EEG electrodes. We set the adjacency matrix A, constructed as a fully-connected graph, to represent the connectivity between each pair of vertices in the graph. Taking EEG electrode channels as vertices (denoted by ch), the $v = ch \times fea \in \mathcal{V}$ could be reorganized to form the vertex set V, and then we set the input graph $G = \{V, A\}$. Inspired by [29], we introduce three EdgeConv blocks. Specifically, for a vertex v_i in V and one of its adjacent vertices $v'_i \in V'$, a transformed feature h_i of v_i is obtained from EdgeConv by

$$h_i = \phi_e(fea'_i - fea_i) \quad (3)$$

where ϕ_e denotes the shared multi-layer perceptron (sMLP) of the eth EdgeConv block. Taking $H_{pooled}^{1\sim3}$ as global max pooled vertex features from three EdgeConv blocks, the feature H extracted by DGCNN, which is also the output feature of our proposed extractor, is obtained from $\varphi(H_{pooled}^1 \oplus H_{pooled}^2 \oplus H_{pooled}^3)$, where φ is linear transformations.

3.4 Dyadic-Subject EEG Network (DSEN)

As shown in Fig. 2, our proposed dyadic-subject EEG network (DSEN) consists of a triplet network and a classifier, in which we first package our source data from a triplet set and a dyadic set as $S = \{S_{tri}, S_{dual}\}$. The triplet source data is set as $S_{tri} = \{X, Y, Z\}$, where X and Y represent dyadic subjects (selected for S_{tri} with ground truth of relationship types), while Z is a subject selected randomly from the other participants, ensuring that Z is a stranger to both X and Y. The S_{dual} is formed as $\{X, Y, L\}$, where L is the set of labels with values of 0 or 1 indicating that the dyadic subject X and Y are strangers or freinds, respectively. To update the parameters of feature extractor $\hat{\theta}_f$ in training, the weight of f via triplet loss, the distance of cosine similarity between anchor (X) and positive (Y) or negative (Z) is calculated on the output features $\{H_X, H_Y, H_Z\}$ from the extractor. Then $\hat{\theta}_f$ is updated with back propagation of $\mathcal{L}_{triplet}$. To train the classifier, we fuse the two features $\{H_X, H_Y\}$ extracted from S_{dual}. For the feature fusion, we integrated the attention mechanism since we perceive H_X and H_Y as sequences of embeddings, and interactions and relationships between them could be captured. To obtain the matrices of query (Q), key (K) and value (V) for both of H_X and H_Y, each of the two features is transformed linearly using learnable matrices W_Q, W_K and W_V by:

$$Q_{X,Y}, K_{X,Y}, V_{X,Y} = W_{Q,K,V} \times H_{X,Y}. \quad (4)$$

Then the attention score S between H_X and H_Y, and the attention-based fusion feature H_{fused} are computed by:

$$S_{X,Y} = \frac{Q_{X,Y} \times K_{Y,X}^T}{c}$$
$$H_{fused} = (softmax(S_X) \times V_Y) \oplus (softmax(S_Y) \times V_X), \quad (5)$$

where c is a constant scale factor that is set to $128^{0.5}$ in our experiments. Subsequently, H_{fused} is used to generate the prediction for stranger and friend relationships of the dyadic subjects by passing through fully connected layers. To

Algorithm 1. training of DSEN

Input: data set S_{tri}, S_{dual}
Output: parameters $\hat{\theta}_f, \hat{\theta}_c$

    calculate the instantaneous amplitude set $\{A_X, A_Y, A_Z\}$ by $\{X, Y, Z\}$ from S_{tri}, S_{dual} via Eq. 1
    while not converge and not reach the max number of iterations **do**
        $fea_{local\text{-}X,Y,Z} = CNN1D_{local}(A_{X,Y,Z})$
        $T_{X,Y,Z} = CNN1D_{global}(fea_{local\text{-}X,Y,Z})$
        $H^1_{X,Y,Z} = EdgeConv_1(T_{X,Y,Z})$
        $H^2_{X,Y,Z} = EdgeConv_2(H^1_{X,Y,Z})$
        $H^3_{X,Y,Z} = EdgeConv_3(H^2_{X,Y,Z})$
        $H_{X,Y,Z} = H^1_{X,Y,Z} \oplus H^2_{X,Y,Z} \oplus H^3_{X,Y,Z}$
        Use $H_{X,Y,Z}$ of $X, Y, Z \in S_{tri}$
        compute $\mathcal{L}_{triplet}$ by Eq. 10 and 11
        update $\hat{\theta}_f$ with $\frac{\partial \mathcal{L}_{triplet}}{\partial \theta_f}$
        use $H_{X,Y}$ of $X, Y \in S_{dual}$
        compute covariance $R_{X,Y,XY}$
        compute $\mathcal{L}_{CCA}$ by Eq. 8 and 9
        get H_{fused} by Eq. 4 and 5
        $preds = Classifier(H_{fused})$
        $\mathcal{L}_{Classification} = Classificationloss(preds, labels)$
        compute $\mathcal{L}_{Combined}$ by Eq. 6
        update $\hat{\theta}_f$ and $\hat{\theta}_c$ with $\frac{\partial \mathcal{L}_{Combined}}{\partial \theta_f}$ and $\frac{\partial \mathcal{L}_{Combined}}{\partial \theta_c}$
    end while
    return $(\hat{\theta}_f, \hat{\theta}_c)$

update both $\hat{\theta}_f$ and the weight of the classifier $\hat{\theta}_c$, we propose a combined loss defined as:

$$\mathcal{L}_{Combined} = \alpha \cdot \mathcal{L}_{Classification} + \beta \cdot \mathcal{L}_{CCA} \qquad (6)$$

where $\mathcal{L}_{Classification}$ denotes the cross-entropy loss, and α, β are the combined factors, which are both set to 1 in our experiments. Algorithm 1 illustrates the complete training procedure of our proposed DSEN framework. Details of the triplet loss and the CCA loss used in our method are presented in the following section.

3.5 Loss Function

CCA Loss. Inspired by [27] that integrates canonical correlation analysis (CCA) with deep neural networks for multimodal language analysis, we introduced the CCA loss to learn the correlation between the dyadic features extracted by the proposed feature extractor. The features H_X, H_Y are normalized through mean centering, which ensures zero mean for each feature, and facilitates further covariance computation. With $H_X^{adapted}, H_Y^{adapted}$, we then calculate the covariance matrices of the two features R_X, R_Y and the cross-covariance R_{XY}. Our objective with CCA loss is to maximize the correlation between the features extracted by the feature extractor f, as shown in the equation:

$$\max_{f} corr(f(A_X), f(A_Y)) \tag{7}$$

the matrix E corresponding to canonical correlation coefficients between the two features is derived as:

$$E = R_X^{-\frac{1}{2}} R_{XY} R_Y^{-\frac{1}{2}}. \tag{8}$$

To update the parameters of f, in (7), CCA loss is defined as:

$$\mathcal{L}_{CCA} = -trace(E^T E)^{\frac{1}{2}} = -\sum s, \tag{9}$$

where s are the singular values of E and derived from singular value decomposition of E.

Triplet Loss. To further train the f to discriminate the relationship type between two subjects, we introduce the triplet loss as an additional criterion that helps f to learn more distinctions by comparing a pair of similar samples against a dissimilar sample. Given an *anchor* sample, a *positive* sample, and a *negative* sample, we formulate the computation of two samples based on cosine similarity as:

$$dist_{p/n} = 1.0 - \frac{anchor \cdot positive/negative}{\|anchor\|^2 \times \|positive/negative\|^2} \tag{10}$$

Then the triplet loss is defined as:

$$\mathcal{L}_{triplet} = ReLU(dist_p - dist_n + margin). \tag{11}$$

where the *margin* is set up to 1 in our experiment.

4 Experiments

4.1 Implementation

To validate the efficiency of our proposed DSEN on discriminating the social relationship type between two subjects, we compared five state-of-the-art methods for processing EEG data on our proposed dyadic-subject EEG dataset. Considering our dyadic data and the task of classifying the relationship, rather than emotion recognition as many of these methods were originally designed for, we implemented these models to extract features from EEG data of each individual separately. We then simply concatenated the features extracted from the EEG of both subjects as feature fusion and fed them through two linear layers for classification. All methods were trained and tested on our dyadic-subject EEG dataset. Noting that our dataset is imbalanced, with 18 pairs of subjects who are strangers and 54 pairs who are freinds to each other, and in order to verify the subject-independence of our model, we partitioned our training and test sets as follows: we selected 15 pairs of subjects from both the strangers and freinds relationships. The data samples from these 30 pairs, segmented using a 2-second window and then concatenated across all 9 video segments, were used to form the training set. The data samples from all remaining subject pairs constituted the test set.

Table 2. Results of different methods applied on our proposed dyadic-subject EEG dataset.

method	accuracy	F1 score
SVM	0.65	0.78
DBN	0.72	0.83
DGCNN	0.61	0.75
LSTM	0.58	0.72
EEGNet	0.67	0.79
SSLAPP	0.77	0.86
DSEN (ours)	**0.86**	**0.92**

4.2 Classification Results

The methods we used for comparison are SVM [28], DGCNN [26], LSTM [4], EEGNet [17], DBN [32], and SSLAPP [19]. Considering that there are currently no deep learning methods designed to handle dyadic EEG data, in our experiments, we selected some representative methods for EEG-based emotion recognition or sleeping stage classification. These methods were utilized as encoders for EEG data. Subsequently, the EEG features extracted from two individuals were concatenated, and an MLP classifier was employed to classify their relationship. In particular, as the traditional SVM model requires the input features to be a one-dimensional vector. Therefore, for the SVM classifier, preprocessing of the data is done that we first flatten each data sample from the dyadic EEG dataset and then compute the Pearson correlation between the dyadic samples over the entire time step range to serve as the artificial ISC features for the SVM. The accuracy and F1 score results are reported in Table 2. Compared to the baseline methods, including CNN-based, GCN-based, RNN-based, and traditional machine learning approach SVM, our method achieved a significant improvement of at least 14% in classification accuracy for determining the relationship type between the two subjects, reaching over 85%. meanwhile, in terms of the F1 score, our proposed DSEN also outperforms other methods by at least 0.09. The results indicate that the mechanism introduced in our model is more suitable for analyzing dyadic-subject EEG data. Compared to those algorithms designed for tasks like emotion recognition for individuals, our method can predict the relationship between two participants more effectively.

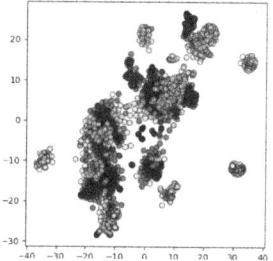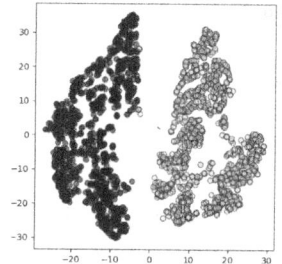

Fig. 3. (a) is the distribution plot of the unprocessed dyadic EEG data in a two-dimensional plane after dimension reduction by t-SNE. (b) is the distribution plot of extracted and fused features from the dyadic EEG data in the plane after dimension reduction by t-SNE.

4.3 Feature Discrimination Comparison

To validate the feature extraction capability of our proposed model for dyadic-subject EEG data, we conducted an experiment to visually compare the features extracted by our feature extractor and fused using our methods, with the unprocessed EEG amplitude data. These features were assessed based on their ability to discriminate different classes of data samples in our dyadic-subject EEG dataset. In the experiment, we employed t-distributed stochastic neighbor embedding (t-SNE) to reduce the dimension of the high-dimensional features. The features were then visualized in a 2D coordinate space. As illustrated in Fig. 3, it is obvious that the feature vectors extracted by our model from the dyadic-subject EEG can effectively distinguish between different social relationship types.

4.4 Ablation Study

In order to validate the efficiency of the specific methods introduced in our framework, we conducted a series of ablation studies. We compared the performance of our full model against various ablated versions, including those without the combined CCA loss, without the triplet loss, and versions where the features extracted from the feature extractor were directly concatenated instead of using an attention mechanism for fusion. The removal of CCA loss showed a notable decrease in performance, indicating its vital role in enhancing the correlation learning between the data of participants. Similarly, excluding the triplet loss led to reduced accuracy, showing its importance in optimizing the feature space for better discrimination and classification. The absence of an attention mechanism resulted in a lower F1 score, illustrating its efficacy in weighting and fusing features from the dyadic participants. Additionally, we also evaluated a version that employs raw EEG data without transforming it to compute the instantaneous amplitude of the EEG waveform. This approach significantly underperformed, demonstrating the necessity of data transformation for effective feature extraction and prediction. The outcomes of these ablation studies, presented in Table 3, indicate that the proposed enhancements in our framework significantly improve the prediction of the relationship between the two participants.

Table 3. Results of ablation studies.

modifications	accuracy	F1 score
without CCA loss	0.81	0.89
without triplet	0.78	0.87
without attention	0.81	0.89
raw EEG data	0.52	0.66
full model	**0.86**	**0.92**

5 Conclusion

In this paper, we introduce a novel research task that aims to explore the relationship between two individuals based on their EEG data. To accomplish this task, we delineated a comprehensive strategy encompassing experiment design, dataset construction, the preliminary statistical analyses that indicate the connection between the social relationship types and the dyadic EEG signals, an initial feature selection, and a DSEN framework designed for processing dyadic-subject EEG data. Our proposed classification algorithm, tailored for dyadic EEG data, incorporates advanced mechanisms such as CCA loss, triplet training, and attention-based feature fusion. By comparing our method with baseline algorithms designed for individual EEG emotion recognition, our results on the proposed dyadic-subject EEG dataset show its effectiveness in predicting interpersonal relationship types. Our ablation studies further emphasize the importance of each component we introduced.

In conclusion, our results demonstrate the potential of our framework in effectively utilizing dyadic EEG data to predict relationships. The introduced mechanisms and strategy have set a benchmark in this novel research area, drawing more attention to the exploration of interpersonal relationship types based on the synchrony of inter-brain within a group using EEG.

Acknowledgments. This work was supported by the Natural Science Foundation of China (U2336214 and 62332019).

References

1. Aftanas, L.I., Reva, N.V., Varlamov, A.A., Pavlov, S.V., Makhnev, V.P.: Analysis of evoked EEG synchronization and desynchronization in conditions of emotional activation in humans: temporal and topographic characteristics. Neurosci. Behav. Physiol. **34**, 859–867 (2004)

2. Al-Nafjan, A., Hosny, M., Al-Ohali, Y., Al-Wabil, A.: Review and classification of emotion recognition based on EEG brain-computer interface system research: a systematic review. Appl. Sci. **7**(12), 1239 (2017)
3. Alarcão, S.M., Fonseca, M.J.: Emotions recognition using EEG signals: a survey. IEEE Trans. Affect. Comput. **10**(3), 374–393 (2019). https://doi.org/10.1109/TAFFC.2017.2714671
4. Alhagry, S., Fahmy, A.A., El-Khoribi, R.A.: Emotion recognition based on EEG using LSTM recurrent neural network. Int. J. Adv. Comput. Sci. Appl. **8**(10) (2017)
5. Astolfi, L., et al.: Neuroelectrical hyperscanning measures simultaneous brain activity in humans. Brain Topogr. **23**(3), 243–256 (2010)
6. Bashivan, P., Rish, I., Heisig, S.: Mental state recognition via wearable EEG. arXiv preprint arXiv:1602.00985 (2016)
7. Ding, Y., Hu, X., Xia, Z., Liu, Y.J., Zhang, D.: Inter-brain EEG feature extraction and analysis for continuous implicit emotion tagging during video watching. IEEE Trans. Affect. Comput. **12**(1), 92–102 (2021). https://doi.org/10.1109/TAFFC.2018.2849758
8. Dornhege, G., Blankertz, B., Curio, G., Muller, K.R.: Boosting bit rates in non-invasive EEG single-trial classifications by feature combination and multiclass paradigms. IEEE Trans. Biomed. Eng. **51**(6), 993–1002 (2004)
9. Duan, R.N., Zhu, J.Y., Lu, B.L.: Differential entropy feature for EEG-based emotion classification. In: 2013 6th International IEEE/EMBS Conference on Neural Engineering (NER), pp. 81–84. IEEE (2013)
10. Gamliel, H.N., Nevat, M., Probolovski, H.G., Karklinsky, M., Han, S., Shamay-Tsoory, S.G.: Inter-group conflict affects inter-brain synchrony during synchronized movements. Neuroimage **245**, 118661 (2021)
11. Ge, Y., Zhao, G., Zhang, Y., Houston, R.J., Song, J.: A standardised database of Chinese emotional film clips. Cogn. Emot. **33**(5), 976–990 (2019)
12. Hakim, U., et al.: Quantification of inter-brain coupling: a review of current methods used in haemodynamic and electrophysiological hyperscanning studies. NeuroImage **280**, 120354 (2023). https://doi.org/10.1016/j.neuroimage.2023.120354, https://www.sciencedirect.com/science/article/pii/S1053811923005050
13. Haslam, N.: Categories of social relationship. Cognition **53**(1), 59–90 (1994)
14. Hazarika, N., Chen, J.Z., Tsoi, A.C., Sergejew, A.: Classification of EEG signals using the wavelet transform. Signal Process. **59**(1), 61–72 (1997)
15. Knight, A.P., Eisenkraft, N.: Positive is usually good, negative is not always bad: the effects of group affect on social integration and task performance. J. Appl. Psychol. **100**(4), 1214–1227 (2015)
16. Koelstra, S., Muhl, C., Soleymani, M., Lee, J.S., Yazdani, A., Ebrahimi, T., Pun, T., Nijholt, A., Patras, I.: DEAP: a database for emotion analysis; using physiological signals. IEEE Trans. Affect. Comput. **3**(1), 18–31 (2012). https://doi.org/10.1109/T-AFFC.2011.15
17. Lawhern, V.J., Solon, A.J., Waytowich, N.R., Gordon, S.M., Hung, C.P., Lance, B.J.: EEGNet: a compact convolutional neural network for EEG-based brain-computer interfaces. J. Neural Eng. **15**(5), 056013 (2018)
18. Le Van Quyen, M., et al.: Comparison of Hilbert transform and wavelet methods for the analysis of neuronal synchrony. J. Neurosci. Methods **111**(2), 83–98 (2001)
19. Lee, H., Seong, E., Chae, D.K.: Self-supervised learning with attention-based latent signal augmentation for sleep staging with limited labeled data. In: Proceedings of the Thirty-First International Joint Conference on Artificial Intelligence, IJCAI, pp. 3868–3876 (2022)

20. Liang, H., Liu, Y., Wang, H., Jia, Z., Center, B.: Teacher assistant-based knowledge distillation extracting multi-level features on single channel sleep EEG. In: Proceedings of the Thirty-Second International Joint Conference on Artificial Intelligence, IJCAI, pp. 3948–3956 (2023)
21. Lotte, F., Bougrain, L., Clerc, M.: Electroencephalography (EEG)-based brain-computer interfaces (2015)
22. Morihisa, J.M., Duffy, F.H., Wyatt, R.J.: Brain electrical activity mapping (beam) in schizophrenic patients. Arch. Gen. Psychiatry **40**(7), 719–728 (1983)
23. Pan, Y., Cheng, X., Zhang, Z., Li, X., Hu, Y.: Cooperation in lovers: an f NIRS-based hyperscanning study. Hum. Brain Mapp. **38**(2), 831–841 (2017)
24. Sharma, A., Rai, J.K., Tewari, R.P.: Epileptic seizure anticipation and localisation of epileptogenic region using EEG signals. J. Med. Eng. Technol. 1–14 (2018)
25. Song, T., Zheng, W., Lu, C., Zong, Y., Zhang, X., Cui, Z.: MPED: a multi-modal physiological emotion database for discrete emotion recognition. IEEE Access **7**, 12177–12191 (2019). https://doi.org/10.1109/ACCESS.2019.2891579
26. Song, T., Zheng, W., Song, P., Cui, Z.: EEG emotion recognition using dynamical graph convolutional neural networks. IEEE Trans. Affect. Comput. **11**(3), 532–541 (2018)
27. Sun, Z., Sarma, P.K., Sethares, W.A., Liang, Y.: Learning relationships between text, audio, and video via deep canonical correlation for multimodal language analysis. In: Proceedings of the AAAI Conference on Artificial Intelligence (2019). https://api.semanticscholar.org/CorpusID:207930647
28. Suykens, J.A., Vandewalle, J.: Least squares support vector machine classifiers. Neural Process. Lett. **9**, 293–300 (1999)
29. Wang, Y., Sun, Y., Liu, Z., Sarma, S.E., Bronstein, M.M., Solomon, J.M.: Dynamic graph CNN for learning on point clouds. ACM Trans. Graph. (TOG) (2019)
30. Zhang, G., Yu, M., Liu, Y.J., Zhao, G., Zhang, D., Zheng, W.: SparseDGCNN: recognizing emotion from multichannel EEG signals. IEEE Trans. Affect. Comput. **14**(1), 537–548 (2023). https://doi.org/10.1109/TAFFC.2021.3051332
31. Zhang, T., Cui, Z., Xu, C., Zheng, W., Yang, J.: Variational pathway reasoning for EEG emotion recognition. In: Proceedings of the AAAI Conference on Artificial Intelligence (2020)
32. Zheng, W.L., Lu, B.L.: Investigating critical frequency bands and channels for EEG-based emotion recognition with deep neural networks. IEEE Trans. Auton. Ment. Dev. **7**(3), 162–175 (2015)
33. Zheng, W.L., Zhu, J.Y., Lu, B.L.: Identifying stable patterns over time for emotion recognition from eeg. IEEE Trans. Affect. Comput. **10**(3), 417–429 (2017)

Effect of Music Training in Neural Responses to Emotional Speech Prosody: Insights from EEG and Brain Network Analysis

Ci-Jun Gao[1,2], Xucheng Liu[2,3,4], Gufeng Jia[2], Hongzhe Yang[2], Wei Tao[2,3,4], Haiyan Wu[1,2], Ruey-Song Huang[2,3], and Feng Wan[2,3,4(✉)]

[1] Department of Psychology, Faculty of Social Sciences, University of Macau, Taipa, Macao
[2] Centre for Cognitive and Brain Sciences, Institute of Collaborative Innovation, University of Macau, Taipa, Macao
fwan@um.edu.mo
[3] Department of Electrical and Computer Engineering, Faculty of Science and Technology, University of Macau, Taipa, Macao
[4] Centre for Artificial Intelligence and Robotics, Institute of Collaborative Innovation, University of Macau, Taipa, Macao

Abstract. Speech prosody conveys crucial information during social communication. However, the neural mechanisms underlying speech prosody in the human brain remain unclear. The primary aim of this study was to explore the impact of long-term musical training on the recognition of emotional speech prosody. Employing a task that isolated prosody recognition without semantic content interference, we recorded behavioral and electrophysiological data. Participants were categorized into two groups: individuals with systematic musical training and those without any formal musical training. Our findings indicate that the musician group exhibited enhanced neural activity in the frontal region, characterized by significant differences in the P50, P200, P3a, P3b, and P600 components. Additionally, a larger Power Spectral Density (PSD) was observed in the frontal, occipital, and parietal regions of musicians, particularly in the theta and alpha frequency bands. Graph theoretical analysis also indicated that musically trained individuals demonstrated an increased local coefficient and more concentrated degree. This study provides a comprehensive analysis of the neural alteration during speech prosody recognition attributable to long-term musical training, incorporating both static and dynamic analyses, with a focus on temporal and spatial dimensions. Our study provides new insights into the neural mechanisms induced by speech prosody.

Keywords: Music Training · Electroencephalogram · Event-related Potential · Brain Network · Power Spectral Density

1 Introduction

Speech prosody plays a crucial role in social communication, conveying non-semantic information such as emotions, which is essential for successful interactions. Given the similarities and shared properties between music and language, numerous studies have explored the effects of musical training on the processing of emotional prosody [17,24]. Musicians are known to have an advantage in recognizing speech information, demonstrated through greater accuracy and efficiency in task performance [27]. In previous studies, investigating musician versus non-musician emotional processing in behavioral tasks, consistent findings have been reported regarding the superior performance of musicians in recognizing auditory information and emotional content [24]. These studies have employed electroencephalography (EEG) to explore the neural mechanisms underlying emotional prosody processing in musicians compared to non-musicians, using stimuli from natural language with manipulated semantic and prosodic emotions. This research illuminates the electrophysiological correlates of processing emotional speech. Specifically, these studies have highlighted the modulation of the N200 and P300 components by emotional prosody, suggesting that emotional information can be processed at an early stage. The processing of vocal emotions appears to be rapid and non-attentional, occurring even when there is no requirement to integrate additional emotional information. Neurological studies indicate that emotional prosody processing is predominantly right-lateralized and is associated with positive ERP components [31]. In musicians, there is evidence of enhanced sensitivity to emotions in speech, potentially due to structural and functional reorganizations induced by music training [17].

In addition to the electrophysiological investigation, the processing of linguistic and emotional prosody in the cortical surfaces and subcortical areas has been studied using fMRI, revealing specific neural regions involved in supra-segmental acoustic information extraction, meaningful acoustic sequence representation, as well as the explicit evaluation of emotional prosody. Implicit processing of emotional prosody seems to engage subcortical regions responsible for automatic emotional reactions, such as the amygdala and the middle temporal gyrus, which integrates information from various sources, allowing complex divisions and interpretations of speech [32]. The general hypothesis is that within the right hemisphere, emotional prosody primarily activates the middle temporal gyrus.

Inspired by the literature, we further investigated the precise neural mechanisms that are related to the musicians' advantage in emotional prosody recognition. There is some evidence that auditory-perceptual enhancements might play a role, which is consistent with the idea of overlapping sensory pathways for processing music and vocal emotion information. Whether this overlap extends to higher-order stages of socioemotional processing remains unclear. EEG and fMRI studies, combined with behavioral tasks, will be valuable in investigating these mechanisms.

In the current study, we utilized EEG analysis to capture both the temporal dynamics and the long-term static features during speech prosody processing. ERP analysis was employed to delineate the neural activity across specific time courses, highlighting the temporal precision in brain responses. Conversely, static brain networks provided insights into the consistent connectivity patterns among different brain regions over extended periods. These networks represent the average connection strength during these intervals and are characterized by their stability over time. The use of static networks is instrumental in enhancing our understanding of the interactions between various brain regions under specific conditions, thus facilitating the exploration of the relationships between brain network structures, cognitive functions, and behavioral performances. Furthermore, the application of Power Spectral Density (PSD) analysis played a pivotal role in interpreting the complex oscillatory activity within the brain. This methodology allows for the decomposition of EEG signals into their constituent frequencies, which is crucial for diagnosing neurological conditions, monitoring brain activity, and exploring cognitive states. Consequently, PSD has become an indispensable tool in neuroscientific research, providing deeper insights into the dynamics of brain function. Finally, we used the graph theoretical analysis to quantify and compare the functional connectivity differences between musician and non-musician groups during speech prosody processing. This approach enabled a structured analysis of the network properties, offering a comprehensive view of how musical training might influence neural connectivity during auditory tasks.

2 Method

2.1 Experiment Design

Thirty subjects were recruited from the university (Age: $M = 22.9; SD = 2.9$). 15 were assigned to the music group with a minimum of three years of professional training on musical instruments, and the remaining 15 who had not undergone any formal music training were assigned to the non-music group. Following previous research on gender differences and handedness in emotion perception and language lateralization [2,14,27], only male and right-handed subjects were included to eliminate the confounding factor.

The stimuli were meaningless words that were created by splitting commonly used two-character words from the Lexicon of Common Words in Contemporary Chinese [12] and reorganizing the characters into new two-character combinations. The purpose of using these meaningless words was to have a control condition that lacked semantic content. To ensure high-quality auditory stimuli, the recordings were performed by a professional voice actress. By using meaningful and meaningless words recorded by the same professional, the study aimed to maintain a controlled and standardized auditory experience for the participants.

For each of the words, they were with either happy or angry prosody, by a professional voice actress in a recording studio. For the senseless words performed in happy prosody, we label them as SH; for the senseless words performed in angry prosody, we label them as SA. The audio stimuli were later edited and analyzed by Praat [3]. The acoustic features of the audio are summarized in the following table.

Figure 1 shows the experiment design for one trial. At the beginning of each trial, a fixation cross was presented in the center of the screen, while participants were required to fix the cross. After 1000 ms of presentation, the auditory

Table 1. Acoustic feature of auditory stimulus.

Mean	SA	SH
Duration (ms)	640.303	642.792
$f0$ (Herz)	431.554	425.755
Intensity (dB)	75.863	73.432

stimulus was presented with the cross remaining on the screen. The participants were allowed to make key responses as soon as the audio was presented, and the screen would turn blank when the response was made. If the participants fail to make any response within 5 s after the stimuli onset, the screen will turn blank automatically and continue to the next trail. After the key response, a blank screen appeared for a duration of 3000–3200 ms. The varied inter trial intervals were adopted to reduce artifacts, as per [27] (Table 1).

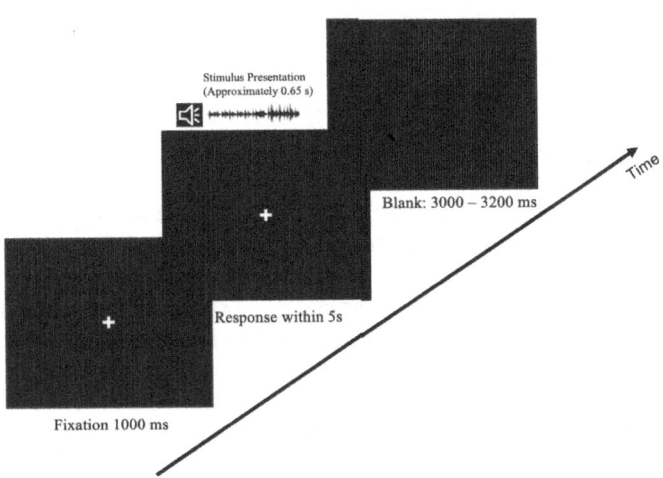

Fig. 1. Illustration of a single trial.

The task was counterbalanced designed in terms of the response key to eliminate the possible effects of the participants' tendency to automatically associate positive or negative emotional information with a particular direction (in the current study: keys on the left or right side) Some previous studies suggest that the mental spatial metaphor of emotional valence and intensity may be related to the people's biological determinants (i.e., hemisphere lateralization, handedness) and cultural determinants (i.e., language) [4,5,15,29]. In this context, to eliminate the confounding factor of directional bias in the groups, for all three tasks, we designed two subversions. In one version, the key 'G' represents positive emotion, and the key 'H' represents negative emotion; in the other version of the task, the associations were reversed. Half of the participants participated in the tasks that required them to respond to positive emotions with the index finger putting on the key 'G', and the other half of the participants took part in the tasks that instructed them to respond to the positive emotion with their index finger placed on the key 'H'. The association for each participant was kept consistent throughout the whole task.

2.2 Data Acquisition

The EEG data recorded using the Biosemi A/D box with the ActiView system (BioSemi BV, Amsterdam, Netherlands) with 64 active Ag/AgCl scalp pin-type electrodes. The electrodes are arranged based on the international standard 10–20 system, ensuring standardized and reproducible positioning of scalp electrodes. This system is based on the relationship between the location of an electrode and the underlying area of the cerebral cortex. Each electrode site was labeled according to the 10–20 system nomenclature, where 'F' stands for frontal, 'T' for temporal, 'C' for central, 'P' for parietal, and 'O' for occipital lobes. The letter 'Z' refers to an electrode placed on the midline. Additional electrodes (e.g., Fpz, Oz) were placed according to the extended 10-10 system to provide higher resolution data of the scalp potential field.

The electrode offset was kept below 40 mV. Different from most of the other EEG recording systems, the Biosemi adopts two additional electrodes as ground electrodes: the Common Mode Sense active electrode (CMS) and the Driven Right Leg passive electrode (DRL), which are placed on the left and right side of 'POz' channel. The EEG signal was sampled at the rate of 2048 Hz.

The behavioral data was recorded and stored locally by the E-Prime 3.0 (Psychology Software Tools, Pittsburgh, PA) and the trigger signals were sent from the E-Prime computer to the ActiView trigger input port via a standard USB to parallel port converter cable.

2.3 Data Preprocessing

The raw signal was band-pass (0.1–40 Hz) filtered, and a notch finite impulse response (FIR) filter was also applied (49–51 Hz) to remove power interference of 50 Hz in brainwave data [26,30]. The data was referenced by average reference. Additionally, the independent component analysis (ICA) was employed to remove the artifacts such as eye blink and other muscular-induced activities to improve the signal-to-noise ratio (SNR) [7,9,13].

2.4 ERP Analysis

Data analyses for the ERP analysis were conducted by the EEGLAB toolbox v2022.0 [6]. According to the previous ERP studies, EEG signals will be segmented into temporal windows that correspond to event-related potentials (ERPs) and the segmentation will be based on event marks that were recorded during experiments, the epochs were chosen as a 200 ms pre-stimulus baseline and an 800 ms post-onset windows [19–24].

Only the EEG recordings of trials that were correctly responded to were included in the analysis. The single subject ERP waveforms and the grand average waveforms from all 64 electrodes were plotted for eye inspection of the centering of peaks and their relevant electrodes. The scalp topographies of the ERP were also plotted for assistance in locating the spatial activation. For the statistical analysis, twenty-two electrodes were grouped into four regions of interest based on their spatial distribution on the scalp: left anterior (AF7, Fp1, F3, FC1, FC3, FC5, C1), left posterior (P5, P7, PO3, PO7, O1), right anterior (F2, F8, FC2, FC4, C2), right posterior (TP8, P8, P10, PO8, O2) [28].

In light of preceding research and insightful observations gleaned from the figures, ten components have been identified for subsequent analysis. The time windows selected for these components are as follows: P50 (25–85 ms), P100 (50–150 ms), N100 (50–150 ms), P200 (150–250 ms), N200 (150–250 ms), P3a (275–375 ms), P3b (300–600 ms), and P600 (500–750 ms). These temporal intervals are strategically chosen and centered on the peak activity of the respective ERP components and will be rigorously analyzed in the forthcoming stages of this research.

The single-trial EEG signal of each condition was averaged to single-subject ERP waveforms and was measured by computing the mean amplitude and peak latency in windows that were suggested by the previous study and the eye inspection of the waveforms and topographies. Data from channels that were selected for the same region of interest were averaged.

2.5 PSD Calculation

For the PSD calculation, all EEG data were filtered and separated into the frequency bands alpha (8–13 Hz), beta (13–30 Hz), theta (3.5–6.75 Hz), and low gamma (31–39.75 Hz).

To analyze EEG data effectively, the initial step involves filtering the data to isolate the desired frequency bands. This preprocessing aims to enhance the signal quality by reducing noise and excluding irrelevant frequencies. Subsequently, the continuous signal is segmented into shorter sections. In the context of Welch's method, these segments often overlap significantly, typically by 50%, to improve the reliability of the spectral estimates. Each segment is then treated with a window function, such as a Hamming or Hann window, to mitigate spectral leakage caused by the discontinuities at the edges of each segment. This step is crucial for obtaining a more accurate frequency domain representation of the EEG signal.

Following the application of the window function, each segment undergoes a Fourier transform, typically implemented using a fast Fourier transform (FFT) algorithm. This transformation converts the time-domain data into its frequency-domain counterpart, facilitating the subsequent analysis. The mathematical representation of FFT in each single segment is as follows:

$$P_k(f) = \frac{1}{N|\omega|^2}|FFT\{x_k(n) \cdot w(n)\}|^2 \tag{1}$$

Following In this equation, $x_k(n)$ is the k-th signal segment, $w(n)$ is the window function, N is the segment length, and $|\omega|^2$ is the window function energy that provides the necessary normalization.

The power spectral density (PSD) of each segment is derived by calculating the modulus squared of the FFT results, which quantifies the power present at each frequency component within the segment. This procedure, known as computing the periodogram, is repeated for each segment.

$$PSD(f) = \frac{1}{K}\sum_{k=1}^{K} P_k(f) \tag{2}$$

To enhance the stability and reduce the variance of the power spectral estimates, the periodograms of all segments are averaged arithmetically. This averaging process is a definitive aspect of Welch's method, providing a more robust estimation of the signal's power spectral density by averaging the spectral estimates across overlapping segments.

By employing this methodical approach, Welch's method significantly improves the accuracy and reliability of power spectral density estimates from EEG data, making it a preferred choice for analyzing physiological signals where precision is paramount.

2.6 Functional Network and Graph Theory

The Pearson's correlation is used here to calculate statistical dependency. The static connectivity was calculated with the discrete non-stationary times series of pairs of the anatomically divided brain regions.

The methodology for analyzing EEG-based static networks involves several stages. Initially, data collection is completed. Following this, data preprocessing and segmentation are conducted to ensure data quality and consistency. In the network construction phase, each EEG electrode is treated as a node. Statistical methods, such as Pearson correlation coefficients, are employed to compute the strength of connectivity between channels, thereby defining the edges between nodes. As illustrated in the following equation, calculates the static connectivity between region X and Y, each has discrete non-stationary time series $X = \{x_i\}_{i=1,2,...,n}$ and $Y = \{y_i\}_{i=1,2,...,n}$. The detail of the calculation is presented as the following:

$$r = \frac{\sum_{i=1}^{n} x_i y_i - n\overline{x}\overline{y}}{\sqrt{\sum_{i=1}^{W} x_i^2 - n\overline{x}^2}\sqrt{\sum_{i=1}^{W} y_i^2 - w\overline{y}^2}} \tag{3}$$

In this equation, n represents the number of total data time points, and the means of X and Y are represented as $\overline{x}$ and $\overline{y}$. Before the calculation, the time series were z-score normalized with 0 as means and 1 as standard deviation.

Subsequently, the constructed network is usually analyzed using graph theory methods. This analysis typically involves a range of graph theory parameters that provide insights into the network's structure and function. These mathematical definitions of complex network measures are shown as follows:

Local Efficiency: Assesses the network's resilience, indicating how well information is exchanged by the neighbors of each node when that node is removed. The equation below shows the calculation of the node i.

$$E_{loc} = \frac{1}{n} \sum_{i \in N} E_{loc,i} = \frac{1}{n} \sum_{i \in N} \frac{\sum_{j,h \in N, j \neq i} a_{ij} a_{ih} [d_{jh}(N_i)]^{-1}}{k_i(k_i - 1)} \tag{4}$$

Degree: One of the most direct and fundamental metrics of the networks, it denotes the number of direct connections linked to each node, which reflects the node's centrality and influence within the network. The following equation shows the calculation of the degree of node i, by calculating the number of non-zero elements in the corresponding row or column of the connectivity adjacency matrix.

$$k_i = \sum_{j \in N} a_{ij} \tag{5}$$

The application of these graph theory metrics allows us to uncover significant aspects of the network's topology and functionality, enhancing the understanding of how brain networks correlate with psychological and behavioral processes.

An adjacency matrix was constructed for each frequency band by using coherence values as edge weights. We applied a proportional threshold to the weighted adjacency matrix to retain the top 10% strongest connections, thereby ensuring network sparsity and comparability across subjects.

We further calculated the graph theoretical measures to characterize the topological properties of the networks. The analysis was conducted using the Brain Connectivity Toolbox in MATLAB [25] The following measures of complex networks were conducted: local coefficient and the degree.

2.7 Statistical Analysis

In the analysis of behavioral data, a two-tailed independent t-test was employed to compare reaction times (RT) across two emotional prosody conditions in two distinct groups.

For the ERP analysis, separate Analyses of Variance (ANOVAs) were conducted across different regions of interest (ROIs), focusing on various ERP components and measures. Specifically, ERP data corresponding to one feature from a selected component within an ROI underwent a 2×2 repeated measures ANOVA. This analysis included two within-subject factors, namely emotional valence (Happy and Angry), and two between-subject factors (Group: Music and Non-Music). This procedure was replicated for each combination of feature, component, and ROI.

3 Results

3.1 Behavioral Results

As Figure 2(a) shows, the two groups completed the tasks with accuracy rates exceeding 90%, and no significant differences were observed(Music: $M = 0.93, SD = 0.24$; Non-music: $M = 0.95$, $SD = 0.22$). The response time data failed to reveal any interaction between group and prosody emotions, however, the independent $t-test$ detected a significant group difference in reaction time as shown in Figure 2(b). Specifically, the music group showed a longer reaction time compared with the non-music group ($F(1, 3328) = 65.64, p < 0.001$).

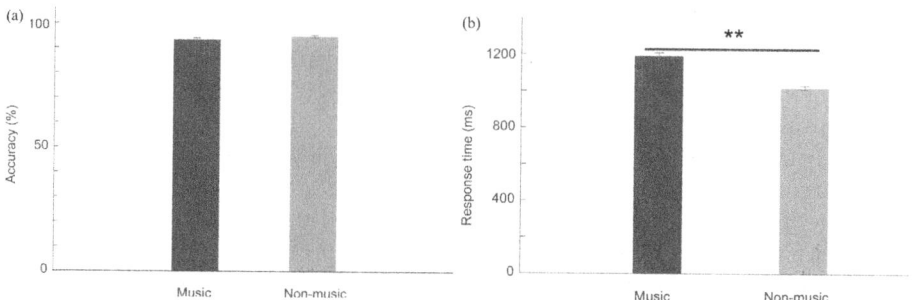

Fig. 2. Response accuracy and response time in music and non-music groups. ** Indicate $p < 0.001$

3.2 ERP Results

Separated two-way repeated ANOVA was conducted for each component to investigate the effects of prosodic emotions and musical experience. The P50, P200, N200, P3a, P3b, and P600 components were focused on the analysis and were quantified as mean amplitude and peak latency.

For the P50 component, in the left anterior region, a significant main effect of musical experience was detected ($F(1, 18) = 4.85, p = 0.034$). Individuals with musical training ($M = 0.292, SD = 0.641$) exhibited greater P50 mean amplitudes compared to their non-musical counterparts ($M = -0.462, SD = 1.311$). The left posterior region also revealed a significant main effect, $F(1, 18) =$

7.75, $p = 0.009$, with the musicians ($M = -0.124, SD = 0.529$) presenting lower mean amplitudes relative to non-musicians ($M = 0.784, SD = 1.262$). In the right posterior region, a significant main effect was observed, $F(1, 18) = 10.41, p = 0.003$. Participants with a background in music exhibited significantly lower mean amplitudes ($M = -0.037, SD = 0.214$) compared to those without musical experience ($M = -0.009, SD = 0.251$). Analysis of P50 peak latency revealed a significant main effect of musical experience was observed in the left anterior region, $F(1, 18) = 5.652, p = 0.023$. This finding indicates a difference in neural timing between the two groups, with the non-musician group displaying a delayed P50 peak ($M = 53.169\,\text{ms}, SD = 10.351\,\text{ms}$) compared to the musician group ($M = 46.032\,\text{ms}, SD = 8.593\,\text{ms}$). Furthermore, the main effect of prosodic emotions on P50 peak latency reached marginal significance, indicating that the emotional modulation of the auditory stimuli has a tendency to exhibit a differential effect on the latency of the P50 response across the sample. Specifically, for the main effect of prosodic emotion, $F(1, 18) = 3.242, P = 0.08$, with a delayed display of peak during the Angry prosody condition ($M = 47.324, SD = 9.872$) than the Happy prosody condition ($M = 52.578, SD = 6.192$). Although these results are only marginally significant, it is consistent with the neuroimaging and behavioral data in previous studies which suggest that the angry stimulus captures attention faster [16].

The results of the P200 component also revealed no significant interaction effects between prosody emotions and musical experience. Similarly, no significant main effect of prosody emotions was observed. However, the analysis did indicate a marginally significant main effect of musical experience on the P200 mean amplitude in the left anterior region ($F(1, 18) = 2.895, p = 0.097$). Specifically, the group with musical experience exhibited a mean amplitude of 1.048 with a standard deviation of 1.217, while the non-musical group had a mean amplitude of 0.198 with a standard deviation of 1.763. A similar result was found in the right anterior region ($F(1, 18) = 3.845, p = 0.058$). Specifically, the group with musical experience exhibited a mean amplitude of 0.67 with a standard deviation of 0.856, while the non-musical group had a mean amplitude of 0.083 with a standard deviation of 0.969. A significant main effect of the group was found in the left posterior region ($F(1, 18) = 7.079, p = 0.011$). Specifically, the group with musical experience exhibited a mean amplitude of -1.48 with a standard deviation of 1.258 while the non-musical group had a mean amplitude of -0.172 with a standard deviation of 1.691.

For the N200 component, contrarily, musical experience exerted a significant main effect in the left posterior region ($F(1, 18) = 7.079, p = 0.011$), with musicians exhibiting a reduced N200 area amplitude ($M = -1.48, SD = 1.258$) compared to non-musicians ($M = -0.172, SD = 1.691$). This suggests that musical training may lead to a more pronounced neural response as reflected by the larger N200 amplitude in musicians. These findings could implicate adaptive neural plasticity in auditory regions as a result of musical expertise.

The 2×2 ANOVA analysis conducted on the P3a mean amplitude revealed no significant interaction effects between prosody emotions and musical experience. However, the analysis did indicate significant main effects of musical experience in specific brain regions. In the left anterior region, a significant main effect of musical experience was found ($F(1, 18) = 8.521, p = 0.006$), suggesting that the music group exhibited a smaller mean amplitude ($M = 0.598, SD = 1.814$) compared to the non-music group ($M = -0.966, SD = 1.498$). The left posterior region also showed a significant main effect of musical experience ($F(1, 18) = 6.212, p = 0.017$), with the music group again having a smaller mean amplitude than the non-music group (Music: $M = -0.028, SD = 1.456$; Non-music: $M = 1.332, SD = 1.832$). Consistent with the findings from other regions, right posterior regions displayed a similar pattern ($F(1, 18) = 7.85, p = 0.008$), where the music group exhibited a smaller P3a mean amplitude ($M = -0.52, SD = 1.692$) compared to the non-music group ($M = 1.193, SD = 2.008$).

For the P3b component, the ANOVA analysis revealed no significant interaction effects between prosody emotions and musical experience. However, the analysis did indicate significant main effects of musical experience in specific brain regions. In the left anterior region, a significant main effect of musical experience was found ($F(1, 18) = 12.812, p = 0.001$), suggesting that the music group exhibited a smaller mean amplitude ($M = -0.289, SD = 2.048$) compared to the non-music group ($M = -2.221, SD = 1.288$). The left posterior region also showed a significant main effect of musical experience ($F(1, 18) = 15.756, p < 0.001$), with the music group again having a smaller mean amplitude than the non-music group (Music: $M = 1.203, SD = 1.429$; Non-music: $M = 3.353, SD = 1.833$). Consistent with the findings from other regions, right posterior regions displayed a similar pattern ($F(1, 18) = 11.014, p = 0.002$), where the music group exhibited a smaller P3b mean amplitude ($M = 0.581, SD = 1.856$) compared to the non-music group ($M = 2.768, SD = 2.16$).

For the P600 mean amplitude, similarly, it indicated no significant interactions between prosodic emotions and musical experience. Additionally, the results showed no main effect of prosodic emotions on P600 mean amplitude. However, there was a significant main effect of musical experience in several brain regions. In the left anterior regions, a substantial main effect of musical experience was observed ($F(1, 18) = 11.084, p = 0.002$). The music group had a mean amplitude significantly lower than the non-music group (Music: $M = -1.428, SD = 2.153$; Non-music: $M = -3.583, SD = 1.865$). A similar effect was found in the left posterior regions ($F(1, 18) = 10.948, p = 0.002$), where the music group exhibited a smaller mean amplitude ($M = 2.775, SD = 1.89$) compared to the non-music group ($M = 5.191, SD = 2.540$). In the right posterior regions, the main effect of musical experience was also significant ($F(1, 18) = 8.164, p = 0.007$). The music group had a lower mean amplitude ($M = 1.804, SD = 2.169$) relative to the non-music group ($M = 4.456, SD = 3.342$). Statistical analysis on P600 peak latency revealed a significant main effect of music experience in the right anterior region ($F(1, 18) = 7.223, p = 0.011$), indicating an earlier P600 peak for the musically trained participants ($M = 572.889\,\text{ms}, SD = 40.348\,\text{ms}$)

compared to the participants without music training ($M = 616.327$ ms, $SD = 60.805$ ms) (Fig. 3).

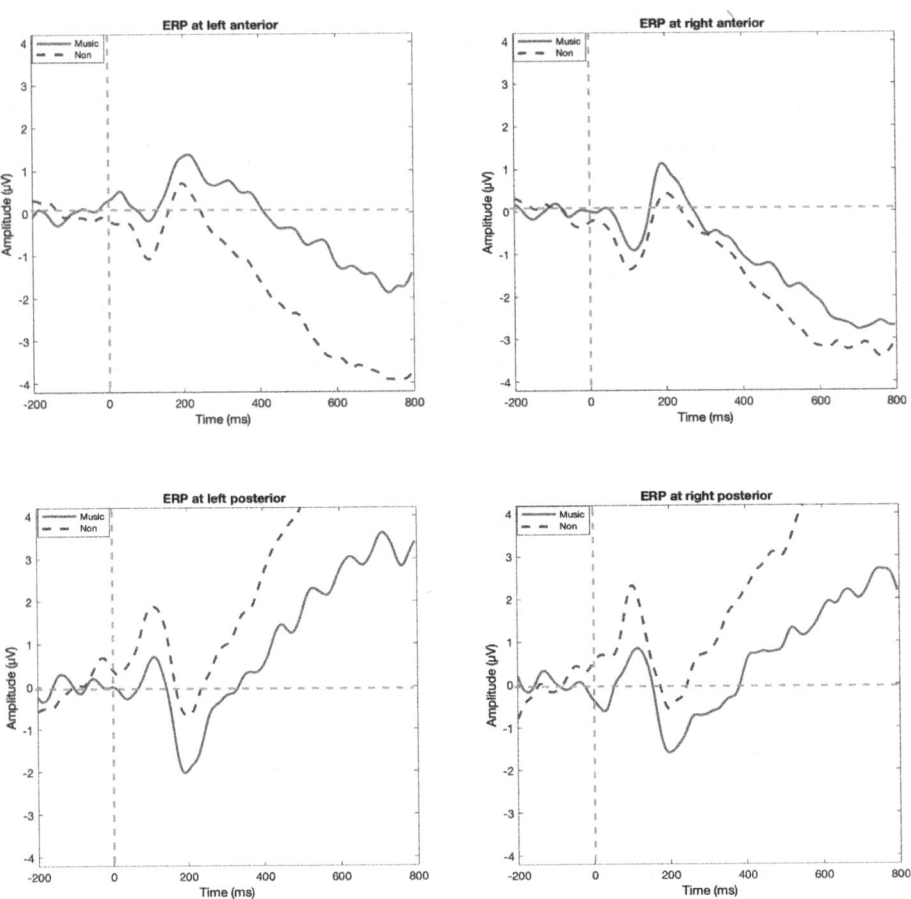

Fig. 3. Grand averaged ERP waveform in four regions of interest: Left Anterior, Right Anterior, Left Posterior, Right Posterior. The red lines indicate the grand averaged ERP waveform of all subjects from the music group, and the blue lines indicate the grand averaged ERP waveform of all subjects from the non-music group. (Color figure online)

3.3 PSD Results

A series of Mann-Whitney U tests were conducted to compare the Power Spectral Density (PSD) between groups of musicians and non-musicians across different brain regions during exposure to emotional prosody conditions, specifically "happy" and "angry". The PSD was analyzed within several frequency bands: theta, alpha, beta, and low gamma. The results revealed statistically significant differences between musicians and non-musicians that varied by both brain region and emotional prosody condition (Fig. 4).

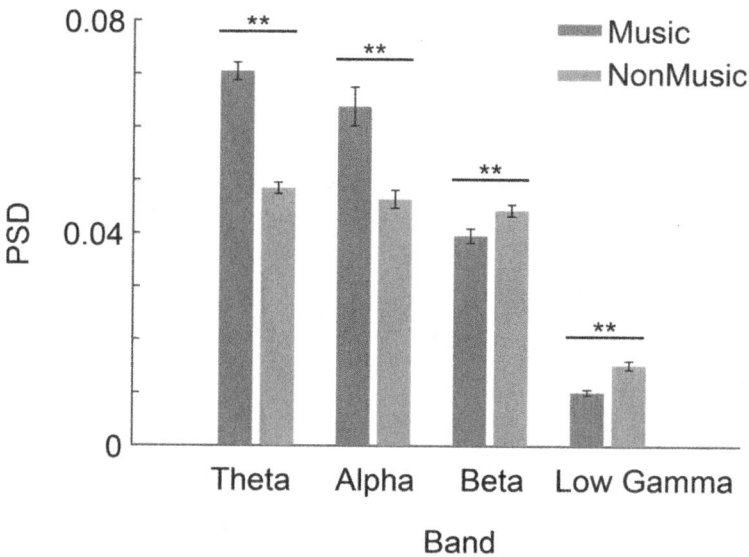

Fig. 4. PSD of between music and non-music groups across four frequency bands.

In the frontal region, significant differences were observed in the Theta, Beta, and Low Gamma frequency bands under both happy and angry prosody conditions. Specifically, for the Angry Prosody condition, the correlations were significant with $r = 0.688 (p < .001)$ for theta band, $r = -0.806 (p < .001)$ for beta band, and $r = -0.848 (p < .001)$ for low gamma band. Similarly, for the Happy Prosody condition, significant correlations were also found $r = 0.806 (p < .001)$ for theta band, $r = -0.812 (p < .001)$ for beta band, and $r = -0.854 (p < .001)$ for low gamma band.

In the central region, notable distinctions were observed in the Theta band under both emotional conditions. Specifically, in the Angry Condition, the correlation was moderate with $r = 0.452$ $(p = 0.033)$, while in the Happy Condition, it was stronger with $r = 0.66$ $(p = 0.001)$. However, significant differences in the Beta band were only detected in the angry prosody condition, with a correlation of $r = -0.434$ $(p = 0.04)$. This suggests that the central region's neural activity in the Beta band is particularly sensitive to anger-related prosodic cues (Table 2).

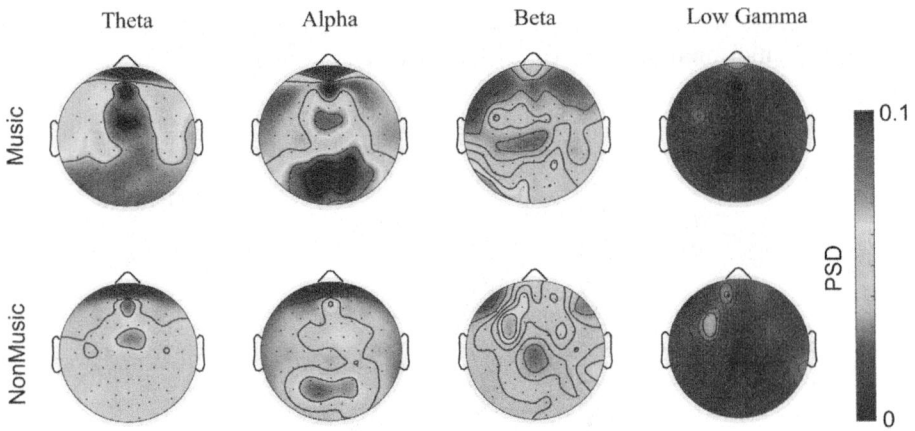

Fig. 5. Topology maps of PSD across four frequency bands. First panel: Music Group; Second panel: Non-music Group.

Table 2. Frontal region PSD analysis with the Mann-Whitney U test.

Frequency Band	Condition	U-statistic	r-value	p-value
Theta	Angry	261	0.688	$p < .001*$
	Happy	281	0.806	$p < .001*$
Alpha	Angry	198	0.316	0.1
	Happy	183	0.227	0.231
Beta	Angry	8	-0.806	$p < .001*$
	Happy	7	-0.812	$p < .001*$
Low Gamma	Angry	1	-0.848	$p < .001*$
	Happy	0	-0.854	$p < .001*$

In the parietal regions, significant differences were detected across several frequency bands under both happy and angry conditions. For the Angry Condition, there was a strong correlation in the Theta band ($r = 0.8, p < .001$) and a moderate correlation in the Alpha band ($r = 0.511, p = 0.007$). For the Happy Condition, the correlations were similarly strong in the Theta band ($r = 0.854, p < .001$) and moderate in the Alpha band ($r = 0.505, p = 0.008$). Additionally, significant differences were observed in the Low Gamma band, but only under the Happy Condition ($r = -0.487, p = 0.01$) (Tables 3 and 4).

Table 3. Central region PSD analysis with the Mann-Whitney U test.

Frequency Band	Condition	U-statistic	r-value	p-value
Theta	Angry	150	0.452	0.033*
	Happy	174	0.66	0.001*
Alpha	Angry	119	0.182	0.385
	Happy	121	0.2	0.344
Beta	Angry	48	−0.434	0.04*
	Happy	55	−0.373	0.085
Low Gamma	Angry	75	−0.2	0.344
	Happy	57	−0.356	0.1

Table 4. Parietal region PSD analysis with the Mann-Whitney U test.

Frequency Band	Condition	U-statistic	r-value	p-value
Theta	Angry	280	0.8	$p < .001*$
	Happy	289	0.854	$p < .001*$
Alpha	Angry	231	0.511	0.007*
	Happy	230	0.505	0.008*
Beta	Angry	193	0.287	0.131
	Happy	195	0.298	0.121
Low Gamma	Angry	95	−0.292	0.126
	Happy	62	−0.487	0.01*

In the temporal regions, significant differences were observed in the Theta band for both happy and angry conditions. Specifically, the correlation was strong in the Angry Condition ($r = 0.832, p = 0.005$) and also robust in the Happy Condition ($r = 0.786, p = 0.009$). Furthermore, significant differences in the Low Gamma band were found exclusively in the Happy Condition ($r = -0.693, p = 0.029$). These results underscore the temporal regions' responsiveness to emotional prosody, with distinct patterns emerging across different frequency bands and emotional states (Table 5).

Table 5. Temporal region PSD analysis with the Mann-Whitney U test.

Frequency Band	Condition	U-statistic	r-value	p-value
Theta	Angry	36	0.832	0.005*
	Happy	35	0.786	0.009*
Alpha	Angry	17	−0.046	0.937
	Happy	15	−0.139	0.756
Beta	Angry	9	−0.416	0.225
	Happy	6	−0.555	0.1
Low Gamma	Angry	8	−0.462	0.17
	Happy	3	−0.693	0.029*

Lastly, the occipital region, although typically linked with visual processing, shows a pronounced neural response to emotional prosody across multiple frequency bands. Significant differences were observed across the Theta, Alpha, and Beta bands for both emotional conditions. Noteworthy correlations were recorded with $r = 0.84(p < .001)$ for Theta, $r = 0.814(p < .001)$ for Alpha, and $r = 0.84(p < .001)$ for Beta in both the Angry and Happy Conditions (Table 6).

Table 6. Occipital region PSD analysis with the Mann-Whitney U test.

Frequency Band	Condition	U-statistic	r-value	p-value
Theta	Angry	64	0.84	$p < .001*$
	Happy	64	0.84	$p < .001*$
Alpha	Angry	63	0.814	$p < .001*$
	Happy	63	0.814	$p < .001*$
Beta	Angry	64	0.84	$p < .001*$
	Happy	64	0.84	$p < .001*$
Low Gamma	Angry	30	−0.053	0.901
	Happy	28	−0.105	0.759

Significant group differences were mainly discovered in the frontal, occipital, and parietal areas, specifically for the alpha and theta frequency bands as Fig. 5 shows.

3.4 Brain Network and Graph Theoretical Analysis Results

For the brain network analysis, we divided the 64 EEG channels into five brain regions based on anatomical structures and analyzed the connectivity based on the five ROIs accordingly - frontal, central, parietal, temporal, and occipital regions. As Fig. 6 shows, there are significant differences in connection strength between the two groups, especially in the connections between Frontal - Central ($r = -0.10, p = 0.004$) and Parietal - Occipital regions ($r = -0.26, p < 0.001$).

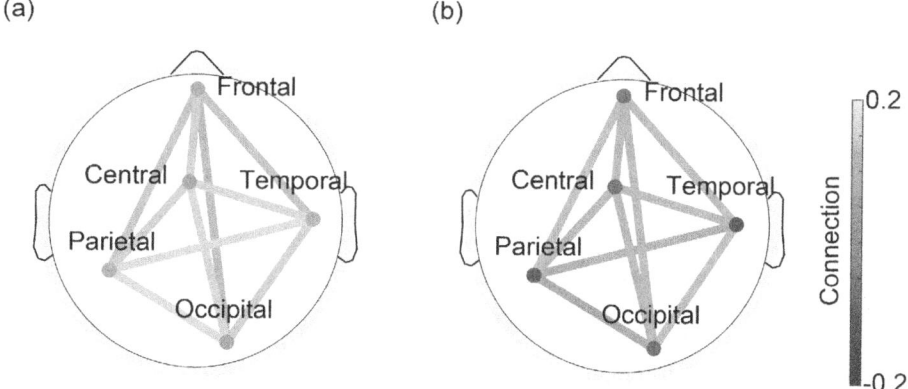

Fig. 6. Brain map of the group averaged ROIs connections. The color bar indicates the connection strength and ranges from -0.2 to 0.2. (a) Music group, (b) Non-music group. (Color figure online)

The results from t-tests revealed a consistently higher local efficiency in the music group compared to the non-music group across all emotional prosody categories. Specifically, under the anger prosody condition, the music group exhibited a mean local efficiency of 0.626 ($SD = 0.016$), which was notably higher than the 0.606 ($SD = 0.017$) observed in the non-music group. The t-test for this condition yielded a t-value of 2.82 with a significant p-value of 0.01. Similarly, in the happy prosody condition, the music group demonstrated a mean local efficiency of 0.62558 ($SD = 0.018$), compared to the non-music group's mean of 0.607 ($SD = 0.015$). The t-value for this condition was 2.71, with a p-value of 0.013.

Kolmogorov-Smirnov (K-S) statistical results indicate that there is a significant difference in the degree distributions between the two groups under both emotional prosody conditions. Specifically, in the anger prosody condition, the K-S statistic is 0.093 with a p-value of 0.002, indicating a significant difference between the groups, and in the happy prosody condition, the K-S statistic is 0.091 with a p-value of 0.003, also indicating a significant difference.

4 Discussion

In this study, we compared the speech prosody recognition processing in the group with music training and the group without formal music training. The ERP analysis first revealed the time course and regional activation differences between the groups. With the ERP analysis results, the music groups show a stronger response than the non-musically trained group in the anterior regions, while the non-music group shows a stronger response than the music group in the posterior regions. Such patterns were found in P50, P200, P3a, P3b, P600. This regional difference could be because of the differential neural engagement of the two groups. The anterior regions are more related to the higher-order cognitive functions such as executive functions, attention, and decision making while the posterior regions are shown to be related to the primary sensory processing. The relatively stronger ERP response of the music group in the anterior region may indicate more employment of higher order cognitive resources [18], while the non-music group may rely on more basic auditory processing strategies due to the lack of specialized training; it is also possible that long-term music training allows the participants from music group assign less resource for auditory processing. As for the non-music group, it requires more effort and neural resources to process the speech emotion conveyed by prosody. Results of P50 peak latency suggest that musical experience is associated with faster auditory processing, as reflected in the earlier P50 peak latency, but only in the left anterior region. This specificity may reflect the influence of musical training on the neural mechanisms underlying auditory perception and temporal processing, which may suggest a prompter reaction of the musicians to auditory cues [11].

The static functional network analysis was employed to uncover the interaction between brain regions over an extended period. This was done to demonstrate how brain networks organize and deploy cognitive resources differently during speech prosody recognition between music and non-music groups.

We calculated the PSD in the whole brain across conditions in two groups respectively, it was found that the musically trained individuals have generally higher PSD than the non-musically trained individuals. We further calculated the PSD in different anatomical brain regions and frequency bands. In theta band, we found a significantly higher PSD in the music group in both angry and happy prosody conditions across five brain regions compared to the non-music group. The alpha and theta frequency bands exhibit the most notable differences between groups across all types of emotional prosody, regardless of semantic content. The alpha frequency band, which is associated with a relaxed mental state and the activation of the default mode network, shows enhanced Power Spectral Density (PSD) in the frontal and occipital regions. This may suggest that musical training may be associated with enhanced neural efficiency in processing emotional prosody, as evidenced by the higher local efficiency values in the music group across different emotional conditions, as well as more efficient information processing and improved attention management. Additionally, increased theta activity in the frontal region may indicate enhanced attentional focus and emotional processing [8,10] (Fig. 7).

We further analyzed the interactions between regions with the functional connectivity results and quantified the connectivity into graph theory parameters. The results of the graph theoretical analysis showed a significant difference in local efficiency between the two groups regardless of the stimuli type. Local efficiency represents the capacity of information transfer in local connections, it

(a)

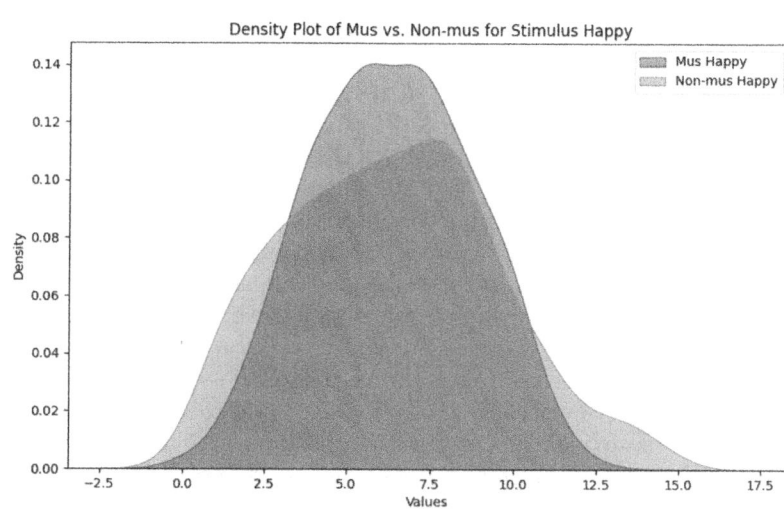

(b)

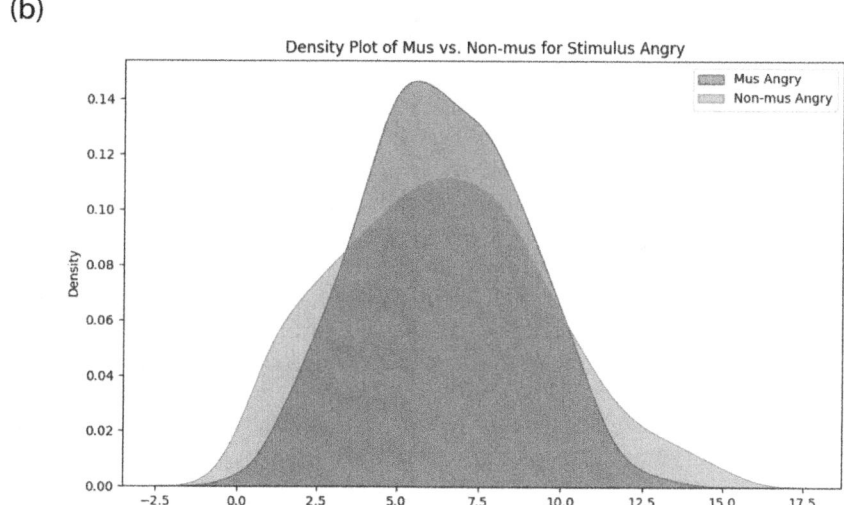

Fig. 7. K-S statistics for the Degree in music and non-music group under different emotional prosody conditions.

quantifies how well information is exchanged among the immediate neighbor of a given node when this node is removed. The high local efficiency observed in the musician group indicates that their network is capable of sustaining robust regional communication when responding to speech prosody. This enhanced connectivity could lead to improvements in several auditory tasks, including the recognition of pitch, timbre, rhythm, and emotional content in both music and speech.

The Kolmogorov-Smirnov (K-S) statistical results indicate that there is a significant difference in the degree distributions between the two groups under both emotional prosody conditions. Specifically, a 'more concentrated' degree distribution in the music group suggests that some regions of the music group-representing specific brain regions-have a highly consistent response of the brain communication for the speech prosody stimuli. The results can be found that in both emotional conditions, the music group shows a more concentrated degree, which may suggest that musically trained individuals may have a denser or more richly interconnected network in the areas of the brain that process emotional prosody [1].

These findings indicate that musical training is associated with distinct neural processing of emotional prosody. The influence of musical expertise on brain activity was evident in conditions of both positive and negative emotional valence, with some effects being consistent across emotions and others being condition-specific.

5 Conclusion

In this study, advanced neural imaging techniques, such as EEG and brain network analysis, were utilized to investigate both dynamic and static processing of emotional prosody. Importantly, we compared the neural processing mechanisms between individuals with long-term musical training and those without any musical background. A thorough analysis of the EEG recordings was conducted, revealing that musical training has a transferable effect on speech prosody processing. This is reflected in neural adaptations characterized by less reliance on primary auditory resources, a more optimized neural network, and more focused information processing. This study provides a comprehensive examination of the neural mechanisms by which music training influences speech prosody processing, offering unique insights for music education and neuroscience. Additionally, it suggests potential applications in rehabilitating language functions in clinical settings.

Acknowledgments. This work was supported in part by The Science and Technology Development Fund, Macau SAR (File no. 0022/2021/APD), The University of Macau Research Committee (MYRG projects 2017-00207-FST and 2022-00197-FST) and Guangdong Basic and Applied Basic Research Foundation (Grant No. 2023A1515010844).

References

1. Bassett, D.S., Bullmore, E.T.: Human brain networks in health and disease. Curr. Opin. Neurol. **22**(4), 340–347 (2009)

2. Beaton, A.A.: The relation of planum temporale asymmetry and morphology of the corpus callosum to handedness, gender, and dyslexia: a review of the evidence. Brain Lang. **60**(2), 255–322 (1997)
3. Boersma, P., Weenink, D.: PRAAT: doing phonetics by computer (version 5.3.51) (2007)
4. Casasanto, D.: Embodiment of abstract concepts: good and bad in right-and left-handers. J. Exp. Psychol. Gen. **138**(3), 351 (2009)
5. Casasanto, D., Jasmin, K.: Good and bad in the hands of politicians: spontaneous gestures during positive and negative speech. PLoS ONE **5**(7), e11805 (2010)
6. Delorme, A., Makeig, S.: Eeglab: an open source toolbox for analysis of single-trial EEG dynamics including independent component analysis. J. Neurosci. Methods **134**(1), 9–21 (2004)
7. Hallez, H., et al.: Removing muscle and eye artifacts using blind source separation techniques in ictal EEG source imaging. Clin. Neurophysiol. **120**(7), 1262–1272 (2009)
8. Herholz, S.C., Zatorre, R.J.: Musical training as a framework for brain plasticity: behavior, function, and structure. Neuron **76**(3), 486–502 (2012)
9. James, C.J., Hesse, C.W.: Independent component analysis for biomedical signals. Physiol. Meas. **26**(1), R15 (2004)
10. Koelsch, S., Siebel, W.A.: Towards a neural basis of music perception. Trends Cogn. Sci. **9**(12), 578–584 (2005)
11. Kraus, N., Chandrasekaran, B.: Music training for the development of auditory skills. Nat. Rev. Neurosci. **11**(8), 599–605 (2010)
12. for Language, N.W.C., Characters: Lexicon of common words in contemporary Chinese (2008)
13. Li, R., Principe, J.C.: Blinking artifact removal in cognitive EEG data using ICA. In: 2006 International Conference of the IEEE Engineering in Medicine and Biology Society, pp. 5273–5276. IEEE (2006)
14. Lin, Y., Ding, H., Zhang, Y.: Gender differences in identifying facial, prosodic, and semantic emotions show category-and channel-specific effects mediated by encoder's gender. J. Speech Lang. Hear. Res. **64**(8), 2941–2955 (2021)
15. Maass, A., Russo, A.: Directional bias in the mental representation of spatial events: nature or culture? Psychol. Sci. **14**(4), 296–301 (2003)
16. Maratos, F.A.: Temporal processing of emotional stimuli: the capture and release of attention by angry faces. Emotion **11**(5), 1242 (2011)
17. Mitchell, R.L., Elliott, R., Barry, M., Cruttenden, A., Woodruff, P.W.: The neural response to emotional prosody, as revealed by functional magnetic resonance imaging. Neuropsychologia **41**(10), 1410–1421 (2003)
18. Moreno, S., Marques, C., Santos, A., Santos, M., Castro, S.L., Besson, M.: Musical training influences linguistic abilities in 8-year-old children: more evidence for brain plasticity. Cereb. Cortex **19**(3), 712–723 (2009)
19. Paulmann, S., Bleichner, M., Kotz, S.A.: Valence, arousal, and task effects in emotional prosody processing. Front. Psychol. **4**, 345 (2013)
20. Paulmann, S., Kotz, S.A.: Early emotional prosody perception based on different speaker voices. NeuroReport **19**(2), 209–213 (2008)
21. Paulmann, S., Seifert, S., Kotz, S.A.: Orbito-frontal lesions cause impairment during late but not early emotional prosodic processing. Soc. Neurosci. **5**(1), 59–75 (2010)
22. Pinheiro, A.P., et al.: Sensory-based and higher-order operations contribute to abnormal emotional prosody processing in schizophrenia: an electrophysiological investigation. Psychol. Med. **43**(3), 603–618 (2013)

23. Pinheiro, A.P., Galdo-Alvarez, S., Rauber, A., Sampaio, A., Niznikiewicz, M., Gonçalves, O.F.: Abnormal processing of emotional prosody in Williams syndrome: an event-related potentials study. Res. Dev. Disabil. **32**(1), 133–147 (2011)
24. Pinheiro, A.P., et al.: Abnormalities in the processing of emotional prosody from single words in schizophrenia. Schizophr. Res. **152**(1), 235–241 (2014)
25. Rubinov, M., Sporns, O.: Complex network measures of brain connectivity: uses and interpretations. Neuroimage **52**(3), 1059–1069 (2010)
26. Saber, M.: Removing powerline interference from EEG signal using optimized fir filters. J. Artif. Intell. Metaheuristics **1**(1), 8–19 (2022)
27. Schirmer, A., Kotz, S.A.: ERP evidence for a sex-specific Stroop effect in emotional speech. J. Cogn. Neurosci. **15**(8), 1135–1148 (2003)
28. Schirmer, A., Kotz, S.A., Friederici, A.D.: Sex differentiates the role of emotional prosody during word processing. Cogn. Brain Res. **14**(2), 228–233 (2002)
29. Shapiro, L.A.: The Routledge Handbook of Embodied Cognition (2014)
30. Smith, E.R.: What do connectionism and social psychology offer each other? J. Pers. Soc. Psychol. **70**(5), 893 (1996)
31. Twist, D.J., Squires, N.K., Spielholz, N.I., Silverglide, R.: Event-related potentials in disorders of prosodic and semantic linguistic processing 1. Cogn. Behav. Neurol. **4**(4), 281–304 (1991)
32. Wildgruber, D., Ackermann, H., Kreifelts, B., Ethofer, T.: Cerebral processing of linguistic and emotional prosody: fMRI studies. Prog. Brain Res. **156**, 249–268 (2006)

Potential Indicator for Continuous Emotion Arousal by Dynamic Neural Synchrony

Guandong Pan[1,3], Zhaobang Wu[1,3], Yaqian Yang[2,3], Xin Wang[2,3,5,6,7], Longzhao Liu[2,3,5,6,7], Zhiming Zheng[2,3,4,5,6,7,8], and Shaoting Tang[2,3,4,5,6,7,8](✉)

[1] School of Computer Science and Engineering, Beihang University, Beijing 100191, China
{pan_gd,wuzhaobang}@buaa.edu.cn
[2] Institute of Artificial Intelligence, Beihang University, Beijing 100191, China
{yangyaqian,wangxin_1993,longzhao,tangshaoting}@buaa.edu.cn, zzheng@pku.edu.cn
[3] Key Laboratory of Mathematics, Informatics and Behavioral Semantics, Beihang University, Beijing 100191, China
[4] Institute of Medical Artificial Intelligence, Binzhou Medical University, Yantai 264003, China
[5] Zhongguancun Laboratory, Beijing 100094, China
[6] Beijing Advanced Innovation Center for Future Blockchain and Privacy Computing, Beihang University, Beijing 100191, China
[7] PengCheng Laboratory, Shenzhen 518055, China
[8] State Key Lab of Software Development Environment, Beihang University, Beijing 100191, China

Abstract. The need for automatic and high-quality emotion annotation is paramount in applications such as continuous emotion recognition and video highlight detection, yet achieving this through manual human annotations is challenging. Inspired by inter-subject correlation (ISC) utilized in neuroscience, this study introduces a novel Electroencephalography (EEG) based ISC methodology that leverages a single-electrode and feature-based dynamic approach. Our contributions are three folds: Firstly, we reidentify two potent emotion features suitable for classifying emotions-first-order difference (FD) an differential entropy (DE). Secondly, through the use of overall correlation analysis, we demonstrate the heterogeneous synchronized performance of electrodes. This performance aligns with neural emotion patterns established in prior studies, thus validating the effectiveness of our approach. Thirdly, by employing a sliding window correlation technique, we showcase the significant consistency of dynamic ISCs across various features or key electrodes in each analyzed film clip. Our findings indicate the method's reliability in capturing consistent, dynamic shared neural synchrony among individuals, triggered by evocative film stimuli. This underscores the potential of our approach to serve as an indicator of continuous human emotion arousal. The implications of this research are significant for advancements in affective computing and the broader neuroscience field, suggesting a streamlined and effective tool for emotion analysis in real-world applications.

Keywords: Emotion Annotation · Inter-subject Correlation · Electroencephalography (EEG)

1 Introduction

Capturing the ground truth of dynamic human emotions is pivotal for advancements in continuous emotion recognition and video highlight detection [3,6,8]. Traditionally, the collection of two-dimensional emotional responses, namely arousal and valence, has relied on subjective human annotations [8,38]. However, the dual demands of video observation and annotation tasks impose a cognitive burden on annotators, which can compromise the quality of the data collected. Additionally, the requirement for conscious rating in such tasks may affect long-term efficiency. These limitations pose a significant scientific challenge: Is there an alternative method to annotate human emotions both unconsciously and efficiently?

The technique of inter-subject correlation (ISC), derived from neuroscience, offers a promising approach to capturing group-level shared neural responses without conscious effort from participants [16–18,29]. This neural synchrony, reflecting the processing of emotional information in response to evocative video content [36], serves as an indicator of variations in human emotional arousal. Traditional ISC research, primarily utilizing functional Magnetic Resonance Imaging (fMRI), has established the reliability of this method across different cinematic experiences [16]. In contrast, electroencephalography (EEG) offers advantages over fMRI in terms of affordability, portability, and high temporal resolution, making it more applicable to real-world tasks [40]. Recent studies have applied EEG to extract ISC, but have predominantly focused on multichannel analyses [9–11], which do not meet the practical needs of applications requiring fewer channels. To address this limitation, we propose a novel feature-based dynamic EEG-based ISC method computed on each individual channel. This approach can autonomously capture the continuous synchronization of brain regions among a population, thereby providing a potential neural-based indicators for representing the ground truth of human dynamic emotion arousal.

In brief, the contributions of our work are as follows:

- We reidentify two robust emotion features that are pivotal in distinguishing emotional states. The first feature, derived from the time domain, is the first-order difference (FD), and the second, from the frequency domain, is differential entropy (DE).
- Utilizing an overall correlation analysis, this study demonstrates the heterogeneous synchronized performance across electrodes. This finding is in alignment with established neural emotion patterns from previous research, thus serving as a validation of our methodology's effectiveness.
- By employing a sliding window correlation technique, we illustrate the consistency of dynamic ISCs across various features or key electrodes within each analyzed film clip. This consistency underscores the reliability of our method in capturing dynamic shared neural synchrony among individuals,

triggered by emotionally evocative film stimuli. Our approach suggests significant potential to serve as an indicator of continuous human emotion arousal.

2 Related Works

2.1 Inter-Subject Correlation

Inter-subject correlation (ISC) quantifies the similarity in brain responses among individuals exposed to **the same naturalistic stimuli**, such as films or stories. In the seminal work [17], Hasson et al. pioneered the exploration of ISC by demonstrating synchronous neural activity across participants during film viewing, not only in primary sensory areas but also in higher associative cortices involved in complex cognitive functions such as face recognition. Over the past decades, the intuitive underlying principles of ISC have facilitated its adoption across a wide range of fields, yielding significant insights into mechanisms of attention [23,33,35], friendship prediction [31], emotion [8,20,30], marketing [2,9], memory processes [5,15], video highlight detection [6], healthcare applications [14], and even interspecies comparisons [27].

Historically, the majority of ISC research utilized functional Magnetic Resonance Imaging (fMRI). However, there has been a gradual shift towards incorporating alternative modalities such as Electroencephalography (EEG) [8,9,25], Electrocorticography (ECoG) [19], Magnetoencephalography (MEG) [24], functional Near-Infrared Spectroscopy (fNIRS) [26], Electrocardiography (ECG) [34], analysis of eye movements [18], and other physiological signals [6]. Notably, Haufe et al. [19] confirmed that fMRI, ECoG, and EEG exhibit comparable repeat-reliability across viewings, highlighting EEG's potential for capturing shared neural responses with its high temporal resolution and suitability for naturalistic settings due to portable equipment.

Nevertheless, a significant gap in the literature exists regarding the practical implementation of ISC using EEG in real-world environments, where the use of a minimal number of electrodes is preferable. While past studies often employed multiple electrodes to enhance analysis performance [9,11,23,35], this approach is less feasible in the real world, potentially limiting the broader application of EEG-based ISC techniques. Addressing this gap by developing methodologies for ISC analysis with fewer electrodes could significantly expand the utility and accessibility of EEG for real-world applications. By isolating the ISC analysis to each individual electrode, we leverage both overall and sliding window correlation techniques to meticulously assess the potential of single-channel data to capture neural synchrony. Results demonstrate the reliability of our methodology in capturing dynamic shared neural synchrony, suggesting the potential to serve as a reliable indicator of continuous emotional arousal. This approach not only simplifies the hardware requirements for real-world applications but also maintains analytical precision, thus significantly advancing the practical deployment of EEG-based ISC techniques.

2.2 EEG Feature Extraction

Traditionally, inter-subject correlation (ISC) techniques have been applied directly to raw brain signals. However, in the realm of EEG-based emotion recognition, a variety of robust feature extraction methods have been developed. These methods have proven effective in distinguishing emotional valences through model training and testing. This presents a compelling case for implementing feature-related ISC analysis.

In 2013, the Differential Entropy (DE) feature was introduced [12], with subsequent studies affirming its efficacy in representing emotional states [40,41,45]. A comprehensive review in 2019 [42] evaluated EEG features across four domains-time, frequency, time-frequency, and spatial-using the sparse linear discriminant analysis (SLDA) method across three distinct datasets (SEED, DREAMER, CAS-THU). This review highlighted that time-domain features, particularly the first-order difference (FD), exhibited superior performance in discriminating emotional valence.

Motivated by these findings, our study begins with a user-dependent comparative analysis of features to ascertain their efficacy in extracting emotional information. Following this preliminary evaluation, DE and FD features are identified as particularly potent and are subsequently selected for further analysis in extracting neural synchrony. This approach aims to enhance the applicability of EEG-based single-channel ISC techniques by leveraging sophisticated feature-based analyses.

3 Materials and Methodology

3.1 Dataset and Preprocessing

The SEED EEG dataset, a publicly available affective dataset, was introduced by Zheng et al. [40]. It comprises EEG recordings from fifteen Chinese participants (seven males and eight females; mean age: 23.27, SD: 2.37) who viewed various Chinese film clips designed to evoke distinct emotional valences: positive, negative, and neutral. Each participant was exposed to the same 15 clips, approximately four minutes each, across three separate sessions on different days. EEG data were captured using a 62-channel setup conforming to the International 10–20 system via the ESI Neuroscan system. For our study, 45 sessions (15 participants, three sessions each) of EEG records are analyzed to extract statistical shared neural responses at the population level for each film clip.

Our analysis utilize the "Preprocessed Data" set, sampled at 200 Hz and bandpass filtered from 0 to 75 Hz. Using MATLAB's EEGLAB toolbox [7], we apply a notch filter between 48 and 52 Hz to eliminate electrical noise. Visual inspections of channel data for each session are used for the interpolation of faulty channels. Additionally, artifacts from eye movements and muscle contractions are isolated and removed through independent component analysis (ICA) implemented in EEGLAB.

3.2 EEG-Based ISC

Feature Extraction. Original features in EEG analysis are derived from the absolute raw values of EEG signals following preprocessing, whereas first-order difference features represent the differences between consecutive sample values. First-order difference features are particularly valuable as they capture the non-linear dynamics of EEG signals, namely the rate of change in voltage, which is closely linked to emotional states [42]. In our ISC methodology, we introduce a variable, *scale*, which aggregates s sample points into a single feature point. This scaling allows for the analysis of shared cognitive information across varying temporal resolutions and reduces the length of the feature vectors, thereby decreasing storage requirements and enhancing computational efficiency.

The equations for the original and first-order difference feature vectors for a single EEG record are defined as [32]:

$$V_O(i) = \frac{1}{s} \sum_{j=i \cdot s}^{(i+1) \cdot s - 1} |x(j)|, 0 \leq i < \lfloor \frac{N}{s} \rfloor, \tag{1}$$

$$V_{FD}(i) = \frac{1}{s} \sum_{j=i \cdot s}^{(i+1) \cdot s - 1} |x(j+1) - x(j)|, 0 \leq i < \lfloor \frac{N}{s} \rfloor, \tag{2}$$

where i represents the index of the feature vector, $x(j)$ deontes the j-th sample point in the EEG record, N is the total number of samples, and s is the scale factor converting s sample points into one feature point.

Additionally, we incorporate the Differential Entropy (DE) feature [12], which provides a quantification of the complexity or uncertainty associated with EEG signal frequency distributions. It has been found that the subbands of EEG signals are nearly subject to Gaussian distribution [12]. The DE feature is calculated over a fixed scale of 200 sample points, corresponding to a 1-second time scale at a 200 Hz sampling rate. We compute DE features using Fast Fourier Transform with a 1-s-long window and no overlapping window. Given that a series of a certain band X follows a Gaussian distribution $N(\mu, \sigma^2)$, where μ is the mean and σ^2 is the variance of the distribution, the DE is defined by the formula:

$$V_{DE}(i) = -\sum_{k=0}^{K-1} p(X_k) \log p(X_k) = \frac{1}{2} \log(2\pi e \sigma^2), 0 \leq i < \lfloor \frac{N}{s} \rfloor \tag{3}$$

where i represents the index of the feature vector, $p(x)$ represents the probability distribution of the EEG frequency amplitude values within the defined window, namely Gaussian distribution. This fixed-scale analysis provides a consistent measure of entropy across all EEG channels, offering insights into the informational content of the signals related to emotions. In our work, we compute DE features for four frequency bands respectively, i.e., δ and θ band (1–7 Hz), α band (8–13 Hz), β band (14–29 Hz), γ band (30–47 Hz).

Similarity Measurement. Pearson correlation coefficient (PCC) is used in ISC similarity measurement that was proposed by Karl Pearson. PCC can measure a linear association or dependence, between two continuous variables [29]. Given two random variables X, Y and their paired N samples $x(i), y(i), i = 1, 2, 3, ...N$, the formula of PCC is as follows:

$$PCC(X,Y) = \frac{Cov(X,Y)}{\sigma_X \cdot \sigma_Y} = \sum_{i=1}^{n} \frac{(x_i - \bar{x})(y_i - \bar{y})}{\sqrt{(x_i - \bar{x})^2}\sqrt{(y_i - \bar{y})^2}} \quad (4)$$

where $Cov(X, Y)$ means the covariance between X, Y, $\bar{x}$ and σ_x are the mean and standard deviation of $x(i)$, $\bar{y}$ and σ_y are the mean and standard deviation of $y(i)$ and N means the number of samples.

Overall and Sliding Window Correlation. Inter-subject correlation (ISC) can be computed using two distinct methodologies: overall correlation and sliding window correlation. The overall correlation method computes a single correlation coefficient, reflecting the global similarity across the entire lengths of two feature vectors. In contrast, the sliding window correlation method segments the feature vectors into smaller, overlapping windows. For each window, a separate correlation is computed, resulting in a sequence that illustrates temporal fluctuations in correlation, providing insights into dynamic synchronization patterns. The length of this sequence is influenced by both the width of the window and the overlap between successive windows, which is set to 1 s in this study.

The choice of window width is crucial and remains a subject of ongoing research [28]. In our analysis, window widths of 10 s and 70 s are utilized to adequately capture brief scenes and longer narrative arcs within the films, respectively. These correlation calculations are performed for each pairwise combination of subjects, across all channels and all films, resulting in a substantial dataset comprising 990 pairs, 62 channels, and 15 films within the SEED dataset. For each film and channel, sliding window correlations are averaged across all pairs of subjects to determine population-level synchronous responses. These correlations are denoted as SW-O-ISC, SW-FD-ISC, and SW-DE-ISC for the original, first-order difference (FD), and differential entropy (DE) feature vectors, respectively. To enhance computational efficiency, the Python library 'taichi' [21] is employed for parallel computation of sliding window ISC.

The mathematical representations for the overall (ζ) and sliding window (ξ) correlations are given by:

$$\zeta_{i,j}^{f,c} = PCC(V_i^{f,c}, V_j^{f,c}), \quad (5)$$

$$\xi^{f,c}(k) = \frac{1}{M^*} \sum_{i,j} PCC(v_i^{f,c}(k), v_j^{f,c}(k)), \quad (6)$$

where i, j represent participant indices ($1 \leq i \leq M, i < j \leq M$, with M being the total number of participants), f, c denote film and channel indices respectively,.

$v_i^{f,c}(k)$ represents the segmented feature vector in the kth window. M^* signifies the total number of pairwise comparisons.

4 Experiments and Results

4.1 EEG Feature Classification and Selection for Effective Emotion Representation

Table 1. User-dependent recognition accuracy(%) of different feature extraction methods on SEED.

Feature	Original	FD	Hjorth	NSI	PSD	DE
Acc	66.18	87.96	83.27	47.09	77.67	82.46

In the field of EEG-based emotion recognition, features in the frequency domain have been widely used, such as power spectral density (PSD) and differential entropy (DE) [1,22,40,45],. However, a recent study [42] have showed that features in the time domain, such as the first difference, exhibit superior discriminative power in differentiating emotional valence compared to their frequency domain counterparts.

To further explore the discriminative capabilities of these features, we conducted an experiment using the SEED dataset, employing the linear kernel Support Vector Machine (SVM) classifier. We select four time-domain features for evaluation: the original feature, the first-order difference (FD) feature, the Hjorth parameter, and the non-stable index (NSI). Additionally, we include PSD and DE from the frequency domain for comparative analysis. A detailed description of these feature extraction methods is available in [42].

The study employs a user-dependent training strategy, configuring one classifier per subject. Each EEG record, approximately four minutes in duration, is segmented into 1-s samples from which features are extracted. A five-fold cross-validation approach is utilized for training and testing the classifiers, with data shuffling prior to model fitting to enhance performance. Hyperparameter optimization is conducted through nested grid search, with the SVM's gamma and C parameters ranging logarithmically from 2^{-4} to 2^4, each fold independently selecting optimal parameters.

Results, summarized in Table 1, reveal that the FD feature achieved the highest classification accuracy at 87.96%, surpassing other features significantly, with a notable 5% improvement over the frequency domain feature, DE. It is important to note, however, that certain features may exhibit enhanced performance with alternative classifiers instead of the linear kernel SVM, such as those employing non-linear kernels. In brief, these results underscore the potential of FD and DE features in effectively capturing nuanced emotional information from EEG data.

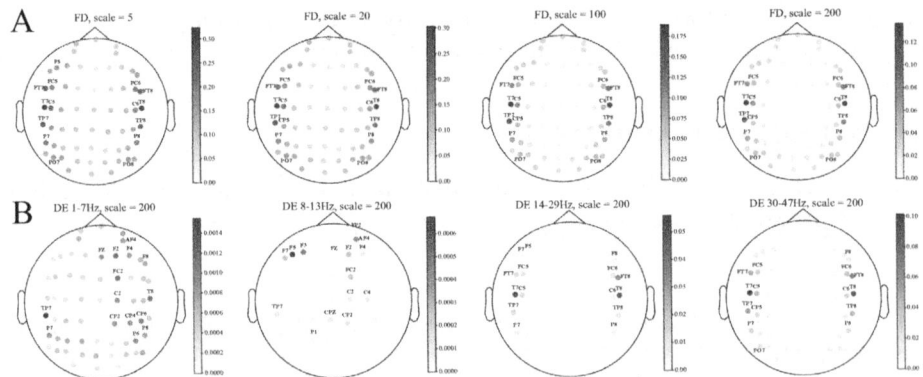

Fig. 1. Heterogeneous synchronized performance of electrodes. A and **B** illustrate the synchronized percentage of each electrode with overall correlation on different scales of FD feature or different bands of DE feature. After Family-Wise Error Rate (FWER) Bonferroni correction for multiple comparison across films, pairwise subjects, and channels three dimensions, we calculate the percentage of significant correlations across films and pairwise subjects using the adjusted p values of correlation ($p < 0.05$).

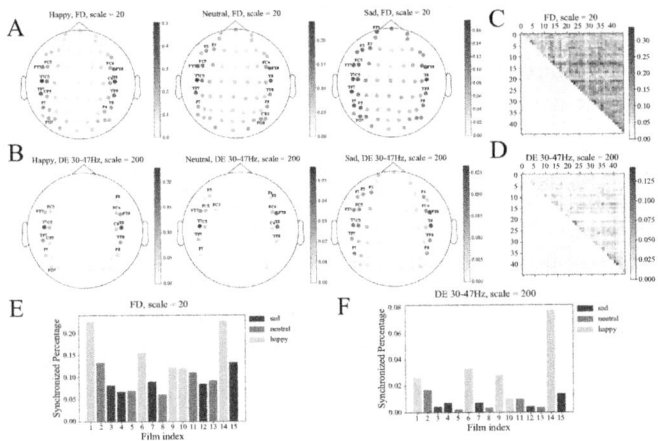

Fig. 2. Additional views of synchronized percentage for overall correlation. A and **B** show the montage views of three emotion valences on FD, scale = 20 and DE 30–47 Hz, scale = 200 respectively. **C** and **D** depict the synchronized percentage on each pairwise subject. **E** and **F** plot the synchronized percentage for each film.

4.2 Heterogeneous Synchronized Percentage of Electrodes on Overall ISC

We apply the overall correlation to EEG features from identical electrodes, assessing their synchrony to preliminarily evaluate the viability of single-electrode EEG-based Inter-Subject Correlation (ISC). First-order Difference

(FD) and Differential Entropy (DE) features are employed to compute ISC. Our findings reveal that electrodes near the temporal lobe exhibit the stronger ability to capture global synchrony, with consistent performance across various scales of the FD feature and different frequency bands of the DE feature.

For each film clip and each pair of subjects, we correlate two complete feature vectors from the same channel to obtain a p value. This process results in a multi-dimensional p-value tensor ($F \times P \times C$), where F denotes films, P indicates pairwise subjects, and C represents channels. Each scale of FD features and each band of DE features correspond to a specific p-value tensor. We apply a Family-Wise Error Rate (FWER) Bonferroni correction to the p-value tensor to address multiple comparisons. Following Hasson et al. [17], we use the 'synchronized percentage' as a metric for synchronization performance, calculated as the proportion of significant correlations ($p < 0.05$) across channels, pairwise subjects, or films. This metric reflects the intensity of shared cognitive activity.

Our analysis across all scales of FD features and the β, γ bands of DE features shows a heterogeneous yet consistent spatial pattern among the electrodes (Fig. 1-A B). Electrodes near the bilateral temporal lobes are notably effective in capturing shared information processing related to emotions, a finding corroborated by other studies in EEG-based emotion recognition [39, 43, 45]. Similar patterns are observed for films with different emotional valences (Fig. 2-A B), underscoring the consistency of these spatial patterns.

Moreover, films with a 'happy' emotional valence tend to evoke stronger shared neural synchrony, suggesting a greater potential of such films to engage audiences similarly (Fig. 2-A B). Additionally, both FD and DE features show synchronization across many pairwise subjects, confirming that the observed overall synchronization is not limited to a few pairs but is rather widespread (Fig. 2-C D). The consistent induction of neural synchrony by each film, albeit with varying potentials linked to different valences, further supports our findings (Fig. 2-E F). Besides, the montage view plots are improved based on mne.channels.DigMontage.plot [13].

4.3 Validating the Reliability of Dynamic ISC

In this section, we employ the sliding window correlation technique (Fig. 3-A) to analyze the dynamic shared neural responses across participants at critical electrodes (FT7, FT8, TP7, TP8, T7, T8). We calculate the mean correlation coefficients derived from the dynamic inter-subject correlation (ISC) of emotion-related features or key electrodes to assess the reliability of our methods. Our findings reveal that the similarity of different dynamic ISCs exhibits significant performance across all film categories and both window sizes, particularly for 'happy' films. This suggests that consistent dynamic synchronous neural responses to stimuli can be effectively captured through single-electrode sliding window correlations, underscoring the potential of this method to represent group-level emotional arousal triggered by the films.

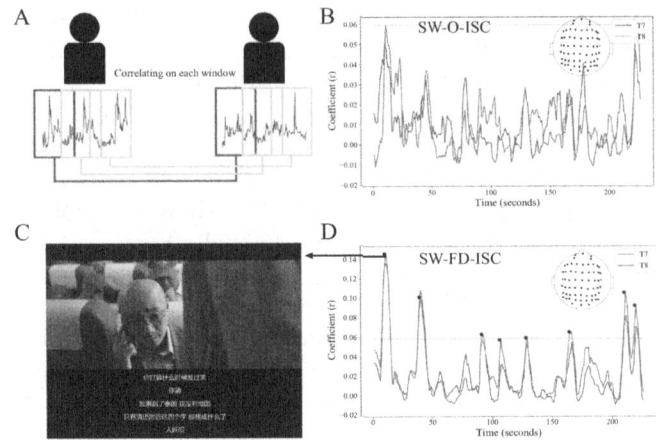

Fig. 3. Case comparison of the dynamic ISCs at electrodes T7 and T8. A. The illustration of computing one dynamic ISC between one pairwise subject. **B** and **D** The example dynamic ISCs of SW-O-ISC and SW-FD-ISC on the electrodes T7, T8 for the first film clip in SEED. **C.** A scene from the first film clip corresponding to the window at the first peak.

We begin with a comparative analysis of the dynamic ISC at electrodes T7 and T8 using sliding window correlations with original inter-subject correlation (SW-O-ISC) and first-difference inter-subject correlation (SW-FD-ISC) at a scale of 20, focusing on the first film from the SEED dataset (Fig. 3-B D). The grey dotted lines in the plots indicate the highest correlation coefficient for SW-O-ISC. Notably, many instances in SW-FD-ISC exceed this benchmark. This comparison demonstrates that SW-FD-ISC provides smoother, stronger, and more consistent responses than SW-O-ISC, highlighting its efficacy in capturing emotion-related dynamic ISC at pivotal electrodes. Additionally, we illustrate a 10-second scene corresponding to the first peak in SW-FD-ISC, which coincides with a humorous moment eliciting significant emotional arousal in the audience (Fig. 3-C).

Further comprehensive comparisons of dynamic ISCs across different emotion-related features or several key electrodes are conducted. In each film clip, various scales of first-difference (FD) features and the β or γ bands of differential entropy (DE) features are employed to extract population-level dynamic synchronous responses at each key electrode. The significance of coefficients in each dynamic ISC is assessed using the Wilcoxon signed-rank test with adjustments for multiple comparisons via the Benjamini-Hochberg False Discovery Rate (FDR) [4]. Only significant coefficients are color-coded, and Z-score normalization is employed prior to comparing the performance of dynamic ISCs. In the first film, significant peaks are consistently observed in dynamic ISCs at window 10, indicating the reliable capture of fine-grained and emotion-related information (Fig. 4-A B). Moreover, dynamic ISCs at window 70 consistently

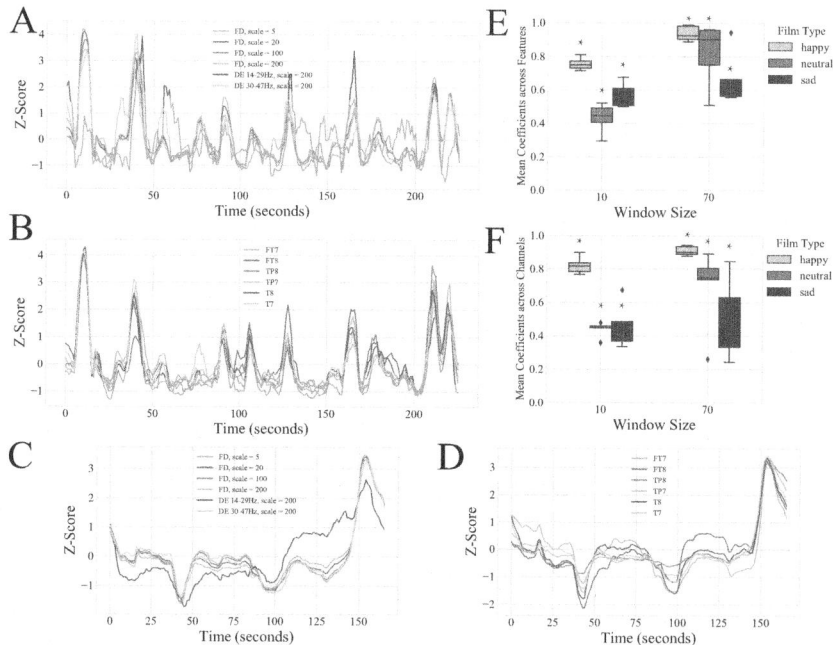

Fig. 4. The consistent dynamic ISCs on both 10 and 70 window sizes. A and **C** show the dynamic ISCs across features on the electrode T7, window size 10 s or 70 s respectively. **B** and **D** illustrate the dynamic ISCs across key electrodes, window size 10 s or 70 s respectively. The significance of coefficients in each dynamic ISC is assessed using the Wilcoxon signed-rank test with adjustments for multiple comparisons via the Benjamini-Hochberg False Discovery Rate (FDR). Only significant coefficients are color-coded. **E** and **F** depict the mean correlation coefficients of dynamic ISCs across features (with the electrode T7) or channels (with FD, scale = 20) under category view. Results with other configurations have similar performance and are not shown for brevity. The significance of each category is tested using one-sample t-tests against a threshold of 0.2.

capture all significant variations in shared responses, suggesting the presence of similar cognitive activities among the audience on a larger scale (Fig. 4-C D).

To assess the performance across all film clips, we calculate the mean correlation coefficients for each individual film clip, categorizing them by movie emotion labels, each category containing five clips. These coefficients are derived by averaging the pairwise correlations of dynamic ISCs across various features or key channels. We evaluate the significance of these means using one-sample t-tests against a threshold of 0.2. The resulting box plots for each category demonstrate significant performance, affirming that the similarity among dynamic ISCs for film clips within each category is substantial (Fig. 4-E F). Particularly noteworthy are films in the 'happy' category, which display mean coefficients approaching 0.8. This suggests that these films are exceptionally effective at engaging the audience and eliciting synchronized neural responses, indicative of strong emotional engagement.

5 Discussion and Conclusion

In this study, we explore the viability of using EEG-based inter-subject correlation (ISC) for automatic and unconscious emotion annotation. With a particular focus on reducing the number of electrodes required, we have developed a novel single-electrode and feature-based dynamic ISC method. The contributions of our work are threefold: (1) We reidentify two effective emotion features, one from the time domain and one from the frequency domain: first-order difference (FD) and differential entropy (DE). (2) We utilize the overall correlation to demonstrate the heterogeneous synchronized performance of electrodes. This performance is consistent with neural emotion patterns identified in previous research, providing a validation of our method's effectiveness. (3) Applying a sliding window correlation, we illustrate the similarity of dynamic ISCs across various features or key electrodes for each film clip. This indicates the reliability of our method in capturing consistent dynamic shared neural synchrony among individuals induced by evocative film clips, highlighting its potential as an indicator of human continuous emotion arousal.

Our findings on the temporal lobe's dominance in neural synchrony are consistent with a breadth of multidisciplinary literature, encompassing neuroscience and affective computing. Neuroscientists, such as Vytal and Hamann [39], using functional Magnetic Resonance Imaging (fMRI), identified distinct activation patterns associated with basic emotions in specific brain regions: happiness primarily activates the right superior temporal gyrus (STG) and left anterior cingulate cortex (ACC), while sadness is primarily associated with the left medial frontal gyrus (medFG). Furthermore, several studies have employed EEG and computational models to delineate spatial distribution patterns of emotional responses. Critical emotion patterns located at the temporal lobe have been observed through various methods, including DE feature frequency bands, weight distributions of deep belief networks [40], average energy distributions [44,45], spatio-temporal feature characteristics by contrastive learning intracranial seizure analysis (CLISA) [37], and gradient visualization of a teacher-student model [43]. These findings collectively underscore the integral role of the temporal lobes in brain emotion functions. The heterogeneous electrode performance observed in our results aligns with these established spatial patterns of emotions.

Acknowledgements. This work is supported by National Key Research and Development Program of China (2021YFB2700300), Program of National Natural Science Foundation of China (62141605, 12201026,12301305)

Disclosure of Interests. The authors have no competing interests.

References

1. Alarcao, S.M., Fonseca, M.J.: Emotions recognition using EEG signals: a survey. IEEE Trans. Affect. Comput. **10**(3), 374–393 (2017). https://doi.org/10.1109/TAFFC.2017.2714671. https://ieeexplore.ieee.org/document/7946165/
2. Barnett, S.B., Cerf, M.: A ticket for your thoughts: method for predicting content recall and sales using neural similarity of moviegoers. J. Consum. Res. **44**(1), 160–18 (2017).https://doi.org/10.1093/jcr/ucw083. https://academic.oup.com/jcr/article/44/1/160/2938969
3. Baveye, Y., Dellandrea, E., Chamaret, C., Chen, L.: LIRIS-ACCEDE: a video database for affective content analysis. IEEE Trans. Affect. Comput. **6**(1), 43–55 (2015). https://doi.org/10.1109/TAFFC.2015.2396531. http://ieeexplore.ieee.org/document/7024148/
4. Benjamini, Y.: Discovering the false discovery rate. J. Roy. Stat. Soc. Series B: Stat. Methodol. **72**(4), 405–416 (2010). https://doi.org/10.1111/j.1467-9868.2010.00746.x. https://rss.onlinelibrary.wiley.com/doi/abs/10.1111/j.1467-9868.2010.00746.x
5. Chen, J., Leong, Y.C., Honey, C.J., Yong, C.H., Norman, K.A., Hasson, U.: Shared memories reveal shared structure in neural activity across individuals. Nat. Neurosci. **20**(1), 115–125 (2017). https://doi.org/10.1038/nn.4450. https://www.nature.com/articles/nn.4450
6. Chênes, C., Chanel, G., Soleymani, M., Pun, T.: Highlight detection in movie scenes through inter-users, physiological linkage. In: Ramzan, N., Van Zwol, R., Lee, J.S., Clüver, K., Hua, X.S. (eds.) Social Media Retrieval, pp. 217–237. Springer, London (2012). https://doi.org/10.1007/978-1-4471-4555-4_10
7. Delorme, A., Makeig, S.: EEGLAB: an open source toolbox for analysis of single-trial EEG dynamics including independent component analysis. J. Neurosci. Methods **134**(1), 9–21 (2004)
8. Ding, Y., Hu, X., Xia, Z., Liu, Y.J., Zhang, D.: Inter-brain EEG feature extraction and analysis for continuous implicit emotion tagging during video watching. IEEE Trans. Affect. Comput. **12**(1), 92–102. https://doi.org/10.1109/TAFFC.2018.2849758. https://ieeexplore.ieee.org/document/8392701/
9. Dmochowski, J.P., Bezdek, M.A., Abelson, B.P., Johnson, J.S., Schumacher, E.H., Parra, L.C.: Audience preferences are predicted by temporal reliability of neural processing. Nat. Commun. **5**(1), 4567 (2014). https://doi.org/10.1038/ncomms5567. https://www.nature.com/articles/ncomms5567
10. Dmochowski, J.P., Ki, J.J., DeGuzman, P., Sajda, P., Parra, L.C.: Extracting multidimensional stimulus-response correlations using hybrid encoding-decoding of neural activity. NeuroImage **180**, 134–146 (2018). https://doi.org/10.1016/j.neuroimage.2017.05.037. https://www.sciencedirect.com/science/article/pii/S1053811917304299
11. Dmochowski, J.P., Sajda, P., Dias, J., Parra, L.C.: Correlated components of ongoing EEG point to emotionally laden attention - a possible marker of engagement? Front. Hum. Neurosci. **6**. https://doi.org/10.3389/fnhum.2012.00112. http://journal.frontiersin.org/article/10.3389/fnhum.2012.00112/abstract
12. Duan, R.N., Zhu, J.Y., Lu, B.L.: Differential entropy feature for EEG-based emotion classification. In: 2013 6th International IEEE/EMBS Conference on Neural Engineering (NER), pp. 81–84 (2013). https://doi.org/10.1109/NER.2013.6695876. ISSN: 1948-3554
13. Gramfort, A., et al.: MEG and EEG data analysis with MNE-python. Front. Neuroinf. **7** (2013). https://doi.org/10.3389/fnins.2013.00267. https://www.frontiersin.org/article/10.3389/fnins.2013.00267

14. Hasson, U., et al.: Shared and idiosyncratic cortical activation patterns in autism revealed under continuous real-life viewing conditions. Autism Res. **2**(4), 220–231 (2009). https://doi.org/10.1002/aur.89. https://www.ncbi.nlm.nih.gov/pmc/articles/PMC2775929/
15. Hasson, U., Furman, O., Clark, D., Dudai, Y., Davachi, L.: Enhanced intersubject correlations during movie viewing correlate with successful episodic encoding. Neuron **57**(3), 452–462 (2007). https://doi.org/10.1016/j.neuron.2007.12.009. https://linkinghub.elsevier.com/retrieve/pii/S0896627307010082
16. Hasson, U., Malach, R., Heeger, D.J.: Reliability of cortical activity during natural stimulation. Trends Cogn. Sci. **14**(1), 40–48 (2009). https://doi.org/10.1016/j.tics.2009.10.011. https://linkinghub.elsevier.com/retrieve/pii/S1364661309002393
17. Hasson, U., Nir, Y., Levy, I., Fuhrmann, G., Malach, R.: Intersubject synchronization of cortical activity during natural vision. Science **303**(5664), 1634–1640 (2004). https://doi.org/10.1126/science.1089506. https://www.science.org/doi/10.1126/science.1089506
18. Hasson, U., Yang, E., Vallines, I., Heeger, D.J., Rubin, N.: A hierarchy of temporal receptive windows in human cortex. J. Neurosci. **28**(10), 2539–2550 (2008). https://doi.org/10.1523/JNEUROSCI.5487-07.2008. https://www.jneurosci.org/lookup/doi/10.1523/JNEUROSCI.5487-07.2008
19. Haufe, S., et al.: Elucidating relations between fMRI, ECoG, and EEG through a common natural stimulus. NeuroImage **179**, 79–91 (2018). https://doi.org/10.1016/j.neuroimage.2018.06.016. https://linkinghub.elsevier.com/retrieve/pii/S1053811918305238
20. Hu, X., Wang, F., Zhang, D.: Similar brains blend emotion in similar ways: Neural representations of individual difference in emotion profiles. Neuroimage **247**, 118819 (2021). https://doi.org/10.1016/j.neuroimage.2021.118819. https://linkinghub.elsevier.com/retrieve/pii/S1053811921010909
21. Hu, Y., Li, T.M., Anderson, L., Ragan-Kelley, J., Durand, F.: Taichi: a language for high-performance computation on spatially sparse data structures. ACM Trans. Graph. **38**(6), 201 (2019)
22. Jenke, R., Peer, A., Buss, M.: Feature extraction and selection for emotion recognition from EEG. IEEE Trans. Affect. Comput. **5**(3), 327–339 (2014). https://doi.org/10.1109/TAFFC.2014.2339834. http://ieeexplore.ieee.org/document/6858031/
23. Ki, J.J., Kelly, S.P., Parra, L.C.: Attention strongly modulates reliability of neural responses to naturalistic narrative stimuli. J. Neurosci. **36**(10), 3092–3101 (2016). https://doi.org/10.1523/JNEUROSCI.2942-15.2016
24. Lankinen, K., Saari, J., Hari, R., Koskinen, M.: Intersubject consistency of cortical MEG signals during movie viewing. NeuroImage **92**, 217–224 (2014). https://doi.org/10.1016/j.neuroimage.2014.02.004. https://linkinghub.elsevier.com/retrieve/pii/S1053811914000950
25. Liu, S., et al.: What makes a good movie trailer?: Interpretation from simultaneous EEG and eyetracker recording. In: Proceedings of the 24th ACM International Conference on Multimedia, pp. 82–86. ACM (2016). https://doi.org/10.1145/2964284.2967187. https://dl.acm.org/doi/10.1145/2964284.2967187
26. Liu, Y., et al.: Measuring speaker-listener neural coupling with functional near infrared spectroscopy. Sci. Rep. **7**(1), 43293 (2017). https://doi.org/10.1038/srep43293. https://www.nature.com/articles/srep43293
27. Mantini, D., et al.: Interspecies activity correlations reveal functional correspondence between monkey and human brain areas. Nat. Method **9**(3), 277–282

(2012). https://doi.org/10.1038/nmeth.1868. https://www.nature.com/articles/nmeth.1868
28. Mokhtari, F., Akhlaghi, M.I., Simpson, S.L., Wu, G., Laurienti, P.J.: Sliding window correlation analysis: modulating window shape for dynamic brain connectivity in resting state. Neuroimage **189**, 655–666 (2019). https://doi.org/10.1016/j.neuroimage.2019.02.001. https://linkinghub.elsevier.com/retrieve/pii/S1053811919300874
29. Nastase, S.A., Gazzola, V., Hasson, U., Keysers, C.: Measuring shared responses across subjects using intersubject correlation. Social Cogn. Affect. Neurosci. **14**(6), 667–685 (2019). https://doi.org/10.1093/scan/nsz037
30. Nummenmaa, L., Glerean, E., Viinikainen, M., Jääskeläinen, I.P., Hari, R., Sams, M.: Emotions promote social interaction by synchronizing brain activity across individuals. Proc. Natl. Acad. Sci. **109**(24), 9599–9604 (2012). https://doi.org/10.1073/pnas.1206095109. https://www.pnas.org/doi/10.1073/pnas.1206095109
31. Parkinson, C., Kleinbaum, A.M., Wheatley, T.: Similar neural responses predict friendship. Nat. Commun. **9**(1), 332 (2018). https://doi.org/10.1038/s41467-017-02722-7. https://www.nature.com/articles/s41467-017-02722-7
32. Picard, R., Vyzas, E., Healey, J.: Toward machine emotional intelligence: analysis of affective physiological state. IEEE Trans. Pattern Anal. Mach. Intell. **23**(10), 1175–1191 (2001). https://doi.org/10.1109/34.954607. http://ieeexplore.ieee.org/document/954607/
33. Poulsen, A.T., Kamronn, S., Dmochowski, J., Parra, L.C., Hansen, L.K.: EEG in the classroom: synchronised neural recordings during video presentation. Sci. Rep. **7**(1), 43916 (2017). https://doi.org/10.1038/srep43916. https://www.nature.com/articles/srep43916
34. Pérez, P., et al.: Conscious processing of narrative stimuli synchronizes heart rate between individuals. Cell Rep. **36**(11), 109692 (2021). https://doi.org/10.1016/j.celrep.2021.109692. https://linkinghub.elsevier.com/retrieve/pii/S2211124721011396
35. Rosenkranz, M., Holtze, B., Jaeger, M., Debener, S.: EEG-based intersubject correlations reflect selective attention in a competing speaker scenario. Front. Neurosci. **15**, 685774 (2021). https://doi.org/10.3389/fnins.2021.685774. https://www.frontiersin.org/articles/10.3389/fnins.2021.685774/full
36. Sachs, M.E., Habibi, A., Damasio, A., Kaplan, J.T.: Dynamic intersubject neural synchronization reflects affective responses to sad music. NeuroImage **218**, 116512 (2019). https://doi.org/10.1016/j.neuroimage.2019.116512. https://www.sciencedirect.com/science/article/pii/S1053811919311036
37. Shen, X., Liu, X., Hu, X., Zhang, D., Song, S.: Contrastive learning of subject-invariant EEG representations for cross-subject emotion recognition (2022). https://doi.org/10.1109/TAFFC.2022.3164516. https://ieeexplore.ieee.org/document/9748967/
38. Soleymani, M., Asghari-Esfeden, S., Fu, Y., Pantic, M.: Analysis of EEG signals and facial expressions for continuous emotion detection. IEEE Trans. Affect. Comput. **7**(1), 17–28 (2015). https://doi.org/10.1109/TAFFC.2015.2436926
39. Vytal, K., Hamann, S.: Neuroimaging support for discrete neural correlates of basic emotions: a voxel-based meta-analysis. J. Cogn. Neurosci. **22**(12), 2864–2885 (2019). https://doi.org/10.1162/jocn.2009.21366. https://direct.mit.edu/jocn/article/22/12/2864/5011/Neuroimaging-Support-for-Discrete-Neural
40. Zheng, W.-L., Lu, B.-L.: Investigating critical frequency bands and channels for EEG-based emotion recognition with deep neural networks. IEEE Trans. Auton.

Mental Dev. **7**(3), 162–175 (2015). https://doi.org/10.1109/TAMD.2015.2431497. http://ieeexplore.ieee.org/document/7104132/
41. Yi, K., Wang, Y., Ren, K., Li, D.: Learning topology-agnostic EEG representations with geometry-aware modeling. Adv. Neural Inf. Process. Syst. **36**, 53875–53891 (2023). https://proceedings.neurips.cc/paper_files/paper/2023/hash/a8c893712cb7858e49631fb03c941f8d-Abstract-Conference.html
42. Yu, M., et al.: A review of EEG features for emotion recognition. Scientia sinica informationis **49**(9), 1097–1118 (2019). https://doi.org/10.1360/N112018-00337. http://engine.scichina.com/doi/10.1360/N112018-00337
43. Zhang, S., Tang, C., Guan, C.: Visual-to-EEG cross-modal knowledge distillation for continuous emotion recognition. Pattern Recogn. **130**, 108833 (2022). https://doi.org/10.1016/j.patcog.2022.108833. https://linkinghub.elsevier.com/retrieve/pii/S0031320322003144
44. Zhao, L.M., Li, R., Zheng, W.L., Lu, B.L.: Classification of five emotions from EEG and eye movement signals: complementary representation properties. In: 2019 9th International IEEE/EMBS Conference on Neural Engineering (NER), pp. 611–614. IEEE (2019). https://doi.org/10.1109/NER.2019.8717055. https://ieeexplore.ieee.org/document/8717055/
45. Zheng, W.L., Zhu, J.Y., Lu, B.L.: Identifying stable patterns over time for emotion recognition from EEG. IEEE Trans. Affect. Comput. **10**(3), 417–429 (2017). https://doi.org/10.1109/TAFFC.2017.2712143. https://ieeexplore.ieee.org/document/7938737/

Exploring EEG-Based Neural Correlates of Multivariate Ordinal Emotion Representations

Xuyang Chen[1,2], Xin Xu[1], Dan Zhang[3], Quanying Liu[1], and Xinke Shen[1(✉)]

[1] Department of Biomedical Engineering, Southern University of Science and Technology, Shenzhen 518055, China
`shenxk@sustech.edu.cn`
[2] Department of Biology, Southern University of Science and Technology, Shenzhen 518055, China
[3] Department of Psychological and Cognitive Sciences, Tsinghua University, Beijing 100084, China

Abstract. Precise measurement of emotion is a key challenge in understanding human emotion and developing emotional artificial intelligence. Most existing studies have regarded participants' emotional annotations in interval form as the ground truth of their emotional experience. However, recent studies suggest that the ordinal forms of emotional annotation better represent the emotional externalization process, offering a promising approach for precise emotion measurement. In this study, we explored the neural basis of multivariate ordinal emotion representations using a video-elicited EEG dataset (n = 123). We conducted inter-situation representational similarity analysis (RSA) and inter-subject RSA to reveal the EEG substrates of emotion variations and individual differences, respectively. Our findings indicate that both inter-situation and inter-subject variations in EEG features are better explained by ordinal emotion representations than by interval ones, supporting the ordinal nature of emotion from a neural perspective. Besides, multivariate ordinal representations showed better inter-subject reliability and higher representational similarity with EEG features compared to univariate counterparts, highlighting the co-occurrence nature of human emotions. Taken together, these findings demonstrate that multivariate ordinal emotion ratings provide a more accurate measure of emotional ground truth, which is crucial for enabling machines to precisely understand and express human emotions.

Keywords: Multivariate Ordinal Emotion Representations · EEG · Representational Similarity Analysis · Emotional Artificial Intelligence

1 Introduction

Understanding human emotions and accurately measuring them is a fundamental challenge in both psychological research and the development of emotional artificial intelligence (AI). Emotional AI, which aims to enable machines to recognize, interpret, and respond to human emotions, holds significant promise for enhancing human-computer interaction, improving mental health diagnostics, and fostering empathetic technologies. The pursuit of precise emotion measurement is crucial for advancing these applications and ensuring that emotional AI can effectively understand and interact with human emotional states.

Current methods for measuring emotions still have notable limitations, primarily due to the prevalent use of interval measurement approaches [1]. Interval data is problematic for subjective constructs like emotions because it can misrepresent the true emotional state. This misrepresentation occurs because emotions are inherently relative rather than absolute, making absolute interval traces inaccurate [2, 3]. Treating interval data as nominal data introduces biases, taking researchers further away from the underlying ground truth. For instance, dichotomized labels from interval data can cause issues where similar samples around the boundary are placed in different classes, leading to inaccuracies. Additionally, interval measurements often show low reliability, especially in dynamic scenarios where emotions change over time [3]. The absolute differences in interval scores lack value and stability [2, 4], as participants' scores largely depend on the relative comparison between current and past experiences, rather than an absolute scale. These shortcomings highlight the need for more accurate methods of emotion annotation.

Recent studies have suggested that ordinal measurement methods may provide a more accurate reflection of emotional experiences [3]. Unlike interval ratings, ordinal ratings align more closely with how humans naturally perceive and express emotions, which are inherently comparative. Empirical studies have demonstrated several advantages of ordinal annotations: 1) Higher inter-rater reliability: Ordinal ratings reduce subjective biases and inconsistencies associated with interval scales, leading to more consistent ratings across different raters. 2) Stability and robustness: Ordinal measures capture the order of emotional intensity without assuming exact differences, making them less susceptible to distortions from extreme rating strategies [5, 6]. 3) Closer alignment with human perception: Ordinal data better reflects the relative nature of emotional experiences, avoiding the pitfalls of misrepresenting emotional states with fixed intervals. These advantages suggest that ordinal emotion representations may better capture the underlying structure of emotional experiences and offer a promising alternative to traditional interval measurement approaches. However, the neural basis of ordinal emotion measurements remains not fully understood.

Understanding the neural correlates of ordinal emotion representations is crucial for validating their use and integrating them into emotional AI systems. Electroencephalography (EEG) provides a valuable tool for investigating these neural bases due to its high temporal resolution and non-invasive nature. By examining the EEG substrates associated with ordinal emotion representations, we can better understand how the brain encodes and differentiates between various emotional states on an ordinal scale.

Besides, while recent research has emphasized the importance of multivariate emotion representation [7, 8], current ordinal measurements are often limited to single or bivariate representations (e.g. [9, 10]). Multivariate emotion representations can capture the complexity and co-occurrence of emotional experiences, allowing for a more nuanced and comprehensive characterization of emotional experiences [11]. For instance, studies have shown that emotional experiences often involve simultaneous activation of multiple emotions, such as feeling both happy and surprised at the same time [12, 15]. However, current ordinal measurement methods typically concentrate on single or bivariate dimensions, which overlooks the rich interplay between different emotional dimensions.

In this study, we investigate the neural basis of multivariate ordinal emotion representations using an EEG dataset with 123 participants. We conducted a comprehensive examination of the reliability and validity of these ordinal representations across eight emotional dimensions using representational similarity analysis (RSA). Our study explores the EEG substrates that correspond to variations in emotional experiences across situation and across subjects. The findings indicate that both inter-situation and inter-subject variations in EEG features are better explained by ordinal emotion representations, supporting the notion that emotions are inherently ordinal from a neural perspective. Additionally, multivariate ordinal representations demonstrated higher inter-subject reliability and greater representational similarity with EEG features than their univariate counterparts.

2 Methods

2.1 Data Acquisition and Processing

The FACED Dataset. To validate the proposed algorithm, we used data from the Finer-grained Affective Computing EEG Dataset (FACED) released by Chen et al. [14]. The FACED dataset aims to provide fine-grained emotional computing data, offering a balanced classification of positive and negative aspects of emotions.

The dataset recorded 32-channel EEG signals from 123 participants, each of whom watched 28 emotion-inducing video materials covering nine emotion categories: amusement, inspiration, joy, tenderness, anger, fear, disgust, sadness, and neutral emotion. Each category of negative and positive emotions had three video clips, while the neutral emotion category had four clips. On average, these video clips were about 66 s long, with durations ranging from 34 to 129 s. After each video clip, subjects were required to report their subjective ex-periences during the video-watching session on 12 items, which included ratings on eight emotion categories: anger, fear, disgust, sadness, amusement, inspiration, joy, tenderness, as well as on four emotion dimensions: valence, arousal, liking, and familiarity. Subjects provided ratings on a continuous scale from 0 to 7 for each item. For the valence item, a rating of 0 indicated "very negative" and a rating of 7 indicated "very positive." For the other items, 0 indicated "not at all" and 7 indicated "very much." The following analysis focused on the multivariate ratings on eight emotion categories.

EEG Data Preprocessing. The EEG data were downsampled to 125 Hz and under-went bandpass filtering from 0.5 to 40 Hz using a fourth-order zero-phase-shift Butterworth filter. Noisy channels were identified manually by examining data variance and abnormal values. On average, 0.06 ± 0.26 noisy channels per trial were found in the dataset. These noisy channels were interpolated using the average of three adjacent channels. Independent component analysis (ICA) was then employed to remove ocular and muscle artifacts, with 2–4 components removed per subject. Finally, the data were re-referenced to the average of two mastoid channels. The preprocessing steps utilized Noisetools and Fieldtrip toolboxes.

2.2 The Computation of Distance and Ordinal Representations

The core hypothesis of ordinal representation is that each subject evaluates emotions in a relative manner. Therefore, we transform all annotations for one subject at a time. To extract ordinal information from the original interval annotations, we first consider the relative differences in each dimension of annotations to obtain the distance representation (Fig. 1b). Then, we remove the magnitude of differences, which is meaningless in ordinal representation, and retain only the sign (corresponding to the greater/lesser relationship in ordinal representation, Fig. 1c). The computation details are as follows.

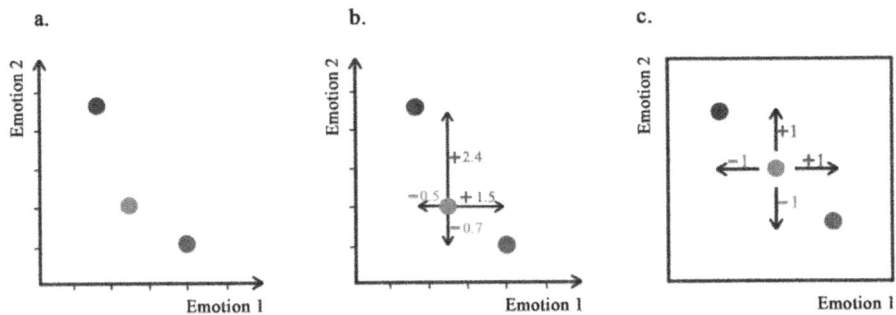

Fig. 1. General idea of extracting ordinal representation from interval annotations. (a). The original annotation in interval scale. (b). Considering the relative rank on each emotion dimension. (c). Only preserving the rank information yields the ordinal representation.

Distance Representation. For a participant's annotations on E emotions across all N situations, $\hat{Y}_1, \hat{Y}_2, \ldots, \hat{Y}_N$, where $\hat{Y}_i \in \mathbb{R}^E$, all annotations forms a network with N nodes and $N \times (N - 1)/2$ edges. Instead of representing the emotion of each trial in the original emotion space, we represent the emotions by its relative position in this network. The position of each node in the network depends only on its relative position with other nodes and is independent of absolute coordinates. These relative positions are represented by a set of distance vectors, so we call this representation as distance representation. Specifically, for $\hat{Y}_i$, its distance representation consists of a set of distance vectors $\{Dist_{ij} | j = 1, 2, \ldots, i, \ldots, N\}$, where:

$$Dist_{ij} = \hat{Y}_i - \hat{Y}_j \tag{1}$$

The vector $Dist_{ij} \in \mathbb{R}^E$ represents the distance between nodes $\hat{Y}_i$ and $\hat{Y}_j$. By concatenating N one-dimensional distance vectors, we obtain the distance representation matrix $D_i \in \mathbb{R}^{E \times N}$.

Ordinal Representation. In ordinal measurement, the specific numerical differences between two annotations are not meaningful. Therefore, we transform the distances into ordinal categories—greater than, less than, or equal to. This transformation removes the emphasis on the actual magnitude of differences and focuses instead on the relative ordering of nodes. This transformation is represented as:

$$O_{i,e,n} = \begin{cases} -1, & \text{if } D_{i,e,n} < -tol \\ 1, & \text{if } D_{i,e,n} > tol \\ 0, & \text{else} \end{cases} \quad (2)$$

$$e = 1, 2, \ldots, E; n = 1, 2, \ldots, N.$$

Here, 1, 0, and −1 are used to represent greater than, equal to, and less than in ordinal measurement. The transformed matrix $O_i \in \mathbb{R}^{E \times N}$ only records the size relationship information from the distance representation, discarding the numerical values. This representation is equivalent to ordinal measurement, so we call it an ordinal representation.

The parameter *tol* represents the error tolerance threshold [13]. In the emotion annotation interface, participants often evaluate the intensity of various emotions by dragging sliders. Even if participants aim to provide the same annotation level, it is unlikely that they will drag the slider to exactly the same value. Therefore, small interval differences are considered equal during transformation. A tolerance of 0.6 was determined by pre-experiment to yield best inter-rater reliability, and was adopted throughout the study.

To further confirm the role of *tol*, we also employed another ordinal representation, denoted as $\tilde{O}$, as a control without using *tol*. $\tilde{O}$ is identical to O in all aspects except for setting *tol* to 0.

2.3 Inter-rater Reliability

Inter-rater reliability measures the degree of agreement among a group of raters [16]. It is widely used as a reliability test for emotion annotations. Here, we test if the inter-rater reliability of ordinal representations are higher than interval representations.

Krippendorff's Alpha. Krippendorff's alpha is a statistical measure used to assess the reliability of agreement among raters. It is particularly useful in situations where multiple raters evaluate a set of items and provides a versatile measure that can handle different types of data (nominal, ordinal, interval, or ratio). It is calculated as

$$\alpha = 1 - \frac{D_O}{D_e}$$

where D_O is the observed disagreement and D_e is the expected agreement. Please refer to [18, 19] for details of the calculation. It produces a value between −1 and 1, where −1 indicates systematical disagreement, 0 indicates no agreement beyond chance and 1 indicates perfect agreement.

The Agreement of Inter-situation Similarity Matrix Across Subjects. Another approach is to measure the consistency of the inter-situation dissimilarity matrix across subjects. A higher consistency indicates that the way a participant rates emotions across different situations is similar to the way other participants rate them.

Here, we compared the inter-rater reliability between univariate and multivariate emotion ratings. For univariate emotion ratings, only a single dimension of emotion ratings or its derived ordinal representations was used to calculate inter-rater reliability. In contrast, for multivariate emotion ratings, all E emotional dimensions were utilized.

2.4 Representational Similarity Analysis

Representational Similarity Analysis (RSA) is a widely used multivariate analysis technique in cognitive neuroscience, aimed at quantifying and comparing information representations in the brain or computational models. RSA characterizes representations in a brain region or computational model by constructing a distance matrix of response patterns, revealing which differences between stimuli are emphasized and which are attenuated [20].

Here, we conduct two types of RSA (Fig. 2). One is based on inter-situation similarity, in which we compare the inter-situation similarity patterns of neural signals with different emotion ratings. This analysis aims to test if the inter-situation differences of emotion ratings have underlying neural substrates. The other is based on inter-subject similarity, where we compare the inter-subject similarity patterns of neural signals with the inter-subject similarity of emotion ratings. This analysis aims to test if the inter-subject discrepancy of the ratings can be reflected in the inter-subject difference of neural activities.

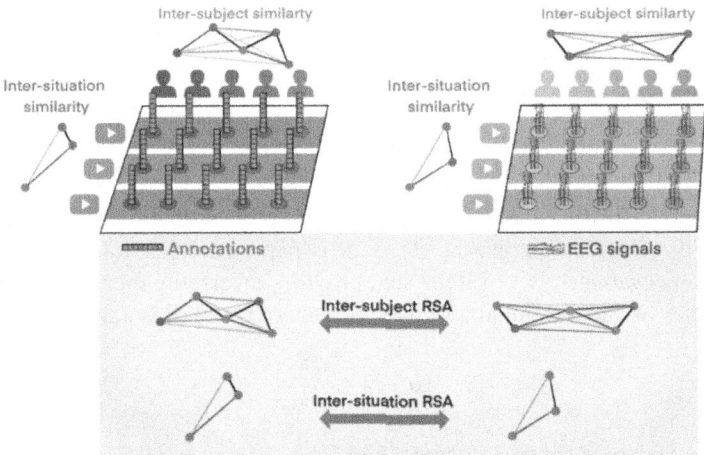

Fig. 2. The overall illustration of inter-subject representational similarity analysis (RSA) and inter-situation RSA.

Inter-situation Representation Similarity Analysis. In the Inter-situation representation similarity analysis, for a specific participant, a set of Inter-situation similarity matrices is generated based on their emotion annotations across all situations. Similarly, another set of inter-situation similarity matrices is generated based on their neural signals across all situations [8].

Inter-situation Similarity Matrix of Emotion Ratings. Here, we denote $S_i^{sit} \in \mathbb{R}^{v \times v}$ standing for the Inter-situation similarity matrix of emotion annotations for participant i. It is computed from the annotations across all situations for that participant, $R_i \in \mathbb{R}^{v \times E}$, where v corresponds to the total number of situations (i.e., videos) in the experiment.

Specifically, $R_i \in \mathbb{R}^{v \times E}$ is L2-normalized along the situation dimension to obtain normalized annotations $R_i^{(1)} \in \mathbb{R}^{v \times E}$. Next, the annotations are multiplied by their own transpose to obtain the Inter-situation similarity matrix of emotion annotations:

$$S_i^{sit} = R_i^{(1)} R_i^{(1)T}, \quad i = 1, 2, \ldots, u$$ (3)

where u is the total number of subjects. Each element of the similarity matrix represents the cosine similarity value between the annotations of two situations (Fig. 3).

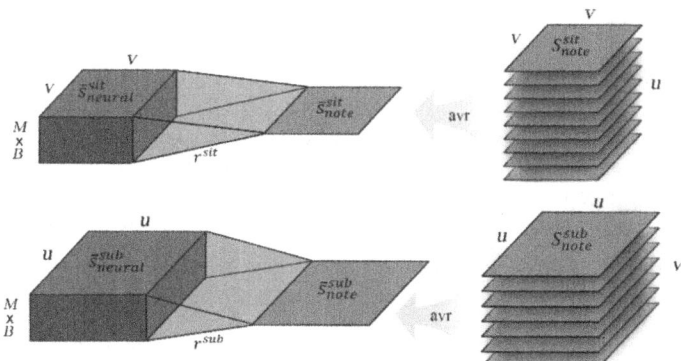

Fig. 3. Pipeline of Inter-situation RSA (above) and inter-subject RSA (below). Blue cubes stands for similarity matrices of EEG, while green squares stands for similarity matrices of annotation.

Inter-situation Similarity Matrix of Neural Signals. The computation process for neural signals is in general similar to the previous part. The Inter-situation similarity matrix of neural signals for participant i, $\tilde{S}_i^{sit} \in \mathbb{R}^{v \times v \times M \times B}$ is computed from the EEG features of that participant, $X_i \in \mathbb{R}^{v \times M \times B}$, where M is the number of channels and B is the number of frequency bands.

Specifically, the signals are filtered into five frequency bands: Delta: 0.5–3 Hz, Theta: 4–7 Hz, Alpha: 8–13 Hz, Beta: 14–29 Hz, Gamma: 30–47 Hz. Next, frequency domain features widely used in emotion analysis are extracted, including Differential Entropy (DE) and Power Spectral Density (PSD). The feature matrix is denoted as $X_i \in \mathbb{R}^{v \times M \times B}$.

Difference between each feature dimension is computed to obtain the inter-situation similarity matrix of neural signals:

$$\tilde{S}^{sit}_{i,v_1,v_2,m,b} = |X_{i,v_1,m,b} - X_{i,v_2,m,b}| \qquad (4)$$
$$i = 1, 2, \ldots, u; m = 1, 2, \ldots, M; b = 1, 2, \ldots, B$$

where $\tilde{S}^{sit}_i \in \mathbb{R}^{v \times v \times M \times B}$ reflects the multi-situation neural response pattern of participant i. Each element of $\tilde{S}^{sit}_i$ corresponds to the difference of an EEG feature between two situations.

Here we compute the average inter-situation similarity matrix of ratings across all participants, obtaining $\overline{s}^{sit\_rating} \in \mathbb{R}^{v \times v}$, and the average inter-situation similarity matrix of EEG features for all participants, obtaining $\overline{s}^{sit\_neural} \in \mathbb{R}^{v \times v \times M \times B}$. For each channel m and frequency band b, we compute the correlation coefficient between its neural similarity matrix and rating similarity matrix:

$$r^{sit}_{m,b} = \text{pearson\_corr}\left(\overline{s}^{sit\_rating}, \overline{s}^{sit\_neural}_{:,:,m,b}\right), \qquad (5)$$
$$m = 1, 2, \ldots, M; b = 1, 2, \ldots, B$$

$r^{sit} \in \mathbb{R}^{M \times B}$ encompasses the inter-situation neural-rating consistency across all channels and frequency bands, representing the neural substrates of different emotion experiences.

Inter-subject Representational Similarity Analysis. The work of inter-subject RSA is generally symmetrical to inter-situation RSA, except that the calculation dimension of similarity matrices is changed from situation dimension to subject dimension. In inter-subject RSA, for a specific situation, a set of inter-subject similarity matrices is generated based on the emotional annotations of all subjects. Another set of inter-subject similarity matrices is generated based on the neural signals of all subjects [21]. The correlation between the two is used to charaterize the correspondence between the inter-subject variance of ratings and inter-subject difference of neural signals.

Inter-subject Similarity Matrix of Emotional Annotations. The inter-subject similarity matrix of situation j is denoted as $S^{sub}_j \in \mathbb{R}^{u \times u}$, where u is the number of subjects. This is calculated based on the rantings of all subjects in this situation $R_j \in \mathbb{R}^{u \times E}$. Similarly, R_j is L2-normalized along the E dimension to obtain the normalized annotation $R^{(1)}_j \in \mathbb{R}^{u \times E}$. Next, the annotation is multiplied by its transpose to obtain the inter-subject similarity matrix of emotional annotations:

$$S^{sub}_j = R^{(1)}_j R^{(1)T}_j, j = 1, 2, \ldots, v \qquad (6)$$

$S^{sub}_j \in \mathbb{R}^{u \times u}$ reflects the multi-subject emotional annotation patterns for situation j. Each element of S^{sub}_j represents the cosine similarity of the annotation vectors given by two subjects for situation j. $S^{sub} \in \mathbb{R}^{v \times u \times u}$ was averaged over the situation dimension to obtain the rating similarity matrix $\overline{s}^{sub\_rating} \in \mathbb{R}^{u \times u}$.

Inter-subject Similarity Matrix of Neural Signals. The inter-subject similarity matrix of neural signals across all situations, $\bar{s}^{sub\_neural} \in \mathbb{R}^{u \times u \times M \times B}$, is obtained from the EEG features of all subjects in all situations, denoted as $X^{total} \in \mathbb{R}^{u \times v \times M \times B}$. For situation j and subjects i_1, i_2, the computation is as follows:

$$\tilde{S}^{sub}_{i_1,i_2,j,:,:} = \left| X^{total}_{i_1,j,:,:} - X^{total}_{i_2,j,:,:} \right| \quad (7)$$
$$i_1, i_2 = 1, 2, \ldots, u; j = 1, 2, \ldots, v$$

Here, $\tilde{S}^{sub}_{i_1,i_2} \in \mathbb{R}^{v \times M \times B}$ represents the neural dissimilarity between subject i_1, i_2 calculated regarding v situations, M channels and B bands. Then average along situation dimension, yielding the average neural similarity matrix $\bar{s}^{sub\_neural} \in \mathbb{R}^{u \times u \times M \times B}$.

The matrices $\bar{s}^{sub\_rating} \in \mathbb{R}^{u \times u}$ and $\bar{s}^{sub\_neural} \in \mathbb{R}^{u \times u \times M \times B}$ reflect the inter-subject variability patterns across all situations in neural activity or ratings, i.e., inter-subject differences. Here, for each channel m and frequency band b, we calculate the correlation coefficient between its neural similarity matrix and the rating similarity matrix:

$$r^{sub}_{m,b} = pearson\_corr(\bar{s}^{sub\_rating}, \bar{s}^{sub\_neural}_{:,:,m,b}) \quad (9)$$
$$m = 1, 2, \ldots, M; \quad b = 1, 2, \ldots, B$$

$r^{sub} \in \mathbb{R}^{M \times B}$ represents the correlation of inter-subject ratings' difference and EEG features' difference.

3 Results

3.1 Inter-rater Reliability

For inter-rater reliability test, ordinal representation $O(0.425 \pm 0.159)$ has a higher Krippendorff's alpha than the original interval ratings $\hat{Y}(0.409 \pm 0.154)$ (Fig. 4a). Besides, the inter-situation similarity matrix of the ordinal representation O showed significantly higher inter-subject consistency than interval ratings $\hat{Y}$ (Fig. 4b). The ordinal representation without tolerance $\tilde{O}$ yields inconsistent performance on these two criteria, indicating the importance of introducing tolerance into ordinal representations.

Multivariate emotion representations exhibit higher inter-subject reliability compared to single-variate representations, regardless of the specific single-variate emotion used and the similarity algorithm applied (Fig. 5).

3.2 Neural Substrates of Ordinal Emotion Representations

In the inter-situation RSA, regardless of which frequency domain features are used, both O and $\tilde{O}$ representations significantly outperform $\hat{Y}$ representations (Fig. 7a, left bars). There's a same situation for the inter-subject RSA (Fig. 7b, left bars).

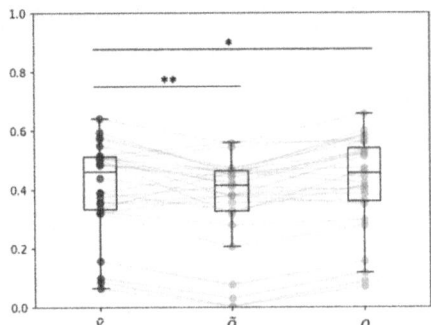
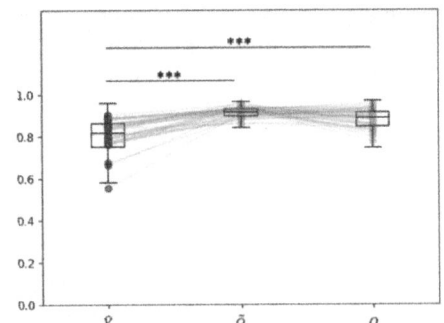

Fig. 4. The reliability result. (a). The reliability according to Krippendorff's alpha under different representations. (b). The pairwise similarity of inter-situation similarity matrices under cosine similarity.

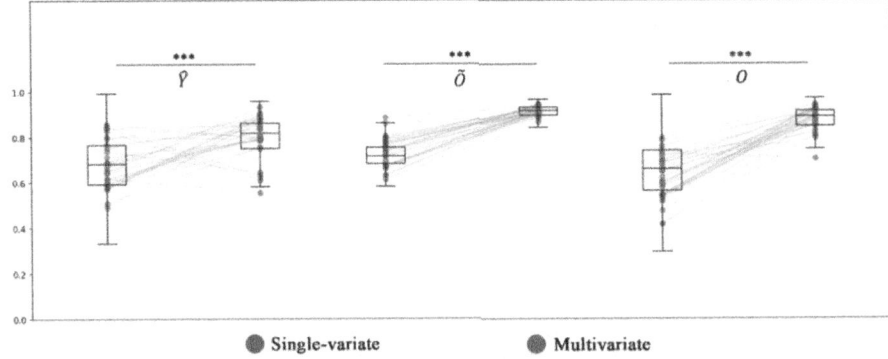

Fig. 5. Inter-subject reliability based on inter-situation RSA of single-variate and multivariate emotion annotation, under 3 different representations. Shown in figure is the case where 'joy' dimension is chosen as the single-variate, and similarity is based on cosine similarity. Under any chosen single-variate and any similarity metric, the results are consistently similar to the shown case.

Multivariate emotion representations consistently exhibit higher correlation with neural signals (Fig. 6). Overall, in the two reliability indicators and two validity indicators used in this study, the ordinal representation consistently outperformed the original annotations. On the other hand, the multivariate representation consistently outperformed the single-variate annotations. This further confirms that significant emotional characteristics of individuals may be hidden within multivariate emotional structures, and multivariate ordinal representation is a promising approach to measure the emotion states.

In both inter-situation RSA and inter-subject RSA, we performed RSA between each EEG component across channels and frequency bands with the emotion annotations. The resulting correlation coefficients represent the consistency between each

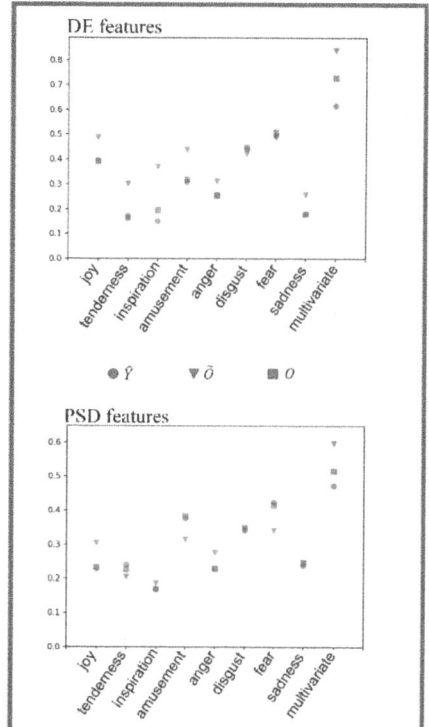 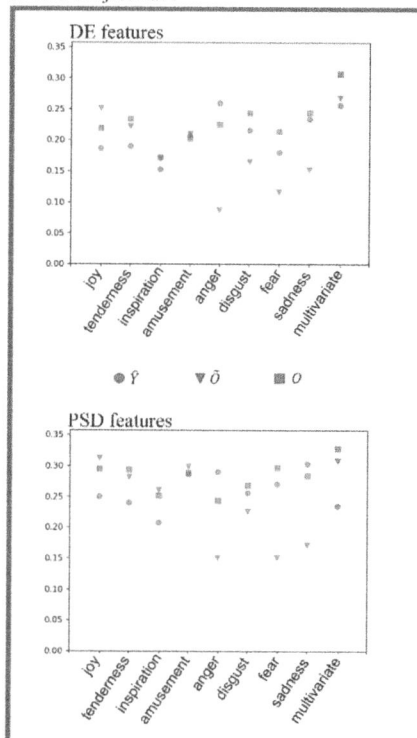

Fig. 6. The maximum correlation coefficient between EEG signals and ratings of single-variate and multivariate emotion annotation, under 3 different representations. (a). validity based on inter-situation RSA. In the upper plot, DE is selected as the EEG feature, while in the lower plot, PSD is used. (b). validity based on inter-subject RSA. In the upper plot, DE is selected as the EEG feature, while in the lower plot, PSD is used.

EEG component and the emotion annotations. Here, we present the spatial distribution of the correlation coefficients for each frequency band using brain topographic maps. These patterns correspond to the neural basis of the emotion annotations under different representations.

Figure 7 shows the spatial patterns of correlation analysis in inter-situation RSA and inter-subject RSA. Regardless of the feature (DE or PSD) or representation used, the brain activities most correlated with emotional representations are concentrated in the temporal area at high frequency bands (beta, gamma bands). Furthermore, under the ordinal representation, the correlation in this region and frequency band is higher than under the interval representation.

a. Inter-situation RSA

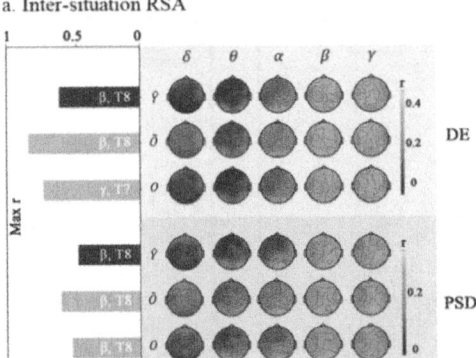

b. Inter-subject RSA

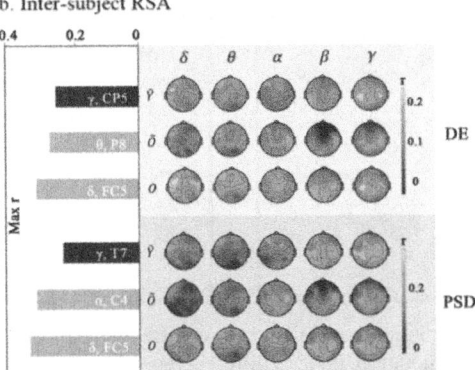

Fig. 7. The neural representation and the maximum correlation coefficients in (a). Inter-situation RSA. (b). Inter-subject RSA. The bars at left indicates the levels of maximum r, and the corresponding frequency bands and channels are shown in the bars. Shown right are the correlation coefficients plotted on the scalp topoplot.

4 Discussion

In this research, we investigated the neural basis of multivariate ordinal emotion representations through the analysis of EEG data from 123 participants who viewed a series of emotion-inducing videos. Our study utilized inter-situation and inter-subject representational similarity analysis (RSA) to uncover the neural correlates of emotional variations and individual differences. The results revealed that ordinal representations of emotions better capture the neural patterns associated with emotional experiences than interval representations. Specifically, both inter-situation and inter-subject variations in EEG features were more accurately reflected by ordinal measures. Additionally, multivariate ordinal representations outperformed univariate ones in terms of inter-subject reliability and alignment with EEG features. These findings highlight the importance of using ordinal scales for emotion measurement, providing a more nuanced understanding of the neural substrates underlying human emotions.

One of the most intriguing findings of this study is the strong neural basis for multivariate ordinal emotion representations compared to their univariate counter-parts. This indicates that considering the complexity and co-occurrence of multiple emotional states provides a more robust measure of emotional experiences, as multivariate representations capture the simultaneous activation of different emotions, reflecting the real-world emotional dynamics more accurately. This aligns with the notion that human emotions are rarely experienced in isolation but rather as complex, intertwined states [12, 15].

The analysis showed that the neural substrates corresponding to ordinal emotion representations are concentrated in specific brain regions, particularly the temporal areas, and are most prominent in the higher frequency bands (beta and gamma). These findings are consistent with previous research indicating that these brain regions and frequency bands are crucial for processing complex emotional stimuli [23, 24]. The higher correlation between ordinal representations and neural activity in these areas suggests that the brain's encoding of emotional information is more closely aligned with ordinal, rather than interval, scales.

This study replicated the findings of previous research that highlighted the higher inter-subject reliability of ordinal emotion representations over interval ones [3, 22]. By extending this analysis to multivariate contexts, we discovered that multivariate ordinal representations exhibit even greater inter-subject reliability than their univariate counterparts. This finding underscores the robustness and consistency of multivariate ordinal scales in capturing emotional experiences across different individuals. The higher inter-subject reliability suggests that multivariate approaches can better account for individual differences in emotional processing, providing a more reliable measure that reflects the true variance in human emotional responses. Moreover, the inter-subject RSA results confirmed that multivariate representations better capture inter-subject variations in neural activities, demonstrating their effectiveness in reflecting nuanced differences in human emotional responses.

These findings collectively demonstrate that multivariate ordinal emotion representations offer a superior approach to capturing the complexity of human emotions. They not only improve the alignment with neural data but also provide a more consistent framework for inter-rater reliability. This has profound implications for the development of emotional AI, suggesting that incorporating ordinal scales and multivariate representations could significantly enhance the accuracy and empathy of AI systems in understanding and responding to human emotions.

Acknowledgments. The authors wish to thank Zongsheng Li for his valuable suggestions. This work was funded in part by the National Key R&D Program of China (2021YFF1200804), Shenzhen Science and Technology Innovation Committee (2022410129, KCXFZ20201221173400001, KJZD20230923115221044).

Disclosure of Interests. The authors have no competing interests.

References

1. Cowie, R., Cox, C., Martin, J.C., Batliner, A., Heylen, D., Karpouzis, K.: Issues in emotion recognition (2011)
2. Kim, M., Pavlovic, V.: Structured output ordinal regression for dynamic facial emotion intensity prediction. In: Daniilidis, K., Maragos, P., Paragios, N. (eds.) Computer Vision–ECCV 2010. Lecture Notes in Computer Science, vol. 6313, pp. 649–662. Springer, Heidelberg (2010). https://doi.org/10.1007/978-3-642-15558-1_47
3. Yannakakis, G.N., Cowie, R., Busso, C.: The ordinal nature of emotions: an emerging approach. IEEE Trans. Affect. Comput. **12**(1), 16–35 (2021)
4. Sethu, V., Provost, E.M., Epps, J., Busso, C., Cummins, N., Narayanan, S.: The ambiguous world of emotion representation. arXiv preprint arXiv:1909.00360 (2019)
5. Neuman, Y., Cohen, Y.: Predicting change in emotion through ordinal patterns and simple symbolic expressions. Mathematics **10**(13), 2253 (2022)
6. Mitsios, M., et al.: Improved Text Emotion Prediction Using Combined Valence and Arousal Ordinal Classification. arXiv preprint arXiv:2404.01805 (2024). https://doi.org/10.48550/arXiv.2404.01805
7. Siemer, M., Mauss, I., Gross, J.J.: Same situation–different emotions: how appraisals shape our emotions. Emotion **7**(3), 592 (2007)
8. Liu, J., Hu, X., Shen, X., Song, S., Zhang, D.: Electrophysiological representations of multivariate human emotion experience. Cogn. Emot. **38**(3), 378–388 (2024)
9. Apicella, A., Arpaia, P., Mastrati, G., et al.: EEG-based detection of emotional valence towards a reproducible measurement of emotions. Sci. Rep. **11**, 21615 (2021). https://doi.org/10.1038/s41598-021-00812-7
10. Yannakakis, G.N., Martínez, H.P.: Grounding truth via ordinal annotation. In: 2015 International Conference on Affective Computing and Intelligent Interaction (ACII), pp. 574–580. IEEE, Xi'an (2015). https://doi.org/10.1109/ACII.2015.7344627
11. Werner-Seidler, A., Hitchcock, C., Hammond, E., et al.: Emotional complexity across the life story: elevated negative emodiversity and diminished positive emodiversity in sufferers of recurrent depression. J. Affect. Disord. **273**(April), 106–112 (2020). https://doi.org/10.1016/j.jad.2020.04.060
12. Ellsworth, P.C., Smith, C.A.: From appraisal to emotion: differences among unpleasant feelings. Motiv. Emot. **12**(3), 271–302 (1988). https://doi.org/10.1007/BF00993115
13. Zoumpourlis, G., Patras, I.: Pairwise ranking network for affect recognition. In: 2021 9th International Conference on Affective Computing and Intelligent Interaction (ACII), pp. 1–8. IEEE (2021)
14. Chen, J., Wang, X., Huang, C., Hu, X., Shen, X., Zhang, D.: A large finer-grained affective computing EEG dataset. Sci. Data **10**(1), 740 (2023)
15. Ellsworth, P.C., Smith, C.A.: Shades of Joy: patterns of appraisal differentiating pleasant emotions. Cogn. Emot. **2**(4), 301–331 (1988). https://doi.org/10.1080/02699938808412702
16. Bartko, J.J., Carpenter, W.T.: On the methods and theory of reliability. J. Nerv. Ment. Dis. **163**(5), 307–317 (1976)
17. Borsboom, D., Mellenbergh, G.J., Van Heerden, J.: The concept of validity. Psychol. Rev. **111**(4), 1061 (2004)
18. Hayes, A.F., Krippendorff, K.: Answering the call for a standard reliability measure for coding data. Commun. Methods Meas. **1**(1), 77–89 (2007)
19. Krippendorff, K.: Testing the reliability of content analysis data: what is involved and why. In: Krippendorff, K., Bock, M.A. (eds.) The Content Analysis Reader. Sage, Thousand Oaks (2009)

20. Kriegeskorte, N., Mur, M., Bandettini, P.A.: Representational similarity analysis-connecting the branches of systems neuroscience. Front. Syst. Neurosci. **2**, 249 (2008)
21. Hu, X., Wang, F., Zhang, D.: Similar brains blend emotion in similar ways: neural representations of individual difference in emotion profiles. Neuroimage **247**, 118819 (2022)
22. Kim, J., Truong, K.P., Charisi, V., Zaga, C., Evers, V., Chetouani, M.: Multimodal detection of engagement in groups of children using rank learning. In: Human Behavior Understanding: 7th International Workshop, HBU 2016, Amsterdam, Proceedings, vol. 7, pp. 35–48. Springer, Heidelberg (2016)
23. Pessoa, L.: On the relationship between emotion and cognition. Nat. Rev. Neurosci. **9**, 148–158 (2008). https://doi.org/10.1038/nrn231724
24. Yang, K., Tong, L., Shu, J., Zhuang, N., Yan, B., Zeng, Y.: High gamma band EEG closely related to emotion: evidence from functional network. Front. Hum. Neurosci.Neurosci. **14**, 89 (2020)

The Co-varying Multimodal Pattern in Treatment-Resistant and Non-treatment-Resistant Schizophrenia

Siyuan Cao[1,2,3], Shuzhan Gao[4,5], Chuang Liang[1,2,3], Vince D. Calhoun[6], Xuyun Wen[1,2,3], Zening Fu[6], Lei Wu[6], Rongtao Jiang[7], Daoqiang Zhang[1,2,3], Shile Qi[1,2,3(✉)], and Xijia Xu[4,5(✉)]

[1] Department of Computer Science and Technology, Nanjing University of Aeronautics and Astronautics, Nanjing, China
shile.qi@nuaa.edu.cn

[2] Key Laboratory of Brain-Machine Intelligence Technology, Ministry of Education, Nanjing University of Aeronautics and Astronautics, Nanjing, China

[3] MIIT Key Laboratory of Pattern Analysis and Machine Intelligence, Nanjing University of Aeronautics and Astronautics, Nanjing, China

[4] Department of Psychiatry, Affiliated Nanjing Brain Hospital, Nanjing Medical University, Nanjing, China

[5] Department of Psychiatry, Nanjing Brain Hospital, Medical School, Nanjing University, Nanjing, China
xuxijia@c-nbh.com

[6] Tri-Institutional Center for Translational Research in Neuroimaging and Data Science (TReNDS) Georgia State University, Georgia Institute of Technology, Emory University, Atlanta, GA, USA

[7] Department of Radiology and Biomedical Imaging, Yale University, New Haven, CT, USA

Abstract. Schizophrenia (SZ) is a severe mental illness, with 20%-40% exhibit an inadequate or poor response to the first-line antipsychotic drugs (treatment-resistant schizophrenia, TR-SZ). However, the neural mechanisms underlying this treatment-resistance in SZ remain unclear. This study aimed to identify the co-varying multimodal pattern that distinguishing among TR-SZ, non-treatment-resistant schizophrenia (NTR-SZ) and healthy controls (HC) by unsupervised fusion, in which fractional amplitude of low-frequency fluctuation (fALFF), fraction anisotropic (FA), and gray matter volume (GMV) were used as fusion input. 63 TR-SZs, 221 NTR-SZs, and 86 healthy controls (HCs) were included in this study. The joint multimodal components (including fALFF_IC3, FA_IC12, GMV_IC9) were identified that group discriminating among TR-SZ, NTR-SZ and HC (TR-SZ < NTR-SZ < HC), constructing the central executive network (CEN) and the salience network (SAN). Moreover, these components were correlated with the cognitive scores that were validated on three independent SZ cohorts.

Keywords: Multimodal Fusion · Schizophrenia · Treatment-resistance · Co-varying multimodal pattern · CEN · SAN

S. Cao and S. Gao—These authors contributed equally to this work.

1 Introduction

Schizophrenia (SZ) is a severe psychiatry characterized by complex psychopathology [1–4], including positive, negative, and cognitive symptoms [5]. It affects around 0.5–1.0% of the population [6], but 20%–40% SZ exhibit inadequate or poor response to first-line antipsychotics, a condition called treatment-resistant schizophrenia (TR-SZ) [7]. TR-SZ can be defined as a situation in which significant improvement of psychopathology and/or other target symptoms has not been demonstrated despite treatment with 2 different antipsychotics from at least 2 different chemical classes (at least 1 should be an atypical antipsychotic) in the previous 5 years at the recommended antipsychotic dosages for a treatment period of at least 2–8 weeks [8]. Patients with TR-SZ are often highly symptomatic and may require extensive periods of hospital care [9] and tend to have a longer duration of untreated psychosis, younger age of onset, and poor pre-morbid functioning [10–12]. This suggests that TR-SZ may be a unique subtype of SZ. To guide therapeutic scheme for SZ earlier and more accurately, it is important to clarify the different mechanisms between TR-SZ and NTR-SZ by combining neuroimaging technology such as magnetic resonance imaging (MRI).

Exploring the brain differences between TR-SZ and NTR-SZ is a challenging and important problem. Researchers have identified abnormal brain regions between TR-SZ and NTR-SZ based on single-modality analysis. Reduced connectivity between ventral striatum and substantia nigra, between dorso caudal putamen and thalamus, and elevated connectivity between dorsal caudate and medial prefrontal cortex were reported in TR-SZ compared to NTR-SZ [13]. Compared to NTR-SZ, TR-SZ showed increased regional homogeneity (ReHo) in the left postcentral gyrus and decreased ReHo in the right angular gyrus [14], the left dorsolateral prefrontal cortex, and right superior parietal cortex [15]. For structural imaging, TR-SZ showed reduced gray matter volume (GMV) in the frontal regions, precentral, postcentral gyrus [16, 17] and hippocampus [18] compared to NTR-SZ. TR-SZ showed less cortical volume [19] and lower cortical thickness [20, 21] in the frontal region compared to NTR-SZ. These previous evidences suggest that TR-SZ and NTR-SZ exhibit fundamental differences in both brain structure and function.

However, most studies used either single-modality analyses or performed a multimodal comparison after separate analyses within each modality. In this case, the cross-information among multiple modalities is either missing or not being fully leveraged to improve neuromarker identification, despite the evidence that such information is highly informative [22–24]. Although a recent study attempted to address this limitation by fusing GMV, fractional amplitude of low-frequency fluctuation (fALFF) and ReHo to identify modality-shared abnormal brain regions in TR-SZ, the sample size was relatively small (15 TR-SZ and 40 NTR-SZs were included) [25]. Other studies focused solely on region of interest (ROI) analysis, which may need additional prior information compared to data-driven fusion analysis. Multimodal fusion can jointly analyze multimodality data to leverage the cross-information, thereby revealing important relationships that cannot be detected by using a single neuroimaging modality [26–29].

In this study, we aimed to identify the co-varying multimodal brain network in TR-SZ and NTR-SZ by fusing fMRI, sMRI and DTI (63 TR-SZs, 221 NTR-SZs, and 86 HCs were included). We focused on the following two goals: (1) to identify the co-varying multimodal brain network that distinguishing among TR-SZ, NTR-SZ and HC

2 Methods and Materials

2.1 Participants and Data Preprocessing

63 TR-SZs (mean ± standard deviation age: 35.6 ± 11.5; gender: 27 male/36 female), 221 NTR-SZs (age: 32.3 ± 9.9; gender: 58 male/163 female) and 86 HCs (age: 29.6 ± 7.1; gender: 40 male/46 female) were collected from the Affiliated Brain Hospital of Nanjing Medical University (NBH). The symptom severity was assessed by the Positive and Negative Syndrome Scale (PANSS). All subjects with TR-SZ met the criteria of World Federation of Societies of Biological Psychiatry (WFSBP) [8]. Participants were informed about the study procedures. Written informed consent was obtained from all participants and their legal guardians. The study was approved by the local Ethics Committee of the Affiliated Nanjing Brain Hospital of Nanjing Medical University (No. KY44, 2011). The fMRI and sMRI data were preprocessed using the SPM12 in an automated analysis pipeline developed at TReNDS (https://trendscenter.org/) and the diffusion tensor imaging (DTI) data was preprocessed by FMRIB Software Library. fALFF from fMRI, fraction anisotropic (FA) from DTI, GMV from sMRI were used as fusion input.

Group differences were observed in age between HC and TR-SZ ($p = 4.63e−04*$), and gender among the three groups ($p = 7.76e−04*$). Thus, age and gender were regressed out prior to fusion analysis. Independent cohorts from Center for Biomedical Research Excellence (COBRE), Function Biomedical Informatics Research Network (FBIRN) and Bipolar and Schizophrenia Network of Intermediate Phenotypes (BSNIP) studies were used for validation.

2.2 Multimodal Fusion

Based on the analytic plan pointed out in the introduction, we conducted the following procedures: (1) multi-set canonical correlation analysis + joint independent component analysis (MCCA+JICA)[30] was performed to identify the co-varying multimodal brain networks that discriminating among TR-SZ, NTR-SZ and HC (Fig. 1a-b); (2) the correlations between the identified brain networks and symptoms/cognition were calculated across three independent SZ cohorts (Fig. 1c).

In unsupervised fusion, we assume that there are n multimodal datasets X_k (subject-by-voxel, $N \times L$), each is a linear mixture of spatial maps C_k (component-by-voxel, $M \times L$) and a corresponding mixing coefficient matrix D_k (subject-by-component, $N \times M$), where k represents the kth modality, as shown in Eq. (1). The number of components M was estimated based on the minimum description length (MDL) criterion [31].

$$X_k = D_k C_k, k = 1,2,\cdots,n \tag{1}$$

MCCA maximize the covariations among mixing matrices of each modality, as shown in Eq. (2).

$$max \sum_{k,j=1}^{n} \|corr(D_k, D_j)\|_2^2 \tag{2}$$

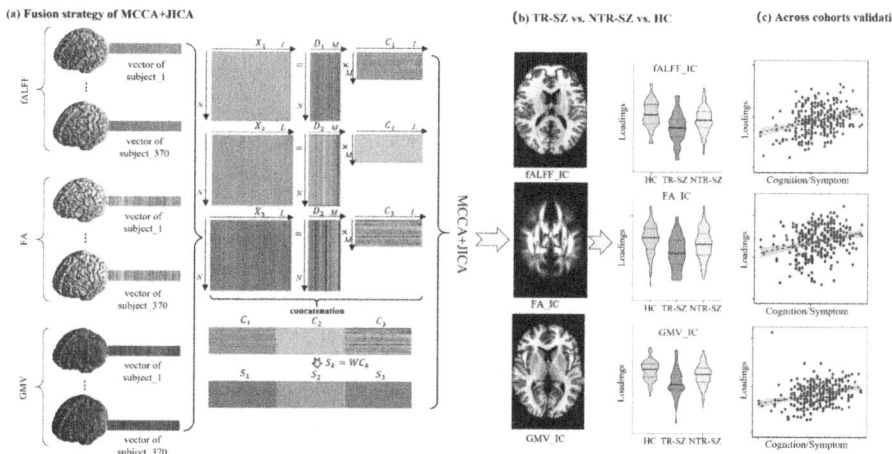

Fig. 1. Analysis framework of this study. (a): The flowchart of the MCCA + JICA fusion approach. (b): Group comparisons of the components' loadings. (c): Correlations between cognition/symptom and the components' loadings in independent cohorts.

Dimension reduction is first performed on X_k with principal component analysis (PCA), thus the signal subspace given by $Y_k = X_k E_k$ (subject-by-component, $N \times M$) is determined to avoid overfitting because of the high dimensionality of data ($L \gg M$). Where E_k contains eigenvectors corresponding to significant (the top M highest) singular values. MCCA is performed on Y_k by maximizing the sum of squared correlations (SSQCOR) among canonical variants (CVs) D_k and then we obtain the spatial maps $C_k = pinv(D_k)X_k$.

Then joint ICA is further performed on the concatenated maps of $[C_1, \cdots, C_M]$, to keep the modality linkage of the potential target components and maximize the spatial independence with the demixing matrix W shared by data from all modalities. Finally, the ICs S_k and their mixing matrices A_k are obtained in Eq. (3–4).

$$W[C_1, \cdots, C_M] = [S_1, \cdots, S_M] \qquad (3)$$

$$X_k = D_k C_k = (D_k W^{-1})S_k = A_k S_k, k = 1,2,\cdots,n \qquad (4)$$

2.3 Across Cohorts Validation

To assess the generalization, the back-reconstruction [32] of the identified brain networks in NBH to independent cohorts was performed based on the linear projection model. Here, take COBRE as an example, the spatial map derived from NBH ($S_{NBH,k}$) was used as the brain map of COBRE to estimate the corresponding mixing matrix for COBRE ($A_{COBRE,k}$) based on Eq. (5). Where $S_{NBH,k}$ and $A_{COBRE,k}$ denote the spatial maps and the corresponding mixing matrix derived by "MCCA + JICA" for COBRE, while $X_{COBRE,k}$ is the input feature matrix. k denotes modality.

$$A_{COBRE,k} = X_{COBRE,k} pinv(S_{NBH,k}) \qquad (5)$$

3 Results

3.1 Multimodal Components

MCCA + JICA was performed on TR-SZ, NTR-SZ and HC to identify the co-varying multimodal brain network that show significant differences among TR-SZ, NTR-SZ and HC (Fig. 2a, M = 32, including fALFF_IC3, FA_IC12 and GMV_IC9). The components identified by MCCA + JICA shown significantly group differences between TR-SZ and NTR-SZ (Fig. 2b, p = 3.8e−04*, p = 6.8e−04*, p = 7.1e−04* for fALFF, FA and GMV separately), between SZ and HC (Fig. 2c, p = 9.4e−06*, p = 3.6e−04*, p = 3.0e−06* for fALFF, FA and GMV separately), among TR-SZ, NTR-SZ and HC (Fig. 2d, p = 1.5e−08*, p = 3.0e−06*, p = 2.1e−09* for fALFF, FA and GMV separately). For fMRI, in the prefrontal cortex (PFC), left precuneus and anterior cingulate cortex (ACC), NTR-SZ showed decreased fALFF compared to HC and increased fALFF compared to TR-SZ. For DTI, in corpus callosum (CC), major forceps and cingulate gyrus, NTR-SZ showed decreased FA compared to HC and increased FA compared to TR-SZ, while in posterior corona radiate, NTR-SZ showed increased FA compared to HC and decreased FA compared to TR-SZ. For sMRI, in ACC, supramarginal gyrus, rolandic operculum, insula and inferior frontal gyrus (IFG), NTR-SZ showed decreased GMV compared to HC and increased GMV compared to TR-SZ, while in right lingual gyrus, NTR-SZ showed increased GMV compared to HC and decreased GMV compared to TR-SZ.

3.2 Associations with Cognition and Symptoms

As is shown in Fig. 3, fALFF_IC3 were positively correlated with cognitive composite scores (including FBIRN: $r = 0.30, p = 8.3e-07$* and BSNIP: $r = 0.23, p = 2.2e-04$*) across two cohorts. Besides, GMV_IC9 were positively correlated with the processing speed (including COBRE: $r = 0.38, p = 3.0e-04$*; FBIRN: $r = 0.38, p = 1.6e-10$* and BSNIP: $r = 0.26, p = 2.4e-05$*) and cognitive composite scores (including COBRE: $r = 0.28, p = 0.012$; FBIRN: $r = 0.37, p = 6.0e-10$* and BSNIP: $r = 0.28, p = 7.1e-06$*) across three cohorts.

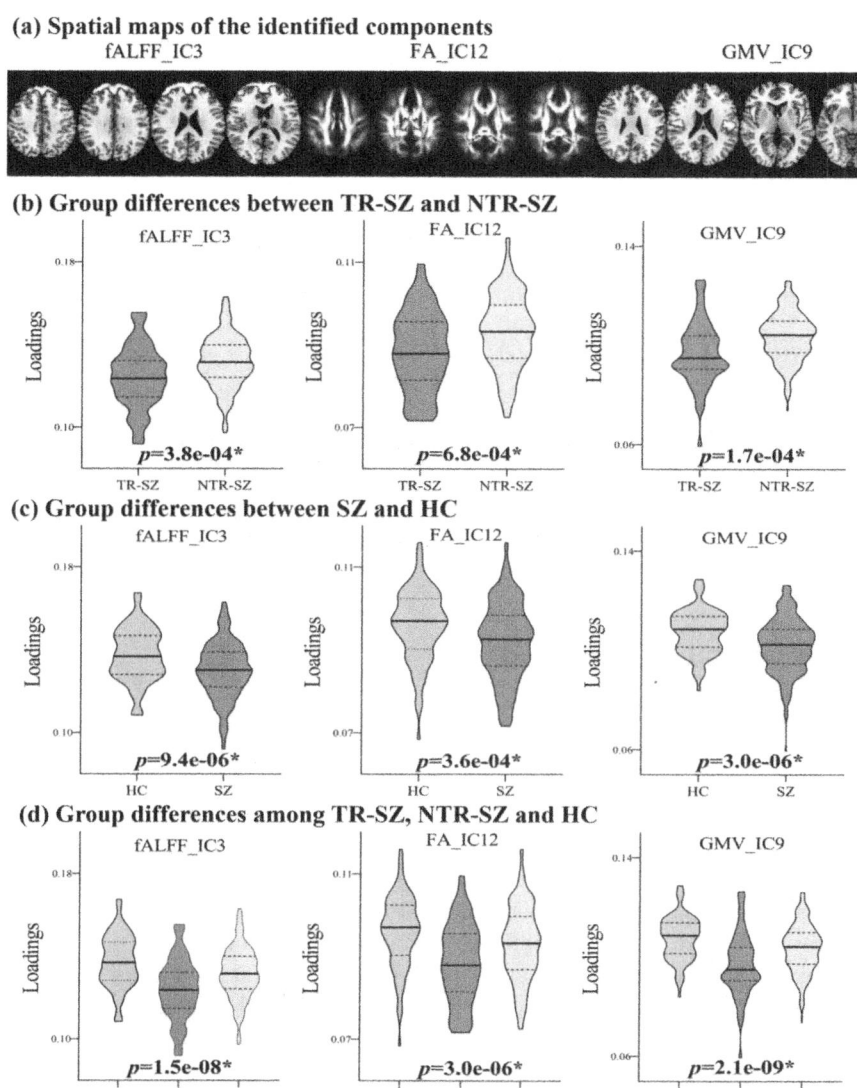

Fig. 2. The identified ICs. (a): The spatial maps of the identified components visualized at |Z| > 2, where the positive Z-values (red regions) means TR-SZ < NTR-SZ < HC while the negative Z-values (blue regions) means TR-SZ > NTR-SZ > HC. Group differences in the components' loadings between TR-SZ and NTR-SZ (b), HC and SZ (c), among HC, TR-SZ and NTR-SZ (d). The black solid/dashed lines in (b-d) represent the median/quartile. *Significance passed false discovery rate (FDR) correction for multiple comparisons. (Color figure online)

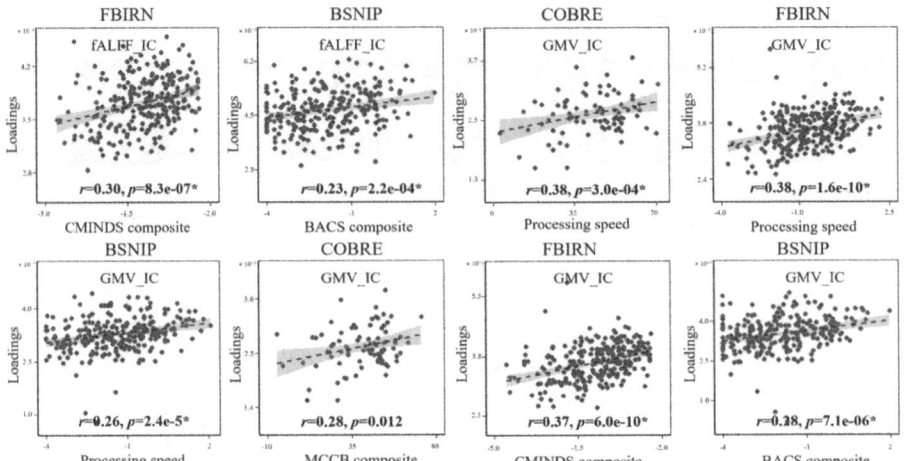

Fig. 3. Correlation analysis between the identified components' loadings in three external cohorts and symptom/cognition. (*) is used to indicate results which pass FDR correction for multiple comparisons.

4 Discussion and Conclusion

This study represents the first attempt to identify a multimodal co-varying network to investigate differences in brain function and structure among TR-SZ, NTR-SZ and HC by fusing three MRI features (fALFF, FA, GMV). Two key points are worth noting. (1) We identified the multimodal co-varying network which show significant differences among TR-SZ, NTR-SZ and HC. The aberrant brain regions in the identified network were mainly in the central executive network (CEN) and the salience network (SAN). (2) The identified joint ICs were positively correlated with cognitive composite score and processing speed, which are replicable across several cohorts.

One major finding was the identification of the multimodal co-varying CEN and SAN showing group differences among TR-SZ, NTR-SZ and HC. Dysfunction in these two brain networks have been widely reported in SZ [33, 34]. The current results further extend the previous findings that CEN and SAN also exists in explaining differential therapeutic resistance in SZ. Specifically, the ACC, an important node of the two networks, is one of the most widely investigated brain regions in SZ, showing differences in function [35], structure [36], and neurometabolites [37] compared to HC. In our study, ACC showed significant differences in both fALFF and GMV components among TR-SZ, NTR-SZ and HC. In addition, decreased fALFF was found in PFC (an integral part of CEN), and decreased GMV in insula (a key node of SAN), in both TR-SZ and NTR-SZ compared to HC. Moreover, CC showed decreased FA in TR-SZ compared to NTR-SZ, in line with a previous study [38]. Overall, the identified multimodal co-varying network can reveal the fundamental differences in both brain function and structure among TR-SZ, NTR-SZ and HC and TR-SZ suffer more severe impairment compared to NTR-SZ in the identified CEN and SAN (TR-SZ < NTR-SZ < HC).

Another finding was the replicable positive correlations between the identified ICs and the cognitive scores, i.e., the lower identified components' loadings, the worse cognitive function. Notably, the abnormal CEN and SAN in the multimodal co-varying network has been demonstrated in previous studies that associated with cognitive function in SZ. The CEN, which has been proved to be active in task-positive tasks due to their effects on attention, working memory and response selection [39]. The SAN, plays a key role in detecting and processing of cognitive, emotional and homeostatic salience events [40]. Moreover, CC plays a fundamental role in integrating information and mediating complex behaviors [41]. These collectively suggest that CEN, SAN and CC may be highly relative with the general cognitive function in SZ and further reveal the brain functional and structural mechanism of cognitive variation in TR-SZ and NTR-SZ.

In summary, this study utilized an unsupervised data-driven fusion method, MCCA + JICA, to identify a co-varying multimodal brain network that differentiating among TR-SZ, NTR-SZ and HC. By fusing three modalities (fALFF, FA and GMV), this approach provides a novel insight into the differential therapeutic resistance between TR-SZ and NTR-SZ. Additionally, the multimodal network deepens the understanding of functional and structural mechanisms underlying the treatment resistance in SZ. To the best of our knowledge, this is the first attempt to identify a co-varying multimodal CEN and SAN to distinguish among TR-SZ, NTR-SZ and HC using a relatively larger sample size.

Acknowledgments. This work was supported by the Natural Science Foundation of Jiangsu Province, China (BK20220889), the National Natural Science Foundation of China (62376124, 82172061 and 81771444), the Key Research and Development Plan of Jiangsu Province, China (BE2023668, BE2022677), and the Fundamental Research Funds for the Central Universities, (NJ2023032).

References

1. Moretti, P.N., et al.: Accessing gene expression in treatment-resistant schizophrenia. Molec. Neurobiol. **55**, 7000–7008 (2018)
2. Qi, S., et al.: Derivation and utility of schizophrenia polygenic risk associated multimodal MRI frontotemporal network. Nat. Commun. **13**(1), 4929 (2022)
3. Liang, C., et al.: Psychotic symptom, mood, and cognition-associated multimodal MRI reveal shared links to the salience network within the psychosis spectrum disorders. Schizophr. Bull. **49**(1), 172–184 (2023)
4. Zhao, C., et al.: Cross-cohort replicable resting-state functional connectivity in predicting symptoms and cognition of schizophrenia. Human Brain Mapp. **45**(7), e26694 (2024)
5. Howes, O.D., Murray, R.M.: Schizophrenia: an integrated sociodevelopmental-cognitive model. Lancet **383**(9929), 1677–1687 (2014)
6. Van Os, J., Kenis, G., Rutten, B.P.: The environment and schizophrenia. Nature **468**(7321), 203–212 (2010)
7. Lerner, V., Libov, I., Kotler, M., Strous, R.D., Psychiatry, B.: Combination of "atypical" antipsychotic medication in the management of treatment-resistant schizophrenia and schizoaffective disorder. Prog. NeuroPsychopharmacol. Biol. Psychiat. **28**(1), 89–98 (2004)
8. Hasan, A., et al.: World federation of societies of biological psychiatry (WFSBP) guidelines for biological treatment of schizophrenia, part 1: update 2012 on the acute treatment of schizophrenia and the management of treatment resistance. **13**(5), 318–378 (2012)

9. McGlashan, T.H.: A selective review of recent North American long-term followup studies of schizophrenia. Schizophr. Bull. **14**(4), 515–542 (1988)
10. Lally, J., et al.: Two distinct patterns of treatment resistance: clinical predictors of treatment resistance in first-episode schizophrenia spectrum psychoses. Psychol. Med. **46**(15), 3231–3240 (2016)
11. Fusar-Poli, P., et al.: Development and validation of a clinically based risk calculator for the transdiagnostic prediction of psychosis. JAMA Psychiat. **74**(5), 493–500 (2017)
12. Kanahara, N., Yamanaka, H., Suzuki, T., Takase, M., Iyo, M.: First-episode psychosis in treatment-resistant schizophrenia: a cross-sectional study of a long-term follow-up cohort. BMC Psychiat. **18**(1), 1–10 (2018)
13. White, T.P., Wigton, R., Joyce, D.W., Collier, T., Fornito, A., Shergill, S.S.: Dysfunctional striatal systems in treatment-resistant schizophrenia. Neuropsychopharmacology **41**(5), 1274–1285 (2016)
14. Gao, S., et al.: Distinguishing between treatment-resistant and non-treatment-resistant schizophrenia using regional homogeneity. Front. Psychiat. **9**, 282 (2018)
15. van Veelen, N.M., Vink, M., Ramsey, N.F., van Buuren, M., Hoogendam, J.M., Kahn, R.S.: Prefrontal lobe dysfunction predicts treatment response in medication-naive first-episode schizophrenia. Schizophr. Res. **129**(2–3), 156–162 (2011)
16. Anderson, V.M., Goldstein, M.E., Kydd, R.R., Russell, B.R.: Extensive gray matter volume reduction in treatment-resistant schizophrenia. Int. J. Neuropsychopharmacol. **18**(7), pyv016 (2015)
17. Quarantelli, M., et al.: Patients with poor response to antipsychotics have a more severe pattern of frontal atrophy: a voxel-based morphometry study of treatment resistance in schizophrenia. BioMed Res. Int. **2014**(1), 325052 (2014)
18. Yang, K., et al.: A multimodal study of a first episode psychosis cohort: potential markers of antipsychotic treatment resistance. Molec. Psychiat. **27**(2), 1184–1191 (2022)
19. Barry, E.F., et al.: Mapping cortical surface features in treatment resistant schizophrenia with in vivo structural MRI. Psychiat. Res. **274**, 335–344 (2019)
20. Itahashi, T., et al.: Dimensional distribution of cortical abnormality across antipsychotics treatment-resistant and responsive schizophrenia. NeuroImage Clin. **32**, 102852 (2021)
21. Zugman, A., et al.: Reduced dorso-lateral prefrontal cortex in treatment resistant schizophrenia. Schizophrenia Res. **148**(1–3), 81–86 (2013)
22. Meng, X., et al.: Predicting individualized clinical measures by a generalized prediction framework and multimodal fusion of MRI data. Neuroimage **145**, 218–229 (2017)
23. Liu, S., et al.: Linked 4-way multimodal brain differences in schizophrenia in a large Chinese Han population. Schizophrenia Bull. **45**(2), 436–449 (2019)
24. Calhoun, V.D., Sui, J.: Multimodal fusion of brain imaging data: a key to finding the missing link (s) in complex mental illness. Biol. Psychiat. Cogn. Neurosci. Neuroimaging **1**(3), 230–244 (2016)
25. Yao, C., et al.: A multimodal fusion analysis of pretreatment anatomical and functional cortical abnormalities in responsive and non-responsive schizophrenia. Front. Psychiat. **12**, 73717 (2021)
26. Qi, S., et al.: Multimodal fusion with reference: searching for joint neuromarkers of working memory deficits in schizophrenia. IEEE Trans. Med. Imaging **37**(1), 93–105 (2017)
27. Qi, S., et al.: Reward processing in novelty seekers: a transdiagnostic psychiatric imaging biomarker. Biol. Psychiat. **90**, 529–539 (2021)
28. Qi, S., et al.: The relevance of transdiagnostic shared networks to the severity of symptoms and cognitive deficits in schizophrenia: a multimodal brain imaging fusion study. Transl. Psychiat. **10**(1), 149 (2020)

29. Qi, S., et al.: Links between electroconvulsive therapy responsive and cognitive impairment multimodal brain networks in late-life major depressive disorder. BMC Med. **20**(1), 477 (2022)
30. Sui, J., et al.: Three-way (N-way) fusion of brain imaging data based on mCCA+ jICA and its application to discriminating schizophrenia. NeuroImage **66**, 119–132 (2013)
31. Li, Y.O., Adalı, T., Calhoun, V.D.: Estimating the number of independent components for functional magnetic resonance imaging data. Human Brain Mapp. **28**(11), 1251–1266 (2007)
32. Qi, S., et al.: Electroconvulsive therapy treatment responsive multimodal brain networks. Hum. Brain Mapp. **41**(7), 1775–1785 (2020)
33. Eisenberg, D.P., Berman, K.F.: Executive function, neural circuitry, and genetic mechanisms in schizophrenia. Neuropsychopharmacology **35**(1), 258–277 (2010)
34. Palaniyappan, L., Liddle, P.F.: Does the salience network play a cardinal role in psychosis? An emerging hypothesis of insular dysfunction. J. Psychiat. Neurosci. **37**(1), 17–27 (2012)
35. Zhou, C., et al.: Altered patterns of the fractional amplitude of low-frequency fluctuation and functional connectivity between deficit and non-deficit schizophrenia. Front. Psychiat. **10**, 680 (2019)
36. Ji, Y., et al.: Genes associated with gray matter volume alterations in schizophrenia. Neuroimage **225**, 117522 (2021)
37. Roberts, R.C., Barksdale, K.A., Roche, J.K., Lahti, A.C.: Decreased synaptic and mitochondrial density in the postmortem anterior cingulate cortex in schizophrenia. Schizophrenia Res. **168**(1–2), 543–553 (2015)
38. McNabb, C.B., Kydd, R., Sundram, F., Soosay, I., Russell, B.R.: Differences in white matter connectivity between treatment-resistant and treatment-responsive subtypes of schizophrenia. Psychiat. Res. Neuroimaging **282**, 47–54 (2018)
39. Sridharan, D., Levitin, D.J., Menon, V.: A critical role for the right fronto-insular cortex in switching between central-executive and default-mode networks. Proc. Natl. Acad. Sci. **105**(34), 12569–12574 (2008)
40. Bressler, S.L., Menon, V.: Large-scale brain networks in cognition: emerging methods and principles. Trends Cogn. Sci. **14**(6), 277–290 (2010)
41. Hinkley, L.B., et al.: The role of corpus callosum development in functional connectivity and cognitive processing (2012)

Investigating the Dynamics of Seizure Neuroactivities Using Hidden Markov Model

Tao Feng[1], Huihang Ke[2], Hui Yao[3], and Chao Wu[1]()

[1] Zhejiang University, Hangzhou 310058, China
chao.wu@zju.edu.cn
[2] ZTE Corporation, Nanjing 210000, China
[3] Zhejiang Post and Telecommunication Construction CO., LTD.,
Hangzhou 310051, China

Abstract. Epilepsy impacts 65 million globally, necessitating advanced research for effective treatment strategies. This study evaluates the HMM-MVAR model's utility in epilepsy, particularly in seizure prediction and brain network analysis. Initial simulations revealed its superiority over traditional HMM and HMM-TDE approaches. Utilizing the Helsinki EEG database, the model achieved a notable average error of 2.759 in seizure timing prediction. Further application on sEEG data from TLE patients in the HUP database identified six distinct state-time transition sequences, providing critical insights into epileptic network dynamics. These results underscore the potential of the HMM-MVAR model in enhancing clinical epilepsy management and understanding its underlying mechanisms.

Keywords: Temporal Lobe Epilepsy · Hidden Markov Model · Brain Network Analysis

1 Introduction

Epilepsy, affecting 65 million people worldwide, presents a spectrum of seizure types and cognitive impacts [11]. Understanding its neurological mechanisms is crucial, as recent studies link seizure events to complex interactions of ion channel mutations causing network hyperexcitability [17]. This knowledge is essential for developing effective therapeutic interventions. However, challenges persist, including precise localization of epileptogenic zones and managing drug-resistant epilepsy (DRE), which affects about one-third of patients [15]. Hurdles such as seizure type heterogeneity, varied treatment responses, and the need for personalized therapeutic approaches must be addressed. Additionally, considering the risk of comorbidities and long-term medication side effects is essential in treatment planning.

In epilepsy research, various techniques like functional Magnetic Resonance Imaging (fMRI) and positron emission tomography(PET) have provided detailed images of brain activity but lack the necessary temporal resolution to fully capture seizures [18]. Magnetoencephalography(MEG) offers improved timing but at higher costs [1]. Electroencephalogram (EEG) remains a primary, non-invasive tool for real-time monitoring despite its spatial resolution limitations [23]. Stereoelectroencephalography (sEEG), though invasive, gives accurate spatial details for seizure origins [8]. Combined with computational methods, these tools enhance our understanding of seizure dynamics and aid in treatment development [13]. Thalamic sEEG is being investigated for its potential in guiding neurostimulation therapies [3].

In clinical reality, EEG and sEEG are the key measurements in understanding seizures. Signal processing methods such as wavelet decomposition reveal the spectral intricacies of seizures [19]. Machine learning methods, such as CNN, enable the automated detection of seizure patterns [2]. Data from EEG and MEG are used to map brain networks, aiding in surgical planning [4]. Deep learning extracts feature without traditional engineering, offering personalized insights [4]. However, challenges remain in feature extraction variability, non-stationarity of signals, and the opacity of deep learning models, which complicates clinical application [12]. The Hidden Markov Model (HMM) is pivotal in epilepsy research, adeptly mapping the brain's electrical activity sequences noted in EEG and sEEG. It unravels the non-visible states and transitions in brain activity, addressing the challenges of signal variability and non-stationarity [2]. HMMs flexibly track the duration of these states, surpassing traditional methods limited by the assumption of signal stationarity [15]. For seizure prediction, HMMs identify crucial preictal changes, providing key temporal insights for early intervention [15]. However, the HMM model is challenged by its geometric state duration assumption, which may not suit the fast-switching episodic nature of seizures. Moreover, converting HMM's abstract states into concrete physiological interpretations remains complex, affecting the practicality of the results. These issues highlight the demand for more refined models or supplementary methods to enhance accuracy in seizure detection and forecasting.

To tackle this problem, the HMM-MAVR model advances epilepsy research by enhancing the traditional HMM with autoregressive features, allowing for a finer grasp of EEG and sEEG data. This sophisticated model captures temporal patterns and dynamic brain states more precisely, aiding in the nuanced interpretation of neural activity and improving seizure management prospects [21]. Through variational Bayesian inference, it refines data analysis, addressing previous HMM limitations and paving the way for more precise clinical applications [20].

In this work, we initially used a set of simulated signals to compare the performance of three models: HMM, (Hidden Markov Model with Time-Delay Embedding) HMM-TDE, and HMM-MVAR. Preliminary conclusions indicated that the HMM-MVAR model better fits the target signal by comparing the minimum free energy of the model outputs. To further verify if the HMM-MVAR model can accurately describe the neurophysiological activity during an epileptic seizure, we used the Helsinki scalp EEG database and designed a subtask to predict the timing of seizures. The HMM-MVAR model showed an average relative error of 2.759 in predicting seizure timing, suggesting it can accurately depict the neurophysiological process of seizures. We then applied the model for sEEG signal analysis in temporal lobe epilepsy (TLE) seizures, constructing the HMM-MVAR model for multiple seizures of a single patient using the HUP intracranial EEG database. We observed six representative state-time transition sequences and conducted a brain network analysis on these states, identifying key nodes and connections within the brain network for each state. This guides exploring biological markers of epilepsy and the pathways of neural electrical activity.

The contribution of this work is threefold:

1. Introduced the application of the HMM-MVAR model for brain state pattern analysis in epileptic seizures, confirming its capability to characterize brain activity in epilepsy through a subtask of seizure timing prediction on a clinical scalp EEG dataset.
2. Employed the model in the stereoelectroencephalographic analysis of temporal lobe epilepsy (TLE) seizures, achieving a state-time course (STC) that delineates six distinctive brain states from the sEEG signals of a single patient across multiple seizures.
3. Clinically, the model facilitated the inference of crucial brain nodes, networks, and neural electrical conduction pathways across multiple seizures in the same patient, offering valuable insights into the neurodynamics of epileptic events, with implications for clinical science.

2 Methodology

2.1 Datasets

Simulation Dataset. To compare the different performances of the Hidden Markov Model (HMM), the HMM-MVAR model, and the HMM-TDE model (Hidden Markov Model with Time-Delay Embedding) on simulated data, we utilize a data simulation platform to generate a 3-channel, 4-state time series with 10 trials. Each trial contains 10,000 data points to mimic actual EEG signals.

Helsinki Dataset. The Helsinki scalp EEG database, compiled by Stevenson et al. in 2018, is a unique public resource featuring multichannel EEG data from 79 full-term newborns, including expert-annotated epileptic seizures, from Helsinki University Hospital's neonatal ICU [16].

HUP Dataset. The HUP intracranial EEG dataset features 58 patients with drug-resistant epilepsy, including subdural and depth electrode recordings and surgical outcomes [7,10]. It provides interictal EEG data in ICBM152 MNI space, detailing epileptogenic zones, patient demographics, and treatment specifics.

2.2 Modelling

Helsinki EEG Data Pre-processing. The Helsinki database uses clinician assessments to filter EEG data, removing segments marked as seizures or with noise. After 1–60 Hz filtering, re-referencing, and artifact removal, data is segmented into 60-s pieces with 0–10 s gaps, aiding in seizure onset prediction.

HUP Data Pre-processing. In preprocessing HUP's EEG, we used a 1–80 Hz filter and a 60 Hz notch filter to minimize interference. Seizure segmentation covered from onset to full signal spread, maxing at 45 s. Thus, we used 45-s segments for both seizure and non-seizure data for consistency.

HMM-MVAR. The Hidden Markov Model (HMM) posits that each observation is associated with a specific state. Fox et al. [7] expanded on HMM by developing the Beta-Process Auto-Regressive Hidden Markov Model (BP-AR-HMM), which integrates autoregressive models with HMM, using Beta Process for initialization. This model was further refined by [21] into the HMM-MVAR, incorporating a probabilistic framework and variational Bayesian inference for more effective parameter estimation. This method, particularly useful for large datasets due to its efficiency, updates hypotheses based on new evidence. It seeks to approximate the posterior distribution $q(z;\theta)$ with $p(z \mid x)$, aiming to minimize the Kullback-Leibler divergence to determine the optimal parameters.

$$\theta^* = \arg\min_{\theta} KL(q(z;\theta)\|p(z \mid x)) \tag{1}$$

The Kullback-Leibler divergence, $KL(q(z;\theta)\|p(z \mid x))$, measures the divergence between the two distributions $q(z;\theta)$ and $p(z \mid x)$. This approach transforms the problem of solving for the posterior distribution into one of parameter optimization. For the HMM-MVAR model specifically, the multi-channel EEG and sEEG signal is denoted as $O_t \in \mathbb{R}^N$, with the hidden states denoted as H_t, where $t = 1,\ldots,T$. Assuming Gaussian noise, the observation model can be represented as:

$$O_t \mid H_t = k \sim \mathcal{N}\left(\sum_{l \in \rho} y_{t-l} W_l^{(k)}, \Sigma^{(k)}\right) \tag{2}$$

In the Eq. (2), ρ represents the lag order of the MVAR model, $W_l^{(k)}$ is the AR coefficients matrix for state k, and $\sum^{(k)}$ is the diagonal noise covariance matrix. The inverse of $\sum^{(k)}$ can be modeled with a Wishart distribution, and if it is diagonal, a Gamma distribution is used.

$$\Sigma^{(k)^{-1}} \sim \text{Wishart}\,(l_0, \mathcal{B}_0) \tag{3}$$

$$\varpi_{ii}^{(k)} \sim \text{Gamma}\,(l_0, \mathcal{B}_0) \tag{4}$$

The Markov dynamic probability could be represented as:

$$P(H_t = k_1 \mid H_{t-1} = k_2) = \Theta_{k_1 k_2} \quad (5)$$

$$P(H_t = k_1) = \eta_k \quad (6)$$

Among them $\Theta_{k_1 k_2}$ and η_k are the HMM-MVAR model parameters to be inferred, modeled using the Dirichlet distribution:

$$\Theta_k = \text{Dir}(v_0) \quad (7)$$

$$\eta_k = \text{Dir}(\xi_0) \quad (8)$$

Among these v_0 and ξ_0 are parameters inferred through the HMM-MVAR training process. The given text establishes a Bayesian hierarchy for controlling model parameters from (1) to (8). The variational Bayesian method assumes additional factorization within the parameter space and requires conjugate prior distributions for parameter inference. This approach utilizes an iterative algorithm that operates on one set of parameters at a time to minimize free energy. The algorithm seeks to minimize a cost function known as free energy, which is the sum of the model's average log-likelihood, the Kullback-Leibler divergence between the actual and factorized distributions, and the negative entropy of the factorized distribution. The overall workflow has been shown in Fig. 1.

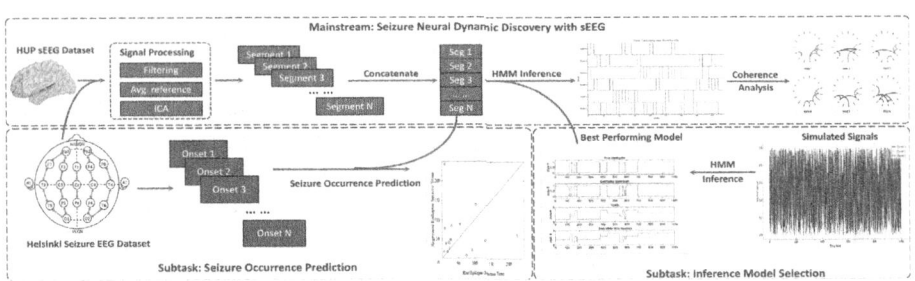

Fig. 1. Overall Workflow of the Proposed Method.

3 Experiments and Results

3.1 Model Selection: with Simulated Signals

We utilized a data simulation platform to generate a three-channel, four-state time series with 10 trials, each containing 10,000 data points, to emulate real brain electrical signals. These signals were then analyzed using three different models. We compared the difference between the estimated parameters of each model and the true parameters of the simulated data. Additionally, we compared the final free energy values of each model to assess their fit and complexity.

Table 1 displays the minimum free energy values estimated by three different models. From the table, it is evident that the HMM-MVAR model yields the lowest final free energy, indicating superior training performance, particularly with sparse data. Consequently, this study selects the HMM-MVAR model as the training model of choice.

Table 1. Predictive Seizure Statistics

Model	Minimum Free Energy
HMM	1.385×10^6
HMM-MVAR	1.291×10^6
HMM-TDE	1.947×10^7

3.2 Model Evaluation: Seizure Prediction with EEG

In the model implementation, the HMM framework utilizes three hidden states, and the MAR model's order is also three. Power spectral density and coherence between channels are estimated using parameters from a third-order MVAR model. A sparsity-enforcing LASSO regression technique based on the L1 norm is employed. This study analyzes epileptic EEG data from Patient Number One in the Helsinki database, partitioning the remaining data into 60-second segments after excluding seizure periods and corrupted segments. Additionally, the time interval from the end of each data segment to the onset of the next seizure is recorded. The segmented data were sequentially connected to form a long sequence, which was then modeled for neurodynamic activity using the HMM-MVAR model. To verify the HMM-MVAR model's decoding ability for epileptic EEG data, we regressed the state time series obtained from the model against the intervals leading to the next seizure. The scatter plot of the regression results is shown in Fig. 2(a). We compared the actual seizure times with the regressed seizure times and documented the relative error information which has show in Table 2.

The connectivity matrix, based on channel coherence, displays only the top 10% of connections by strength. The graph reveals that the connections with the highest coherence are concentrated in the patient's left temporal lobe, which aligns with the lesion labels provided by the database. The three hidden states decoded by the model correspond to brain functional connectivity circular graphs as shown in Fig. 3.

The graph displays the functional connectivity between channels for each state. From the circular graph, we can discern substantial differences in brain functional connections among the three states, indicating that the HMM model has effectively classified brain functional connectivity as reflected by the EEG signals. Notably, channels with dense connections in the circular graph are concentrated in the patient's left temporal lobe, suggesting that during epileptic

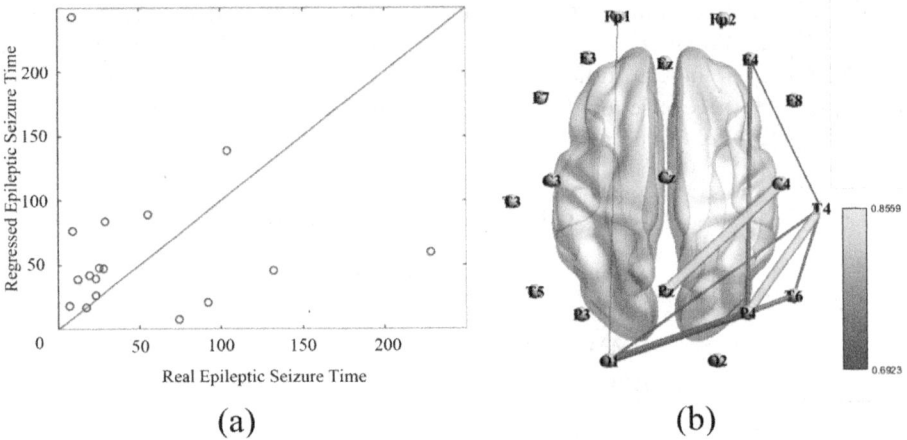

Fig. 2. (a) Epileptic Seizure Prediction Regression; (b) Example Brain Functional Connectivity Graph at the Electrode Level for State 1.

seizures, normal brain activity is suppressed, and the functional connection strength between foci of epilepsy is greater. This experiment also conducted a statistical analysis of the circular graphs of brain functional connectivity between states, with the complex network characteristics of the three states presented in Table 3.

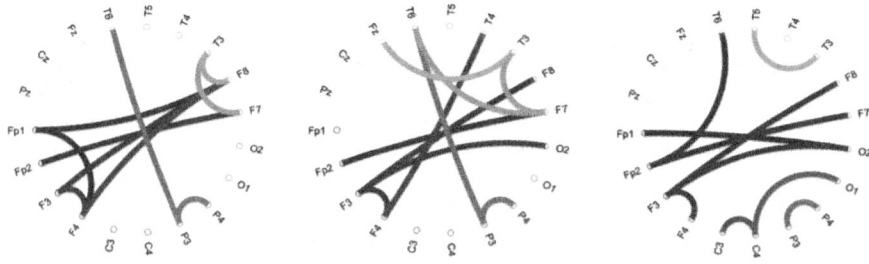

Fig. 3. Circular Graphs of Brain Functional Connectivity for Three States.

The experiment analyzed four complex network indicators: clustering coefficient, characteristic path length, local efficiency, and global efficiency [14]. It is apparent from the graph that all four indicators for State 2 are significantly lower than those for States 1 and 3, indicating that the brain functional connectivity in State 2 has lower efficiency of information transfer and robustness against local node damage compared to States 1 and 3, which are common characteristics of brain activity during epileptic seizures. Therefore, the frequency of occurrence of State 2, which has poor information transfer efficiency, can serve as a basis for predicting epileptic seizures.

Table 2. Predictive Seizure Statistics

ID	Actual Time	Predicted Time	Relative Error
1	7	18.345	1.621
2	9	242.389	25.932
3	9	76.605	7.512
4	12	38.879	2.240
5	17	17.142	0.008
6	19	42.052	1.213
7	23	39.437	0.715
8	23	26.401	0.148
9	25	47.788	0.912
10	28	47.274	0.688
11	29	83.995	1.896
12	55	89.171	0.621
13	74	7.887	0.893
14	92	20.910	0.773
15	104	138.672	0.333
16	132	45.535	0.655
17	228	59.887	0.737

Table 3. Complex Network Analysis of Brain Functional Connectivity Circular Graphs

	State 1	State 2	State 3	Mean
Clustering Coefficient	0.764	0.667	0.755	0.729
Char. Path Length	0.789	0.706	0.771	0.755
Local Efficiency	0.764	0.667	0.755	0.729
Global Efficiency	0.803	0.731	0.771	0.768

3.3 Model Application: Neural Dynamics of TLE

We analyzed sEEG data from patients with focal epilepsy in the left temporal lobe, selecting those with minimal artifact channels and high-quality electrodes. Electrode placement was based on clinical symptoms, EEG, and neuroimaging, tailored to individual patient needs. Due to variability in electrode configurations, we focused on individual analyses, choosing patients like HUP144 with multiple seizures for robustness. HUP144 had 111 good channels, data from five seizures, and two interictal recordings. Using MNE software, we mapped HUP144's electrodes, mainly in the left hemisphere due to the left temporal focus, to ICBM152 MNI space and assigned them to brain regions using Freesurfer [6], creating a functional brain partitioning map (Fig. 4(a)). To focus on key epileptic areas, 31 channels from rods RA, RB, and RC in HUP144's dataset were used, centered around the left hippocampus as shown in Fig. 4(b). This selection aimed to capture the core features of the patient's temporal lobe epilepsy. The experiment employed a consistent setup with six HMM states and a fifth-order MVAR model for spectral and coherence analysis.

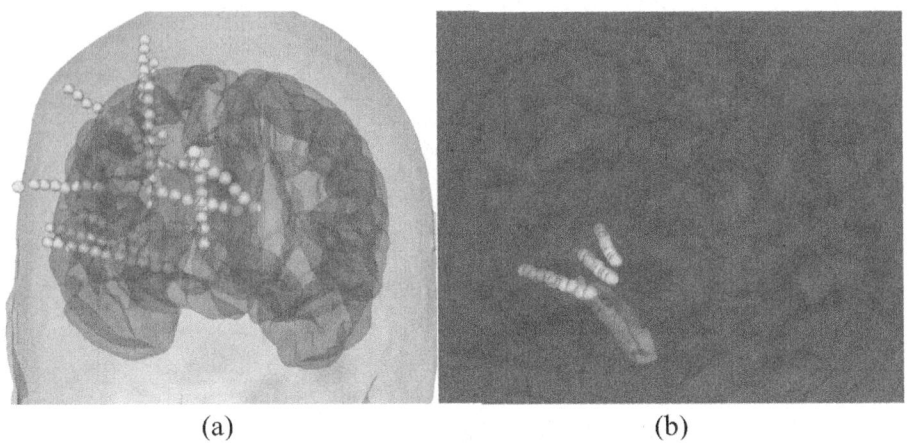

Fig. 4. (a) Electrode Placement Map for Patient HUP144; (b) Localization Map of the 31 Channel Electrodes Enrolled for Patient HUP144.

Post-exclusion, 111 channels remained, with data segmented into six groups (resected, normal, and full 31-channel set for ictal and interictal phases) for individual HMM-MVAR modeling. Resected channels were those surgically removed, and normal channels were the non-resected ones on the same rods, expected to have stronger connections due to their proximity, as illustrated in Fig. 4(b). Analysis across these groups revealed distinct connectivity patterns, underscoring HMM's capability to differentiate functional connections, as depicted in Fig. 5.

Figure 6 shows that during seizures, connectivity in the Ep_resect group is focused in the lower half, near the seizure onset zone (soz) channels RA1-RA9. In contrast, the Ip_resect group's interictal connectivity is more uniform, indicating stabilized EEG signals and normal brain function outside of seizures. This highlights the model's ability to capture the dynamic changes in brain activity associated with epilepsy. Complex network analysis offers a fresh perspective on the brain's structure-function nexus, treating it as an intricate web of neural connections. This approach, leveraging cutting-edge neurotechnological methods, sheds light on cognition, development, and pathology by evaluating network characteristics such as clustering and efficiency. Our research employs this technique for epilepsy prediction and lesion localization, utilizing data from six groups. Results, presented in Table 4, focus on network clustering and global efficiency, key indicators of interconnectivity and information transfer.

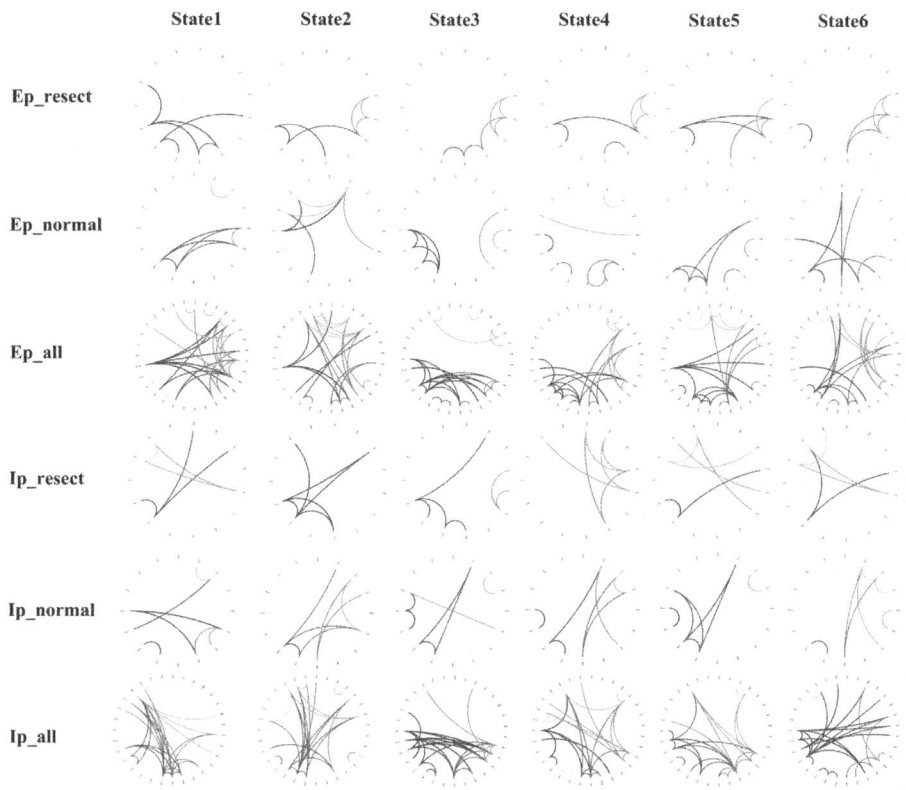

Fig. 5. Circular Graphs of Brain Functional Connectivity for Six Experimental Groups.

Our study shows distinct brain activity patterns through fluctuations in key network metrics, with interictal states displaying higher clustering coefficients and efficiencies than ictal states, indicating seizures' disruptive effects on brain function. Particularly, the Ep_resect group exhibits more pronounced seizure characteristics, with minimal signal differences between normal and surgically altered channels due to their proximity. Interictal lesion areas show notably tighter connectivity, especially in the Ip_resect group. Seizures lead to reduced information transfer efficiency, with EEG data revealing a rapid spread of epileptic activity, suppressing normal brain functions and leading to significant patient impacts.

4 Discussion

Through the application of the HMM-MVAR model to the analysis of sEEG signals during TLE seizures, we can draw the following conclusions: Firstly, in the analysis of the temporal state sequences of brain functional connectivity, the

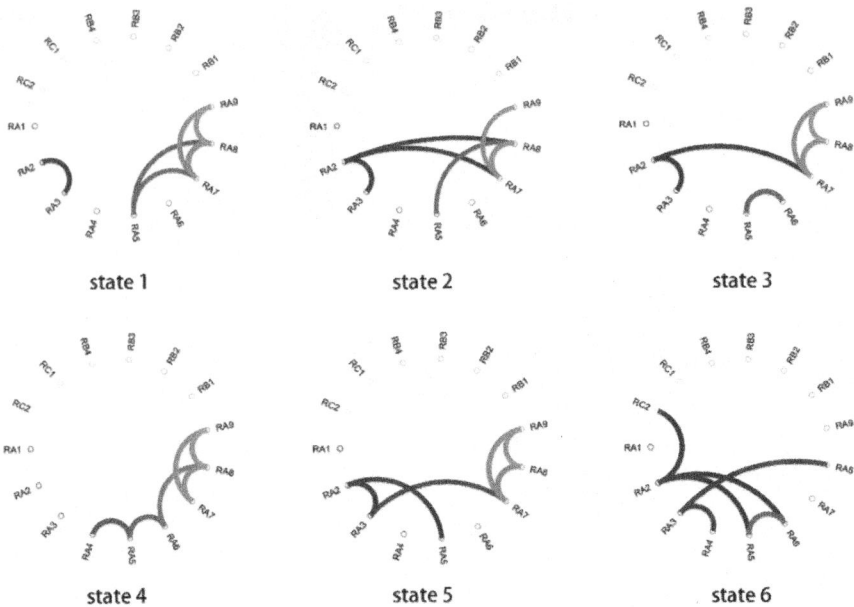

Fig. 6. Ep-resect Brain Functional Connectivity Map.

state time series obtained by this method can differentiate the brain functional connections under different states, with each state showing significant differences in areas of dense connectivity. This indicates that the model can effectively identify different states of brain functional connectivity within the EEG signals, reflecting the model's effective classification capabilities [9]. Secondly, regarding the specificity of brain networks during and between epileptic seizures, during seizure episodes, enhanced functional connectivity is observed in specific areas, particularly in channels corresponding to the epileptogenic zone. In contrast, a more uniform distribution of connectivity is observed during the interictal periods. These differences reveal significant changes in brain activity during seizures and a relatively stable state of brain networks between seizures [5]. Thirdly, in the transition of brain connectivity state sequences in the epileptogenic zone, network characteristics exhibit considerable fluctuations across different states, particularly during seizures, where there is a decline in all measured network characteristics. Notably, electrodes corresponding to the epileptogenic zone show an increase in brain network metrics during the interictal period, underscoring the distinctive role of the epileptogenic zone in brain function outside of seizure episodes [22].

Table 4. Complex Network Metrics for Brain Functional Networks

Measurements	Conditions	State 1	State 2	State 3	State 4	State 5	State 6	Mean
Clustering Coefficient	Ep_resect	0.481	0.334	0.373	0.435	0.466	0.811	0.484
	Ep_normal	0.506	0.457	0.333	0.392	0.885	0.478	0.509
	Ep_all	0.422	0.396	0.375	0.413	0.601	0.797	0.501
	IP_resect	0.897	0.846	0.880	0.303	0.885	0.661	**0.745**
	IP_normal	0.352	0.832	0.727	0.774	0.357	0.514	0.593
	Ip_all	0.734	0.301	0.660	0.877	0.364	0.202	0.523
Characteristic Path Length	Ep_resect	0.500	0.371	0.402	0.458	0.495	0.813	0.507
	Ep_normal	0.526	0.477	0.358	0.418	0.886	0.496	0.527
	Ep_all	0.439	0.419	0.399	0.432	0.610	0.799	0.516
	IP_resect	0.898	0.847	0.882	0.326	0.886	0.666	**0.751**
	IP_normal	0.371	0.836	0.733	0.779	0.376	0.525	0.603
	Ip_all	0.737	0.322	0.665	0.879	0.381	0.223	0.534
Global Efficiency	Ep_resect	0.509	0.395	0.416	0.464	0.510	0.813	0.518
	Ep_normal	0.535	0.485	0.379	0.436	0.886	0.500	0.537
	Ep_all	0.447	0.436	0.417	0.442	0.611	0.799	0.525
	IP_resect	0.898	0.847	0.882	0.342	0.886	0.666	**0.754**
	IP_normal	0.381	0.836	0.733	0.779	0.387	0.526	0.607
	Ip_all	0.737	0.341	0.665	0.879	0.391	0.246	0.543
Local Efficiency	Ep_resect	0.481	0.334	0.373	0.435	0.466	0.811	0.484
	Ep_normal	0.506	0.457	0.333	0.392	0.885	0.478	0.509
	Ep_all	0.422	0.396	0.375	0.413	0.601	0.797	0.501
	IP_resect	0.897	0.846	0.880	0.303	0.885	0.661	**0.745**
	IP_normal	0.352	0.832	0.727	0.774	0.357	0.514	0.593
	Ip_all	0.734	0.301	0.660	0.877	0.364	0.202	0.523

Compared to past studies, this work presents the following advancements: Firstly, it analyzes epileptic EEG signals from the perspective of state time series, taking into account the transitions between different brain states during epileptic seizures. Secondly, the model aggregates three fundamental states of infantile epileptic seizures and six fundamental states of intractable TLE. Thirdly, it analyzes the patterns of each brain state from the perspective of brain networks, with the ability to trace back to the epileptogenic zone through key network nodes. Nonetheless, this work still has the following limitations: Firstly, the HMM-MVAR model has only been analyzed for two types of epilepsy, infantile epilepsy and intractable TLE, and the generalizability of the model needs to be verified. Secondly, the output of the HMM model relies on high-quality sEEG signal sampling, with signal quality having a significant impact on the results. Future research can focus on the generalizability of the model and its compatibility with other brain signals. In summary, this method has shown good fit for the characteristics of epileptic brain signals and has led to new discoveries in the sEEG dataset, demonstrating the potential for clinical practice.

5 Conclusion

In conclusion, the HMM-MVAR model applied to sEEG signals during TLE seizures offers valuable insights into brain connectivity dynamics. This work advances the field by analyzing state time series of epileptic EEG signals, considering transitions between brain states during seizures. Additionally, it identifies fundamental states of epileptic seizures and traces brain network patterns, linking back to the epileptogenic zone. These findings have implications for understanding epileptic brain dynamics and may inform clinical practice.

Acknowledgements. This work was supported by the National Science and Technology Major Project (2021ZD0110505, 2021YFC3340300), National Natural Science Foundation of China (U19B2042), the Zhejiang Provincial Key Research and Development Project (2023C01043, 2022C03106), University Synergy Innovation Program of Anhui Province (GXXT-2021-004), Academy Of Social Governance Zhejiang University, Fundamental Research Funds for the Central Universities (226-2022-00064).

References

1. Bagić, A.I., Knowlton, R.C., Rose, D.F., Ebersole, J.S.: American clinical magnetoencephalography society clinical practice guideline 1. J. Clin. Neurophysiol., 1 (2011)
2. Batista, J., Pinto, M.F., Tavares, M., Lopes, F., Oliveira, A., Teixeira, C.: Eeg epilepsy seizure prediction: the post-processing stage as a chronology. Sci. Rep. **14**, 407 (2024)
3. Bernabei, J.M., Litt, B., Cajigas, I.: Thalamic stereo-eeg in epilepsy surgery: Where do we stand? Brain **146**, 2663–2665 (2023)
4. Cao, M., Vogrin, S.J., Peterson, A.D.H., Woods, W., Cook, M.J., Plummer, C.: Dynamical network models from eeg and meg for epilepsy surgery-a quantitative approach. Front. Neurol. **13** (2022)
5. Englot, D.J., Konrad, P.E., Morgan, V.L.: Regional and global connectivity disturbances in focal epilepsy, related neurocognitive sequelae, and potential mechanistic underpinnings. Epilepsia **57**, 1546–1557 (2016)
6. Fischl, B.: Freesurfer. NeuroImage **62**, 774–781 (2012)
7. Fox, E., Jordan, M., Sudderth, E., Willsky, A.: Sharing features among dynamical systems with beta processes. In: Bengio, Y., Schuurmans, D., Lafferty, J., Williams, C., Culotta, A. (eds.) Advances in Neural Information Processing Systems, vol. 22. Curran Associates, Inc. (2009)
8. Gonzalez-Martinez, J., Bulacio, J., Alexopoulos, A., Jehi, L., Bingaman, W., Najm, I.: Stereoelectroencephalography in the "difficult to localize" refractory focal epilepsy: early experience from a North American epilepsy center. Epilepsia **54**, 323–330 (2013)
9. Kramer, M.A., Eden, U.T., Kolaczyk, E.D., Zepeda, R., Eskandar, E.N., Cash, S.S.: Coalescence and fragmentation of cortical networks during focal seizures. J. Neurosci. **30**, 10076–10085 (2010)
10. Lancaster, J.L., et al.: Bias between mni and talairach coordinates analyzed using the icbm-152 brain template. Human Brain Mapp. **28**, 1194–1205 (2007)

11. Moshé, S.L., Perucca, E., Ryvlin, P., Tomson, T.: Epilepsy: new advances. Lancet **385**, 884–898 (2015)
12. Nagahama, Y., et al.: Outcome of stereo-electroencephalography with single-unit recording in drug-refractory epilepsy, pp. 1–10 (2023)
13. Nourmohammadi, A., et al.: Passive functional mapping of receptive language cortex during general anesthesia using electrocorticography. Clin. Neurophysiol. **147**, 31–44 (2023)
14. Sporns, O.: Structure and function of complex brain networks. Dial. Clin. Neurosci. **15**, 247–262 (2013)
15. Srinivasan, S., Dayalane, S., Mathivanan, S.K., Rajadurai, H., Jayagopal, P., Dalu, G.T.: Detection and classification of adult epilepsy using hybrid deep learning approach. Sci. Rep. **13**, 17574 (2023)
16. Stevenson, N.J., Tapani, K., Lauronen, L., Vanhatalo, S.: A dataset of neonatal eeg recordings with seizure annotations. Sci. Data **6**, 190039 (2019)
17. Stöber, T.M., Batulin, D., Triesch, J., Narayanan, R., Jedlicka, P.: Degeneracy in epilepsy: multiple routes to hyperexcitable brain circuits and their repair. Commun. Biol. **6**, 479 (2023)
18. Tatum, W., et al.: Clinical utility of eeg in diagnosing and monitoring epilepsy in adults. Clin. Neurophysiol. **129**, 1056–1082 (2018)
19. Urigüen, J.A., Garcia-Zapirain, B.: Eeg artifact removal-state-of-the-art and guidelines. J. Neural Eng. **12**, 031001 (2015)
20. Vidaurre, D., Myers, N.E., Stokes, M., Nobre, A.C., Woolrich, M.W.: Temporally unconstrained decoding reveals consistent but time-varying stages of stimulus processing. Cerebral Cortex **29**, 863–874 (2019)
21. Vidaurre, D., Quinn, A.J., Baker, A.P., Dupret, D., Tejero-Cantero, A., Woolrich, M.W.: Spectrally resolved fast transient brain states in electrophysiological data. NeuroImage **126**, 81–95 (2016)
22. Wilke, C., Worrell, G., He, B.: Graph analysis of epileptogenic networks in human partial epilepsy. Epilepsia **52**, 84–93 (2011)
23. Zijlmans, M., Jacobs, J., Zelmann, R., Dubeau, F., Gotman, J.: High-frequency oscillations mirror disease activity in patients with epilepsy. Neurology **72**, 979–986 (2009)

Suppressing Seizure via Optimal Electrical Stimulation to the Hub of Epileptic Brain Network

Zhichao Liang[1], Guanyi Zhao[1], Yinuo Zhang[1], Weiting Sun[1], Jingzhe Lin[1], Jialin Wang[2], and Quanying Liu[1(✉)]

[1] Department of Biomedical Engineering, Southern University of Science and Technology, Shenzhen, China
`liuqy@sustech.edu.cn`
[2] Shenzhen Middle School, Shenzhen, China

Abstract. The electrical stimulation to the seizure onset zone (SOZ) serves as an efficient approach to seizure suppression. Recently, seizure dynamics have gained widespread attendance in its network propagation mechanisms. Compared with the direct stimulation to SOZ, other brain network-level approaches that can effectively suppress epileptic seizures remain under-explored. In this study, we introduce a platform equipped with a system identification module and a control strategy module, to validate the effectiveness of the hub of the epileptic brain network in suppressing seizure. The identified surrogate dynamics show high predictive performance in reconstructing neural dynamics which enables the model predictive framework to achieve accurate neural stimulation. The electrical stimulation on the hub of the epileptic brain network shows remarkable performance as the direct stimulation of SOZ in suppressing seizure dynamics. Underpinned by network control theory, our platform offers a general tool for the validation of neural stimulation.

Keywords: Epilepsy · Network-Coupled Dynamics · System Identification · Seizure Control · Control Node Selection

1 Introduction

Episodes of focal epilepsy engage networks that operate at various spatiotemporal scales [1–3]. The epileptogenic network (EN), formed by the brain regions that are actively involved in the initiation of epileptic seizures, has gained widespread attention [4,5]. The abnormal activities within specific nodes of epileptogenic networks typically show cascade reactions. Evidence has been reported that EN directly impacts the success of epilepsy surgical interventions [6]. Therefore, it is hypothesized that appropriate electrical stimulation to EN may help suppress seizures.

Z. Liang and G. Zhao—Equal contribution.

Nearly one-third of epilepsy patients are drug-resistant, which means medication cannot effectively control their seizures [7]. Although epilepsy surgery is considered as primary treatment for drug-resistant epilepsy, it has significant risks, such as irreversible structural damage to the brain. Alternatively, electrical stimulation, as an advanced technique for neurodiagnostic and therapeutic purposes, has the merits of safety and flexibility. Some pioneer studies have reported that neural stimulation effectively suppressed epileptic seizures by desynchronizing the epileptogenic networks [8–10]. In clinical practice, stimulation targets are typically selected within the EN or some other connected brain areas [11]. However, the regulatory strategies and effects of these regions lack theoretical and experimental quantitative validation [12]. More importantly, these stimulation strategies are highly dependent on the clinician's experience rather than systematic quantitative design [13]. There is an urgent need to model optimal stimulation strategies to provide personalized treatment plans for patients [14].

The human brain is a complex network-coupled dynamical system that the information transits within the network [15]. Due to the physical constraints of directly stimulating the target area, we aim to develop a feasible and optimal control strategy at the network level. In the past few years, the network control theory has provided a theoretical framework for the understanding of the controllability of specific nodes [16,17]. Therefore, from a network perspective, especially in EN, it is worth studying how to select control nodes and whether the optimal control strategy applied to the selected control nodes can effectively suppress the capture dynamics.

In this study, we aim to design effective control strategies based on network control theory to suppress epileptic seizures in neural activity. To this end, we use a linear dynamical model to reconstruct the neural dynamics of seizure propagation, enabling it to serve as a virtual platform for accurately predicting epileptic activities. Based on this surrogate model, combined with Model Predictive Control (MPC) theory, we calculate the optimal control strategies for different control nodes. The main contributions of this study are summarized as follows.

- We present an optimal control framework for seizure suppression, which contains a data-driven system identification module and a control strategy design module (Fig. 1).
- We demonstrate that the control strategy on the hub of the epileptic brain network, as well as on the SOZ, can effectively suppress seizures with evidence from simulation (Fig. 3) and from the real iEEG data (Fig. 4).

2 Related Work

Some electrical stimulation techniques have been used for seizure suppression, including deep brain stimulation (DBS) [18], responsive neurostimulation (RNS) [19], vagus nerve stimulation (VNS) [19] and some non-invasive transcranial stimulations [20,21]. Past research has employed empirical stimulation parameters for long-term stimulation in epilepsy patients [22,23]. A major challenge to applying electrical stimulation for seizure suppression lies in choosing the stimulation sites and designing the optimal stimulation parameters. Recent studies have proposed the Koopman-MPC framework [24] and unscented Kalman filter [25] to optimize the stimulation parameters, such as frequency, intensity, and timing of stimulation. However, theoretical and computational work on choosing stimulation sites is still missing.

Clinically, common targets for epilepsy electrical stimulation include the cerebellum, centromedian thalamus, hippocampus, anterior nucleus of the thalamus, motor cortex, caudate, subthalamic nucleus, and the epileptogenic foci [26]. Currently, the selection of stimulation targets is predominantly based on the typical regions of epileptic lesions, a method that lacks objective quantitative metrics and heavily relies on the physician's experiential judgment, making it challenging to tailor personalized treatment strategies for different patients. Therefore, there is a critical need for the development of objective quantitative indices to assess different stimulation nodes quantitatively, thereby optimizing stimulation strategies further.

The consumption of energy during state transition with brain network control theory characterizes the feasibility or difficulty of reaching the target state. Especially in the feasibility of state transition of cognitive function [16,17,27]. Gu et al. proposed a brain network control strategy, modeling the brain using a linear dynamical model to calculate the energy cost of transitions between different brain states from a network perspective [17]. By quantitatively calculating the optimal energy input required for specific nodes, the nodes that have stronger driving forces on brain states can be identified [27]. Some other studies have applied network control theory to understand the worm movement [28], as well as to guide electrical stimulation [29].

3 Method

In this section, we introduce our framework for suppressing seizures via electrical stimulation to the hub of the epileptic brain network (see Fig. 1). The framework is composed of two parts: (1) Learning the surrogate neural dynamics of seizure-like activity (**System Identification**). (2) Selecting the control node and calculating the optimal stimulation policy to suppress seizure (**Control Strategy**).

3.1 Problem Definition

To address the control problem in suppressing seizure dynamics, we need to design an accurate prediction model and a valid control strategy. We treat the neural dynamics of seizure propagation as a dynamical system, which can be approximated with a linear dynamical system or more advanced surrogate models in a data-driven fashion (Fig. 1b). The reconstructed dynamics require the capability of fast online learning and accurate prediction. The control strategy is designed under the guidelines of the surrogate dynamics with the consideration of control node selection and the corresponding optimal stimulation (Fig. 1c). Meanwhile, we need to consider the safety of electrical stimulation. We introduce the range of minimum or maximum stimulation parameters in the optimization problem of obtaining the optimal control strategy.

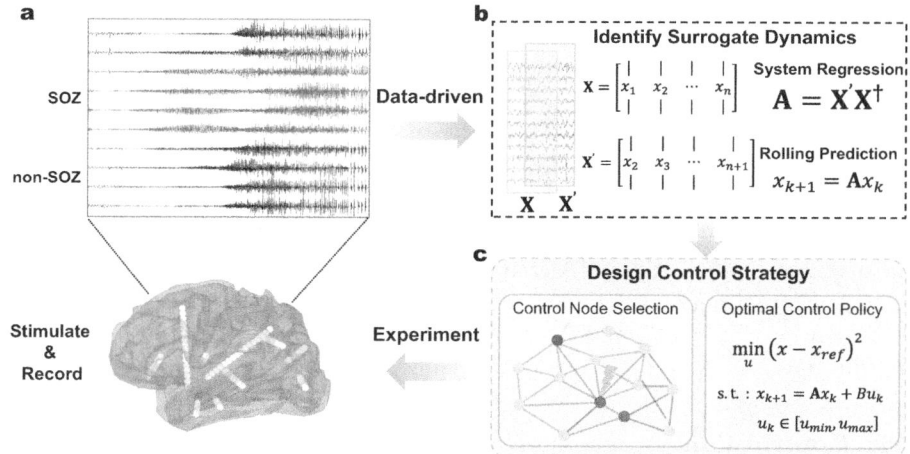

Fig. 1. The optimal control framework for seizure suppression, including a system identification module and an optimal control strategy module. **a**, Brain nodes in SOZ (red lines) and non-SOZ (black lines) of a subject are pre-determined by clinicians and the real iEEG data is from the Epilepsy-iEEG-Multicenter-Dataset. **b**, Identifying the system dynamics via reconstructing the neural dynamics. **c**, Designning the optimal control strategy, considering two factors: control node selection (left) and the optimal control policy (right). The dark nodes indicate potential control nodes. The optimal inputs u to the control nodes are designed by minimizing the error between the surrogate dynamics and the reference dynamics and control energy.

3.2 Dynamic Mode Decomposition in Identifying Seizure Dynamics

Dynamic mode decomposition (DMD) has been widely applied in extracting the spatio-temporal dynamic modes of resting state and task-related fMRI BOLD signals [30] or predicting behaviour [31]. In our study, to investigate the local neural dynamics from data, we applied DMD to reconstruct seizure dynamics, as shown in Fig. 1b.

From the observed seizure data, we can obtain a sequence of vector data $\{x_1, x_2, \ldots, x_n, x_{n+1}\}$, where $x_i \in \mathcal{R}^m$ representing the activity of m recording electrodes at time step i. We can construct the following two matrices X and X' with:

$$X = \begin{bmatrix} | & | & & | \\ x_1 & x_2 & \cdots & x_n \\ | & | & & | \end{bmatrix} \quad \text{and} \quad X' = \begin{bmatrix} | & | & & | \\ x_2 & x_3 & \cdots & x_{n+1} \\ | & | & & | \end{bmatrix} \quad (1)$$

where X and $X' \in \mathcal{R}^{m \times n}$, n is the length of time window (Fig. 1b).

We can identify the seizure dynamics by using the DMD algorithm, which aims to identify the linear operator A by approximating the linear dynamics of

$$x_{t+1} = Ax_t, \tag{2}$$

written in terms of these data matrices as

$$X' \approx AX. \tag{3}$$

The best fit of linear operator A can be solved by minimizing the Frobenius norm of $\|X' - AX\|_F$:

$$A = X'X^\dagger \tag{4}$$

where $X^\dagger$ is the pseudo-inverse of X.

In DMD theory, to address potential high-dimensional problems and perform dimension reduction, X is subjected to Singular Value Decomposition with $X = U\Sigma V^*$, where $*$ denotes complex-conjugate transpose. Then the pseudo-inverse of X is $X^\dagger = V\Sigma^{-1}U^*$, and $X' = AU\Sigma V^*$, where $U \in R^{m \times m}$ and $V \in R^{n \times n}$ are unitary.

3.3 Control Policy

Model predictive control is an optimization-based control framework. It minimizes the objective function in a finite prediction horizon with the control inputs and states transition constraints. In this study, we design the control strategy by considering the control node selection and its optimal control policy under the model predictive control framework. The standard objective function of MPC is defined as follows,

$$\min_{u} \sum_{t=1}^{T_p} (x_t - x_{ref})^T Q_x (x_t - x_{ref}) + \sum_{t=1}^{T_c} u_t^T Q_u u_t \tag{5}$$
$$\text{s.t.} \quad x_{t+1} = f(x_t, u_t)$$
$$u_t \in [u_{\min}, u_{\max}]$$

where x_t represents the system state and x_{ref} is the reference signal, u_t is the input of the system, Q_x is the positive definite weighted matrix for penalizing the deviance and Q_u is a non-negative matrix for regularizing the amplitudes of control inputs. T_p and T_c are the length of the predictive horizon and control horizon, respectively.

In control engineering, the method to solve the optimization problem in Eq. (5) is chosen based on the system dynamics $f(\cdot)$. If $f(\cdot)$ is a nonlinear function of x_t and u_t, it becomes a nonlinear model predictive control problem and can be solved by a nonlinear programming solver with the primal-dual interior method. If $f(\cdot)$ is a linear function of x_t and u_t (i.e., $x_{t+1} = Ax_t + Bu_t$), the optimization problem can be solved by a linear quadratic programming solver directly. In both cases, the optimization problem in Eq. (5) can also be solved by other gradient descent methods (i.e., Adam, SGD).

3.4 Simulation Platform for Seizure Propagation

We initially applied the network-coupled Jansen-Rit model to simulate the dynamics of seizure propagation. The standard Jansen-Rit model delineates a cortical column that is composed of the complex functional interaction among excitatory, inhibitory interneurons, and pyramidal neurons (Fig. 2a) [32]. As shown in Fig. 2a, the functional dynamics of these three populations contain two modules:

(1) The conversion of average pulse density into average postsynaptic membrane potential (PSP), which is described as the impulse response function for both excitatory and inhibitory interneurons:

$$h_E(t) = \begin{cases} Aate^{-at}, & t \geq 0 \\ 0, & t < 0, \end{cases} \quad (6)$$

and

$$h_I(t) = \begin{cases} Bbte^{-bt}, & t \geq 0 \\ 0, & t < 0, \end{cases} \quad (7)$$

where the constants A and B determine the maximum amplitude of the PSPs for excitatory (EPSPs) and inhibitory (IPSPs) responses respectively; a and b are the inverse time constants for the excitatory and inhibitory postsynaptic potentials respectively.

(2) Then it translates the postsynaptic membrane potentials back into average pulse density, characterized by a sigmoid function:

$$S(v, r) = \frac{\zeta_{max}}{1 + e^{r(\theta - v)}}, \quad (8)$$

where ζ_{max} denotes the maximum firing rate, r the slope of the function, and θ the half maximal response of the neuron population. Additionally, pyramidal neurons are stimulated externally by $p(t)$.

Considering the information conversion with other interconnection cortical columns, we can model the neural dynamics of each cortical column as

$$\dot{x}_{0,i}(t) = y_{0,i}(t)$$
$$\dot{y}_{0,i}(t) = Aa\left[S(x_{1,i}(t) - x_{2,i}(t))\right] - 2ay_{0,i}(t) - a^2 x_{0,i}(t)$$
$$\dot{x}_{1,i}(t) = y_{1,i}(t) + u_i(t))$$
$$\dot{y}_{1,i}(t) = Aa\left[p(t) + C_2 S(C_1 x_{0,i}(t) + g \sum_{j=1}^{N} M_{i,j} x_{3,j}\right] - 2ay_{1,i}(t) - a^2 x_{1,i}(t)$$
$$\dot{x}_{2,i}(t) = y_{2,i}(t)$$
$$\dot{y}_{2,i}(t) = Bb\left[C_4 S(C_3 x_{0,i}(t), r_2)\right] - 2by_{2,i}(t) - b^2 x_{2,i}(t)$$
$$\dot{x}_{3,i}(t) = y_{3,i}(t)$$
$$\dot{y}_{3,i}(t) = A\bar{a}\left[S(x_{1,i}(t) - x_{2,i}(t))\right] - 2\bar{a}y_{3,i}(t) - \bar{a}_i^2 x_{3,i}(t)$$
$$(9)$$

where x_0, x_1, and x_2 represent the outputs of the PSP blocks for pyramidal, excitatory, and inhibitory neurons, respectively. x_3 is the output for long-range interactions of pyramidal neurons. The constants C_1, C_2, C_3, and C_4 scale the connectivity among the neural populations. $M_{i,j}$ is the structural connections between cortical column i and j (Fig. 2b). g is the strength of global connections. $u_i(t)$ is the input of node i. The detailed information about the Jansen-Rit neural dynamics is summarized in study [32,33].

Considering that the hippocampus is a common SOZ in focal epilepsy, in our simulation we set the hippocampus as SOZ (Fig. 2a, bottom). We set the amplitude of PSPs for excitatory interneurons with $A = 7.8$ for a seizure onset zone and $A = 6.8$ for others. The seizure dynamics propagate to the whole brain through the structural connection (Fig. 2d). The high-frequency abnormal discharges are shown in the SOZ, and the low-frequency dynamics occur in the propagation regions with strong SC (Fig. 2d).

3.5 Real iEEG Data from an Epileptic Patient

To study whether the control strategy performs well on the real data, we conduct experiments on the real neural dynamical data from an epileptic patient. The real iEEG data is from an open dataset (Epilepsy-iEEG-Multicenter-Dataset,https:// openneuro.org/datasets/ds004100/versions/1.1.3). Each subject recorded iEEG signals for multiple complete seizure events (from 120 s before seizure onset to 60 s after seizure termination) at a sampling rate exceeding 512 Hz. To keep consistent with the JR simulation, we selected the patient (i.e., Subject HUP-146) with SOZ located in the hippocampus and with successful surgical outcomes, implying that the SOZ was correctly labeled by medical experts. In total 122-channel iEEG signals were recorded from this patient (Fig. 1 a).

To be noted, we did not directly apply electrical stimulation to the patient. Instead, we first learn the parameters of the dynamical model in a data-driven fashion, by fitting the model with the real iEEG data. Then we conduct the virtual stimulation experiment on the identified dynamical model, identical to the experiments run on the simulation platform.

3.6 Experiment Protocol

The selection of control nodes determines the control matrix B. For the control experiments on the simulation platform, we consider 4 types of control nodes to apply virtual stimulation, including (1) *SOZ hub*, i.e., hippocampus, (2) *SC hub*, the hub node with the strongest structural connection to SOZ, (3) *FC hub*, the hub node with the strongest functional connectivity with SOZ, (4) *Top weighted degree*, the node that connects to SOZ with the highest degree. Specifically, in the simulation platform, the hippocampus in the left hemisphere (LH) is the defined SOZ, the thalamus in LH is the hub node with the strongest structural connection and functional connection to the hippocampus, and the precuneus in RH has the highest degree of the whole brain and connects to SOZ. Towards these 4 types of control nodes in simulation, we solved the corresponding control problem formulated in Eq. (5) separately. The optimized control strategy u^* of Eq. (5) was then applied to the network-coupled Jansen-Rit dynamics.

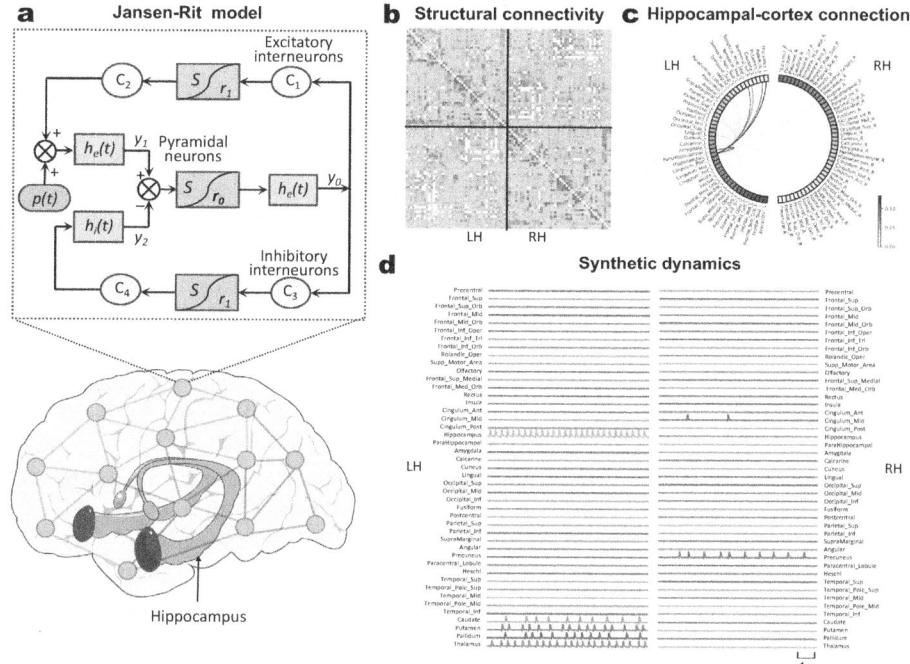

Fig. 2. A simulation of the propagation process of seizure dynamics on the whole brain. **a**, The neural dynamics of each node in the whole brain can be simplified as Jansen-Rit dynamics. The neural dynamics between interconnected regions are propagated through the white matter fiber connections. **b**, An example of the whole brain structural connectivity. **c**, The connection that links to the hippocampus. **d**, The simulated neural dynamics of the whole brain. The hippocampus is predefined as an SOZ with high-frequency abnormal discharge, while other regions with abnormal discharge are considered as propagation regions. The regions with normal discharge are non-seizure regions.

For the control experiment on the dynamical model from the real iEEG data, we consider 3 types of control nodes, including (1) *SOZ*, in total 11 nodes which are predefined by medical experts, (2) *FC hub*, the node with the highest FC with SOZ, (3) *Top weighted degree*, the top-11 highest degrees of FC. The covariance-based functional connectivity is estimated on the first 2 s of iEEG during the ictal period. In our experiment, it is not a direct electrical stimulation to the patient. The dimensionless variable $u(t)$ in the surrogate dynamics $x(t+1) = Ax(t) + Bu(t)$ has no physical meaning. We can obtain the optimized u^* of Eq. (5) iteratively. The optimized control strategy was applied to the surrogate system learned from iEEG data, rather than the real brain.

4 Results

4.1 Suppressing Seizure on Simulation Platform

In the simulation platform, we validate the sufficiency of seizure suppression with the hub control node on a network-coupled Jansen-Rit dynamics. We select one of the cortical columns (Hippocampus, with $A = 7.8$ in Eq. (2)) as a SOZ to simulate the seizure propagation process. The neural dynamics are propagated through the white matter fiber tracts (the structural connections are projected in the AAL atlas). We also define the reference dynamics with $A = 6.8$ in Eq. (2) of the whole brain (Fig. 3 a).

We simulate the neural dynamics for 2.5 s with a 100 Hz sampling rate (Fig. 3 a). We inserted inputs on the 4 types of pre-selected control nodes respectively (Fig. 3 b). The results show that the optimal control strategy on SOZ, SC and FC hub have better control effects in suppressing seizure propagation than the node with the highest degree. When inserting optimal strategy on the hippocampus, both seizure-like activity on SOZ and PZ are suppressed. The optimal strategy on the thalamus suppressed the seizure-like activity on PZ and weakened the activity on SOZ.

To quantify the control effect, we further calculate the average absolute error between neural dynamics and reference after applying optimal stimulation. As well as the decrease in firing rate, *i.e.*, the ratio of the time points where the amplitude is suppressed to the total number of time points with high amplitude during epileptic discharge. It can be observed that when stimulating the hippocampus, the tracking error is the smallest and the inhibition rate of discharge reaches 100%.

The control effects on stimulating the thalamus showed a smaller tracking error with a global decrease rate of 57%. The last type of control nodes have almost no control effect on suppressing epilepsy (see Fig. 3 c). Considering the safety of electrical stimulation, to achieve the effect of reducing epileptic seizures while requiring lower power stimulation, we analyzed the control energy in Fig. 3 d. The result shows that the energy of electrical stimulation applied to the hippocampus and the thalamus is much lower than the Precuneus. Interestingly, we found that the optimal input for the thalamus appears as a periodic pulse sequence. Further spectral analysis revealed that the primary frequency is concentrated around 5 Hz (see Fig. 3 e). This low-frequency, low-energy stimulation pattern may be valuable for designing electrical stimulation in real-world applications.

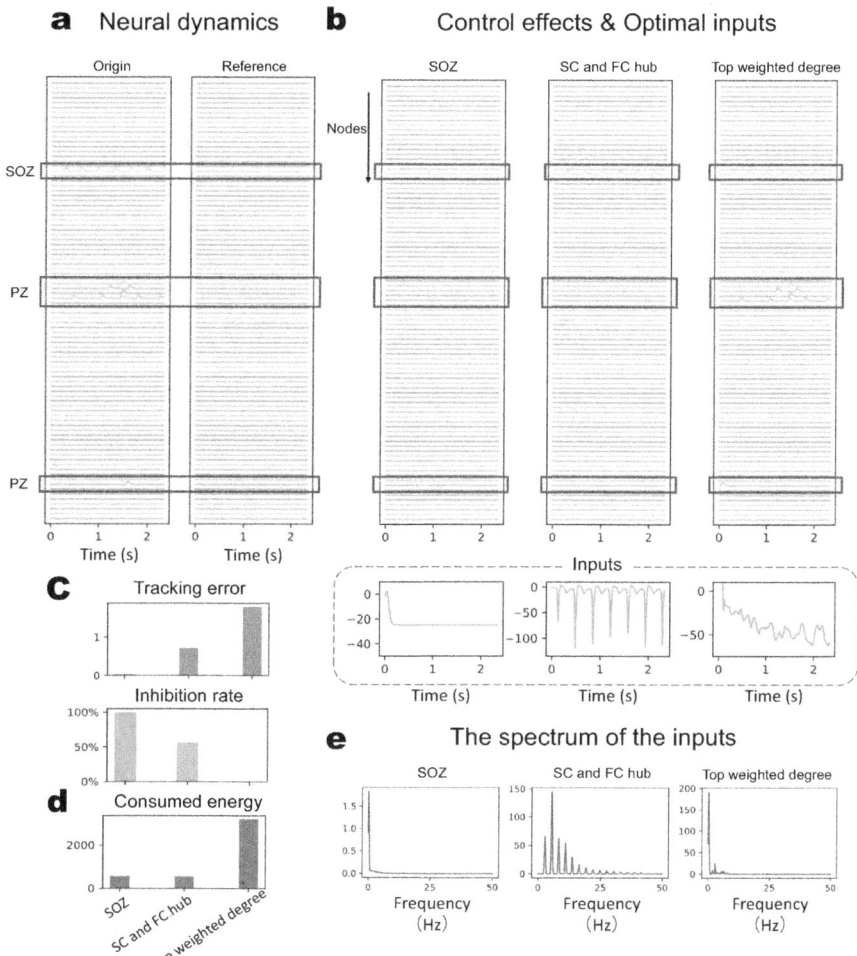

Fig. 3. Control effects on the JR simulation platform. We input the optimized electrical stimulation to the pre-selected control node and illustrate the control effects. **a**, The original seizure dynamics (left) and reference neural dynamics (right). **b**, Control effects (top) and the optimal inputs (bottom). From left to right, the control node is set to the hippocampus (SOZ), thalamus (SC and FC hub), and precuneus (Top weighted degree), respectively. The results show that stimulating the hippocampus or thalamus can successfully suppress seizure propagation. **c**, The mean absolute error of system dynamics and reference dynamics. **d**, The average energy consumed to control the system. **e**, The spectrum of optimal inputs on each pre-selected node.

4.2 Suppressing Real Seizure Dynamics

We further validate the efficiency of seizure suppression on real data. We first validate the prediction performance of the DMD framework $x(t+1) = Ax(t)$ in predicting seizure dynamics. During the experiment, the hyperparameter of the length of the time window is $time\_length = 512$, and the length of prediction is $n\_horizon = 5$. In Fig. 4 a, the result shows sufficiency prediction accuracy on real data with explained variance $EV = 0.771$. Then we conduct the virtual stimulation experiment on the identified system with the consideration of control node selection and designing optimal control strategy. Control node selection determines the control matrix B, which is a binary diagonal matrix with the index controlled or not.

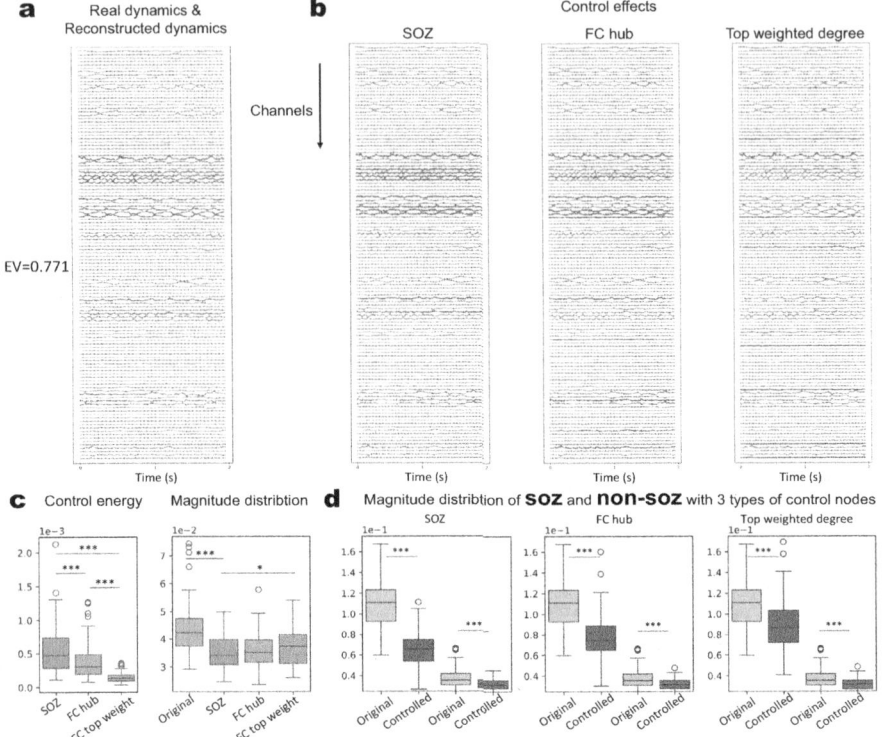

Fig. 4. Control effects on real data. We input the optimized virtual electrical stimulation to the pre-selected control node and illustrate the control effects on the identified model $x_{k+1} = Ax_k + Bu_k$. **a**, The original seizure dynamics (gray line) and the reconstructed neural dynamics (blue dot line). **b**, Control effects with optimal inputs to three types of control nodes (left: SOZ, middle: FC hub, right: top degree of FC). **c**, Distribution of control energy (left) and magnitude (right). **d**, Magnitude distribution of soz (blue-paired) and non-soz (green-paired) with 3 types of control nodes. (∗ for p-value< 0.05 and ∗ ∗ ∗ for p-value$< 1e^{-4}$)

Then we conduct experiments equipped with the system identification module on real data. The control effects of stimulating 3 types of control nodes (i.e.,

SOZ, FC hub, FC top weight) are shown in Fig. 4 b respectively. The results show that both SOZ and FC hubs can suppress seizures. To quantify the control effects, we calculated the energy of the input sequences and the amplitude of the neural signals influenced by the inputs, respectively. The quantitative results are shown in Fig. 4c. We show that the control to SOZ requires the highest input energy, while it has the maximal neural magnitude reduction, indicating effective suppression of seizures. On the contrary, the inputs to FC top weight are inefficient in suppressing neural activity. These two control nodes showed significant statistical differences in both control energy and amplitude distribution (p-value$< 1e^{-4}$ of two-sided t-test). There are no significant differences between controlling SOZ and controlling the FC hub in amplitude reduction of neural activity.

Furthermore, we compared the control effects on SOZ and non-SOZ given the optimal inputs with 3 types of control node selection, as shown in Fig. 4 d. The results show that both the SOZ and non-SOZ decrease the amplitude of neural activity compared with the original neural activity with significant statistical differences (p-value$< 10^{-4}$ of two-sided t-test). However, the control effects with optimal inputs on control nodes that are related to SOZ are sufficient to suppress seizure dynamics.

5 Discussion and Conclusion

From Methodology Perspective. Our study presented a general platform based on system identification and brain network control theory for suppressing seizure dynamics. We applied the dynamic mode decomposition to iteratively update the neural dynamics online, which guarantees high predictive performance in reconstructing neural dynamics. The platform was validated with both synthetic data and real iEEG data from the Epilepsy-iEEG-Multicenter-Dataset. With the high predictive performance, we further conduct model predictive control with the identified surrogate model in inferring the optimal control strategy under the consideration of different types of control nodes. However, the sparsity of sEEG affects the system identification of the whole brain. Lou et al. proposed the framework of fusion multi-modal neural dynamics to reconstruct the whole brain neural dynamics using sEEG-EEG data, which can improve the performance of system identification [34].

From Clinical Perspective. Validating the effectiveness of different stimulation strategies on neural treatment is the core question in clinical neuroscience. Our work delves into the optimal electrical stimulation to the hub of the epileptic brain network, focusing on two key aspects: control node selection and inferring optimal inputs. Selecting the control node enables the validation of the feasibility of conducting the electrical stimulation to reach target states. Moreover, the optimal inputs are key to realizing the objective with less consumed energy.

In summary, we proposed a platform from the control theory perspective for validating the efficiency of different control strategies (control node selection) in suppressing seizures. The hub of the epileptic brain network serves as the alternative control node for realizing seizure suppression.

Acknowledgement. The authors gratefully acknowledge Prof. Xiang Liao, Dr. Liang Chen and Dr. Chen Yao for their insightful discussions, comments and suggestions. This work was funded in part by the National Key R&D Program of China (2021YFF1200804), Shenzhen Science and Technology Innovation Committee (2022410129, KCXFZ20201221173400001, KJZD20230923115221044).

References

1. Laufs, H.: Functional imaging of seizures and epilepsy: evolution from zones to networks. Curr. Opin. Neurol. **25**(2), 194–200 (2012)
2. Bartolomei, F., Guye, M., Wendling, F.: Abnormal binding and disruption in large scale networks involved in human partial seizures. EPJ Nonlinear Biomed. Phys. **1**(1), 1–16 (2013). https://doi.org/10.1140/epjnbp11
3. Zaveri, H.P., Pincus, S.M., Goncharova, I.I., Duckrow, R.B., Spencer, S.S.: Large scale brain networks in epilepsy. In: Advanced Signal Processing Algorithms, Architectures, and Implementations XVIII, vol. 7074, pp. 250–259. SPIE (2008)
4. Bartolomei, F., Lagarde, S., Wendling, F., McGonigal, A., Jirsa, V., Guye, M., Bénar, C.: Defining epileptogenic networks: contribution of seeg and signal analysis. Epilepsia **58**(7), 1131–1147 (2017)
5. Wang, H.E., et al.: Delineating epileptogenic networks using brain imaging data and personalized modeling in drug-resistant epilepsy. Sci. Transl. Med. **680**, eabp8982 (2023)
6. Burns, S.P., et al.: Network dynamics of the brain and influence of the epileptic seizure onset zone. Proc. Natl. Acad. Sci. **111**(49), E5321–E5330 (2014)
7. Rocha, L.L., Lazarowski, A., Cavalheiro, E.A.: Pharmacoresistance in Epilepsy: From Genes and Molecules to Promising Therapies. Springer, Heidelberg (2023). https://doi.org/10.1007/978-3-031-36526-3
8. de Castro Medeiros, D., Moraes, M.F.D.: Focus on desynchronization rather than excitability: a new strategy for intraencephalic electrical stimulation. Epilepsy Behav. **38**, 32–36 (2014)
9. Tao, Yu., et al.: High-frequency stimulation of anterior nucleus of thalamus desynchronizes epileptic network in humans. Brain **141**(9), 2631–2643 (2018)
10. Scherer, M., et al.: Desynchronization of temporal lobe theta-band activity during effective anterior thalamus deep brain stimulation in epilepsy. Neuroimage **218**, 116967 (2020)
11. Sisterson, N.D., Kokkinos, V.: Neuromodulation of epilepsy networks. Neurosurg. Clin. **31**(3), 459–470 (2020)
12. Schaper, F.L., et al.: Mapping lesion-related epilepsy to a human brain network. JAMA Neurol. **80**(9), 891–902 (2023)
13. Piper, R.J., et al.: Towards network-guided neuromodulation for epilepsy. Brain **145**(10), 3347–3362 (2022)
14. Xia, Z., Li, W., Liang, Z., Lou, K., Liu, O.: Controlling network-coupled neural dynamics with nonlinear network control theory. arXiv preprint arXiv:2405.06971 (2024)

15. Luo, Z., Liang, Z., Xu, C., Zhou, C., Liu, Q.: Mapping the whole-brain effective connectome with excitatory-inhibitory causal relationship. arXiv preprint arXiv:2301.00148 (2022)
16. Gu, S., et al.: Controllability of structural brain networks. Nat. Commun. **6**(1), 8414 (2015)
17. Gu, S., et al.: Optimal trajectories of brain state transitions. NeuroImage **148**, 305–317 (2017)
18. Halpern, C.H., Samadani, U., Litt, B., Jaggi, J.L., Baltuch, G.H.: Deep brain stimulation for epilepsy. In: Neuromodulation, pp. 639–649. Elsevier (2009)
19. Morrell, M.J.: Responsive cortical stimulation for the treatment of medically intractable partial epilepsy. Neurology **77**(13), 1295–1304 (2011)
20. Simula, S., et al.: Transcranial current stimulation in epilepsy: a systematic review of the fundamental and clinical aspects. Front. Neurosci. **16**, 909421 (2022)
21. Wang, M., Lou, K., Liu, Z., Wei, P., Liu, Q.: Multi-objective optimization via evolutionary algorithm (movea) for high-definition transcranial electrical stimulation of the human brain. Neuroimage **280**, 120331 (2023)
22. Roa, J.A., et al.: Long-term outcomes after responsive neurostimulation for treatment of refractory epilepsy: a single-center experience of 100 cases. J. Neurosurg. **1**(aop), 1–8 (2023)
23. Khambhati, A.N., Shafi, A., Rao, V.R., Chang, E.F.: Long-term brain network reorganization predicts responsive neurostimulation outcomes for focal epilepsy. Sci. Transl. Med. **13**(608), eabf6588 (2021)
24. Liang, Z., Luo, Z., Liu, K., Qiu, J., Liu, Q.: Online learning koopman operator for closed-loop electrical neurostimulation in epilepsy. IEEE J. Biomed. Health Inf. **27**(1), 492–503 (2022)
25. Chang, S., et al.: A data driven experimental system for individualized brain stimulation design and validation. IEEE Trans. Neural Syst. Rehabil. Eng. **29**, 1848–1857 (2021)
26. Fisher, R.S., Velasco, A.L.: Electrical brain stimulation for epilepsy. Nat. Rev. Neurol. **10**(5), 261–270 (2014)
27. Liang, Z., Zhang, Y., Wu, J., Liu, Q.: Reverse engineering the brain input: network control theory to identify cognitive task-related control nodes. arXiv preprint arXiv:2404.16357 (2024)
28. Yan, G., et al.: Network control principles predict neuron function in the caenorhabditis elegans connectome. Nature **550**(7677), 519–523 (2017)
29. Stiso, J., et al.: White matter network architecture guides direct electrical stimulation through optimal state transitions. Cell Rep. **28**(10), 2554–2566 (2019)
30. Casorso, J., Kong, X., Chi, W., Van De Ville, D., Yeo, B., Liégeois, R.: Dynamic mode decomposition of resting-state and task fmri. NeuroImage **194**, 42–54 (2019)
31. Ikeda, S., Kawano, K., Watanabe, S., Yamashita, O., Kawahara, Y.: Predicting behavior through dynamic modes in resting-state fmri data. Neuroimage **247**, 118801 (2022)
32. Jansen, B.H., Rit, V.G.: Electroencephalogram and visual evoked potential generation in a mathematical model of coupled cortical columns. Biol. Cybern., 357–366 (1995)

33. Coronel-Oliveros, C., Cofré, R., Orio, P.: Cholinergic neuromodulation of inhibitory interneurons facilitates functional integration in whole-brain models. PLoS Comput. Biol. **17**(2), e1008737 (2021)
34. Lou, K., Li, J., Barth, M., Liu, Q.: A data-driven framework for whole-brain network modeling with simultaneous eeg-seeg data. In: International Conference on Intelligent Information Processing, pp. 329–342. Springer, Heidelberg (2024). https://doi.org/10.1007/978-3-031-57808-3_24

AI for Brain Technology

SVFormer: A Direct Training Spiking Transformer for Efficient Video Action Recognition

Liutao Yu[1], Liwei Huang[1,2], Chenlin Zhou[1], Han Zhang[1,3], Zhengyu Ma[1(✉)], Huihui Zhou[1(✉)], and Yonghong Tian[1,2]

[1] AI Department, Pengcheng Laboratory, Shenzhen, China
{yult,zhoucl,mazhy,zhouhh}@pcl.ac.cn
[2] National Key Laboratory for Multimedia Information Processing, School of Computer Science, Peking University, Beijing, China
huanglw20@stu.pku.edu.cn, yhtian@pku.edu.cn
[3] Faculty of Computing, Harbin Institute of Technology, Harbin, China
23B303002@stu.hit.edu.cn

Abstract. Video action recognition (VAR) plays crucial roles in various domains such as surveillance, healthcare, and industrial automation, making it highly significant for the society. Consequently, it has long been a research spot in the computer vision field. As artificial neural networks (ANNs) are flourishing, convolution neural networks (CNNs), including 2D-CNNs and 3D-CNNs, as well as variants of the vision transformer (ViT), have shown impressive performance on VAR. However, they usually demand huge computational cost due to the large data volume and heavy information redundancy introduced by the temporal dimension. To address this challenge, some researchers have turned to brain-inspired spiking neural networks (SNNs), such as recurrent SNNs and ANN-converted SNNs, leveraging their inherent temporal dynamics and energy efficiency. Yet, current SNNs for VAR also encounter limitations, such as nontrivial input preprocessing, intricate network construction/training, and the need for repetitive processing of the same video clip, hindering their practical deployment. In this study, we innovatively propose the directly trained SVFormer (Spiking Video transFormer) for VAR. SVFormer integrates local feature extraction, global self-attention, and the intrinsic dynamics, sparsity, and spike-driven nature of SNNs, to efficiently and effectively extract spatio-temporal features. We evaluate SVFormer on two RGB datasets (UCF101, NTU-RGBD60) and one neuromorphic dataset (DVS128-Gesture), demonstrating comparable performance to the mainstream models in a more efficient way. Notably, SVFormer achieves a top-1 accuracy of 84.03% with ultra-low power consumption (21 mJ/video) on UCF101, which is state-of-the-art among directly trained deep SNNs, showcasing significant advantages over prior models.

Keywords: Video action recognition · Spiking transformer · SVFormer · Direct training · Energy efficiency

1 Introduction

Video is becoming a prevalent and indispensable medium to convey information in daily life, which captures movements, actions, and events over time. Video action recognition (VAR) is an important aspect of video understanding, focusing on automatically identifying actions or activities from videos. This capability is invaluable for various applications, including surveillance, healthcare, entertainment, sports, education, industrial automation, and beyond. Nevertheless, video processing presents greater challenges compared to images, given the necessity to model temporal dynamics within large data volume and heavy information redundancy introduced by the temporal dimension.

As artificial neural networks (ANNs) have shown great success in various computer vision tasks, a lot of studies based on convolutional neural networks (CNNs) [9] or vision transformers (ViTs) [46] have emerged for VAR in recent years. In the early stages, CNNs are the mainstream approaches to VAR, including 2D-CNNs [13,56,68] and 3D-CNNs [7,22,52]. The decomposition of 3D-CNNs [42,53,60], as well as the combination of 2D-CNNs and 3D-CNNs [17], are adopted to reduce the computation cost. However, CNNs struggle to learn long-range dependency between patches, due to the inductive bias of convolution. To overcome this challenge, network models based on ViTs are becoming popular and showing good performance for VAR in recent years [46], including TimeSformer [3], ViViT [2], MViT [12], video swin transformer [35], and so on. To further improve the performance, methods combining convolution and self-attention [33,34,36], or using self-supervised pretraining [51,55,59] are proposed. Nevertheless, the huge computation cost of current ANNs for VAR still limits their practical deployment, especially in power-constrained situations.

The brain-inspired spiking neural networks (SNNs) have garnered significant attention for their potential in temporal information processing and energy efficiency [44], thus are ideal candidates for processing videos. Previous studies show that SNNs exhibit good performance on various tasks, such as object recognition/detection/tracking, robotics control and so on [20,61,70]. In recent years, recurrent SNNs (RSNNs) [8,41] and ANN-converted SNNs [65,67] have been applied to VAR. However, they also encounter limitations in practical application: RSNNs usually need nontrivial input preprocessing and network construction/training, as well as long simulation steps [8,41]; ANN-converted SNNs are based on well-trained ANNs, and need to process the same video clip several times to accomplish the task [65,67].

In this study, we innovatively propose the directly trained SVFormer (Spiking Video transFormer) for VAR, aiming to address the challenges outlined above. SVFormer processes a video clip frame-by-frame without the need for complex input processing, and can be trained end-to-end through the surrogate gradient method, enabling straightforward incremental learning and facilitating practical deployment. SVFormer integrates local feature extraction, global self-attention, and the intrinsic dynamics, sparsity, and spike-driven nature of SNNs, to efficiently and effectively extract spatio-temporal features. Besides, we incorporate parametric LIF neuron [16], a local-global-fusion operation and a novel time-independent batch normalization into SVFormer, contributing to its good performance. We first evaluate SVFormer on the classical UCF101 dataset, and achieve a top-1 accuracy of 84.03% with ultra-low power consumption (21 mJ/video), which is state-of-the-art among directly trained deep SNNs, showing significant advantages over previous models. To validate its generalizability, SVFormer is then evaluated on a larger RGB datasets (NTU-RGBD60) and a neuromorphic dataset (DVS128-Gesture), both demonstrating comparable performance to the mainstream models in a more efficient way. These results showcase that the directly trained SVFormer is an effective and efficient model for VAR.

2 Related Work

2.1 ANNs for VAR

In the early stages, CNNs including 2D-CNNs and 3D-CNNs, are the mainstream approaches to VAR [9]. Some studies applied two-stream models to process RGB frames and optical flows in two separate CNNs, with a late fusion operation in deeper layers [10,18,19,27]. Another line of approach adopts 2D-CNNs to extract frame-level features, and then models temporal information of the input sequence in various ways, such as the consensus module in TSN [56], the bag of features in TRN [68], the pointwise convolutions across frames in TAM [13], and so on. Moreover, to better model the spatio-temporal features, many variants of 3D-CNNs have been introduced, such as C3D [52], I3D [7] and ResNet3D [22]. To mitigate the high computational burden brought by 3D convolutions, some studies tried to decompose the 3D convolution into 2D spatial convolution and 1D temporal convolution, such as P3D [42], S3D [60] and R(2+1)D [53]; or to use a combination of 2D-CNNs and 3D-CNNs, such as SlowFast [17]. However, CNNs mainly extract local features due to limited receptive fields, ignoring long-range dependency across patches.

Recently, ViTs outperform CNNs in many visual tasks due to their enhanced ability to capture long-range dependencies [11,21]. Some researchers proposed different variants based on ViTs for VAR [46]. The classical TimeSformer enables spatio-temporal feature extraction directly from a sequence of frame-level patches, and explores different space-time self-attention schemes [3]. ViViT is a pure-transformer based model, which factorizes the spatial- and temporal-dimensions of the input to handle the long sequences of tokens [2]. To reduce the computation cost, Fan et al. introduced a multi-scale pyramid of features and a pooling self-attention mechanism in MViT [12]; and Liu et al. proposed an inductive bias of locality with shifted window-based self-attention in a video swin transformer [35]. Besides, the combination of convolution and self-attention is also adopted for efficiently learning video representations [33,34,36]. Furthermore, assisted with large-scale self-supervised pretraining, great improvements have been achieved for VAR, such as in VideoMAE [51,55] and MaskFeat [59]. Although current models have realized high classification accuracy for VAR, the excessive data volume and heavy information redundancy introduced by the temporal dimension limit their practical deployment in power-constrained situations.

2.2 SNNs for VAR

Considering the intrinsic dynamics and energy efficiency of SNNs, they are ideal candidates for VAR and have been attempted in some recent work [8,41,57,65,67]. Panda and Srinivasa [41] developed a Driven-Autonomous based reservoir recurrent

SNN to recognize video actions from limited training examples, which is not trivial to train, demonstrating a top-1 accuracy of 81.3% on the UCF101 dataset with pre-extracted multi-scan spike sequences of 300 time steps as input to the model. Further, Chakraborty and Mukhopadhyay [8] presented a heterogeneous recurrent SNN (HRSNN) with unsupervised learning for VAR on several RGB datasets (KTH, 94.32%; UCF101, 77.53%) and one neuromorphic dataset (DVS128-Gesture, 96.54%), using a similar input preprocessing method as [41] and a nontrivial initialization method. Wang et al. [57] progressively trained a two-stream hybrid network (TSRNN) consisting of CNN, RNN, and a novel spiking module, where spiking signals correct the memory of RNN, achieving competitive performance on UCF101 (94.4%) and HMDB51 (69.9%) with both RGB frames and optical flows as input. Zhang et al. [67] constructed a two-stream deep recurrent SNN model through a hybrid ANN-to-SNN conversion method combining channel-wise normalization and tandem learning, obtaining an accuracy of 88.46% on the UCF101 dataset with 200 time steps. Recently, You et al. [65] proposed an improved ANN-to-SNN conversion framework (scalable dual threshold mapping) to mitigate three types of conversion errors (unevenness error, clipping error and quantization error), and obtained ultra-low-latency SNNs based on the SlowFast model backbone, achieving high accuracy on UCF101 (92.94%) and HMDB51 (67.71%) with carefully selected hyperparameters. We observe that current SNNs for VAR have limitations for incremental training and practical application, such as complicated model construction/conversion, nontrivial input preprocessing, and long simulation time steps. Therefore, we directly train a novel spiking transformer for efficient and effective VAR in this paper, which is more suitable for practical deployment.

3 Preliminary and Methodology

In this section, we introduce the proposed SVFormer in detail. Firstly, we briefly introduce the basics of SNNs. Then, we describe the overview of SVFormer and explain the key components such as patch embedding (PE) module, local feature extractor (LFE), global self-attention (GSA) module, local pathway (LP), classification head (CH) and so on. Moreover, we introduce the method to calculate theoretical energy consumption for model inference.

3.1 Spiking Neural Networks

The brain-inspired SNNs are regarded as the third generation of neural networks [37], which innately exhibit temporal dynamics and communicate through binary spikes/events, attracting considerable attention and achieving significant advancements in recent years. SNNs offer powerful computation capability in virtue of their biological plausibility, temporal processing property with intrinsic dynamics, and energy efficiency with event-driven nature, thus considered as promising alternatives to conventional ANNs [44]. Specifically, the communication through sparse binary spikes allow SNNs to adopt low-power accumulate (AC) operations in convolution or linear layers, in place of power-hungry multiply-and-accumulate (MAC) operations when implemented on neuromorphic hardware, leading to high energy efficiency [29,30,40,64].

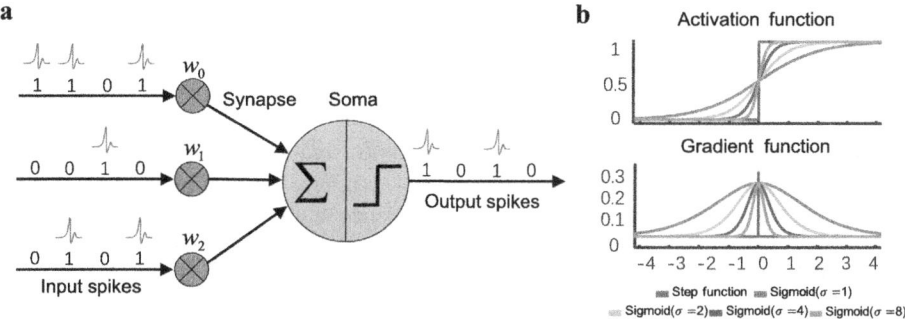

Fig. 1. (a) The scheme of a spiking neuron, of which the input and output are both binary spikes. (b) Sigmoid function approximates the Heaviside activation function of a spiking neuron, and the derivative of it can be utilized to calculate gradients during backpropagation.

The basic units of SNNs are spiking neurons, like ReLUs in ANNs. Leaky Integrate-and-Fire (LIF) neuron model (Fig. 1a) is one of the most commonly adopted neuron models in SNNs [15,63,66,69,71]. The dynamics of a LIF neuron are described as:

$$H[t] = V[t-1] + \frac{1}{\tau}\left(X[t] - (V[t-1] - V_{reset})\right), \tag{1}$$

$$S[t] = \Theta\left(H[t] - V_{th}\right), \tag{2}$$

$$V[t] = H[t]\left(1 - S[t]\right) + V_{reset}S[t], \tag{3}$$

where τ is the membrane time constant, $X[t]$ is the input current at time step t, V_{reset} is the reset potential, V_{th} is the firing threshold. Equation (1) describes the update of membrane potential. Equation (2) describes the spike generation process, where $\Theta(v)$ is the Heaviside step function: if $H[t] \geq V_{th}$ then $\Theta(v) = 1$, meaning a spike is generated; otherwise $\Theta(v) = 0$. $S[t]$ represents whether a neuron fires a spike at time step t. Equation (3) describes the resetting process of membrane potential, where $H[t]$ and $V[t]$ represent the membrane potential before and after the evaluation of spike generation at time step t, respectively. To improve the temporal representation ability of spiking neurons, trainable parameters are incorporated. For example, inspired by heterogeneous neurons in the brain, Fang et al. proposed Parametric LIF (PLIF) neuron [16] by using trainable membrane time constant as follows:

$$H[t] = V[t-1] + k(a)\left(X[t] - (V[t-1] - V_{reset})\right), \tag{4}$$

where $k(a) = \frac{1}{1+exp(-a)} \in (0,1)$, $\tau = \frac{1}{k(a)}$, and a is the trainable parameter.

There are mainly two methods to obtain well-performing deep SNNs: ANN-to-SNN conversion and direct training through surrogate gradients. In ANN-to-SNN conversion, a pretrained ANN is converted to an SNN by replacing the ReLU activation layers with spiking neurons and using scaling operations like weight normalization and threshold balancing [5,6,25,38,45,58]. To mitigate conversion error, the converted SNNs usually suffer from long simulation time steps, which causes high computational cost in practice. In direct training, SNNs are unfolded over the temporal dimension like RNNs and trained with backpropagation through time [32,39,48]. Due to the non-differentiability of the spike generation process, the surrogate gradient method is employed for backpropagation [15,16,31,39]. Specifically, the forward propagation utilizes Heaviside step function to generate spikes (Eq. (2)), which can be approximated by differentiable functions like sigmoid and arctan functions, and the derivative of them are adopted for gradient calculation during backpropagation. Figure 1b illustrates the application of sigmoid function to calculate back-propagated gradients. Direct training with surrogate gradients achieves good performance with few time steps, especially on image classification tasks, thus greatly promotes the development of deep SNNs [15,16,20,62,63,70,71].

3.2 The Proposed SVFormer

Overview. The overall framework of our proposed SVFormer is shown in Fig. 2a. Inspired from the biological brains, efficient and effective multiscale hierarchical modular structures have been widely adopted in DNNs, showing great potential for various tasks [26,28,33,62]. Thus, the backbone of SVFormer is strategically structured in a hierarchical manner, comprising four stages and one local pathway, if not specified. Each of the first two stages consists of one patch embedding (PE) module and several local feature extractors (LFEs), while each of the last two stages consists of one PE module and several global self-attention (GSA) modules. The number of LFEs or GSAs in each stage is indicated as S_i, and the number of channels (or embedding dimension) for each stage is represented as C_i. If not specified, $S = [1, 1, 3, 1]$, meaning that there are one LFE in the first two stages, three GSAs in the third stage, and one GSA in the last stage; and $C = [128, 256, 384, 512]$, meaning that the embedding dimensions for four stages are 128, 256, 384, and 512, respectively. Given a batch of video sequences $I \in \mathbb{R}^{B \times T \times 3 \times H_{in} \times W_{in}}$, T is the length of one video sequence (usually a temporal downsampling of the original video sample), B is the batch size, and H_{in}/W_{in} is the height/width of the video frame. Firstly, we reshape the input into $I \in \mathbb{R}^{T \times B \times 3 \times H_{in} \times W_{in}}$, and feed one video frame into SVFormer in each time step, where T is also the total simulation time steps of SVFormer. After the backbone extracts spatio-temporal information from the video input, the classification head (CH) makes a decision about the action category of the input.

Patch Embedding Module. The PE modules play roles for feature extraction, channel dimension expansion, and patch embedding, raising channel dimension of the feature map, while downsampling the feature map in spatial dimension. The first PE module consists of a convolution (Conv) layer and a batch-normalization (BN) layer, to extract features and spatially downsample the input frame before encoding them into spikes. Other PE modules consist of a spiking neuron (SN) layer, a Conv layer, and a BN layer. For the sake of brevity, the combination of a Conv layer and a BN layer is termed as ConvBN. The computation in the last three PE modules can be expressed as $X_{PEout} = ConvBN(SN(X_{PEin}))$, and the SN operation is omitted in the first PE module as previously described.

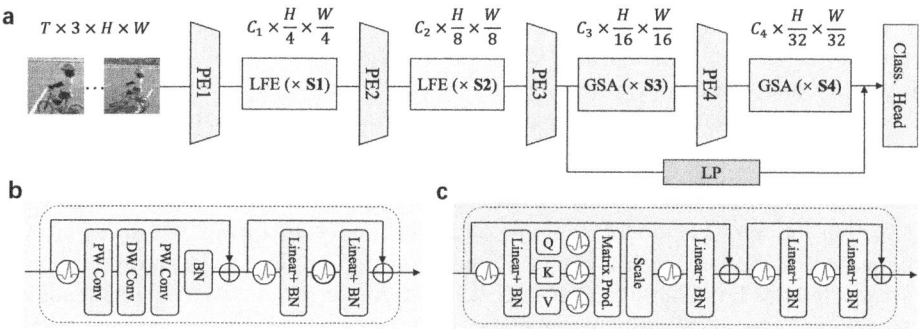

Fig. 2. (a) The overall structure of the hierarchical SVFormer, which includes four stages and one local pathway in default. (b) Structure of one local feature extractor (LFE). (c) Strcuture of one global self-attention (GSA) module.

Local Feature Extractor. As shown in Fig. 2b, one LFE is composed of three cascaded Conv layers and a multi-layer perception (MLP) module, designed for extracting local spatial features. Both parts adopt residual learning with membrane shortcut [62] to avoid gradient vanishing and to improve the model performance. To reduce the number of parameters, we adopt the style of PWConv-DWConv-PWConv like the MobileNet block [24], where PWConv (DWConv) represents pointwise (depthwise) convolution. PWConv is responsible for fusing information across the channel dimension, while DWConv with a relatively large kernel size (5×5 if not specified) for extracting spatial information in each channel. Besides, there is an SN layer ahead of the three Conv layers to transform the feature map into spikes. The MLP module consists of two SN-Linear-BN motifs as usual, which first expands the channel dimension by a ratio ($r = 2$ if not specified) and then reduces it back to the original size, improving the model's representation capability. The combination of a Linear layer and a BN layer is termed as LinearBN. The computation in a LFE can be expressed as: $X = X_{LFEin} + BN(PWConv(DWConv(PWConv(SN(X_{LFEin})))))$, and $X_{LFEout} = X + MLP(X)$, where $MLP(X) = LinearBN(SN(LinearBN(SN(X))))$.

Global Self-Attention Module. As shown in Fig. 2c, a GSA module is composed of one spiking self-attention (SSA) module and one MLP module. Here, we adopt the SSA module which has been proven effective and efficient in [71], and use the same MLP module as that in a LFE. The computation in an SSA module is formulated as follows:

$$A = SN_A(LinearBN_A(SN(X_{SSAin}))), \quad A \in \{Q, K, V\}, \quad X_{SSAin} \in \mathbb{R}^{T \times B \times N \times C}, \quad (5)$$

$$X_{SSAout} = X_{SSAin} + LinearBN(SN(QK^TV * s)), \quad X_{SSAout} \in \mathbb{R}^{T \times B \times N \times C}. \quad (6)$$

In an SSA module, an SN layer first transforms the input into spikes, then three parallel Linear-BN-SN motifs generate the spike-form Q, K, and V tensor, based on which the self-attention score $QK^TV * s$ is calculated through energy-efficient AC operations, where $s = 1/\sqrt{64} = 0.125$ is a predefined scaling factor. The self-attention score is then transformed by a Linear-BN-SN motif before being added to X_{SSAin}.

In the above two equations, N is the number of tokens, which is the multiplication between the height H and width W of the current feature map, i.e. $N = H * W$. Taken together, the computation in a GSA module can be expressed as: $X_{GSAout} = X_{SSAout} + MLP(X_{SSAout})$, where X_{SSAout} is calculated through Eq. (5) and (6).

Local Pathway. To better utilize the features of different scales, we add a local pathway (LP) after the third PE module, and the output of the local pathway will be fused with the output of the last GSA module, as shown in Fig. 2a. The local pathway consists of cascaded SN-DWConv-PWConv-BN layers to further extract local spatial feature in an economical way regarding parameters, and outputs a tensor with the same shape as the output of the last GSA module, which will then be concatenated along the channel dimension before sent into the classification head. The computation of the local pathway can be expressed as $LP(X) = BN(PWConv(DWConv(SN(X))))$, where X means the input of the local pathway.

Classification Head. To make the most of the extracted spatio-temporal features from the backbone, we adopt a classification head (CH) similar to that in [66], which makes a learnable weighted sum of the feature map across temporal and spatial dimension, rather than simply averaging across them. Specifically, for the fused input $X_{CHin} \in \mathbb{R}^{T \times B \times 2C_4 \times H_{out} \times W_{out}}$, we first use an SN layer to convert the feature map into spikes, and reshape it into $X_{CHinS} \in \mathbb{R}^{B \times 2C_4 \times T \times H_{out} \times W_{out}}$. Then, we operate X_{CHinS} with a depthwise 3D convolution layer and a BN layer, i.e. $X_{STF} = 3DConvBN(X_{CHinS})$, which only slightly increases the number of parameters. Further, $X_{STF} \in \mathbb{R}^{B \times 2C_4 \times 1 \times 1 \times 1}$ will be squeezed and operated with a linear layer to generate the classification results. Taken together, the computation in the classification head can be formulated as $Y = CH(X_{CHin}) = Linear(3DConvBN(SN(X_{CHin})))$, where the reshape and squeeze operations are omitted for brevity.

Time-Independent Batch Normalization. For the sake of efficiency, a BN layer in an SNN is usually executed in a parallel way, because the calculation is not time-dependent. The common approach is to reshape the input $X \in \mathbb{R}^{T \times B \times C \times H \times W}$ into $X \in \mathbb{R}^{TB \times C \times H \times W}$, and adopt $nn.BatchNorm2d$ to implement batch normalization (TDBN) [66,69,71]. Obviously, this approach utilizes the feature maps of all time steps equally and simultaneously, which is unreasonable because one cannot use future information for calculation in the current time step. Therefore, we proposed the novel time-independent batch normalization (TIBN) method here. Specifically, we reshape the input into $X \in \mathbb{R}^{T \times BC \times H \times W}$, and adopt $nn.BatchNorm2d$ to implement batch normalization, which treats each time step as a channel and thus independently utilizes information from different time steps.

Overall Computation Process. Based on the above contents, the computation of SVFormer can be summarized as follows.

$$X_{PE_i out} = PE_i(X_{PE_i in}), \quad i \in \{1, 2, 3, 4\} \tag{7}$$

$$X_{LFE_{i,j}out} = LFE_{i,j}(X_{LFE_{i,j}in}), \quad i \in \{1,2\}, \quad j \in \{1,..,S_i\} \tag{8}$$

$$X_{GSA_{i,j}out} = GSA_{i,j}(X_{GSA_{i,j}in}), \quad i \in \{3,4\}, \quad j \in \{1,..,S_i\} \tag{9}$$

$$X_{CHin} = Concate(X_{GSA_{4,S_4}out}, LP(X_{PE_3out})) \tag{10}$$

$$Y = CH(X_{CHin}), \quad Y \in \mathbb{R}^{B \times \#cls} \tag{11}$$

In the above equations, i indicates the stage index, H and W are the height and width of the intermediate feature maps, and $\#cls$ is the number of categories. PE, LFE, GSA, LP and CH represent the computation of patch embedding module, local feature extractor, global self-attention module, local pathway and classification head, respectively. The input shapes of PE, LFE, GSA, LP and CH are $X_{PE_i in} \in \mathbb{R}^{T \times B \times C_{i-1} \times H \times W}$, $X_{LFE_{i,j}in} \in \mathbb{R}^{T \times B \times C_i \times H \times W}$, $X_{GSA_{i,j}in} \in \mathbb{R}^{T \times B \times C_i \times H \times W}$, $X_{PE_3out} \in \mathbb{R}^{T \times B \times C_3 \times H \times W}$, and $X_{CHin} \in \mathbb{R}^{T \times B \times 2C_4 \times H_{out} \times W_{out}}$, respectively. The input to the first PE module is the reshaped video frames, i.e. $X_{PE_1in} = I \in \mathbb{R}^{T \times B \times 3 \times H_{in} \times W_{in}}$. Besides, it should be noted that both the LFE and GSA module do not change the shape of the feature map.

3.3 Theoretical Calculation of Energy Consumption

The theoretical energy consumption of an SNN is usually calculated through multiplication between the number of MAC/AC operations and the energy consumption of each operation on predefined hardware [40,63,66,71]. The number of synaptic operations (SOPs) are calculated as follows:

$$SOP^l = fr^{l-1} \times FLOP^l \tag{12}$$

where fr^{l-1} is the firing rate of spiking neuron layer $l-1$. $FLOP^l$ refers to the number of floating-point MAC operations (FLOPs) of layer l, and SOP^l is the number of spike-based AC operations (SOPs). Assuming the MAC and AC operations are performed on the 45 nm hardware [23], i.e. $E_{MAC} = 4.6pJ$ and $E_{AC} = 0.9pJ$, the energy consumption of SVFormer can be calculated as follows:

$$E_{SVFormer} = E_{AC} \times \left(\sum_{i=2}^{N} SOP^i_{Conv/LN} + \sum_{j=1}^{M} SOP^j_{SSA} \right) + E_{MAC} \times (FLOP^1_{Conv}). \tag{13}$$

$FLOP^1_{Conv}$ represents the FLOPs of the first layer before encoding input frames into spikes, $SOP_{Conv/LN}$ represents the SOPs of a convolution or linear layer, and SOP_{SSA} represents the SOPs of an SSA module. N is the total number of convolution layers and linear layers, and M is the number of SSA modules. During model inference, several cascaded linear operation layers such as convolution, linear and BN layers, can be fused into one single linear operation layer [54,66], still enjoying the AC-type operations with a spike-form input tensor.

4 Experimental Results

In this section, we evaluate the performance of SVFormer on two RGB video datasets (UCF101 [50] and NTU-RGBD60 [47]), as well as a neuromorphic dataset (DVS128-Gesture [1]). We directly train the proposed SVFormer from scratch based on the surrogate gradient method using SpikingJelly [14], which is a popular deep learning framework for building and training SNNs. We compare the performance and energy consumption of SVFormer with existing SNNs or conventional ANNs to demonstrate its effectiveness and efficiency. Besides, we conduct ablation studies to show the effects of some network modules or simulation setups.

4.1 Datasets and Experimental Setup

UCF101 includes 101 action classes with a total of 13,320 videos, which were collected from YouTube [50]. These YouTube videos are recorded in unconstrained environments with cluttered background, camera motion, various illumination conditions, and beyond. These videos have a frame rate of 25 fps and a spatial resolution of 320×240. The length of each video sample ranges from 1.06 s to 71.04 s, of which the average is 7.21 s. In this study, we adopt the first official training-testing split.

NTU-RGBD60 contains 60 kinds of actions with a total of 56,880 samples collected with three different camera angles, which include depth, 3D skeleton, RGB and infrared sequences [47]. This study only adopts RGB videos for action recognition. Both cross-subject (C-Subject) and cross-view (C-View) splits are evaluated in this study. The C-Subject split divides the training set and testing set according to the person ID, while the C-View split divides the samples by the camera ID, as in [47].

DVS128-Gesture contains 11 different classes of gestures collected from 29 individuals under 3 different illumination conditions, which is a neuromorphic dataset collected by dynamic vision sensors [1]. The spatial resolution of DVS128-Gesture is 128×128. We first integrate the stream of events into a sequence of frames as the model input, as usually done [16,63,71].

For both RGB video datasets, the input size of the network is 224×224 in both training and testing phase. Without specification, the batch size is 64, distributed across 4 Nvidia V100 GPUs. The number of training epochs is 600, with a warming-up-then-cosine-decay schedule of learning rate, of which the base value is set empirically to 0.006. For the DVS128-Gesture dataset, the input size of the network is 128×128. We applied a 3-stage model for this small dataset. The batch size is set to 16, and the number of training epochs is 600. The learning rate is set empirically to 0.005 and decayed with a cosine schedule. For all three datasets, AdamW is applied as the optimizer. Moreover, the implementation is based on the SlowFast repository [17] and Uniformer repository [33], and common data augmentation methods like crop, flip, and random erase are applied in the training phase.

4.2 Comparison Results

UCF101 is a popular dataset in the VAR field, which is also commonly applied when using SNNs for VAR. Hence, we evaluate the proposed SVFormer on UCF101 and compare the performance to previous SNNs. The results are listed in Table 1. The SVFormer-base ($S = [1, 1, 3, 1]$, $C = [128, 256, 384, 512]$) is the default network introduced in Sect. 3.2, achieving a top-1 accuracy of 84.03% on the UCF101 dataset, which is state-of-the-art among directly trained deep SNNs for VAR. And we evaluate four modifications of the base model: the shallower SVFormer-ss ($S = [1, 1, 2, 1]$, $C = [128, 256, 384, 512]$), the thinner SVFormer-st ($S = [1, 1, 3, 1]$, $C = [64, 128, 256, 512]$), the deeper SVFormer-dp ($S = [1, 2, 4, 2]$, $C = [128, 256, 384, 512]$) and the wider SVFormer-wd ($S = [1, 1, 3, 1]$, $C = [128, 256, 512, 768]$). All these modifications demonstrate lower accuracy compared to the base model. Three listed recurrent SNNs need nontrivial input preprocessing and 300 simulation time steps to perform the task [8,41], which is unsuitable for practical deployment. The recent ANN-converted SNN, SlowFast-SDM-cv [65], only needs four simulation time steps to achieve a comparable accuracy (92.94%) to its ANN compartment, which is critically dependent on the cautious choice of hyperparameters in the conversion process. Besides, it requires a well-trained ANN as basis and need to repeatedly process the same video clip for four times, thus not friendly for incremental training and not economic for deployment in practice. Further, we directly train two recently published well-performing deep SNNs, SGLFormer [66] and Meta-SpikeFormer [62], on the UCF101 dataset, and the results are inferior to the SVFormer-base model.

To validate the generalizability of the proposed SVFormer, we evaluate SVFormer on a large RGB dataset (NTU-RGBD60) and a neuromorphic dataset (DVS128-Gesture). To the best of our knowledge, SVFormer is the first SNN model that has ever been assessed on the NTU-RGBD60 dataset. Hence, we compare it to two recently published well-performing ANNs tested on RGB frames [43,49]. The results in Table 1 show that SVFormer's accuracy is slightly lower, which is acceptable for the substantial savings of energy consumption. For DVS128-Gesture, We applied the SVFormer-3stg model with one local stage and two global stages, where $S = [1, 2, 1]$ and $C = [64, 128, 256]$. At the same time, we exclude the local pathway for it. The accuracy of the SVFormer-3stg model is 97.92%, which is comparable to mainstream SNNs but with significantly fewer parameters (Table 1).

As described in Sect. 3.1, a main strength of SNNs is energy efficiency, which can be attributed to the sparsity of spikes and the spike-driven communication in SNNs. In this section, we calculate the average theoretical energy consumption of the SVFormer-base model in inference for one video clip from UCF101, with the method described in Sect. 3.3. Firstly, we calculate the number of MAC operations (FLOPs) of convolution and linear layers; then, we count the average firing rate of each spiking neuron layer, and convert corresponding MAC operations (FLOPs) to AC operations (SOPs) by Eq. (12); finally, we calculate the theoretical energy consumption by Eq. (13). The average firing rate of each spiking neuron layer is shown in Fig. 3 (a, b), which indicates the sparsity of spikes during model inference. The FLOPs of the SVFormer-base model's ANN counterpart is 229.163G if executed for 16 repetitions (as 16 time steps in the SNN version), of which the energy consumption is 1054.148 mJ, where all the spiking neurons are replaced by ReLUs. According to the above method, the remaining FLOPs of the SVFormer-base model is 0.700G, and the SOPs is 20.760G, thus the energy consumption is 21.904 mJ. Obviously, the energy cost of SVFormer-base is much lower

Table 1. Comparison results of top-1 accuracy on UCF101, DVS128-Gesture and NTU-RGBD60. † indicates our implementations. The best-performing results are highlighted in bold. SVFormer-base is SOTA of directly trained deep SNNs on UCF101.

Dataset	Model	Param (M)	Time Steps	Top-1 Acc (%)
UCF101	RSNN-reservoir-DA [41]	40.40	300	81.30
	RSNN-HeNHeS-STDP [8]	–	300	77.53
	RSNN-HeNB-BP [8]	–	300	84.32
	RSNN2s-tandem-cvt [67]	–	200	88.46
	SlowFast-SDM-cvt [65]	–	4	**92.94**
	SGLFormer-8-384 [66] †	11.76	16	74.70
	Meta-SpikeFormer [62] †	13.81	16	83.66
	SVFormer-base	13.80	16	**84.03**
	SVFormer-ss	12.53	16	83.61
	SVFormer-st	8.77	16	80.15
	SVFormer-dp	17.72	16	80.25
	SVFormer-wd	24.07	16	81.23
DVS128-Gesture	ConvNet-PLIF [16]	–	20	97.57
	RSNN-HeNHeS-STDP [8]	–	100	96.54
	RSNN-HeNB-BP [8]	–	100	98.12
	Spikformer-2-256 [71]	2.57	16	98.30
	Spikingformer-2-256 [69]	2.57	16	98.30
	SD-Transformer-2-256 [63]	2.57	16	**99.30**
	SGLFormer-3-256 [66]	2.17	16	98.60
	SVFormer-3stg	1.88	16	97.92
NTU-RGBD60	DVANet (ANN) [49]	–	–	93.40(CS)/**98.20**(CV)
	π-ViT (ANN) [43]	–	–	**94.00**(CS)/97.90(CV)
	SVFormer-base	13.76	16	88.12(CS)/94.68(CV)

Table 2. Performance of SVFormer-base on the UCF101 dateset under different noise conditions.

Noise condition	Null	Gaussian noise (a)			Salt-and-pepper noise (P)		
	–	0.1	0.5	1	0.1	0.2	0.3
Top1-acc (%)	84.03	82.26	77.13	64.76	75.36	66.32	55.17

than its ANN counterpart, validating the energy efficiency of SNNs. Further, the ratio of the energy cost of SVFormer-base to that of its ANN counterpart is 1.99% (21:1054), which is much lower than that of SlowFast-SDM-cv (98:128 = 76.56%) [65], showing advantages of directly trained SNNs compared to ANN-converted ones.

Moreover, to test the robustness of the model, we evaluate the trained SVFormer-base model with noisy frames from the UCF101 dataset. Here, we applied two commonly adopted noise types for images, i.e. Gaussian noise and salt-and-pepper noise. For the Gaussian type, we add noise with zero mean and different level of standard deviation $\sigma = a * \sigma_{ori}$ to the original frame, where σ_{ori} indicates the *std* of the original frame. For the salt-and-pepper noise, we randomly transform a pixel into the highest or lowest value of the current frame with a predefined probability P. Figure 4 demonstrates the effects of different level of noise with an exemplar frame. The results in Table 2 demonstrate that SVFormer exhibits resilience to a moderate level of noise, sustaining commendable performance. However, when confronted with excessive noise that markedly degrades frame quality, there is a substantial decline in the model's performance.

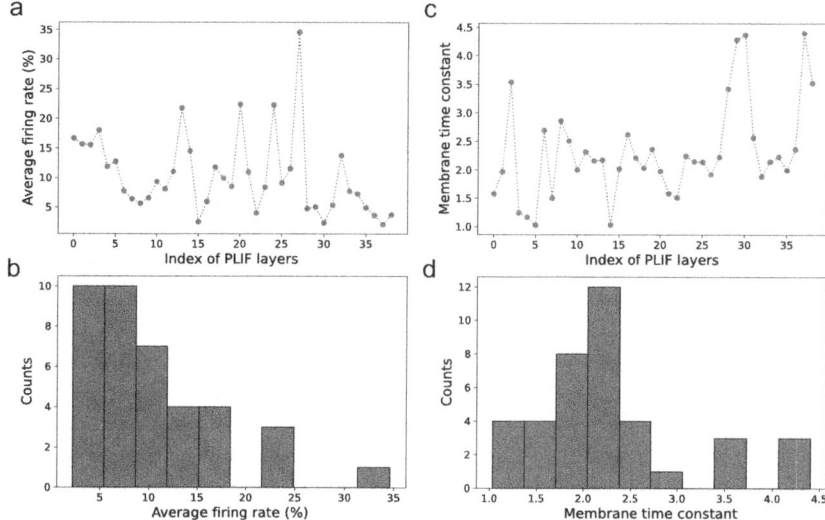

Fig. 3. Average firing rates (a, b) during model inference on the UCF101 dataset, and learned membrane time constants (c, d), for all PLIF layers, showing the trend of both variables as the network goes deeper (a, c), and the corresponding histograms (b, d).

Fig. 4. (a) An exemplar frame with different level of Gaussian noise ($N(0, \sigma)$). (b) An exemplar frame with different level of salt-and-pepper noise, where P means the sprinkling probability.

4.3 Ablation Studies

In this subsection, we conduct some ablation studies based on the UCF101 dataset to show the effects of some network modules or simulation setups. The results are listed in Table 3, where batch size are adjusted according to the model size. First, classification accuracy of a SVFormer without the local pathway is 83.29%, which is slightly lower than the original 84.03%, showing that the local pathway improves the model's representational power for VAR. Second, classification accuracy of a SVFormer with LIF neurons is only 80.89%, demonstrating that learnable membrane time constants of spiking neurons (PLIF) enhance the model's temporal processing capability. As shown in Fig. 3 (c, d), the learned membrane time constants are variable, helping to better represent temporal information, which is also consistent with heterogeneous neurons observed in biological brains. Besides, the membrane time constants of the deeper layers are on average larger than those of the shallower layers, implying that deeper layers integrate spatio-temporal information for longer duration. Third, without using time-independent batch normalization to utilize information from different time steps independently, the model' classification accuracy decrease from 84.03% to 80.65%, indicating the effectiveness of the proposed time-dependent BN. Fourth, we apply different number of frames (i.e. $T=8$ or $T=24$) as network input to test the effects of input video length, where T is the number of frames sampled from the original video sample. And the simulation duration of the SNN model is adjusted to 8 or 24 steps accordingly. Both 8-frame and 24-frame inputs exhibit lower accuracies compared to the original 16-frame input, indicating the necessity of finding a suitable input length to balance accuracy and computational cost.

Additionally, when utilizing SNNs to process time sequences such as videos, addressing the alignment of temporal resolution becomes a critical concern that warrants attention. In this work, we mainly adopt the frame-by-frame approach, i.e. the input length (number of frames) is equal to the simulation duration (number of time steps) of the SNN model, which processes one frame per time step. Here, we also test the clip-by-clip approach, where the SNN model processes a clip of frames in each time step. Firstly, we need to modify the 2D-Conv and 2D-BN layers in the model into 3D ones accordingly, to process a 3D video clip. Then, we separate a sampled video sequence with 16 or 20 frames into 4 parts uniformly, meaning that the model runs 4

Table 3. Ablation studies on the UCF101 dateset.

Architecture	Param (M)	Batch size	Top-1 Acc (%)
SVFormer Base ($T=16$)	13.80	64	84.03
Base - Local pathway	12.32	64	83.29
Base + (PLIF → LIF)	13.80	64	80.89
Base + (TIBN → TDBN)	13.35	64	80.65
Base + ($T=8$)	12.76	64	81.26
Base + ($T=24$)	14.84	40	81.81
Base + 3D clip input (16/4)	12.69	56	74.20
Base + 3D clip input (20/4)	12.89	40	74.09

time steps and processes one clip with 4 or 5 frames in each time step. Finally, we train the model from scratch and test its performance. The results in Table 3 demonstrate that the frame-by-frame approach is superior to the clip-by-clip approach for the UCF101 dataset.

5 Conclusion

In this paper, we propose the directly trained SVFormer, which integrates local feature extraction, global self-attention, and the intrinsic dynamics, sparsity, and spike-driven nature of SNNs, effectively and efficiently learning spatio-temporal representation for VAR. We evaluate SVFormer on two RGB datasets (UCF101, NTU-RGBD60) and a neuromorphic dataset (DVS128-Gesture). The experimental results demonstrate that SVFormer achieves comparable performance to mainstream models for VAR tasks in a more efficient way. Specifically, SVFormer achieves state-of-the-art (top-1 accuracy of 84.03%) among directly trained deep SNNs on UCF101, with ultra-low power consumption (21 mJ/video). These results verify SVFormer's strong representation capability, showing that the multiscale spatio-temporal feature-extraction characteristic endows it with great potential as a backbone for diverse video tasks when equipped with properly designed task heads. However, there are certain limitations associated with this pioneering endeavor. On the one hand, the recognition accuracy lags behind the state-of-the-art of traditional ANNs. One possible solution is to try large-scale self-supervised pretraining for SNNs, which has already shown great success for ANNs [51,55,59]. On the other hand, although a video clip is processed frame-by-frame in SVFormer, the length of the clip is predefined for the ease of network implementation, meaning that the model needs to process all frames before making a decision. One can try to achieve speed-accuracy tradeoff in the model as a biological brain [4], i.e. the model may stop at any step if a decision is made based on predefined criteria, which is more flexible and efficient.

Acknowledgements. This study was supported by grants 62206141 and 62236009 from the National Natural Science Foundation of China, and grant PCL2021A13 from Pengcheng Laboratory.

Disclosure of Interests. The authors have no competing interests to declare that are relevant to the content of this article.

References

1. Amir, A., et al.: A low power, fully event-based gesture recognition system. In: Proceedings of the IEEE/CVF Conference on Computer Vision and Pattern Recognition, pp. 7388–7397 (2017). https://doi.org/10.1109/cvpr.2017.781
2. Arnab, A., Dehghani, M., Heigold, G., Sun, C., Lučić, M., Schmid, C.: Vivit: a video vision transformer. In: Proceedings of the IEEE/CVF International Conference on Computer Vision, pp. 6836–6846 (2021). https://doi.org/10.1109/iccv48922.2021.00676

3. Bertasius, G., Wang, H., Torresani, L.: Is space-time attention all you need for video understanding? In: Proceedings of the 38th International Conference on Machine Learning, vol. 139, pp. 813–824 (2021)
4. Bogacz, R.: Speed-accuracy tradeoff. In: Encyclopedia of Computational Neuroscience, pp. 3225–3228. Springer, Heidelberg (2022). https://doi.org/10.1007/978-1-4614-6675-8_319
5. Bu, T., Fang, W., Ding, J., Dai, P., Yu, Z., Huang, T.: Optimal ann-snn conversion for high-accuracy and ultra-low-latency spiking neural networks. In: International Conference on Learning Representations (2022). https://doi.org/10.48550/arXiv.2303.04347
6. Cao, Y., Chen, Y., Khosla, D.: Spiking deep convolutional neural networks for energy-efficient object recognition. Int. J. Comput. Vision **113**(1), 54–66 (2014). https://doi.org/10.1007/s11263-014-0788-3
7. Carreira, J., Zisserman, A.: Quo vadis, action recognition? A new model and the kinetics dataset. In: Proceedings of the IEEE/CVF Conference on Computer Vision and Pattern Recognition, pp. 6299–6308 (2017). https://doi.org/10.1109/cvpr.2017.502
8. Chakraborty, B., Mukhopadhyay, S.: Heterogeneous recurrent spiking neural network for spatio-temporal classification. Front. Neurosci. **17**, 994517 (2023). https://doi.org/10.3389/fnins.2023.994517
9. Chen, C.F.R., Panda, R., Ramakrishnan, K., Feris, R., Cohn, J., Oliva, A., Fan, Q.: Deep analysis of cnn-based spatio-temporal representations for action recognition. In: Proceedings of the IEEE/CVF Conference on Computer Vision and Pattern Recognition, pp. 6165–6175 (2021). https://doi.org/10.1109/cvpr46437.2021.00610
10. Chéron, G., Laptev, I., Schmid, C.: P-CNN: pose-based cnn features for action recognition. In: Proceedings of the IEEE/CVF International Conference on Computer Vision, pp. 3218–3226 (2015). https://doi.org/10.1109/iccv.2015.368
11. Dosovitskiy, A., et al.: An image is worth 16×16 words: transformers for image recognition at scale. In: International Conference on Learning Representations (2021). https://doi.org/10.48550/arXiv.2010.11929
12. Fan, H., et al.: Multiscale vision transformers. In: Proceedings of the IEEE/CVF International Conference on Computer Vision, pp. 6824–6835 (2021). https://doi.org/10.1109/iccv48922.2021.00675
13. Fan, Q., Chen, C.F.R., Kuehne, H., Pistoia, M., Cox, D.: More is less: learning efficient video representations by temporal aggregation modules. Adv. Neural Inf. Process. Syst. **32**, 2264–2273 (2019). https://doi.org/10.48550/arXiv.1912.00869
14. Fang, W., et al.: Spikingjelly: an open-source machine learning infrastructure platform for spike-based intelligence. Sci. Adv. **9**(40), eadi1480 (2023). https://doi.org/10.1126/sciadv.adi1480
15. Fang, W., Yu, Z., Chen, Y., Huang, T., Masquelier, T., Tian, Y.: Deep residual learning in spiking neural networks. Adv. Neural Inf. Process. Syst. **34**, 21056–21069 (2021). https://doi.org/10.48550/arXiv.2102.04159
16. Fang, W., Yu, Z., Chen, Y., Masquelier, T., Huang, T., Tian, Y.: Incorporating learnable membrane time constant to enhance learning of spiking neural networks. In: Proceedings of the IEEE/CVF International Conference on Computer Vision, pp. 2661–2671 (2021). https://doi.org/10.1109/iccv48922.2021.00266
17. Feichtenhofer, C., Fan, H., Malik, J., He, K.: Slowfast networks for video recognition. In: Proceedings of the IEEE/CVF International Conference on Computer Vision, pp. 6202–6211 (2019). https://doi.org/10.1109/iccv.2019.00630
18. Feichtenhofer, C., Pinz, A., Wildes, R.P.: Spatiotemporal residual networks for video action recognition. Adv. Neural Inf. Process. Syst. **29**, 3468–3476 (2016). https://doi.org/10.5555/3157382.3157486

19. Feichtenhofer, C., Pinz, A., Wildes, R.P.: Spatiotemporal multiplier networks for video action recognition. In: Proceedings of the IEEE/CVF Conference on Computer Vision and Pattern Recognition, pp. 4768–4777 (2017). https://doi.org/10.1109/cvpr.2017.787
20. Guo, Y., Huang, X., Ma, Z.: Direct learning-based deep spiking neural networks: A review. Front. Neurosci. **17**, 1209795 (2023). https://doi.org/10.3389/fnins.2023.1209795
21. Han, K., et al.: A survey on vision transformer. IEEE Trans. Pattern Anal. Mach. Intell. **45**(1), 87–110 (2022). https://doi.org/10.1109/tpami.2022.3152247
22. Hara, K., Kataoka, H., Satoh, Y.: Can spatiotemporal 3d cnns retrace the history of 2d cnns and imagenet? In: Proceedings of the IEEE/CVF Conference on Computer Vision and Pattern Recognition, pp. 6546–6555 (2018). https://doi.org/10.1109/cvpr.2018.00685
23. Horowitz, M.: 1.1 computing's energy problem (and what we can do about it). In: 2014 IEEE International Solid-State Circuits Conference Digest of Technical Papers, pp. 10–14. IEEE (2014). https://doi.org/10.1109/isscc.2014.6757323
24. Howard, A.G., et al.: Mobilenets: efficient convolutional neural networks for mobile vision applications. arXiv preprint arXiv:1704.04861 (2017). https://doi.org/10.48550/arXiv.1704.04861
25. Hunsberger, E., Eliasmith, C.: Spiking deep networks with lif neurons. arXiv preprint arXiv:1510.08829 (2015). https://doi.org/10.48550/arXiv.1510.08829
26. Jiao, L., Gao, J., Liu, X., Liu, F., Yang, S., Hou, B.: Multiscale representation learning for image classification: a survey. IEEE Trans. Artif. Intell. **4**(1), 23–43 (2021). https://doi.org/10.1109/tai.2021.3135248
27. Karpathy, A., Toderici, G., Shetty, S., Leung, T., Sukthankar, R., Fei-Fei, L.: Large-scale video classification with convolutional neural networks. In: Proceedings of the IEEE/CVF Conference on Computer Vision and Pattern Recognition, pp. 1725–1732 (2014). https://doi.org/10.1109/cvpr.2014.223
28. Kruger, N., et al.: Deep hierarchies in the primate visual cortex: what can we learn for computer vision? IEEE Trans. Pattern Anal. Mach. Intell. **35**(8), 1847–1871 (2013). https://doi.org/10.1109/tpami.2012.272
29. Kundu, S., Datta, G., Pedram, M., Beerel, P.A.: Spike-thrift: towards energy-efficient deep spiking neural networks by limiting spiking activity via attention-guided compression. In: Proceedings of the IEEE/CVF Winter Conference on Applications of Computer Vision, pp. 3953–3962 (2021). https://doi.org/10.1109/wacv48630.2021.00400
30. Kundu, S., Pedram, M., Beerel, P.A.: Hire-snn: harnessing the inherent robustness of energy-efficient deep spiking neural networks by training with crafted input noise. In: Proceedings of the IEEE/CVF International Conference on Computer Vision, pp. 5209–5218 (2021). https://doi.org/10.1109/iccv48922.2021.00516
31. Lee, C., Sarwar, S.S., Panda, P., Srinivasan, G., Roy, K.: Enabling spike-based backpropagation for training deep neural network architectures. Front. Neurosci. **14**, 119 (2020). https://doi.org/10.3389/fnins.2020.00119
32. Lee, J.H., Delbruck, T., Pfeiffer, M.: Training deep spiking neural networks using backpropagation. Front. Neurosci. **10**, 508 (2016). https://doi.org/10.3389/fnins.2016.00508
33. Li, K., et al.: Uniformer: unified transformer for efficient spatiotemporal representation learning. In: International Conference on Learning Representations (2022). https://doi.org/10.48550/arXiv.2201.04676
34. Li, K., et al.: Uniformer: unifying convolution and self-attention for visual recognition. IEEE Trans. Pattern Anal. Mach. Intell. (2023). https://doi.org/10.1109/tpami.2023.3282631
35. Liu, Z., et al.: Video swin transformer. In: Proceedings of the IEEE/CVF Conference on Computer Vision and Pattern Recognition, pp. 3202–3211 (2022). https://doi.org/10.1109/cvpr52688.2022.00320
36. Liu, Z., et al.: Convtransformer: a convolutional transformer network for video frame synthesis. arXiv preprint arXiv:2011.10185 (2020). https://doi.org/10.48550/arXiv.2011.10185

37. Maass, W.: Networks of spiking neurons: the third generation of neural network models. Neural Netw. **10**(9), 1659–1671 (1997). https://doi.org/10.1016/s0893-6080(97)00011-7
38. Meng, Q., Xiao, M., Yan, S., Wang, Y., Lin, Z., Luo, Z.Q.: Training high-performance low-latency spiking neural networks by differentiation on spike representation. In: Proceedings of the IEEE/CVF Conference on Computer Vision and Pattern Recognition, pp. 12444–12453 (2022). https://doi.org/10.1109/cvpr52688.2022.01212
39. Neftci, E.O., Mostafa, H., Zenke, F.: Surrogate gradient learning in spiking neural networks: bringing the power of gradient-based optimization to spiking neural networks. IEEE Signal Process. Mag. **36**(6), 51–63 (2019). https://doi.org/10.1109/msp.2019.2931595
40. Panda, P., Aketi, S.A., Roy, K.: Toward scalable, efficient, and accurate deep spiking neural networks with backward residual connections, stochastic softmax, and hybridization. Front. Neurosci. **14**, 653 (2020). https://doi.org/10.3389/fnins.2020.00653
41. Panda, P., Srinivasa, N.: Learning to recognize actions from limited training examples using a recurrent spiking neural model. Front. Neurosci. **12**, 126 (2018). https://doi.org/10.3389/fnins.2018.00126
42. Qiu, Z., Yao, T., Mei, T.: Learning spatio-temporal representation with pseudo-3d residual networks. In: Proceedings of the IEEE/CVF International Conference on Computer Vision, pp. 5533–5541 (2017). https://doi.org/10.1109/iccv.2017.590
43. Reilly, D., Das, S.: Just add π! pose induced video transformers for understanding activities of daily living. In: Proceedings of the IEEE/CVF Conference on Computer Vision and Pattern Recognition, pp. 18340–18350 (2024). https://doi.org/10.48550/arXiv.2311.18840
44. Roy, K., Jaiswal, A., Panda, P.: Towards spike-based machine intelligence with neuromorphic computing. Nature **575**(7784), 607–617 (2019). https://doi.org/10.1038/s41586-019-1677-2
45. Rueckauer, B., Lungu, I.A., Hu, Y., Pfeiffer, M., Liu, S.C.: Conversion of continuous-valued deep networks to efficient event-driven networks for image classification. Front. Neurosci. **11**, 682 (2017). https://doi.org/10.3389/fnins.2017.00682
46. Selva, J., Johansen, A.S., Escalera, S., Nasrollahi, K., Moeslund, T.B., Clapés, A.: Video transformers: a survey. IEEE Trans. Pattern Anal. Mach. Intell. (2023). https://doi.org/10.1109/tpami.2023.3243465
47. Shahroudy, A., Liu, J., Ng, T.T., Wang, G.: NTU RGB+D: a large scale dataset for 3d human activity analysis. In: Proceedings of the IEEE/CVF Conference on Computer Vision and Pattern Recognition, pp. 1010–1019 (2016). https://doi.org/10.1109/cvpr.2016.115
48. Shrestha, S.B., Orchard, G.: Slayer: spike layer error reassignment in time. Adv. Neural Inf. Process. Syst. **31**, 1419–1428 (2018). https://doi.org/10.48550/arXiv.1810.08646
49. Siddiqui, N., Tirupattur, P., Shah, M.: Dvanet: disentangling view and action features for multi-view action recognition. In: Proceedings of the AAAI Conference on Artificial Intelligence, vol. 38, pp. 4873–4881 (2024). https://doi.org/10.1609/aaai.v38i5.28290
50. Soomro, K., Zamir, A.R., Shah, M.: UCF101: a dataset of 101 human actions classes from videos in the wild. arXiv preprint arXiv:1212.0402 (2012). https://doi.org/10.48550/arXiv.1212.0402
51. Tong, Z., Song, Y., Wang, J., Wang, L.: Videomae: masked autoencoders are data-efficient learners for self-supervised video pre-training. Adv. Neural Inf. Process. Syst. **35**, 10078–10093 (2022). https://doi.org/10.48550/arXiv.2203.12602
52. Tran, D., Bourdev, L., Fergus, R., Torresani, L., Paluri, M.: Learning spatiotemporal features with 3d convolutional networks. In: Proceedings of the IEEE/CVF International Conference on Computer Vision, pp. 4489–4497 (2015). https://doi.org/10.1109/iccv.2015.510
53. Tran, D., Wang, H., Torresani, L., Ray, J., LeCun, Y., Paluri, M.: A closer look at spatiotemporal convolutions for action recognition. In: Proceedings of the IEEE/CVF Conference on Computer Vision and Pattern Recognition, pp. 6450–6459 (2018). https://doi.org/10.1109/cvpr.2018.00675

54. Waeijen, L., Sioutas, S., Peemen, M., Lindwer, M., Corporaal, H.: Convfusion: a model for layer fusion in convolutional neural networks. IEEE Access **9**, 168245–168267 (2021). https://doi.org/10.1109/access.2021.3134930
55. Wang, L., et al.: Videomae v2: scaling video masked autoencoders with dual masking. In: Proceedings of the IEEE/CVF Conference on Computer Vision and Pattern Recognition, pp. 14549–14560 (2023). https://doi.org/10.1109/cvpr52729.2023.01398
56. Wang, L., et al.: Temporal segment networks: towards good practices for deep action recognition. In: Leibe, B., Matas, J., Sebe, N., Welling, M. (eds.) ECCV 2016. LNCS, vol. 9912, pp. 20–36. Springer, Cham (2016). https://doi.org/10.1007/978-3-319-46484-8_2
57. Wang, W., Hao, S., Wei, Y., Xiao, S., Feng, J., Sebe, N.: Temporal spiking recurrent neural network for action recognition. IEEE Access **7**, 117165–117175 (2019). https://doi.org/10.1109/access.2019.2936604
58. Wang, Y., Zhang, M., Chen, Y., Qu, H.: Signed neuron with memory: towards simple, accurate and high-efficient ann-snn conversion. In: Proceedings of the Thirty-First International Joint Conference on Artificial Intelligence, pp. 2501–2508 (2022). https://doi.org/10.24963/ijcai.2022/347
59. Wei, C., Fan, H., Xie, S., Wu, C.Y., Yuille, A., Feichtenhofer, C.: Masked feature prediction for self-supervised visual pre-training. In: Proceedings of the IEEE/CVF Conference on Computer Vision and Pattern Recognition, pp. 14668–14678 (2022). https://doi.org/10.1109/cvpr52688.2022.01426
60. Xie, S., Sun, C., Huang, J., Tu, Z., Murphy, K.: Rethinking spatiotemporal feature learning: speed-accuracy trade-offs in video classification. In: Proceedings of the European Conference on Computer Vision, pp. 305–321 (2018). https://doi.org/10.1007/978-3-030-01267-0_19
61. Yamazaki, K., Vo-Ho, V.K., Bulsara, D., Le, N.: Spiking neural networks and their applications: a review. Brain Sci. **12**(7), 863 (2022). https://doi.org/10.3390/brainsci12070863
62. Yao, M., et al.: Spike-driven transformer v2: meta spiking neural network architecture inspiring the design of next-generation neuromorphic chips. In: International Conference on Learning Representations (2024). https://doi.org/10.48550/arXiv.2404.03663
63. Yao, M., Hu, J., Zhou, Z., Yuan, L., Tian, Y., Xu, B., Li, G.: Spike-driven transformer. Adv. Neural Inf. Process. Syst. **36** (2023). https://doi.org/10.48550/arXiv.2307.01694
64. Yin, B., Corradi, F., Bohté, S.M.: Accurate and efficient time-domain classification with adaptive spiking recurrent neural networks. Nat. Mach. Intell. **3**(10), 905–913 (2021). https://doi.org/10.1038/s42256-021-00397-w
65. You, H., et al.: Converting artificial neural networks to ultra-low-latency spiking neural networks for action recognition. IEEE Trans. Cogn. Dev. Syst. (2024). https://doi.org/10.1109/tcds.2024.3375620
66. Zhang, H., et al.: Sglformer: spiking global-local-fusion transformer with high performance. Front. Neurosci. **18**, 1371290 (2024). https://doi.org/10.3389/fnins.2024.1371290
67. Zhang, J., Wang, J., Di, X., Pu, S.: High-accuracy and energy-efficient action recognition with deep spiking neural network. In: International Conference on Neural Information Processing, pp. 279–292. Springer, Heidelberg (2022). https://doi.org/10.1007/978-3-031-30108-7_24
68. Zhou, B., Andonian, A., Oliva, A., Torralba, A.: Temporal relational reasoning in videos. In: Proceedings of the European Conference on Computer Vision, pp. 803–818 (2018). https://doi.org/10.1007/978-3-030-01246-5_49
69. Zhou, C., et al.: Spikingformer: spike-driven residual learning for transformer-based spiking neural network. arXiv preprint arXiv:2304.11954 (2023). https://doi.org/10.48550/arXiv.2304.11954

70. Zhou, C., et al.: Direct training high-performance deep spiking neural networks: a review of theories and methods. Front. Neurosci. **18**, 1383844 (2024). https://doi.org/10.3389/fnins.2024.1383844
71. Zhou, Z., et al.: Spikformer: when spiking neural network meets transformer. In: International Conference on Learning Representations (2023). https://doi.org/10.48550/arXiv.2209.15425

BL-BERT: Extracting Body Language from Behavior Sequences in Freely Moving Mice

Yaning Han[1,2,3,4], Zhiwei Jiang[1,3,4], Furong Ju[1,3,4], Liping Wang[1,3,4], Quanying Liu[5], and Pengfei Wei[1,3,4(✉)]

[1] CAS Key Laboratory of Brain Connectome and Manipulation, Shenzhen-Hong Kong Institute of Brain Science, Shenzhen Institute of Advanced Technology, Chinese Academy of Sciences, Shenzhen 518055, China
pf.wei@siat.ac.cn
[2] University of Chinese Academy of Sciences, Beijing 100049, China
[3] Guangdong Provincial Key Laboratory of Brain Connectome and Behavior, The Brain Cognition and Brain Disease Institute, Shenzhen Institute of Advanced Technology, Chinese Academy of Sciences, Shenzhen 518055, China
[4] Key Laboratory of Brain Cognition and Brain-Inspired Intelligence Technology, The Brain Cognition and Brain Disease Institute, Shenzhen Institute of Advanced Technology, Chinese Academy of Sciences, Shenzhen 518055, China
[5] Department of Biomedical Engineering, Southern University of Science and Technology, Shenzhen 518055, China

Abstract. Artificial intelligence-driven techniques for pose estimation and behavior analysis represent a significant advancement in the elucidation of precise movement trajectories and behavioral components in freely moving animals. However, comprehending the underlying temporal dynamics of these trajectories remains challenging due to the dearth of effective analytical instruments. In order to decipher the nuanced body language inherent in behavioral dynamics at a sequential level, we introduce BL-BERT, a computational framework rooted in Bidirectional Encoder Representation from Transformers (BERT). This framework discerns stereotypical behavior sequences exhibited by freely moving mice, elucidating behavioral dynamics in a linguistically comprehensible manner. BL-BERT discerns salient behavior sequences from input behavior modules, as evidenced by its performance on a custom dataset of interactions among free-moving mice. Diverging from conventional Markov models, BL-BERT unfolds the recurrent structure of behavior sequences, rendering it more interpretable. BL-BERT offers a novel way to apprehend the hierarchical organization of intricate animal behaviors, with promising prospects for widespread applicability across various behavioral paradigms.

Keywords: Computational neuroethology · behavior sequence · Transformer model

1 Introduction

Animal behavior exhibits a hierarchical organization encompassing poses, movements, ethograms, sub-ethomes, and ethomes [1]. Recent strides in artificial intelligence (AI)-driven methodologies for animal behavior quantification have significantly enhanced accessibility to the diverse layers of this organizational structure. Notable approaches such as DeepLabCut [2], SLEAP [3], and ADPT [4] leverage deep neural networks to track animal poses without physical markers. Meanwhile, methodologies like MoSeq [5], Behavior Atlas [6], and VAME [7] employ unsupervised techniques to decompose continuous poses into discrete movements or behavioral modules. However, methodologies explicitly targeting quantification at the upper echelon of ethograms are infrequent. The semantic richness of animal behavior stretches from ethograms to ethomes [1], underscoring the importance of comprehending the significance of animal behavior sequences beyond mere poses and movements.

The primary constituent of an ethogram resides in behavior sequences, which serve as informative manifestations of animal communication [8–11]. These well-organized behavior sequences are represented in the sequential coding of the neural circuits [12]. Traditionally, the Markov model has been employed to extract temporal relationships among behavioral modules, such as feeding behavior [13], circadian behavior [6], and avoidance behavior [5]. Nevertheless, emerging evidence suggests that behaviors may exhibit strongly non-Markovian characteristics [14]. Consequently, reliance on the Markov model could lead to diminished accuracy in behavior sequence identification. Moreover, disentangling the recurrent structure of transitions between behavioral states into distinct sequences proves challenging. These factors collectively contribute to inaccuracies in mapping behavior sequences to neural activities.

Recent advancements in Transformer-based large language models have yielded notable strides in language comprehension [15, 16]. The Transformer, an artificial intelligence model incorporating a self-attention mechanism, has proven adept at discerning semantic nuances within token sequences [15]. Given that behavior sequences can be construed as akin to animal body language, there exists potential to leverage Transformer-based large language models for uncovering the inherent semantic structures within these sequences. The self-attention mechanism inherent in the Transformer architecture offers a solution to the challenge of modeling non-Markovian dynamics [17]. Additionally, unlike the Markov model, the Transformer learns semantics directly from sequences, thus obviating the need to disentangle recurrent structures within behavior sequences.

Inspired by these large language models, we developed a Bidirectional Encoder Representation from Transformers (BERT [18])-based Body Language (BL) extraction computational framework termed BL-BERT. BL-BERT offers a more intuitive depiction of behavior sequence fingerprints compared to the Markov model. Through validation of our customized dataset, BL-BERT demonstrates resilience in handling noise within animal behavior sequences, prioritizing the extraction of meaningful sequential patterns. Our discoveries highlight the potential of AI-driven BL-BERT in fostering a deeper understanding of animal behavior sequences and the underlying neural mechanisms orchestrating them.

2 Results

2.1 Markov Model Fails to Decouple the Recurrent Structure in Temporal Behavior Dynamics

We create a customized dataset to investigate the behavior sequences in free-moving mice (Fig. 1). Now that the behavior sequences can be regarded as body language [9], our dataset is constructed based on the free-social interaction behavior paradigm [19]. This involves placing two mice within an open field setting (Fig. 1A), with one of them equipped with a neural recording device to enable identity differentiation. 15 trials of mice social interaction are collected for subsequent analysis. The time length of each trial is 5 min. Behavioral capture is facilitated by the MouseVenue3D system [20, 21], which employs four multi-view cameras to comprehensively record mouse behavior. Subsequently, raw video footage from the four cameras undergoes behavior analysis utilizing the methodology outlined in Behavior Atlas [6, 22]. Two-dimensional (2D) poses are tracked from each camera angle using DeepLabCut [2]. The subsequent stages encompass 3D pose reconstruction, 3D behavior decomposition, behavior module clustering, and behavior phenotype identification, following the procedures delineated in Behavior Atlas (Fig. 1B). The examples of identified 3D behavior phenotypes are illustrated in Fig. 1C.

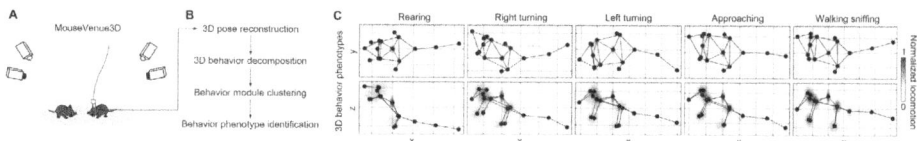

Fig. 1. 3D behavior data acquisition and dataset preparation. **A**, MouseVenue3D for the behavior data acquisition. **B**, The dataset preparation steps for later analysis. **C**, The illustration of 3D poses with identification.

The behavior modules can be regarded as the states in the framework of the Markov model [6]. Optimization of the behavior state transition matrix is feasible through the free-social interaction dataset above (Fig. 2A). The undirected graph visually depicts inherent patterns within behavior state transitions (Fig. 2B). To enhance the clarity of these patterns, a directed graph employing force-directed placement is utilized [23] (Fig. 2C). Furthermore, a control group comprising mice confined within cages is incorporated to underscore intrinsic behavior state transition patterns during free-social interactions. The directed graph reveals that only the node labeled "right_wiggle_groom" exhibits a linear sequence structure, while other behavior states display consistently recurrent patterns. Nonetheless, effective communication within social interactions necessitates clear initiation points, suggesting that recurrent structures may stem from inherent limitations of the Markov model. Consequently, the utilization of the Markov model for representing behavior sequences proves imprecise due to these recurrent structures.

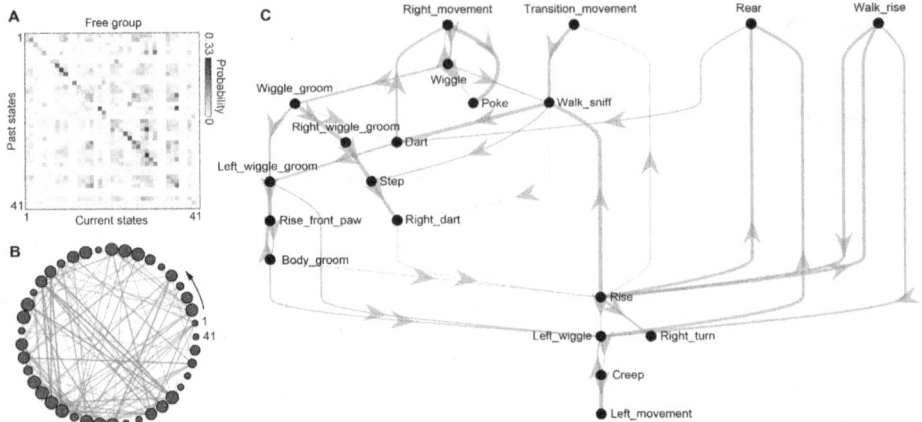

Fig. 2. The behavior state transition representation of the Markov model. **A**, The transition matrix of behavior modules. **B**, The undirected graph of behavior state transition of 41 behavior modules. **C**, The directed graph of behavior state transitions with significant differences in the free group. (t-test or Wilcoxon rank sum test, $n = 15$).

2.2 BL-BERT Bridges the Gap Between Behavior Modules and Behavior Sequences

The method for extracting behavior sequences ought to emphasize the interconnections spanning a series of behavior modules, rather than solely concentrating on transitions between adjacent modules. That is the reason why the development of BL-BERT is undertaken (Fig. 3).

Initially, the behavior modules delineated by BeA (Fig. 3A) are transformed into sequences specific to each mouse. These individual behavior sequences are subsequently resampled into segments (Fig. 3B), facilitating their organization into a dataset for training AI models. Concurrently, the behavior modules are tokenized into discrete behavior tokens analogous to words in natural language. Following the blueprint of BERT's architecture, the tokenized behavior sequences undergo random masking. The Transformer model within BERT is then tasked with processing the masked behavior sequences, with its output representing the unaltered behavior sequences (Fig. 3C). Upon completion of Transformer training, the Monte Carlo method is applied to predict the behavior sequences of each mouse. This involves the repeated random masking of behavior sequences one hundred times, followed by input into the Transformer. Subsequently, the congruent outputs between prediction and unaltered sequences are retained, signifying the presence of meaningful body language cues.

2.3 BL-BERT is Robust in Focusing on Meaningful Behavior Sequences

To ensure the best mask percentage and verify the performance of BL-BERT, we first set the gradient for mask percentage ranges from 0% to 90% (Fig. 4A). Although the 15% mask of BERT has been validated to be important for the performance of tasks [18], the more appropriate percentages of masks could further improve the performance of BERT

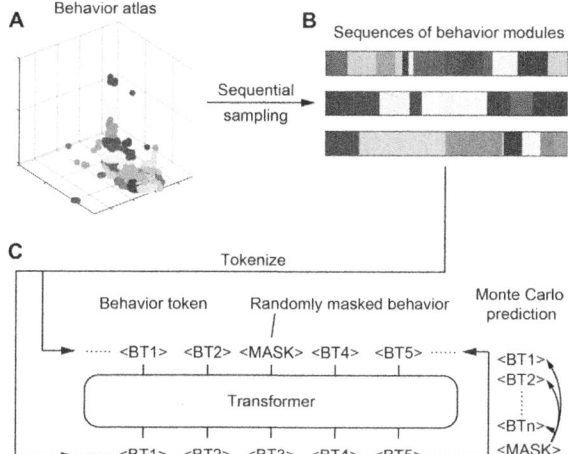

Fig. 3. The framework of BL-BERT. **A**, The behavior atlas of behavior modules. **B**, The sequences of behavior modules transformed from the behavior atlas. **C**, The BERT part of BL-BERT.

[24]. Because the structure of animal body language could be different from human language, the percentages of masks should be chosen correctly. Initial findings indicate a direct correlation between mask percentage augmentation and heightened loss, coupled with a reduction in precision. Notably, precision metrics exhibit a discernible inflection point at approximately 30% mask coverage. To uphold the efficacy of body language extraction, a mask percentage of 20% is determined as the most optimal threshold.

Furthermore, two specific scenarios, namely shuffled sequences and erroneous mask assignments, are employed to validate the robustness of BL-BERT against noise inherent in behavior sequences (Fig. 4B). BL-BERT demonstrates an ability to converge to shuffled sequences but precision levels notably diminish compared to unshuffled counterparts. This observation suggests that certain behavior sequences may not inherently rely on sequential semantics for conveying information. Despite this, BL-BERT exhibits insensitivity towards such non-semantic sequences, as inferred from the predicted precision metrics. Additionally, results of incorrect mask assignments highlight BL-BERT's incapacity to conform to noise, with approximately 80% precision affirming its fidelity solely to accurate representations.

2.4 BL-BERT Quantifies the Behavior Sequence Fingerprints of Free-Moving Mice

Using BL-BERT, the behavior sequences of the free-moving mice are extracted (Fig. 5). Similar to the Markov model analysis, the control group is also used to highlight the intrinsic behavior sequence patterns of the free-social interaction. The behavior sequence fractions with significant differences between the two groups are illustrated and sorted using hierarchical clustering (Fig. 5A). To depict the intricate behavioral sequences, we extend the analysis from the static behavioral module fingerprint, as outlined in [24], to encompass a dynamic behavioral sequence fingerprint (Fig. 5B). Although the temporal

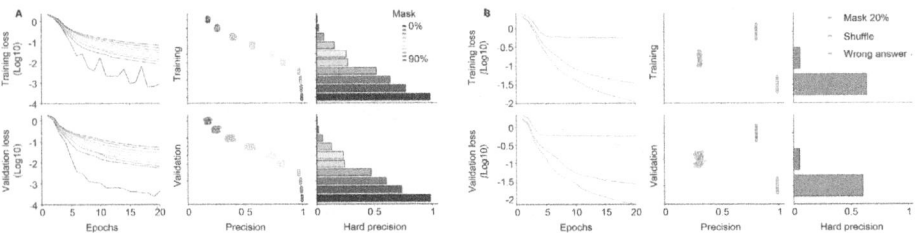

Fig. 4. The parameter optimization and performance evaluation of BL-BERT. **A**, The performance comparison of BL-BERT using different percentages of masks. The top row, the loss, precision, and hard precision in training. The bottom row, the loss, precision, and hard precision in validation. **B**, The performance comparison of BL-BERT using shuffled sequences and wrong answers of masks. The top row, the loss, precision, and hard precision in training. The bottom row, the loss, precision, and hard precision in validation.

dynamics of the behavior sequences are flexible, the fingerprint is more intuitive than the visualization of the directed graph of the Markov model. It is much easier to locate the starting points of each behavior sequence. Additionally, the different font sizes of the behavior modules illustrate their importance in the behavior sequences.

3 Discussion

We devised BL-BERT intending to autonomously identify significant behavior sequences. BL-BERT effectively quantifies the sequential behavior structures of free-moving mice. Furthermore, BL-BERT translates behavior sequences into visually intuitive fingerprints, facilitating the understanding from a human language view angle. Notably, BL-BERT addresses the shortcomings of traditional Markov-based models, particularly concerning the identification of starting points and handling non-Markovian sequences. Collectively, these findings underscore the utility of BL-BERT as a robust tool for behavior sequence extraction.

The extracted behavior sequences show large variabilities across individuals. It will be more comprehensible in analogy with the human language. The sentences with the same meaning could have different structures because of the various habits of the linguistic organization. This view angle suggests that even though mice are reared in the experimental environment, individual differences in behavior can not be ignored. When the observation scale of behavior increases, the variances behind the behavior are revealed. Tracing back to the brain, there should be more individualized neural coding patterns in the top-down control of behavior organizations from modules to sequences.

Recent machine learning-based behavior analysis methods focus on understanding poses from the behavior module level. For example, BeA dynamically decomposes the behavior modules using a dynamic time alignment kernel as the metric [6]. The behavior modules are further clustered into a behavior atlas using the uniform manifold approximation and projection [25] and identified by manual annotations. Although these methods like BeA give a name to each segment of poses, the meaning of behavior is still hard to understand such as the exploratory versus social behaviors. At the behavior module level, the components of exploratory and social behaviors are similar to each

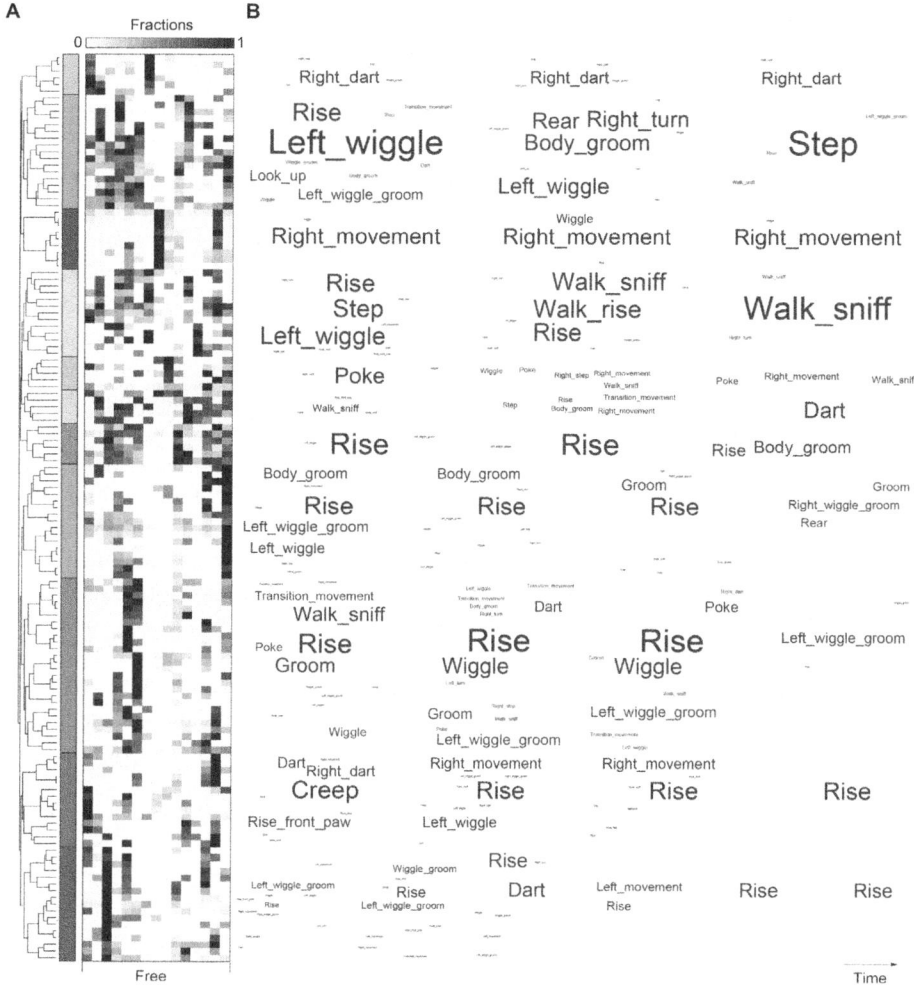

Fig. 5. The behavior sequence fingerprints from BL-BERT. **A**, The cluster gram of behavior sequences of free-moving mice. **B**, The behavior sequence fingerprint of free-moving mice. Blue words represent that the behavior sequences are significantly higher in the free group (t-test or Wilcoxon rank sum test, $n = 15$). The size of the words represents the probability of the behavior modules in a behavior sequence.

other like walking, running, turning, and rearing. Even if the module proportion of exploratory and social behaviors can be different, the sequential structures are impossible to reflect in these proportions. Hence, BL-BERT should be regarded as the next step after the behavior decomposition to quantify the sequential features of behavior.

BL-BERT's applicability extends beyond the detection of social interaction behavior sequences. Other natural behaviors, such as hunting and defensive behaviors, also exhibit stereotypical sequence patterns [26–28]. For example, defensive escape behavior includes threat detection, escape initiation, escape execution, and escape termination stages [28]. These stages are modulated by the superior colliculus, amygdala, periaqueductal gray, and other brain networks [28]. Sequential behavior needs the coordination of these brain circuits. Identifying more meaningful sequential behavior can promote the understanding of the high-order information integration of the brain. The neural representation of sequential behavior should be more abstract than the behavior modules. Analyzing neural codings across hierarchical behaviors can contribute to the interpretation of the organizing structure of latent variables behind neural activities [29, 30].

Consequently, BL-BERT holds promise for widespread utilization. While certain studies have explored the neural underpinnings of behavior sequences, their efficacy has been constrained by the limitations of the Markov model [12, 31]. In contrast, BL-BERT offers a superior alternative, enabling the elucidation of behavior sequences with enhanced performance and strong correlations with neural recordings. Moreover, automated behavior sequence identification mitigates human bias in mapping behaviors to neural activities.

Due to its reliance on deep learning techniques, BL-BERT necessitates considerable computational resources, potentially limiting its implementation within traditional wet lab settings. Another limitation of BL-BERT lies in its interpretability, which hinges upon the annotations assigned to behavior modules. The definition and categorization of these modules significantly influence the interpretation of behavior sequences derived from BL-BERT. Establishing a standardized dataset could serve as a viable solution to obtain consistent and interpretable behavior sequences when employing BL-BERT. AI tools have been applied to decipher the behavior modules from poses in the most recent years [7, 32], which drives the building of a standardized dataset of behavior modules in the near future. The dataset would be essential to fill the gap between behavior modules and body language.

4 Methods

4.1 Animals

Ten male C57BL/6J mice (8–10 weeks old) were used in the experiment. The mice were housed at 5 mice per cage under a 12 h light-dark cycle at 22–25 °C with 40–70% humidity, and were allowed to access water and food ad libitum (Shenzhen Institutes of Advanced Technology, Shenzhen, China). Five of them accepted the surgery for the neural recording of two-photon microscopy. The procedure and experiences of imaging have been described in previous studies [20, 33, 34]. All husbandry and experimental procedures were approved by the Animal Care and Use Committee at the Shenzhen Institute of Advanced Technology, Chinese Academy of Sciences.

4.2 The Process of Behavior Modules

The 3D behavior modules of the mice with neural recording are reconstructed, decomposed, and annotated following the procedure of Behavior Atlas [6]. 41 behavior modules in total are identified. They are "walking rising", "right stepping", "body grooming type 1", "body grooming type 2", "body grooming type 3", "nose grooming", "face grooming type 1", "left movement", "right movement", "poking", "grooming type 1", "looking up", "creeping", "transition movement type 1", "stepping", "darting", "standing grooming", "left wiggling type 1", "transition movement type 2", "transition movement type 3", "rising front paw", "right wiggling grooming", "rearing", "grooming type 2", "left wiggling type 2", "walking sniffing type 1", "right darting", "wiggling", "rising", "left turning", "left wiggling grooming", "pausing type 1", "pausing type 2", "right turning", "approaching", "walking sniffing type 2", "pausing type 3", "face grooming type 2", "wiggling grooming", "left grooming", "face grooming type 3". The behavior modules are shorted in the visualization of the behavior sequence fingerprints for concision.

4.3 Training, Validation, and Prediction of BL-BERT

BL-BERT is trained on one NVIDIA GeForce RTX 3090 GPU. The layers of the Transformer encoder and decoder are set to 6. The batch size is set to 200. The changes in batch size from 16 to 200 influence the training time but not the precision. The number of epochs is set to 20. The number of accumulation steps is set to 10. The initial learning rate is set to 1.0. The learning rate is changed during training using the warm-up strategy [35]. The learning rate is decreased by the inverse square root of the step number. The turning point of step number is 4000. Other configurations and model structures of BL-BERT are the same as the Annotated Transformer [35]. 200,000 behavior sequences are randomly resampled from the raw behavior data. Each behavior sequence includes 60 behavior modules. 80% and 20% of the 200,000 behavior sequences are split separately into training and validation datasets. The prediction of BL-BERT is based on greedy decoding [35]. The token with the highest probability as its next token is selected as the output. The output length is set to 60.

4.4 Statistics

Before hypothesis testing, data were first tested for normality by the Shapiro–Wilk normality test and for homoscedasticity by the F test. For normally distributed data with homogeneous variances, parametric tests were used; otherwise, non-parametric tests were used. All the analyses of variance (ANOVA) have been corrected by the recommended options of Prism v.8.0.

Acknowledgments. This work was supported in part by the National Natural Science Foundation of China (grant no. T2394530 to P.W), the Research Fund for International Senior Scientists (grant no. T2250710685 to P.W.), National Natural Science Foundation of China (grant no. 32222036 to P.W.), STI2030-Major Projects (grant no. 2021ZD0203900 to P.W.), the National Key R&D Program of China (grant no. 2018YFA0701403 to P.W.), Shenzhen Key Basic Research Project (grant no. JCYJ20220818100805013 to P.W), SIAT Distinguished Young Scholars (E4G024 to P.W.), and the Key Area R&D Program of Guangdong Province (grant no. 2018B030338001 to P.W.).

Disclosure of Interests. The authors have no competing interests to declare that are relevant to the content of this article.

References

1. Gris, K.V., Coutu, J.P., Gris, D.: Supervised and unsupervised learning technology in the study of rodent behavior. Front. Behav. Neurosci. **11**(July), 1–6 (2017). https://doi.org/10.3389/fnbeh.2017.00141
2. Mathis, A., et al.: DeepLabCut: markerless pose estimation of user-defined body parts with deep learning. Nat. Neurosci. **21**, 1281–1289 (2018). https://doi.org/10.1038/s41593-018-0209-y
3. Pereira, T.D., et al.: SLEAP: a deep learning system for multi-animal pose tracking. Nat. Methods **19**(4), 486–495 (2022). https://doi.org/10.1038/s41592-022-01426-1
4. Tang, G., Han, Y., Liu, Q., Wei, P.: Anti-drift pose tracker (ADPT): a transformer-based network for robust animal pose estimation cross-species. bioRxiv, p. 2024.02.06.579164 (2024). https://doi.org/10.1101/2024.02.06.579164
5. Wiltschko, A.B., et al.: Mapping sub-second structure in mouse behavior. Neuron **88**(6), 1121–1135 (2015). https://doi.org/10.1016/j.neuron.2015.11.031
6. Huang, K., et al.: A hierarchical 3D-motion learning framework for animal spontaneous behavior mapping. Nat. Commun. **12**(1) (2021). https://doi.org/10.1038/s41467-021-22970-y
7. Luxem, K., et al.: Identifying behavioral structure from deep variational embeddings of animal motion. Commun. Biol. **5**(1) (2022). https://doi.org/10.1038/s42003-022-04080-7
8. Hasegawa, M., Ohtani, N., Ohta, M.: Dogs' body language relevant to learning achievement. Animals **4**(1) (2013). https://doi.org/10.3390/ani4010045
9. Ebbesen, C.L., Froemke, R.C.: Body language signals for rodent social communication. Curr. Opin. Neurobiol. **68**, 91–106 (2021). https://doi.org/10.1016/J.CONB.2021.01.008
10. Wemelsfelder, F., Hunter, A.E., Paul, E.S., Lawrence, A.B.: Assessing pig body language: agreement and consistency between pig farmers, veterinarians, and animal activists. J. Anim. Sci. **90**(10) (2012). https://doi.org/10.2527/jas.2011-4691
11. Taubert, J., et al.: A broadly tuned network for affective body language in the macaque brain. Sci. Adv. **8**(47) (2022). https://doi.org/10.1126/sciadv.add6865
12. Markowitz, J.E., et al.: The striatum organizes 3D behavior via moment-to-moment action selection. Cell **174**(1), 44-58.e17 (2018). https://doi.org/10.1016/j.cell.2018.04.019
13. Liu, Q., et al.: An iterative neural processing sequence orchestrates feeding. Neuron **111**(10) (2023). https://doi.org/10.1016/j.neuron.2023.02.025
14. Alba, V., Berman, G.J., Bialek, W., Shaevitz, J.W.: Exploring a strongly non-Markovian animal behavior (2020). https://arxiv.org/abs/2012.15681v1. Accessed 29 Mar 2024
15. Vaswani, A., et al.: Attention is all you need (2017)
16. Floridi, L., Chiriatti, M.: GPT-3: its nature, scope, limits, and consequences. Minds Mach. **30**(4) (2020). https://doi.org/10.1007/s11023-020-09548-1
17. Tang, B., Matteson, D.S.: Probabilistic transformer for time series analysis. In: Advances in Neural Information Processing Systems (2021)
18. Devlin, J., Chang, M.W., Lee, K., Toutanova, K.: BERT: pre-training of deep bidirectional transformers for language understanding. In: NAACL HLT 2019 - 2019 Conference of the North American Chapter of the Association for Computational Linguistics: Human Language Technologies - Proceedings of the Conference, vol. 1, pp. 4171–4186 (2018). https://arxiv.org/abs/1810.04805v2. Accessed 23 Apr 2023
19. Han, Y., et al.: Multi-animal 3D social pose estimation, identification and behaviour embedding with a few-shot learning framework. Nat. Mach. Intell. **6**, 48–61 (2024). https://doi.org/10.1038/s42256-023-00776-5
20. Han, Y., et al.: MouseVenue3D: a markerless three-dimension behavioral tracking system for matching two-photon brain imaging in free-moving mice. Neurosci. Bull. **38**(3), 303–317 (2022). https://doi.org/10.1007/S12264-021-00778-6/TABLES/2

21. Han, Y., Huang, K., Chen, K., Wang, L., Wei, P.: An automatic three dimensional markerless behavioral tracking system of free-moving mice. In: 2021 IEEE 11th Annual International Conference on CYBER Technology in Automation, Control, and Intelligent Systems, CYBER 2021, pp. 306–310 (2021). https://doi.org/10.1109/CYBER53097.2021.9588299
22. Liu, N., Han, Y., Ding, H., Huang, K., Wei, P., Wang, L.: Objective and comprehensive re-evaluation of anxiety-like behaviors in mice using the Behavior Atlas. Biochem. Biophys. Res. Commun. **559**, 1–7 (2021). https://doi.org/10.1016/j.bbrc.2021.03.125
23. Fruchterman, T.M.J., Reingold, E.M.: Graph drawing by force-directed placement. Softw. Pract. Exp. **21**(11) (1991). https://doi.org/10.1002/spe.4380211102
24. Wettig, A., Gao, T., Zhong, Z., Chen, D.: Should you mask 15% in masked language modeling? In: EACL 2023 - 17th Conference of the European Chapter of the Association for Computational Linguistics, Proceedings of the Conference (2023). https://doi.org/10.18653/v1/2023.eacl-main.217
25. McInnes, L., Healy, J., Melville, J.: UMAP: uniform manifold approximation and projection for dimension reduction (2018). http://arxiv.org/abs/1802.03426
26. Shang, C., et al.: A subcortical excitatory circuit for sensory-triggered predatory hunting in mice. Nat. Neurosci. **22**(6) (2019). https://doi.org/10.1038/s41593-019-0405-4
27. Tseng, Y.T., Schaefke, B., Wei, P., Wang, L.: Defensive responses: behaviour, the brain and the body. Nat. Rev. Neurosci. **24**(11) (2023). https://doi.org/10.1038/s41583-023-00736-3
28. Evans, D.A., Stempel, A.V., Vale, R., Branco, T.: Cognitive control of escape behavior. Trends Cogn. Sci. **23**(4), 334–348 (2019). https://doi.org/10.1016/j.tics.2019.01.012
29. Schneider, S., Lee, J.H., Mathis, M.W., Lee, J.H.: Learnable latent embeddings for joint behavioural and neural analysis. Nature **617**(7960), 360–368 (2023). https://doi.org/10.1038/s41586-023-06031-6
30. Han, Y., Hou, X., Han, C.: Latent embeddings: an essential representation of brain–environment interactions. Brain-X **1**(3) (2023). https://doi.org/10.1002/brx2.40
31. Klaus, A., Martins, G.J., Paixao, V.B., Zhou, P., Paninski, L., Costa, R.M.: The spatiotemporal organization of the striatum encodes action space. Neuron **95**(5), 1171-1180.e7 (2017). https://doi.org/10.1016/j.neuron.2017.08.015
32. Goodwin, N.L., et al.: Simple behavioral analysis (SimBA) as a platform for explainable machine learning In behavioral neuroscience. Nat. Neurosci. **27**, 1411–1424 (2024). https://doi.org/10.1038/s41593-024-01649-9
33. Zong, W., et al.: Fast high-resolution miniature two-photon microscopy for brain imaging in freely behaving mice. Nat. Methods **14**(7), 713–719 (2017). https://doi.org/10.1038/nmeth.4305
34. Zong, W., et al.: Miniature two-photon microscopy for enlarged field-of-view, multi-plane and long-term brain imaging. Nat. Methods (2021). https://doi.org/10.1038/s41592-020-01024-z
35. Rush, A.: The annotated transformer (2019). https://doi.org/10.18653/v1/w18-2509

Benchmarking Neural Decoding Backbones Towards Enhanced On-Edge iBCI Applications

Zhou Zhou[1], Guohang He[1], Zheng Zhang[1,2(✉)], Luziwei Leng[2], Qinghai Guo[2], Jianxing Liao[2], Xuan Song[1], and Ran Cheng[1(✉)]

[1] Southern University of Science and Technology, Shenzhen 518055, China
ranchengcn@gmail.com
[2] Advanced Computing and Storage Lab, Huawei Technologies Co., Ltd., Shenzhen 518055, China
zhangzheng147@huawei.com

Abstract. Traditional invasive Brain-Computer Interfaces (iBCIs) typically depend on neural decoding processes conducted on workstations within laboratory settings, which prevents their everyday usage. Implementing these decoding processes on edge devices, such as the wearables, introduces considerable challenges related to computational demands, processing speed, and maintaining accuracy. This study seeks to identify an optimal neural decoding backbone that boasts robust performance and swift inference capabilities suitable for edge deployment. We executed a series of neural decoding experiments involving non-human primates engaged in random reaching tasks, evaluating four prospective models, Gated Recurrent Unit (GRU), Transformer, Receptance Weighted Key Value (RWKV), and Selective State Space model (Mamba), across several metrics: single-session decoding, multi-session decoding, new session fine-tuning, inference speed, calibration speed, and scalability. The findings indicate that although the GRU model delivers sufficient accuracy, the RWKV and Mamba models are preferable due to their superior inference and calibration speeds. Additionally, RWKV and Mamba comply with the scaling law, demonstrating improved performance with larger data sets and increased model sizes, whereas GRU shows less pronounced scalability, and the Transformer model requires computational resources that scale prohibitively. This paper presents a thorough comparative analysis of the four models in various scenarios. The results are pivotal in pinpointing an optimal backbone that can handle increasing data volumes and is viable for edge implementation. This analysis provides essential insights for ongoing research and practical applications in the field.

Keywords: Neural decoding · Brain-computer interfaces · Deep neural networks

---

Z. Zhou and G. He—Contribute equally to this work.

1 Introduction

Advancements in invasive Brain Computer Interfaces (iBCIs) have demonstrated promising results across various applications, including speech decoding [10,27,28], prosthesis control [7,31], neurological disorders rehabilitation [5,15,21,22] and more. Accurate decoding the brain activities is crucial for the success of these applications. Previous efforts have focused on employing adaptive filters such as Kalman Filters [7,29,30] or traditional machine learning models such as Recurrent Neural Networks (RNNs) [23,27]. However, with the expansion of the available neural data, significant progress has been made using Transformer-based architectures. Models such as Neural Data Transformer (NDT1) [32] leverage multi-session, multi-task and multi-subject neural data, yielding improved decoding performance and enhanced generalization capabilities with unseen data.

Limitations still exist among these methods. Despite the advantages of RNNs for handling long-term dependency, their inherent serial dependency significantly affect the model's inference speed [32]. Meanwhile, it remains unclear whether scaling up GRU model size with data volume improves neural decoding accuracy. Transformers facilitate parallel computation and adhere to the scaling laws [12], but the increase in model size and sequence length leads to quadratic growth in model complexity ($O(n^2)$), requiring a dramatic escalation in computational resources in order to fit in edge-device for portable BCI applications in daily use.

Models such as Receptance Weighted Key Value model (RWKV) [18] and Selective State Space model (Mamba) [8] have been designed utilizing linear attention mechanisms that offer reduced temporal and spatial complexity compared to traditional transformers. These models have demonstrated competitive performance in natural language processing and computer vision tasks [8,16,18], but it remains unclear which model is most suitable as the backbone for neural decoding.

This paper investigates whether recent advancements in model architectures can enhance neural decoding. Instead of benchmarking against state-of-the-art (SoTA) architectures, we compare the RWKV and Mamba models with the GRU and Transformer models in terms of computational efficiency and decoding accuracy. We have designed a series of experiments to assess various parameters: decoding accuracy, adaptiveness to new sessions, inference time, and scalability trends on model size, to identify an optimal neural decoding backbone. To the best of our knowledge, this work might be the first effort to investigate linear attention mechanisms in neural decoding, targeting fast and low-power inference on edge devices.

2 Related Work

2.1 Neural Decoding

Neural decoding primarily relied on adaptive filters or traditional machine learning methods such as Kalman Filters [7,29,30], Wiener Filters [11] or SVM [24].

However, with the advent of deep learning, particularly the emergence of large-scale models, there has been a significant shift in neural decoding approaches. Deep learning models facilitate automated feature learning, reducing the impact of subjective factors and greatly improving decoding accuracy and efficiency. Recurrent neural networks and Transformers have now found more applications in neural decoding tasks [20]. Contemporary applications of brain decoding technologies extend to medical rehabilitation, assistive communication, and human-computer interaction [19].

2.2 RWKV

Transformer has precipitated as disruptive revolution, particularly due to its widespread application of attention mechanisms across multiple domains. However, a significant issue arises as the memory and computational complexity of the Transformer grows quadratically with increasing sequence length. Concurrently, RNNs exhibit linear growth in memory and computational demands but are significantly outperformed by Transformer due to limitations in parallelization and scalability [18]. To address this challenge, Bo Peng et at. have proposed the RWKV, which integrates the efficient parallel training advantages of Transformer with the effective inference comparable to that of similarly scaled Transformer, underscoring its potential and effectiveness in handling large-scale sequence data [2].

2.3 Mamba

The state space model is a mathematical framework used to describe the evolution of systems over time. It employs state vectors to represent the current state of the system and uses state transition equations and observation equations to correlate the changes between system states and the relationship with observed data [9]. Mamba is an enhanced approach based on the structured state space model S4, integrating the recurrent structure of recurrent neural networks and the parallel characteristics of convolution neural networks. This approach excels in capturing long-term dependencies in sequential data and facilitates efficient parallel computation. By combining structured state space models with deep learning techniques, Mamba can handle sequential data more effectively, exhibiting higher modeling capability and predictive performance. Mamba has demonstrated superior performance in various domains, including language modeling, DNA sequence modeling, audio modeling and generation [8].

3 Methods

The system architecture is shown in Fig. 1. The raw neural recording from the Utah array is processed into spike count bins and decoded using GRU, Transformer, RWKV and Mamba as four different backbones. The decoded output is compared with the ground truth motion activities. The detailed workflow is given below.

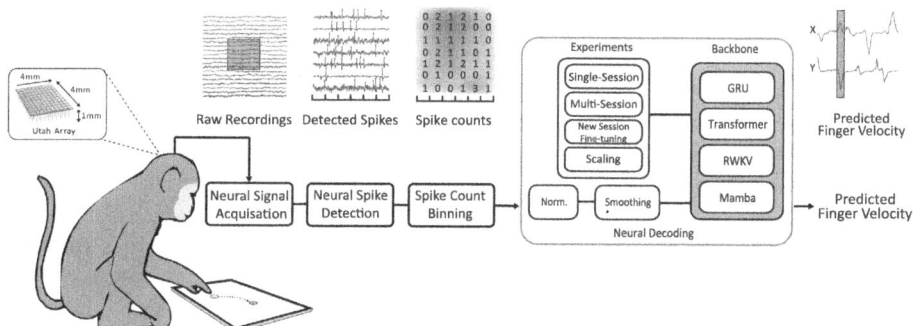

Fig. 1. Raw neural signals were recorded from the primary motor cortex (M1) area of a monkey using a 96-channel Utah microelectrode array during random reach tasks. Spike activity detected from these neural signals was binned temporally across the 96 channels. The resulting matrix of spike counts served as inputs for various methods after normalization and smoothing, and the outputs were the predicted finger velocities along the x and y axes. Experiments conducted under different scenarios facilitated comparisons of predictive accuracy, inference speed, and scalability among the four types of backbone models.

3.1 Data Processing

Datasets. The dataset from [17] is used in this study, which includes a rich collection of neural and behavioral data recorded from nonhuman primates engaged in a random target reaching task. This task requires the subject controlling a ticker to move the computer cursor and reach a series of randomly distributed targets displayed on screen in succession. During the execution of the task, the neural activities from primary motor cortex (M1) and primary sensory cortex (S1) are collected using Utah array, and the position of the subject's hand kinematic trajectories are recorded using motion tracking systems.

The neural recordings in this dataset consists of extracellular spike recordings, with the event times of threshold crossings sorted into discrete units. The recordings collected from subject Indy are used in this studies in total 30 sessions. The kinematic measurements contain the x and y coordinates of the subject's fingertip and cursor position as it reaches out, as well as the x and y coordinates of the set targets, both sampled at a frequency of 250 Hz.

Data Processing. In this studies, we only used recordings collected from the M1 cortex. We partitioned each session of the recorded task into multiple temporal bins with duration of 10 ms. Due to the sampling rate of 250 Hz, the sampling frequency is increased to 1000 Hz using linear interpolation. Within these bins, we quantified the number of spike events (threshold crossings) for each neural recording channel, thereby capturing the discrete neural firing patterns over time. It is worth noting that we use the unsorted spike events known as multiunit activities. In practice, spike sorting can be require too much computation for on-chip processing while using the sorted single-unit activities only bring limited decoding accuracy improvement as shown in [25].

The cursor's velocity is used to characterize the kinematics of the reaching movement. The binned spike event and cursor velocity were temporally aligned, normalized and smoothed with a Gaussian smoothing operation, which attenuates high-frequency noise and elucidate the underlying signal trends following [14].

In the experiments, the input to the model is denoted as $Spk \in \mathbb{R}^{S \times C}$, where S represents the timesteps used for each prediction and C denotes the number of channels. The ground truth denoted as $Vel \in \mathbb{R}^{S \times 2}$, which represents the finger speed in x and y axis at each timestep.

3.2 Backbone Models

GRU. Proposed by Cho et at. [4], GRU is a variant of the Recurrent Neural Network (RNN), specifically designed to address the challenges of gradient explosion and gradient vanishing in training. GRU achieves this by employing update gates and reset gates, which selectively update useful information and capture of long-term dependencies within time series data. GRU can be characterized by the following formulations [6]:

$$h_t^j = (1 - z_t^j) \odot h_{t-1}^j + z_t^j \odot \tilde{h}_t^j \tag{1}$$

$$z_t^j = \sigma(W_z x_t + U_z h_{t-1}^j) \tag{2}$$

$$\tilde{h}_t^j = \tanh(W x_t + U(r_t \odot h_{t-1}^j)) \tag{3}$$

$$r_t^j = \sigma(W_r x_t + U_r h_{t-1}^j) \tag{4}$$

The reset gate (r) is a gating mechanism that modulates the flow of information from the previous activation, allowing the model to discard irrelevant past state information, thus mitigating the vanishing gradient problem. The update gate (z) determines the extent to which the unit updates its activation, or hidden state (h). It controls the degree of information transfer from the previous state to the current state, enabling the model to capture long-term dependencies. The activation (h), commonly referred to as the hidden state, captures the learned information at the current time step and is recursively influenced by past activation. In our work, we employ hidden size $d_h = 256$. The candidate activation $(\tilde{h})$ is a proposed update to the hidden state, which incorporates new input while being modulated by the reset gate to potentially discard the irrelevant previous state.

Transformer. The foundational mechanism of the Transformer is its self-attention mechanism, which enables the model to dynamically adjust the weighting of input data, such as tokens or sequence elements, based on their contextual relevance [26]. Unlike GRUs, which process data sequentially, Transformers handle input in parallel during the training phase, significantly expediting the training process.

We employ the classic Multihead Scaled Dot-Product Attention mechanism along with an encoder-decoder architecture. Unlike traditional approaches that transform input vectors of vocabulary tokens through an embedding layer to embed feature dimensions, we directly take the spike matrix $x \in \mathbb{R}^{S \times C}$ as the input for both encoder and decoder and treat the channel dimension C of the input spike matrix as the feature dimension and project the feature dimension to the model hidden dimension following Eq. 5 and this projection is also used in RWKV and Mamba model.

$$A = f(x) = Wx + b \qquad (5)$$

$$B = E[\text{positions}] \qquad (6)$$

$$input = \text{Dropout}(A + B) \qquad (7)$$

$$output = \text{Decoder}(\text{Encoder}(input), input) \qquad (8)$$

The function f represents a linear mapping layer, where W and b denote the weights and biases of the input layer, respectively. E corresponds to the positional embedding matrix, from which an embedding vector is selected for each positional index. Where $input \in \mathbb{R}^{S \times d_{\text{model}}}$ is the input to the encoder and decoder and $output \in \mathbb{R}^{S \times 2}$ is the predicted x and y axis velocity. In the encoder and decoder, the attention is implemented as below:

$$\text{Attention}(Q, K, V) = \text{softmax}(\frac{QK^T}{\sqrt{d_k}})V \qquad (9)$$

$$\begin{aligned}\text{MultiHead}(Q, K, V) &= \text{Concat}(\text{head}_1, \ldots, \text{head}_h)W^O \\ \text{where head}_i &= \text{Attention}(Q', K', V')\end{aligned} \qquad (10)$$

Where the Q, K, V are calculated following Eq. 5 with independent weights and zero bias, the parameter matrix $W^O \in \mathbb{R}^{hC \times d_{\text{model}}}$ and $Q' \in \mathbb{R}^{S \times h \times d_q}$, $K' \in \mathbb{R}^{S \times h \times d_k}$, $V' \in \mathbb{R}^{S \times h \times d_v}$. Here, we employ $h = 2$ heads, $d_q = d_k = d_v = \frac{d_{\text{model}}}{h}$ and $d_{model} = 128$.

Given the limited variance in input data patterns, the data is processed through two separate attention heads. The system comprises three layers each of encoders and decoders, culminating in the prediction of velocities in the x and y axes.

RWKV. Unlike most RNNs, RWKV is a recurrent model combines the efficient parallelizable training of transformers with the fast inference time. RWKV reformulates the attention mechanism with a variant of linear attention, replacing traditional dot-product token interaction with more effective channel-directed attention [18]. It mitigates the memory bottleneck and quadratic scaling issues inherent in Transformers through efficient linear scaling. It also preserves the ability for parallelized training and ensures robust scalability.

$$r_t = W_r(\mu_r \odot x_t + (1 - \mu_r) \odot x_{t-1}) \quad (11)$$
$$k_t = W_k(\mu_k \odot x_t + (1 - \mu_k) \odot x_{t-1}) \quad (12)$$
$$v_t = W_v(\mu_v \odot x_t + (1 - \mu_v) \odot x_{t-1}) \quad (13)$$

R encodes historical information, activated via a Sigmoid function and incorporating a forgetting mechanism. W signifies the positional weight decay vector, a trainable parameter within the model. The terms K and V function analogously to the key and value in Transformer architectures. Distinct from traditional models where x is simply the embedding of the current token, in the RWKV, x is calculated as the weighted sum of the embeddings of the current token and the previous token.

$$wkv_t = \frac{\sum_{i=1}^{t-1} e^{-(t-1-i)w+k_i} \odot v_i + e^{u+k_t} \odot v_t}{\sum_{i=1}^{t-1} e^{-(t-1-i)w+k_i} + e^{u+k_t}} \quad (14)$$

Equation 14 functions similarly to an attention mechanism, representing position t as a learnable weighted sum of past content. In RWKV, w is treated as a channel-wise vector that adjusts according to the relative position, requiring the training of only a single parameter vector w. u is designated for individual processing of the current token's position, serving to circumvent any potential degradation of w.

Mamba. In contrast to the quadratic scaling observed with traditional models, Mamba demonstrates a throughput up to five times faster than the Transformer and exhibits linear scaling with sequence length [8]. Unlike RNNs, which compress all information into a hidden space and struggle with long-term memory issues, Mamba introduces a selective state-space model. This model offers the benefits of a linear recurrent network, enhanced by mechanisms for rapid training and effective context retention. Improvements in Mamba's Structured State Spaces (SSM) include a selection mechanism that filters out irrelevant information while enabling indefinite memory retention, and a hardware-aware algorithm optimized for GPU memory layouts to facilitate hardware acceleration. This ensures efficient computation cycling without extending the state unnecessarily, thus enhancing performance.

The SSM Mamba consists of the following two equations:

$$x_t = f(x_{t-1}, u_t, w_t) \quad (15)$$
$$y_t = h(x_t, v_t) \quad (16)$$

Equation 15 represents the state transition equation, describing how the system state evolves over time. Here, x_t denotes the system state at time step t, u_t represents the control input, w_t is the process noise, and f is the state transition function. Equation 16 is the observation equation, y_t represents the observation data at time step t, v_t denotes the observation noise, and h is the observation function. The concept of selectivity in Mamba allows the model to selectively remember or forget information at each time step.

4 Experiments and Key Results

4.1 Experiment Settings

To evaluate the capabilities of different backbone models across various dimensions, four distinct experiments were established: single-session, multi-session, new session finetuning, and scaling experiments (set timestpes as 128, 1024, 128 and 1024 respectively). A total of 30 sessions, collected over different days from the same subject, were used. All neural recordings from these 30 days were divided into training and testing datasets with an 8:2 ratio, consistently applied across all experiments.

Single-Session Experiment: This experiment assessed the ability of the backbone models to perform effectively on small datasets. Each of the four models was trained independently on data from individual sessions, with recording lengths varying from 360 s to 3363 s.

Multi-Session Experiment: This experiment focused on the models' capacity to extract deep latent representations from neural recordings with input feature shifting overtime. A unified model was trained using training sets from all sessions. Over time, the quality of the recordings degraded due to scar tissue encapsulation around the implants, leading to increased noise levels and a decrease in detected neural firing rates from over 20 Hz to below 10 Hz. Additionally, the neurons observed on different channels changed over time. Various training strategies were explored to help models adapt to these shifting input features.

New Session Finetuning Experiment: This experiment tested the models' ability to generalize and adapt to unseen data. Models were initially trained with datasets from the first 25 days, and then incrementally finetuned using datasets from the last five days (10 s per iteration). This setup mirrors practical scenarios for BCI calibration on new days, where a shorter calibration time is often critical. The aim was to identify the model that could quickly return to acceptable performance levels, making it more suitable for real-world use outside the laboratory.

Scaling Experiment: This experiment investigated whether increased model size could enhance performance. The scaling law has been a key principle in designing large language models [13], but its applicability in neural decoding remains unexplored.

Table 1. Parameter counts and hyperparameters of models

Model	Parameters	Epochs Single(Multi)	Layers	Embedding Size
GRU	272k	30(50)	1	256
Transformer	316k	50(50)	3	128
RWKV	294k	30(50)	2	88
Mamba	306k	30(50)	2	144

The same hyperparamter settings are used for all experiments except the scaling experiment, with details on their parameter counts and hyperparameters presented in Table 1. The requirement for the Transformer model to undergo 50 epochs may be attributed to its attention mechanism, which necessitates numerous iterations to effectively optimize attention weights. Additionally, the design of the Transformer, which processes entire sequences simultaneously, may contribute to slower convergence rates during training [26].

The R^2 is used to evaluate the neural decoding performance following [1,33]. R^2 typically ranging from 0 to 1, an R-square value of 0 indicates that the model fails to explain any variance in the dependent variable, while a value of 1 indicates a perfect fit if the model to the data. The formula for calculating R^2 is as follows:

$$RSS = \sum_{i=1}^{n}(y_i - \hat{y}_i)^2 \tag{17}$$

$$TSS = \sum_{i=1}^{n}(y_i - \bar{y})^2 \tag{18}$$

$$R^2 = 1 - \frac{RSS}{TSS} \tag{19}$$

where RSS is the residual sum of squares (the sum of the squares of the differences between actual (y_i) and predicted values $(\hat{y}_i)$), and TSS is the total sum of squares (the sum of the squares of the differences between actual values and the mean of the observed values $(\bar{y})$). Table 2 summarizes the evaluation results of four models in different experiments.

4.2 Single-Session Experiment

As shown in Table 2, the RWKV model excels in the single-session experiment, surpassing the GRU model by 0.02 in R^2. However, both the Mamba and Transformer models score below 0.7, indicating that these models are less effective when dataset sizes are limited.

In terms of inference time processing 1280 ms of neural data (1 batch), as detailed in Table 2, shows varying performance among the models. The GRU model requires the longest processing time due to its sequential processing nature. The Transformer model also exhibits relatively long inference times due to its computationally intensive operations. In contrast, the RWKV and Mamba models demonstrate significant advantages in inference speed over both the GRU and Transformer models.

Specifically, RWKV, which is a recurrent neural network devoid of an attention mechanism, avoids the computational overhead associated with computing attention matrices. This model incorporates Token Shift and Channel Mix mechanisms to optimize position encoding and channel blending, thereby enhancing both efficiency and performance. On the other hand, Mamba achieves rapid inference and maintains linear scalability with sequence length through dynamic and selective retention or dismissal of information based on input. Its streamlined and homogeneous architecture, coupled with a selective state space, markedly boosts inference speed.

4.3 Multi-session Experiment

In the multi-session experiment, we explored three different data partitioning strategies during training to identify the most effective approach for aiding models to learn as input features shifted. These strategies are as follows:

- Random partitioning: Batches are randomly selected from random sessions to be fed into the model.
- Sequential partitioning: Data batches are fed into the model in a sequential order, day by day.
- Random session partitioning: Sessions are selected randomly, but within each selected session, data batches are fed sequentially.

The random training strategy results in significantly higher stability and decoding accuracy of the model compared to the other two strategies. Although the data are strongly time-correlated, this approach of random input enhances gradient diversity, reduces cyclic biases in data appearance, and helps prevent overfitting.

Sequential training resulted in limited improvement over the single-session experiment for both the GRU and RWKV models. Although these models can memorize historical information, sequential training may still lead to catastrophic forgetting, thereby only marginally enhancing performance compared to the single-session results. In contrast, the Mamba model demonstrated a significant improvement, nearly 0.1 increase in R^2, over the single-session experiment. This suggests that Mamba's selective state space mechanism is more effective at preserving useful information and handling long-term dependencies compared to the gating mechanisms of GRU or the RWKV model in neural decoding.

However, the random session training strategy failed to provide a diverse training gradient and the data order could not convey long-term dependencies, resulting in underfitting of the model.

Another observation during the multi-session training is the difficulty in achieving convergence with the Transformer model, which required careful tuning of its hyper-parameters. In contrast, the other models exhibited less sensitivity to training hyper-parameter settings.

4.4 Fine-Tuning the Model on New Sessions

As shown in Table 2, among the four models, GRU achieved the highest average R^2 score over 5 days of fine-tuning on new sessions, reaching 0.773. The RWKV and Mamba models scored 1–2% lower, while the Transformer model recorded the lowest score at 0.748. Regarding zero-shot performance, we only saw RWKV achieved an R^2 of 0.7 in one session out of five. On average, none of the models achieved adequate zero-shot performance.

Table 2. Experiments results on all models

Experiment	Indicator	GRU	Transformer	RWKV	Mamba
Single-session	Average R^2	0.715	0.633	**0.717**	0.660
	Inference time/s	0.941	0.822	**0.303**	0.434
Multi-session	**Random train**	**0.838**	0.720	0.812	0.810
	Sequence train	0.749	0.523	0.726	**0.752**
	Random session	0.560	0.314	0.600	0.556
Fine-tuning	Average R^2	**0.773**	0.748	0.763	0.756
	Recovery time/s	214	-	202	**178**
	Zero shot	0.4811	0.383	0.452	0.370
Scaling	Max R^2	0.846	-	0.843	**0.851**
	Increment	0.010	-	0.031	**0.041**

The results from the finetuning experiment indicate that all models are capable of surpassing their performance when trained solely on single-session data. This demonstrates that despite variations in firing rates and neuron-channel mappings over time, the models can distill useful information to enhance neural decoding. The quality of the base model significantly influences the effectiveness of the finetuned model. However, the backbone model alone does not provide zero-shot capability, suggesting that additional architectural designs or training strategies are necessary to enhance the models' adaptability to input feature shifts and improve zero-shot performance.

In terms of the data length required to achieve an acceptable R^2 score of 0.7 through fine-tuning, Mamba outperformed both RWKV and GRU. This superior performance likely stems from Mamba's enhanced ability to resolve long-term dependencies, which facilitates its calibration to unseen data more effectively. Consequently, Mamba emerges as a more viable option for real-world deployment in practical BCI applications due to its robust adaptability.

4.5 Scaling Analysis

In multi-session training, the parameter count for the models we used is approximately 300k. To explore whether increasing the model size could enhance its decoding performance, we examined the improvements achieved by increase the number of layers in GRU, RWKV and Mamba models (Transformer can fail to converge in many cases and is

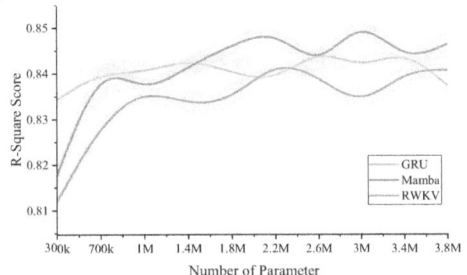

Fig. 2. Scaling parameter counts for the models range from 300k to 3.8M with error

therefore ignored.). The variation in the model's decoding R^2 scores as a function of the parameter count of these models, ranging from 300K to 3M, is illustrated in the Fig. 2.

With an increase in model parameters, the R^2 scores for Mamba and RWKV show significant improvement, reaching 0.843 and 0.851 respectively. This represents increases of 0.031 and 0.041 over their 300k parameter models. In contrast, the GRU model demonstrates only a mild improvement of 0.01 when parameters are increased, and further scaling leads to a declining trend in performance. Despite its gate mechanisms to mitigate the vanishing gradient problem, GRU's inherent sequential processing nature restricts its scalability and limits its efficiency in handling large-scale sequence tasks.

Conversely, RWKV and Mamba exhibit superior scalability and computational efficiency, outperforming GRU. This advantage is largely due to their innovative structural designs and optimization strategies that effectively address the limitations typically associated with recurrent neural networks and traditional Transformers.

While performance gains for RWKV and Mamba level off as model size increases, this plateau is mainly attributable to the limited size of the current dataset. However, with the rapid advancement of BCI technology and the anticipated increase in available data, it is reasonable to predict that RWKV or Mamba could serve as robust backbones for neural decoding in future applications.

5 Discussion

5.1 Suggestions on Model Selection

Each of the four models evaluated demonstrates distinct strengths and weaknesses. The GRU model achieves the secondary prediction accuracy in single session experiment and best predictive accuracy on multi-session experiment on the dataset used in this work. However, its inference time and calibration recovery time is constrained by its inherent serial structure. In contrast, the RWKV and Mamba model have significantly faster inference and calibration recovery time. Additionally, both RWKV and Mamba adhere to the scaling law, demonstrating a gradual improvement in predictive accuracy as model sizes increase. Mamba eventually achieves an R^2 of 0.851 when scale up to 3M, hitting the highest score among all models in different experiment settings. It also becomes compatible with the SoTA neural decoding model POYO [3], trained on a much larger dataset tested on the same task. The Transformer model, however, lags in nearly all performance metrics and is difficult to converge in our experiments.

Consequently, Mamba or RWKV could be suitable backbones for future neural decoding tasks, especially with an increasing amount of available neural recordings. Their scalability and linear computational complexity can significantly enhance decoding performance without the need for excessive computational resources, making them preferable for wearable devices used daily. For BCI applications, this choice can also lead to reduced training times, faster response times, and quicker calibration speeds. However, for studies involving a limited amount of data and those not sensitive to response times, RNN models like GRU or LSTM may suffice to provide high decoding performance in most use cases.

5.2 Limitation and Future Works

One significant challenge within the BCI field is achieving long-term stable neural decoding. Unfortunately, none of the four models can provide long-term stable decoding capabilities without finetuning, based on our experiments. While this work utilizes only one dataset, introducing a more diverse dataset could enable the model to learn a broader array of data features, thereby enhancing its robustness in practical applications.

The degradation in long-term decoding performance is primarily due to input feature shifting [33]. To manage potential data drift over prolonged periods, continuous or online learning strategies could be implemented, allowing the model to continually adapt to new data. From a computational-saving perspective, instead of full parameter updating, tuning only the input and output layers or employing some transfer learning strategies might better accommodate input variations with less computational overhead.

New training strategies can also be explored to guide the model in learning useful latent representations. By implementing a weighted loss scheme that prioritizes recent sessions chronologically, our preliminary results have already shown notably improved zero-shot outcomes.

Additionally, the backbone models in this study were only trained on a random track task with one subject. The adaptation across different tasks and subjects also needs to be carefully evaluated in future studies.

6 Conclusion

This study has conducted a comprehensive comparison of GRU, Transformer, RWKV, and Mamba models in the context of neural decoding for random reach tasks. RWKV and Mamba, which demonstrate faster inference speeds, lower computational complexity, better scalability compared to GRU and Transformer, emerge as preferred choices for deployment on wearable devices. This detailed evaluation of the various strengths and weaknesses of each model not only highlights their individual capabilities but also establishes a robust foundation for future advancement on model architecture. The insights gained from this work guide the development of more efficient and effective neural decoding architecture, paving the way for enhanced performance in practical applications.

Acknowledgement. This work was supported in part by Guangdong Natural Science Funds for Distinguished Young Scholar under Grant 2024B1515020019.

References

1. Ahmadi, N., Constandinou, T.G., Bouganis, C.S.: Robust and accurate decoding of hand kinematics from entire spiking activity using deep learning. J. Neural Eng. **18**(2), 026011 (2021)
2. Alam, M.M., Raff, E., Biderman, S., Oates, T., Holt, J.: Recasting self-attention with holographic reduced representations. In: ICML (2023)
3. Azabou, M., et al.: A unified, scalable framework for neural population decoding. In: Advances in Neural Information Processing Systems 36: Annual Conference on Neural Information Processing Systems, NeurIPS (2023)
4. Bahdanau, D., Cho, K., Bengio, Y.: Neural machine translation by jointly learning to align and translate. In: 3rd International Conference on Learning Representations, ICLR (2015)
5. Cheng, N., et al.: Brain-computer interface-based soft robotic glove rehabilitation for stroke. IEEE Trans. Biomed. Eng. **67**(12), 3339–3351 (2020)
6. Chung, J., Gülçehre, Ç., Cho, K., Bengio, Y.: Empirical evaluation of gated recurrent neural networks on sequence modeling. CoRR 1412.3555 (2014)
7. Gilja, V., et al.: A high-performance neural prosthesis enabled by control algorithm design. Nat. Neurosci. **15**(12), 1752–1757 (2012)
8. Gu, A., Dao, T.: Mamba: linear-time sequence modeling with selective state spaces. arXiv preprint arXiv:2312.00752 (2023)
9. Gu, A., Goel, K., Ré, C.: Efficiently modeling long sequences with structured state spaces. In: ICLR. OpenReview.net (2022)
10. Heelan, C., Lee, J., O'Shea, R., Lynch, L., Brandman, D.M., Truccolo, W., Nurmikko, A.V.: Decoding speech from spike-based neural population recordings in secondary auditory cortex of non-human primates. Commun. Biol. **2**(1), 1–12 (2019)
11. Hochberg, L.R., et al.: Neuronal ensemble control of prosthetic devices by a human with tetraplegia. Nature **442**(7099), 164–171 (2006)
12. Ivgi, M., Carmon, Y., Berant, J.: Scaling laws under the microscope: predicting transformer performance from small scale experiments. In: Goldberg, Y., Kozareva, Z., Zhang, Y. (eds.) Findings of the Association for Computational Linguistics: EMNLP, pp. 7354–7371. Association for Computational Linguistics (2022)
13. Kaplan, J., et al.: Scaling laws for neural language models. CoRR 2001.08361 (2020)
14. Keshtkaran, M.R., et al.: A large-scale neural network training framework for generalized estimation of single-trial population dynamics. Nat. Methods **19**(12), 1572–1577 (2022)
15. Lazarou, I., Nikolopoulos, S., Petrantonakis, P.C., Kompatsiaris, I., Tsolaki, M.: EEG-based brain-computer interfaces for communication and rehabilitation of people with motor impairment: a novel approach of the 21st century. Front. Hum. Neurosci. **12**, 14 (2018)
16. Liu, Y., et al.: Vmamba: visual state space model. CoRR 2401.10166 (2024)
17. Makin, J.G., O'Doherty, J.E., Cardoso, M.M., Sabes, P.N.: Superior arm-movement decoding from cortex with a new, unsupervised-learning algorithm. J. Neural Eng. **15**(2), 026010 (2018)
18. Peng, B., et al.: RWKV: reinventing RNNs for the transformer era. In: Findings of the Association for Computational Linguistics: EMNLP, pp. 14048–14077. Association for Computational Linguistics (2023)
19. Rapeaux, A.B., Constandinou, T.G.: Implantable brain machine interfaces: first-in-human studies, technology challenges and trends. Curr. Opin. Biotechnol. **72**, 102–111 (2021)

20. Roy, Y., Banville, H.J., Albuquerque, I., Gramfort, A., Falk, T.H., Faubert, J.: Deep learning-based electroencephalography analysis: a systematic review. CoRR 1901.05498 (2019)
21. Stanslaski, S., et al.: Design and validation of a fully implantable, chronic, closed-loop neuromodulation device with concurrent sensing and stimulation. IEEE Trans. Neural Syst. Rehabil. Eng. **20**(4), 410–421 (2012)
22. Stanslaski, S., et al.: A chronically implantable neural coprocessor for investigating the treatment of neurological disorders. IEEE Trans. Biomed. Circuits Syst. **12**(6), 1230–1245 (2018)
23. Sussillo, D., et al.: A recurrent neural network for closed-loop intracortical brain-machine interface decoders. J. Neural Eng. **9**(2), 026027 (2012)
24. Taghizadeh-Sarabi, M., Daliri, M.R., Niksirat, K.S.: Decoding objects of basic categories from electroencephalographic signals using wavelet transform and support vector machines. Brain Topogr. **28**(1), 33–46 (2015)
25. Todorova, S., Sadtler, P., Batista, A., Chase, S., Ventura, V.: To sort or not to sort: the impact of spike-sorting on neural decoding performance. J. Neural Eng. **11**(5), 056005 (2014)
26. Vaswani, A., et al.: Attention is all you need. In: Advances in Neural Information Processing Systems 30: Annual Conference on Neural Information Processing Systems, pp. 5998–6008 (2017)
27. Willett, F.R., Avansino, D.T., Hochberg, L.R., Henderson, J.M., Shenoy, K.V.: High-performance brain-to-text communication via handwriting. Nature **593**(7858), 249–254 (2021)
28. Wilson, G.H., et al.: Decoding spoken English from intracortical electrode arrays in dorsal precentral gyrus. J. Neural Eng. **17**(6), 066007 (2020)
29. Wu, W., et al.: Neural decoding of cursor motion using a Kalman filter. In: Advances in Neural Information Processing Systems, vol. 15 (2002)
30. Wu, W., Hatsopoulos, N.G.: Real-time decoding of nonstationary neural activity in motor cortex. IEEE Trans. Neural Syst. Rehabil. Eng. **16**(3), 213–222 (2008)
31. Xu, H., Han, Y., Han, X., Xu, J., Lin, S., Cheung, R.C.: Unsupervised and real-time spike sorting chip for neural signal processing in hippocampal prosthesis. J. Neurosci. Methods **311**, 111–121 (2019)
32. Ye, J., Pandarinath, C.: Representation learning for neural population activity with neural data transformers. arXiv preprint arXiv:2108.01210 (2021)
33. Zhang, Z., Constandinou, T.G.: Firing-rate-modulated spike detection and neural decoding co-design. J. Neural Eng. **20**(3), 036003 (2023)

Enhanced Local Attention with Deep Neural Networks for EEG Decoding

Wei Tao[1,2,3], Jiawo Ye[1], Chio-In Ieong[4], Haiyan Wu[2,5], and Feng Wan[1,2,3(✉)]

[1] Department of Electrical and Computer Engineering, Faculty of Science and Technology, University of Macau, Taipa, Macau
fwan@um.edu.mo
[2] Centre for Cognitive and Brain Sciences, Institute of Collaborative Innovation, University of Macau, Taipa, Macau
[3] Centre for Artificial Intelligence and Robotics, Institute of Collaborative Innovation, University of Macau, Taipa, Macau
[4] Guangdong Institute of Intelligence Science and Technology, Zhuhai, China
[5] Department of Psychology, Faculty of Social Sciences, University of Macau, Taipa, Macau

Abstract. Channel-wise attention mechanisms have significantly improved deep learning-based decoding in brain-computer interfaces (BCIs). However, these methods often fail to fully utilize spatial and temporal dynamics, focusing instead on individual channel enhancements to the detriment of broader EEG signal dynamics. To address these limitations, we introduce the Enhanced Local Attention (ELA) module, seamlessly integrated into deep neural networks that enhances EEG decoding performance. It captures long-range dependencies from deep features across two dimensions through a streamlined architecture. The ELA module employs adaptive 1D convolution for precise localization of spatial and temporal information without reducing dimensions. Additionally, it utilizes group normalization to enhance feature representation by normalizing features across groups. Its lightweight design enables easy integration into existing deep learning frameworks. Comprehensive evaluations on the BCI-IV2b dataset highlight the ELA module's superior performance. Specifically, it increases the decoding accuracy of EEG-Net in the motor imagery task from 81.89% to 85.97%. Furthermore, when integrated with ADFCNN, the ELA module achieves an accuracy of 88.13%, consistently outperforming other state-of-the-art methods.

Keywords: Attention Mechanism · Deep Learning · Brain-Computer Interfaces

1 Introduction

Electroencephalography (EEG) has emerged as a pivotal technology in the field of neuroscience due to its non-invasive nature, cost-effectiveness, and high temporal resolution [4]. Widely used in both research and clinical settings, EEG's

---

W. Tao and J. Ye—Authors contributed equally.

ability to monitor real-time brain activity enables a broad spectrum of applications, from enhancing cognitive performance and predicting epileptic seizures to enabling control over computer interfaces through imagined movements [16,19,20]. This versatility makes EEG indispensable in advancing our understanding of neural dynamics and developing practical solutions that enhance the quality of life for individuals with neurological disorders and for the general population [8]. EEG-based brain-computer interfaces (BCIs) are a transformative area that promises new modes of interaction between the human brain and external devices [2,23,28].

In recent years, deep learning (DL) has significantly enhanced the field of EEG decoding, primarily due to its robust automatic feature learning capabilities, as highlighted by Craik et al. [5]. Among various models, Convolutional Neural Networks (CNNs) excel at decoding and analyzing EEG signals [11,12,17]. They utilize weight-sharing techniques to effectively manage the computational demands associated with high-dimensional input data. Schirrmeister et al. introduced two CNN variants, ShallowConvNet and DeepConvNet, demonstrating that network depth significantly affects performance [17]. Building on this foundation, Kim et al. refined the motor imagery classification capabilities of M-ShallowConvNet by adjusting the kernel size of temporal convolutions and convolutional strides based on the original framework [11]. Additionally, EEGNet, developed by Lawhern et al., employs depthwise and separable convolutions, enabling it to generalize across various BCI paradigms and maintain robust performance, even with limited training data [12]. However, the inherent limitations of EEG data, such as limited availability and a low signal-to-noise ratio (SNR), require more sophisticated network architectures to fully exploit the information within EEG signals. To address this, Mane et al. introduced FBCNet, which incorporates Filter Bank Common Spatial Patterns (FBCSP) to capture distinct features across various frequency bands, thus enhancing feature extraction [15]. Moreover, Dai et al. developed HS-CNN, utilizing multiple convolutional scales to capture a comprehensive range of EEG signal features, significantly boosting classification accuracy [6]. Despite these innovations, CNN-based methods often struggle to capture the long-range dependencies inherent in the complex spatial and temporal patterns of EEG signals, highlighting a crucial area for further research and development in EEG-based BCIs.

To dynamically prioritize the most informative features of EEG data, attention mechanisms have significantly improved the capabilities of deep neural network models for EEG decoding. Among the various types of attention mechanisms, multi-head attention, introduced by Vaswani et al. in the Transformer architecture [25], allows models to attend to different representational subspaces at different positions, facilitating a comprehensive analysis of the entire EEG dataset [30]. Self-attention has been shown to be effective in capturing intricate temporal dependencies within EEG signals, as demonstrated by Tao et al. in their work on emotion-based BCI systems [23]. Channel-wise attention specifically targets the relevance of different EEG channels, which is crucial due to the spatial diversity of brain activity captured across various electrodes [3].

This method enables models to identify and prioritize information from the most significant channels, thus enhancing their ability to interpret complex neural signals. For instance, Tao et al. utilized channel-wise attention to capture dependencies among different electrodes [23], while Su et al. dynamically assigned different weights to EEG channels through a similar mechanism [21]. However, existing methods employing channel-wise attention often utilize the Squeeze-and-Excitation (SE) block, which uses two-dimensional global pooling to compress spatial dimensions into channel dimensions. Although the SE block adeptly encodes inter-channel relationships, it overlooks the spatial positional information within the feature maps, a crucial aspect for fully leveraging the spatial characteristics of EEG data. This limitation highlights the necessity for developing more sophisticated attention frameworks that integrate both channel-wise and spatial attention to enhance the decoding accuracy and interpretability of EEG-based models [7].

To address these issues, this paper introduces an Enhanced Local Attention (ELA) mechanism, which is integrated into deep neural networks to improve EEG decoding performance. Our main contributions are:

- We introduce the ELA module, which uniquely combines adaptive 1D convolution with group normalization. This module not only refines feature representation but also effectively localizes spatial and temporal dynamics without the need for dimension reduction. The ELA module efficiently captures long-range dependencies across channels and time points, essential for decoding intricate EEG patterns.
- The ELA module is both lightweight and efficient, significantly reducing computational demands while delivering high performance. Designed for easy integration into existing deep neural network frameworks, it offers a seamless, plug-and-play enhancement for deep learning models in BCI applications.
- Extensive validation on the BCI-IV2b dataset has demonstrated the ELA module's superior performance. The module significantly improved the decoding accuracy of EEGNet in the motor imagery task from 81.89% to 85.97%. When implemented with ADFCNN, it achieved an exceptional accuracy of 88.13%, surpassing the current state-of-the-art methods and setting a new benchmark in the field.

2 Related Work

Attention mechanisms originated in the field of computer vision, primarily focused on the extraction of spatial features. This focus significantly shifted with the introduction of SCA-CNN [3], which incorporated channel-wise attention along with spatial attention, substantially enhancing image recognition capabilities. Subsequently, the SE block [7] refined the extraction of channel information through efficient squeeze and excitation operations, gaining widespread recognition for its computational efficiency. Expanding on this foundation, the Convolutional Block Attention Module (CBAM) [29] effectively merged channel and spatial features using a combination of average and maximum pooling techniques. The Efficient Channel Attention (ECA) module [27] further advanced this integration by implementing 1-D convolution to enable local cross-channel interactions, thus reducing computational complexity compared to the SE block. Inspired by these developments, researchers have investigated the application of channel and spatial attention mechanisms in EEG decoding.

In EEG applications, MBEEGSE utilized three SE blocks in a tri-branched CNN architecture for MI-EEG decoding, enhancing the adaptability of channel-wise feature responses [1]. Similarly, Tao et al. merged spatial and channel features using a CBAM block within a CNN framework for 2D EEG image decoding in error-related potential BCI systems [22]. Furthermore, MSDAN altered the CBAM structure by integrating a convolution block and a soft-thresholding module to eliminate non-essential features for sleep staging [26]. Additionally, MBSTCNN-ECA-LightGBM employed the ECA module in a four-dimensional MI-EEG signal context, customizing kernel sizes to optimize cross-channel interactions [9].

However, the global pooling operations of the SE and ECA blocks might not effectively capture local spatial and temporal dependencies. To address these challenges, our work introduces the Enhanced Local Attention (ELA) module, designed to establish robust long-term dependencies across spatial and temporal dimensions through adaptive 1-D convolutions and to enhance feature generalization through group normalization. Paired with advanced CNN architectures, this module has undergone rigorous testing through a series of experiments on EEG decoding, demonstrating its efficacy in enhancing the precision of EEG interpretation.

3 Method

3.1 Enhanced Local Attention

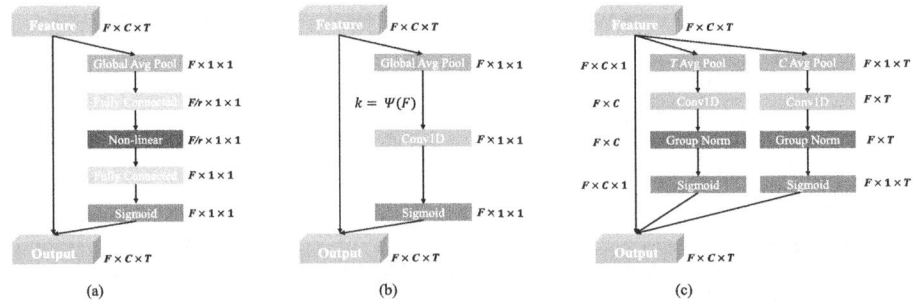

Fig. 1. Architectural comparison of three attention mechanisms: Squeeze-and-Excitation (SE) Block (a), Efficient Channel Attention (ECA) (b), and Enhanced Local Attention (ELA) (c). The diagram illustrates the key components and data flow within each module. "C Avg Pooling" and "T Avg Pooling" refer to 1D-average pooling performed across the channel and temporal dimensions, respectively.

Given EEG signals $\mathbf{X} \in \mathbb{R}^{C \times T}$, where C represents the number of EEG electrode nodes and T denotes the number of sampling points, deep neural networks apply convolution operations to extract features, resulting in an output with the shape $F \times C \times T$. Here, F represents the number of feature maps. As depicted in Fig. 1, various channel-wise attention mechanisms are employed to capture long-range dependencies among the features. The Squeeze-and-Excitation (SE) block leverages global average pooling (GAP) to compress global spatial information into a channel descriptor, aggregating features across all spatial dimensions and recalibrating the model's responses based on this global information. Conversely, the Efficient Channel Attention (ECA) refines this approach by introducing a fast 1D convolution after global pooling, facilitating the capture of local channel-wise dependencies without significantly increasing computational costs.

However, GAP may not effectively capture local spatial and temporal dependencies in EEG features. Therefore, we propose an Enhanced Local Attention (ELA) mechanism that processes the features from spatial and temporal dimensions, respectively. Additionally, we employ the adaptive 1D convolution to manage long-range dependencies across these two dimensions effectively. Besides, group normalization, which boosts the feature learning capability by normalizing multiple feature channels in groups, is employed to accommodate the intricate features extracted.

Mathematically, instead of employing GAP, our method performs separate average pooling for channel and temporal dimensions, enabling effective dual-branch convolution processing to fuse the resultant feature maps, thereby preserving channel dimensions and enhancing feature representation. The pooling operations can be defined as:

$$z_c = \frac{1}{C} \sum_{i=0}^{C-1} f(c, i), \tag{1}$$

$$z_t = \frac{1}{T} \sum_{j=0}^{T-1} f(t, j), \tag{2}$$

where f represents the features extracted from deep neural networks with a shape $F \times C \times T$, F is the number of feature maps, z_c and z_t denote the average feature maps along channel and temporal dimensions, respectively.

Following this, adaptive 1D convolution assesses dependencies among channel and temporal information by dynamically selecting the kernel size based on the number of feature channels:

$$k = \psi(D_z) = \left| \frac{\log_2(D_z)}{\gamma} + \frac{b}{\gamma} \right|_{odd}, \tag{3}$$

where $|t|\,odd$ indicates the nearest odd number of t, D_z represents the dimension of the average feature maps z_c and z_t, and γ and b are set to 2 and 1, respectively.

The processed feature maps then undergo group normalization (GN), which contrasts with the more common batch normalization (BN) and is especially advantageous in scenarios with small batch sizes or in cases where discrepancies in batch sizes between training and testing sets could impact model performance. GN's ability to normalize groups of channels aids in effectively handling the enhanced features. Following GN, a sigmoid activation function scales the attention weights between 0 and 1, with the final step being the multiplication of the enhanced features by the original inputs to produce the outputs. This sophisticated approach ensures the preservation of information integrity and improves the model's overall stability and performance.

3.2 ELA Module for Deep Models

We have developed two network frameworks, incorporating EEGNet and ADFCNN, respectively, integrating the proposed ELA block. The architectures of these neural networks are presented in Tables 1 and 2. To effectively harness temporal and spatial features from raw electroencephalogram signals, we strategically position the ELA module after the temporal convolution layer, which typically outputs the richest feature information.

Table 1. EEGNet with Enhanced Local Attention (ELA)

Module	Layer	Output Shape
Temporal Convolution	Standard Conv2D	$F \times C \times T$
ELA	Avg Pool (T, C)	$F \times C \times 1, F \times 1 \times T$
	Conv1D (C, T)	$F \times C, F \times T$
	Group Norm (C, T)	$F \times C, F \times T$
	Sigmoid (C, T)	$F \times C, F \times T$
	Dot Product	$F \times C \times T$
Spatial Convolution	Depthwise Conv2D	$F \times 1 \times T$
Point-wise Convolution	Separable Conv2D	$F \times 1 \times T/4$
	Average Pooling	$F \times 1 \times T/32$
Classifier	Flatten	$F \times T/32$
	Dense	$N \times 1$

F = Number of convolution kernel, N = Number of classes

In the EEGNet-ELA configuration, EEGNet first performs temporal convolution on the input features. The resultant feature maps are then processed by the ELA module, enhancing both channel and temporal features through one-dimensional convolution and group normalization. These enhanced features are integrated with the outputs of the temporal convolution, and then proceed through spatial and temporal convolution layers before reaching the classification module for prediction.

Similarly, in the ADFCNN-ELA architecture, the ELA module assigns attention weights to the temporal convolution outputs. The output from the ELA module is then fed into the spatial convolution layer. ADFCNN comprises two branches: Branch I employs separable spatial convolution followed by point-wise convolution and then a pooling layer, while Branch II utilizes standard

Table 2. The ADFCNN-ELA Architecture

Module	Layer	Output Shape
Branch-I	Temporal Convolution	$F \times C \times T$
	ELA	$F \times C \times T$
	Separable Spatial Convolution	$F \times 1 \times T$
	Point-wise Convolution	$F \times 1 \times T$
	Average Pooling	$F \times 1 \times T_1$
Branch-II	Temporal Convolution	$F \times C \times T$
	ELA	$F \times C \times T$
	Standard Spatial Convolution	$F \times 1 \times T$
	Average Pooling	$F \times 1 \times T_2$
Fusion	Concatenation	$F \times 1 \times (T_1 + T_2)$
	Attention-based	$F \times 1 \times (T_1 + T_2)$
Classifier	Dense	$N \times 1$

T_1 = Dimension of features learned by Branch I,
T_2 = Dimension of features learned by Branch II

spatial convolution prior to pooling. Subsequently, features from both branches are merged by a fusion module that incorporates ADFCNN's attention mechanism and are then processed by the classification module for final output. This methodical approach ensures that both frameworks leverage the advantages of ELA to enhance feature representation and improve classification accuracy.

4 Experiments and Results

4.1 Dataset and Preprocessing

In this study, we evaluated the effectiveness of the proposed model using the widely recognized public dataset BCI-IV2b. Details of the dataset are provided below:

BCI competition IV 2b dataset (BCI-IV2b) [13]: This dataset comprises EEG recordings from nine subjects at a sampling rate of 250 Hz, using three electrodes positioned at C3, Cz, and C4 according to the international 10–20 system. Two MI tasks (left-hand vs. right-hand) were performed, with each trial lasting 4 s. The first two sessions consisted of 400 trials each, while the remaining three sessions contained 320 trials each. The initial two sessions were conducted without feedback, while the subsequent three sessions included feedback.

During preprocessing, each EEG trial was represented as a matrix $\mathbf{X} \in \mathbb{R}^{C \times T}$, where C denotes the number of EEG electrode nodes and T denotes the number of sampling points. To standardize the dataset, the sampling rate was adjusted by down-sampling the raw EEG data from 1000 Hz to 250 Hz.

4.2 Simulation Setup

In our comparative experiments, we meticulously tuned hyperparameters such as training epochs, batch size, learning rate, and weight decay to ensure optimal performance across various deep learning models. The learning rate was set to 0.001 and weight decay to 0.075 for each model. We standardized the number of training epochs and batch size at 1000 and 16, respectively. All models employed the Adam optimizer and the cross-entropy loss function. The models were implemented in Python using the PyTorch framework and trained on an NVIDIA Tesla V100 GPU.

For performance evaluation, we adopted a cross-session evaluation methodology. Specifically, we applied uniform five-fold cross-validation on data from the first session, and the reported results reflect the average performance across the five models when tested on the second session's data. Furthermore, we utilized the Kappa coefficient as an evaluation metric to assess classification accuracy. This metric effectively eliminates result randomness, providing a robust measure of a model's classification performance, as higher Kappa values indicate better performance. The Kappa coefficient is mathematically defined as:

$$kappa = \frac{1}{N_v} \sum_{i=1}^{N_v} \frac{P_i - P_c}{1 - P_c} \qquad (4)$$

where N_v is the number of cross-validations, which equals 5 in this study; P_i is the average classification accuracy of the i-th cross-validation; and P_c indicates the random classification accuracy for binary classification tasks, set at 0.5 here.

4.3 Effect of Group Normalization on ELA Module

To examine the impact of group normalization on the efficacy of the ELA module in EEG decoding, a comparative analysis was conducted using EEGNet as a baseline. This investigation assessed the ELA module across various GN configurations (GN = 1, GN = 2, GN = 4, and GN = 8), as well as a variant without GN, with each model evaluated using the same EEG dataset and focusing on mean accuracy and kappa scores to measure performance improvements.

Table 3. Comparative Performance of EEGNet and EEG-ELA with Different Group Normalization Settings on Mean Accuracy and Kappa Scores

Model	Mean Accuracy (%)	Mean Kappa (%)
EEGNet	81.89 ± 10.95	63.79 ± 21.90
ELA (w/o GN)	82.14 ± 9.66	64.29 ± 19.32
ELA (GN = 1)	85.67 ± 9.75	71.34 ± 19.51
ELA (GN = 2)	**85.97 ± 9.58**	**71.95 ± 19.17**
ELA (GN = 4)	85.91 ± 10.04	71.82 ± 20.09
ELA (GN = 8)	85.41 ± 10.04	70.83 ± 20.09

As shown in Table 3, the baseline EEGNet achieved a mean accuracy of 81.89% and a kappa score of 63.79%. The ELA module without GN exhibited modest improvements in both metrics, confirming the inherent advantages of the ELA architecture. Notably, integrating group normalization into the ELA module substantially enhanced its performance. With a single group (GN = 1), the mean accuracy rose to 85.67% and the kappa score to 71.34%, indicating that even minimal group normalization significantly improved performance. However, further increases in group number (GN = 2, GN = 4, and GN = 8) showed a performance peak at GN = 2, where the mean accuracy reached 85.97% and the kappa score was 71.95%. Beyond this point, performance slightly decreased with larger group divisions (GN = 4 and GN = 8), suggesting that excessive partitioning may introduce inefficiencies or overfitting.

These results underscore the necessity of optimizing the GN setting to maximize the decoding performance of the ELA module. The optimal performance observed at GN = 2 demonstrates that moderate group normalization can significantly enhance the network's generalization capabilities without adversely affecting computational efficiency. This balance is crucial for leveraging the full potential of the ELA module in practical EEG decoding applications.

4.4 Comparisons with Different Attention Mechanisms

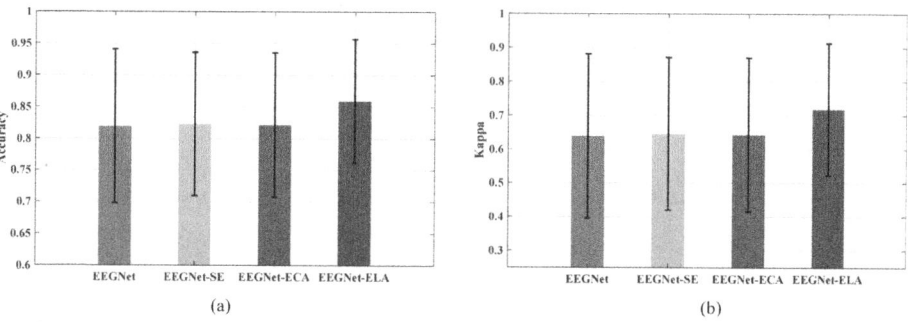

Fig. 2. (a) Accuracy comparison of EEGNet variants with various attention mechanisms, (b) Kappa coefficient comparison of EEGNet variants with various attention mechanisms.

To elucidate the effectiveness of various attention mechanisms within EEG decoding architectures, we conducted a comparative analysis of several EEGNet variants, each incorporating a different attention mechanism: Squeeze-and-Excitation (SE), Efficient Channel Attention (ECA), and Enhanced Local Attention (ELA). The primary aim was to assess the impact of these attention mechanisms on the accuracy and reliability of EEG decoding, as measured by the kappa coefficient.

Figure 2 displays the results of this experimental comparison. Panel (a) shows the accuracy achieved by each EEGNet variant, while panel (b) delineates their kappa coefficients. The conventional EEGNet, used as the baseline, demonstrated that incorporating attention mechanisms generally improves both accuracy and kappa scores. Notably, the EEGNet-ELA variant exhibited the most significant enhancements in both metrics, indicating the ELA mechanism's superior capacity for feature refinement in EEG decoding tasks.

This study highlights the substantial potential of attention mechanisms to enhance the performance of neural network architectures used in EEG decoding, particularly the ELA mechanism, which excels in boosting both accuracy and interpretability of EEG decoding. These results strongly support the integration of sophisticated attention mechanisms into the development of more precise and robust EEG-based BCIs.

4.5 Attention Visualization

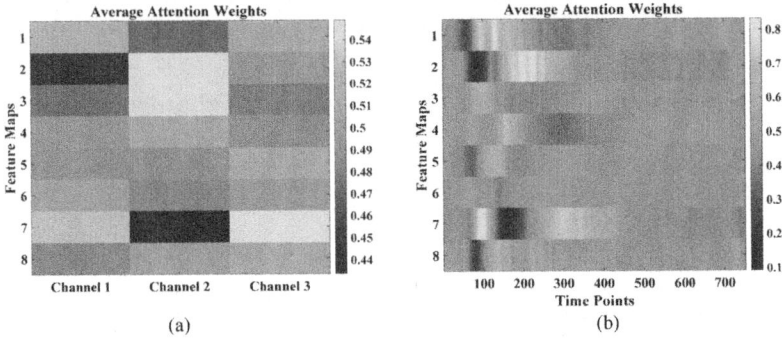

Fig. 3. Attention Visualization of Features Refined by ELA on Spatial and Temporal Dimensions for Subject 5.

To study the efficacy of the ELA module in the context of motor imagery tasks, we conducted a detailed attention visualization analysis, shown in Fig. 3. This analysis aimed to illustrate how the ELA mechanism refines attention weights across both spatial and temporal dimensions for Subject 5 during motor imagery tasks.

Figure 3(a) displays the average attention weights assigned to the three EEG channels (Channel 1, Channel 2, and Channel 3), demonstrating the differential prioritization of channel-specific information by the ELA module, which is crucial for accurate decoding of motor imagery signals. Figure 3(b) extends this analysis to the temporal domain, illustrating the distribution of attention weights across various time points. The color gradients indicate the intensity of attention weights, with warmer colors representing higher values.

This comprehensive visualization underscores the ELA module's ability to dynamically focus on pertinent spatial and temporal features within the EEG data, which is essential for effectively capturing the neural patterns associated with imagined movements. The refined attention weights highlight the ELA module's capability to enhance the discriminative power of the EEGNet by emphasizing critical dependencies, thereby significantly improving the precision and reliability of motor imagery decoding. These insights affirm the module's utility in capturing long-range dependencies and enhancing the overall performance of EEG-based models in motor imagery tasks.

4.6 Feature Visualization

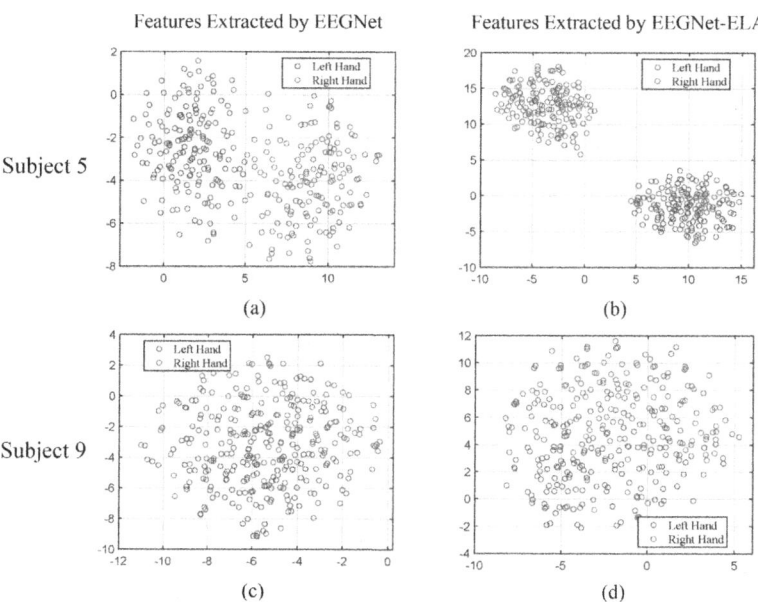

Fig. 4. t-SNE Visualization of EEG Features for Two Different Subjects Extracting from EEGNet and EEGNet-ELA

To explore the impact of the ELA module on feature representation in EEG decoding, we utilized t-distributed Stochastic Neighbor Embedding (t-SNE), a technique for visualizing the distribution and separation of features. This study examined EEG features extracted by two variants of EEGNet—the standard version and the EEGNet-ELA—focusing on features related to two motor imagery tasks: left-hand and right-hand movements across various subjects.

Figure 4 presents the t-SNE visualizations for two subjects, labeled as Subject 5 and Subject 9. Panels (a) and (c) illustrate features from the standard EEGNet, while panels (b) and (d) show those from EEGNet-ELA. Notably, EEGNet-ELA achieves a clearer distinction between the two classes of movements than the baseline EEGNet. Specifically, in panels (b) and (d), the clusters corresponding to different hand movements are more cohesive and distinctly separated, indicating that the ELA mechanism significantly enhances feature discrimination.

This visualization study confirms the qualitative enhancements in feature extraction provided by the ELA mechanism and highlights its potential to improve the accuracy and robustness of EEG-based classification systems. The distinct separation of features relevant to different motor tasks underscores the ELA-enhanced network's improved capability to differentiate between closely related EEG signals, a crucial advancement for the development of more effective BCIs.

4.7 Comparisons Using Different CNNs

In a comprehensive evaluation aimed at assessing the performance of CNNs in EEG decoding, our study integrated the ELA mechanism with another CNN architecture, ADFCNN, and compared the outcomes with several notable related works. Table 4 presents the results of this comparison. It includes benchmarks against traditional and state-of-the-art CNN models such as Deep ConvNet, EEGNet, and various custom configurations like ShallowConvNet and M-ShallowConvNet. Notably, the integration of ELA with EEGNet (EEGNet+ELA) and with Adaptive Feature CNN (ADFCNN+ELA) demonstrated a significant improvement in average accuracy, achieving 85.97% and 88.13%, respectively. These results surpass the performance of conventional models and other recent innovations, highlighting the effectiveness of the ELA mechanism in enhancing the discriminative power of neural networks for EEG signal classification.

Table 4. Comparison of Classification Accuracy Across Different Methods on BCI-IV2b Dataset

Related Work	Method	Average Accuracy (%)
Schirrmeister et al. 2017 [17]	Deep ConvNet	80.00
Lawhern et al. 2018 [12]	EEGNet	81.89
Liu et al. 2022 [14]	SHCNN	83.49
Song et al. 2022 [18]	Conformer	84.63
Schirrmeister et al. 2017 [17]	Shallow ConvNet	84.83
Tang et al. 2023 [10]	MSHCNN	85.25
Our work	EEGNet+ELA	**85.97**
Tao et al. 2023 [24]	ADFCNN	87.81
Our work	ADFCNN+ECA	88.03
Our work	ADFCNN+SE	88.06
Our work	ADFCNN+ELA	**88.13**

Our study underscores the significant role of the ELA module in boosting classification accuracy through advanced feature extraction and representation.

This demonstrates its potential for seamless integration with various deep learning models. The comparative analysis not only confirms the superior performance achieved by incorporating the ELA module but also supports further investigation into adaptive and attention-based enhancements. Such modifications aim to augment the capabilities of neural network models for BCI applications, potentially leading to more effective and robust BCI systems.

5 Conclusion

The Enhanced Local Attention (ELA) module introduced in this study significantly advances EEG decoding capabilities within deep learning frameworks. Detailed evaluations conducted on the BCI-IV2b dataset confirm that integrating adaptive 1D convolution and group normalization within the ELA architecture markedly enhances spatial and temporal feature localization and captures long-range dependencies. These enhancements facilitate superior feature discrimination and enable straightforward integration with prevalent CNN architectures such as EEGNet and ADFCNN, achieving notable improvements in classification accuracy. The ELA module demonstrates superior performance in EEG decoding and underscores the utility of sophisticated, attention-based enhancements in neural network models for brain-computer interface applications. This research paves the way for future innovations aimed at enhancing the accuracy, robustness, and accessibility of BCI technologies.

Acknowledgements. This work was supported in part by The Science and Technology Development Fund, Macau SAR (File no. 0045/2019/AFJ and 0022/2021/APD), The University of Macau Research Committee (MYRG projects 2017-00207-FST and 2022-00197-FST) and Guangdong Basic and Applied Basic Research Foundation (Grant No. 2023A1515010844).

Funding Information. This work is funded in part by University of Macau (File no. MYRG2022-00197-FST, MYRG-GRG2024-00285-FST, MYRG-CRG2024-00048-FST-ICI), by The Science and Technology Development Fund, Macau SAR (File no. 0085/2023/AMJ), and by Guangdong Basic and Applied Basic Research Foundation (Grant No. 2023A1515010844).

References

1. Altuwaijri, G.A., Muhammad, G., Altaheri, H., Alsulaiman, M.: A multi-branch convolutional neural network with squeeze-and-excitation attention blocks for EEG-based motor imagery signals classification. Diagnostics **12**(4), 995 (2022)
2. Chaudhary, U., Birbaumer, N., Ramos-Murguialday, A.: Brain-computer interfaces for communication and rehabilitation. Nat. Rev. Neurol. **12**(9), 513–525 (2016)
3. Chen, L., et al.: SCA-CNN: spatial and channel-wise attention in convolutional networks for image captioning. In: Proceedings of the IEEE Conference on Computer Vision and Pattern Recognition, pp. 5659–5667 (2017)

4. Cohen, M.X.: Where does EEG come from and what does it mean? Trends Neurosci. **40**(4), 208–218 (2017)
5. Craik, A., He, Y., Contreras-Vidal, J.L.: Deep learning for electroencephalogram (EEG) classification tasks: a review. J. Neural Eng. **16**(3), 031001 (2019)
6. Dai, G., Zhou, J., Huang, J., Wang, N.: HS-CNN: a CNN with hybrid convolution scale for EEG motor imagery classification. J. Neural Eng. **17**(1), 016025 (2020)
7. Hu, J., Shen, L., Sun, G.: Squeeze-and-excitation networks. In: Proceedings of the IEEE Conference on Computer Vision and Pattern Recognition, pp. 7132–7141 (2018)
8. Jeunet, C., N'Kaoua, B., Lotte, F.: Advances in user-training for mental-imagery-based BCI control: psychological and cognitive factors and their neural correlates. Prog. Brain Res. **228**, 3–35 (2016)
9. Jia, H., et al.: A model combining multi branch spectral-temporal CNN, efficient channel attention, and LightGBM for MI-BCI classification. IEEE Trans. Neural Syst. Rehabil. Eng. **31**, 1311–1320 (2023)
10. Jia, Z., Lin, Y., Wang, J., Yang, K., Liu, T., Zhang, X.: MMCNN: a multi-branch multi-scale convolutional neural network for motor imagery classification. In: Hutter, F., Kersting, K., Lijffijt, J., Valera, I. (eds.) ECML PKDD 2020. LNCS (LNAI), vol. 12459, pp. 736–751. Springer, Cham (2021). https://doi.org/10.1007/978-3-030-67664-3_44
11. Kim, S.J., Lee, D.H., Lee, S.W.: Rethinking CNN architecture for enhancing decoding performance of motor imagery-based EEG signals. IEEE Access **10**, 96984–96996 (2022)
12. Lawhern, V.J., Solon, A.J., Waytowich, N.R., Gordon, S.M., Hung, C.P., Lance, B.J.: EEGNet: a compact convolutional neural network for EEG-based brain-computer interfaces. J. Neural Eng. **15**(5), 056013 (2018)
13. Leeb, R., Brunner, C., Müller-Putz, G., Schlögl, A., Pfurtscheller, G.: BCI competition 2008-Graz data set B. Graz University of Technology, Austria, pp. 1–6 (2008)
14. Liu, C., et al.: SincNet-based hybrid neural network for motor imagery EEG decoding. IEEE Trans. Neural Syst. Rehabil. Eng. **30**, 540–549 (2022)
15. Mane, R., Robinson, N., Vinod, A.P., Lee, S.W., Guan, C.: A multi-view CNN with novel variance layer for motor imagery brain computer interface. In: 2020 42nd Annual International Conference of the IEEE Engineering in Medicine & Biology Society (EMBC), pp. 2950–2953. IEEE (2020)
16. Pfurtscheller, G., Neuper, C.: Motor imagery and direct brain-computer communication. Proc. IEEE **89**(7), 1123–1134 (2001)
17. Schirrmeister, R.T., et al.: Deep learning with convolutional neural networks for EEG decoding and visualization. Hum. Brain Mapp. **38**(11), 5391–5420 (2017)
18. Song, Y., Zheng, Q., Liu, B., Gao, X.: EEG conformer: convolutional transformer for EEG decoding and visualization. IEEE Trans. Neural Syst. Rehabil. Eng. **31**, 710–719 (2023)
19. Stacey, W.C., Litt, B.: Technology insight: neuroengineering and epilepsy—designing devices for seizure control. Nat. Clin. Pract. Neurol. **4**(4), 190–201 (2008)
20. Stern, Y., et al.: Brain networks associated with cognitive reserve in healthy young and old adults. Cereb. Cortex **15**(4), 394–402 (2005)
21. Su, E., Cai, S., Xie, L., Li, H., Schultz, T.: STAnet: a spatiotemporal attention network for decoding auditory spatial attention from EEG. IEEE Trans. Biomed. Eng. **69**(7), 2233–2242 (2022)

22. Tao, T., Gao, Y., Jia, Y., Chen, R., Li, P., Xu, G.: A multi-channel ensemble method for error-related potential classification using 2D EEG images. Sensors **23**(5), 2863 (2023)
23. Tao, W., et al.: EEG-based emotion recognition via channel-wise attention and self attention. IEEE Trans. Affect. Comput. **14**(1), 382–393 (2020)
24. Tao, W., et al.: ADFCNN: attention-based dual-scale fusion convolutional neural network for motor imagery brain-computer interface. IEEE Trans. Neural Syst. Rehabil. Eng. **32**, 154–165 (2024)
25. Vaswani, A., et al.: Attention is all you need. In: Advances in Neural Information Processing Systems, vol. 30 (2017)
26. Wang, H., et al.: A novel sleep staging network based on multi-scale dual attention. Biomed. Signal Process. Control **74**, 103486 (2022)
27. Wang, Q., Wu, B., Zhu, P., Li, P., Zuo, W., Hu, Q.: ECA-net: efficient channel attention for deep convolutional neural networks. In: Proceedings of the IEEE/CVF Conference on Computer Vision and Pattern Recognition, pp. 11534–11542 (2020)
28. Wong, C.M., et al.: Online adaptation boosts SSVEP-based BCI performance. IEEE Trans. Biomed. Eng. **69**(6), 2018–2028 (2021)
29. Woo, S., Park, J., Lee, J.Y., Kweon, I.S.: CBAM: convolutional block attention module. In: Proceedings of the European Conference on Computer Vision (ECCV) (2018)
30. Xie, J., et al.: A transformer-based approach combining deep learning network and spatial-temporal information for raw EEG classification. IEEE Trans. Neural Syst. Rehabil. Eng. **30**, 2126–2136 (2022)

Mirror Contrastive Loss Based Sliding Window Transformer for Subject-Independent Motor Imagery Based EEG Signal Recognition

Jing Luo(✉) ⓘ, Qi Mao, Weiwei Shi, Zhenghao Shi, Xiaofan Wang, Xiaofeng Lu, and Xinhong Hei(✉)

Shaanxi Key Laboratory for Network Computing and Security Technology and Human–Machine Integration Intelligent Robot Shaanxi University Engineering Research Center, School of Computer Science and Engineering, Xi'an University of Technology, Xi'an, Shaanxi, China
{luojing,heixinhong}@xaut.edu.cn

Abstract. While deep learning models have been extensively utilized in motor imagery based EEG signal recognition, they often operate as black boxes. Motivated by neurological findings indicating that the mental imagery of left or right-hand movement induces event-related desynchronization (ERD) in the contralateral sensorimotor area of the brain, we propose a Mirror Contrastive Loss based Sliding Window Transformer (MCL-SWT) to enhance subject-independent motor imagery-based EEG signal recognition. Specifically, our proposed mirror contrastive loss enhances sensitivity to the spatial location of ERD by contrasting the original EEG signals with their mirror counterparts—mirror EEG signals generated by interchanging the channels of the left and right hemispheres of the EEG signals. Moreover, we introduce a temporal sliding window transformer that computes self-attention scores from high temporal resolution features, thereby improving model performance with manageable computational complexity. We evaluate the performance of MCL-SWT on subject-independent motor imagery EEG signal recognition tasks, and our experimental results demonstrate that MCL-SWT achieved accuracies of 66.48% and 75.62%, surpassing the state-of-the-art (SOTA) model by 2.82% and 2.17%, respectively. Furthermore, ablation experiments confirm the effectiveness of the proposed mirror contrastive loss. A code demo of MCL-SWT is available at https://github.com/roniusLuo/MCL_SWT.

Keywords: Motor imagery · EEG · Brain-computer interface · Event-related Desynchronization · Mirror Contrastive Loss · Sliding Window Transformer

1 Introduction

Brain-Computer Interface (BCI) establishes a direct communication channel between the brain and a computer, facilitating interaction and communication between cognitive processes and external devices, thus fostering the integration of biological and artificial intelligence [1–3]. Among various paradigms, Motor Imagery Brain-Computer Interface (MI-BCI) stands as a cornerstone, allowing users to manipulate external devices or perform specific tasks by mentally simulating movements. MI-BCI holds promising prospects in motor function rehabilitation [4]. Motor Imagery Electroencephalography (MI-EEG) captures EEG signals during motor imagery tasks, representing non-invasive, endogenous brain activity characterized by user-friendly operation, simplicity, flexibility, non-invasiveness, and minimal environmental requirements [5]. Accurate recognition of EEG signals is paramount for the development of robust subject-independent Motor Imagery (MI) Brain-Computer Interface (BCI) systems.

Presently, the primary focus of researchers in the realm of motor imagery EEG signal recognition algorithms lies in single-subject BCI systems, necessitating individual modeling of the target subject and yielding fruitful research outcomes [6]. However, these algorithms typically entail individualized calibration procedures, involving the collection of sufficient individual EEG signals and model adjustments [7]. This augments system complexity and calibration duration, thereby diminishing system usability and convenience. In contrast, subject-independent BCI systems endeavor to tackle the issue of inter-subject generalization, enabling BCI systems to better accommodate multiple subjects and expand the horizons of BCI technology applications [8]. Nonetheless, the spatial variance in event-related desynchronization/event-related synchronization (ERD/ERS) phenomena across different subjects presents significant research challenges [9]. Consequently, enhancing the model's ERD/ERS localization capability is imperative for bolstering subject-independent motor imagery EEG recognition performance.

Although deep learning models have been widely applied in recent years for MI based EEG signal recognition, they often function as black boxes and struggle to precisely localize ERD/ERS, a crucial factor in motor imagery recognition. Furthermore, while transformer-based approaches have demonstrated promising capabilities in extracting discriminative features from EEG signals [10], the computational complexity of the global multi-head self-attention mechanism in transformer models increases quadratically with the length of the input sequence. Currently, the majority of transformer models used in MI-EEG recognition employ short input sequences, thus limiting the temporal resolution of the extracted features.

To address the issues mentioned above, we propose a mirror contrastive loss-based sliding window transformer (MCL-SWT) to enhance subject-independent motor imagery-based EEG signal recognition.

The main contributions of this paper are as follows:

(1) Motivated by neurological findings indicating that the mental imagery of left or right-hand movement induces event-related desynchronization (ERD) in the contralateral sensorimotor area of the brain, the MCL is proposed to enhance sensitivity to the spatial location of ERD by contrasting the original EEG signals with their mirror counterparts—mirror EEG signals generated by interchanging the channels of the left and right hemispheres of the EEG signals.
(2) A temporal SWT based subject-independent MI-EEG signal recognition model is proposed to achieve high time resolution of feature with affordable computational complexity. Specifically, the self-attention scores are calculated in temporal windows, and with the windows sliding along the temporal dimension of the EEG signal, the information from different temporal windows can interact with each other.

(3) The experimental results on subject-independent MI based EEG signal recognition demonstrate the effectiveness of MCL-SWT method in subject-independent MI-EEG classification tasks. Parameter sensitivity experiments have shown robustness of the MCL-SWT model, and the ablation study has validated the effectiveness of MCL.

2 Related Work

2.1 CNN-Based MI-EEG Signal Recognition Method

In recent years, compared to traditional manually designed feature extraction methods, end-to-end feature extraction and classification approaches based on deep neural networks have demonstrated exceptional performance in the domain of motor imagery brain-computer interface (MI-BCI) [6]. Dai et al. introduced a hybrid-scale convolutional network architecture aimed at extracting temporal features of EEG signals across various convolutional scales for EEG-based motor imagery classification [11]. Yang et al. proposed a dual-branch 3D convolutional neural network for the three-dimensional representation of EEG data related to motor imagery, leveraging separate branches for temporal and spatial feature learning to circumvent mutual interference between these features. Furthermore, their framework introduced central loss to further enhance the decoding accuracy of motor imagery EEG [12]. Schirrmeister et al. delved into the impact of different CNN architecture designs on decoding MI-EEG signals, demonstrating superior performance of their proposed Deep ConvNet and Shallow ConvNet compared to alternative methods [13]. Lawhern et al. introduced a compact convolutional neural network dubbed EEGNet, showcasing its versatility across four BCI paradigms and its superior performance over other methods [14]. Mane et al. proposed a filter-bank convolutional network (FBCNet) for MI classification, leveraging multiple bandpass filters, deep convolutional layers for spatial information extraction, and innovative temporal aggregation techniques, outperforming existing methods on EEG signals from both healthy subjects and stroke patients [15]. Zhang et al. devised a weighted convolutional siamese network (WCSN) based on metric learning for feature representation of EEG signals, enhancing decoding accuracy by learning low-dimensional embeddings and implementing an adaptive training strategy to tackle non-stationarity between sessions [16]. Their method achieved significantly better decoding results on both limb neurorehabilitation and healthy subject datasets compared to state-of-the-art approaches. Hou et al. introduced a novel deep learning framework based on graph convolutional neural networks (GCNs) to enhance the recognition performance of raw EEG signals across various motor imagery tasks by capturing functional topological relationships of EEG channels [17]. Their approach involved constructing the Laplacian graph of EEG channels and employing GCNs-Net for feature extraction, followed by dimension reduction and final prediction through fully connected softmax layers.

2.2 Multi-subject MI-EEG Signal Recognition Method

When implementing a single-subject brain-computer interface (BCI) system on a new subject, conducting experiments to collect EEG data for calibration becomes necessary.

This process is time-consuming and labor-intensive, significantly raising the practical application challenges of BCI systems. Consequently, researchers have begun exploring multi-subject motor imagery brain-computer interfaces.

Kwon et al. devised a pioneering framework based on deep convolutional neural networks for spectrum-space feature representation, tailored for multi-subject zero-calibration motor imagery brain-computer interface (MI-BCI). Leveraging a filter bank with multiple frequency bands in conjunction with mutual information and convolutional neural networks, this method constructs generalized features from diverse subjects and frequency bands, exhibiting remarkable performance on the Open BMI dataset [8]. Luo et al. introduced a twin-cascade softmax convolutional neural network for a multi-subject motor imagery BCI. To mitigate subject variability, they employed a cascaded softmax structure comprising subject identification and motor imagery recognition layers. Through joint optimization of subject identification and motor imagery recognition costs during model training, they achieved simultaneous subject identification and motor imagery recognition [18]. Hermosilla et al. developed a novel shallow convolutional neural network model for motor imagery classification, incorporating two convolutional layers for temporal and spatial feature extraction. Through single-subject and multi-subject experiments on three benchmark datasets, their approach surpassed state-of-the-art techniques [19]. Luo et al. proposed a shallow Transformer model for EEG signal decoding in motor imagery tasks. Utilizing multi-head self-attention layers with a global receptive field, they detected and utilized discriminative segments across multiple subjects EEG signals, enhancing classification accuracy. Furthermore, they improved performance through ensemble learning-based network structure and mirror EEG signal construction [5].

2.3 Attention Mechanism-Based MI-EEG Signal Recognition Method

The attention mechanism allocates varying attention weights to different data or feature subsets, enabling the model to prioritize key areas, thereby acquiring detailed information about the target of interest while suppressing irrelevant information [20]. The self-attention mechanism calculates attention weights between each position and other positions to determine their significance in processing. It dynamically learns relationships across different positions in a sequence and addresses long-range dependencies [21]. Consequently, numerous researchers have integrated the attention mechanism with convolutional neural networks (CNNs) to construct models. Zhang et al. designed a convolutional recurrent attention model (CRAM), which uses CNN to encode the spatiotemporal information of EEG signal and establishes a recurrent attention mechanism to explore temporal dynamics between different time periods [22]. Altaheri et al. developed an attention-based temporal convolutional network (ATCNet) model for MI-EEG signal classification. ATCNet, a domain-specific and interpretable deep learning model, highlights valuable features in motor imagery EEG data using a multi-head self-attention mechanism and extracts advanced temporal features with a time convolutional network [23]. Wen et al. designed a CNN-based model architecture for end-to-end training and classification, incorporating a spatial-spectrum-temporal (SST) attention mechanism to adaptively extract the most expressive features from EEG data. Additionally, they proposed a 3D Densely Connected Cross-Stage-Partial Network to segment extracted feature maps, reducing gradient loss and enhancing the model's representational capacity and robustness [24]. Amin et al. proposed a hybrid deep learning model architecture. Firstly, they employed the attention-inception convolutional neural network to extract spatial contextual features, which is crucial for learning the dynamic characteristics of EEG signal. Then, they utilized bidirectional long-short-term memory (Bi-LSTM) to learn temporal features [25]. Li et al. devised a temporal-spectral-based squeeze-and-excitation feature fusion network (TS-SEFFNet) for decoding motor imagery EEG signals. Their model incorporates a deep-temporal convolution block to extract high-dimensional information from EEG data, a multi-spectral convolution block for powerful spectral feature extraction, and a squeeze-and-excitation feature fusion block based on attention mechanism to enhance decoding performance [26]. Dong-Hee Ko proposed an attention-based deep learning approach to extract spatio-spectral features based on significant frequency bands for each subject. The method comprises three parts: extracting spatio-temporal features based on multiple frequency bands, utilizing sub-band attention to identify important frequency bands, and implementing an attention-based bidirectional long short-term memory network to extract time dynamic features [27]. Fan et al. proposed a new network structure for motor imagery EEG classification, called QNet. It includes a newly designed attention module (3D-Attention Module, 3D-AM) for learning attention weights of EEG channels, time points and feature maps. QNet uses a two-branch structure to learn more characteristics. After merging the dual branches, bilinear vectors are obtained. Finally, a fully connected layer is used as classifier [28]. Tao et al. proposed a novel solution called attention-based dual-scale fusion convolutional neural network (ADFCNN), which jointly extracts spectral and spatial information of EEG signal at different scales and provides new insights into integrating effective information from different scales through self-attention [29]. Li et al. introduced the depth-shallow attention multi-frame fusion network (DSA-MFNet) specifically designed for classifying motor imagery EEG signals. DSA-MFNet comprises the depth-shallow attention module for advanced feature extraction and the multi-frame fusion module for exploring inherent temporal variations in EEG data through multi-frame segmentation and recombination techniques [30].

In addition to embedding attention mechanism into convolutional neural network, transformer based models were proposed and applied in EEG decoding [31]. Besides the natural language processing field, successful vision model like Vision Transformer and Swin Transformer were proposed [32, 33]. Google introduced a novel and streamlined network architecture known as the transformer model, specifically designed for sequence modeling and prominently featuring a self-attention mechanism. Departing from traditional convolutional and recurrent layers, the transformer model relies on a multi-head self-attention mechanism. This architecture has garnered significant success in natural language processing and machine translation tasks, demonstrating superiority in handling long-range dependencies and capturing global contextual information [31]. Subsequently, the computer vision community began adopting transformer models. Notably, the vision transformer model was proposed, which segments images into fixed-size patches and feeds them into an enhanced transformer model, outperforming CNN models. Leveraging multi-head self-attention mechanism and positional encoding, this model adeptly captures both global and local image information, enabling effective representation and processing of images with remarkable results [32]. Another innovative visual transformer model, the swin transformer, was introduced. This model strategically processes global information at lower resolutions using a local-partitioned attention mechanism and gradually integrates higher-resolution local information, thereby reducing computational and memory costs while maintaining accuracy. Additionally, it introduces a cross-window communication mechanism, focusing attention within offset windows to enhance computational efficiency and perception, facilitating information flow and feature representation within the network [33].

In summary, attention-related models have rapidly developed and demonstrated excellent representational capabilities, being widely applied in the field of brain computer interface, providing new insights for recognizing motor imagery EEG signal across multi-subject.

2.4 Loss Function Applied in MI-EEG Signal Recognition Method

The loss function, serving as the optimization objective for training neural network models, has garnered significant attention and research focus within the realm of motor imagery recognition. Zhang et al. augmented the loss function by integrating regularization terms for acquired weights, employing squeeze-and-excitation modules to derive weights of EEG channels based on their contributions to EEG classification. They also devised an automated channel selection strategy. To fully exploit spatiotemporal characteristics, they proposed a convolutional neural network that notably outperformed traditional classification methods [34]. Autthasan et al. introduced a multi-task learning model termed MIN2Net, adept at extracting meaningful features from EEG data sans high-complexity preprocessing. Excelling in multi-subject motor imagery EEG classification, MIN2Net achieves end-to-end training through amalgamation of an autoencoder, deep metric learning, and supervised classifier. The autoencoder module aids in feature extraction from EEG data and furnishes discriminative patterns for diverse classes. The deep metric learning module aims to enhance feature discriminative power by refining distance measurement learning, while the supervised classifier module utilizes a standard softmax classifier to categorize latent vectors of input EEG signals. To derive a compact and distinctive latent representation from EEG signals, the model simultaneously minimizes reconstruction loss function, cross-entropy loss function, and triple loss function during optimization [35].

3 Method

This section provides a detailed description of the MCL and SWT model.

3.1 Notations and Definitions

The original EEG signal is defined as $X \in R^{T \times C}$, where T is the number of sampling points and C is the number of EEG channels. The raw EEG signal X serves as the input of the MCL-SWT model, with a batch size of B, resulting in an input data dimension of $B \times 1 \times T \times C$.

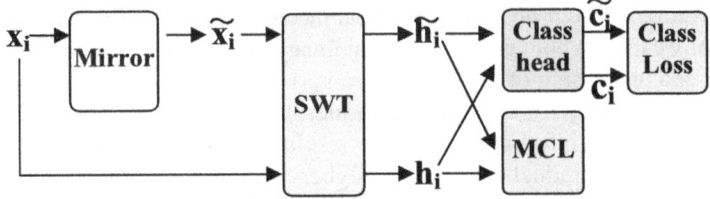

Fig. 1. SWT Model training framework based on MCL

3.2 Mirror Contrastive Loss

In EEG signals, imagining movements of the left or right hand can elicit the ERD phenomenon in the sensory-motor area on the contralateral hemisphere of the brain, and the ERS phenomenon on the same side of the brain. The accurate identification and localization of ERD/ERS are crucial criteria for MI classification. However, existing deep learning models only operate as black boxes and struggle to precisely locate the ERD/ERS phenomenon, leading to suboptimal results. To address this challenge and improve ERD/ERS localization ability of MI recognition model, we propose the MCL in this section. The framework of MCL based SWT model is illustrated in Fig. 1.

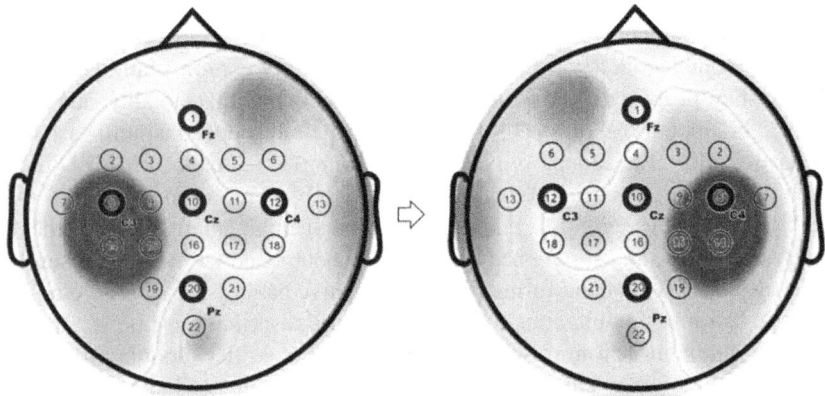

Fig. 2. Swapping left and right EEG channels to construct negative samples

Mirror EEG Signal. To begin, the mirror EEG signal is generated by interchanging the channels of the left hemisphere and the right hemisphere of the EEG signals, as illustrated in Fig. 2. For instance, data from the original C3 channel is transposed to the C4 channel, and data from the C4 channel is transferred to the C3 channel. The electrode positions of the mirror EEG signals mirror those of the original EEG signals, hence the term 'mirror EEG signals'. As the ERD/ERS phenomenon in the mirror EEG signal manifests on the opposite side compared to the original EEG, and the corresponding left and right hand motor imagery label of the mirror EEG signal is designated as the opposite label of the original EEG signal. For instance, if the original EEG is labeled as left hand, then the label for the mirror EEG is right hand.

During the training phase, both mirror and original EEG signals are utilized in model training. By employing this sensible data augmentation technique, the number of training samples is doubled, thereby facilitating the model training process.

MCL. Consequently, the amalgamation of original and mirror EEG signals forms a negative sample pair, which can be leveraged in contrastive loss to bolster the capacity to pinpoint ERD/ERS phenomena through the contrast of deep features from negative sample pairs. Contrastive loss, as described in [40], optimizes the feature distributions during model training by encouraging the model to bring the features from positive sample pairs closer together while pushing features from negative sample pairs apart. The successful construction of positive and negative sample pairs is an essential factor in contrastive loss. While contrastive loss has been successfully applied in deep learning, it typically requires a large number of negative samples to effectively train the model and discern differences between samples.

In this paper, we propose a mirror contrastive loss to enhance the model's sensitivity to the location of ERD/ERS by contrasting the original EEG signals with their mirror EEG signal counterparts. This modification aims to improve the model's spatial awareness of ERD/ERS occurrences. Simultaneously, features less pertinent to the differences in EEG signals between the left and right sides of the brain are subdued. Furthermore, this approach effectively increases the number of negative sample pairs, thereby enhancing the effectiveness of contrastive learning loss.

In the MCL, the samples belonging to the same motor imagery task category are utilized as positive pairs, while the samples belonging to the different motor imagery task category are utilized as negative pairs. Specifically, Mirror Contrastive Learning (MCL) is applied to sample pairs from the original EEG signal and sample pairs from the original EEG signal and its mirror counterpart as:

$$L_d = w_o \sum_{i,j \in \text{OE}} g_{ij} D_{ij} + w_m \sum_{i \in \text{ME}, j \in \text{OE}} g_{ij} D_{ij} \tag{1}$$

where D_{ij} represents the Euclidean distance between sample pairs, consisting of sample i and sample j. The variable g_{ij} takes the value of 1 for positive sample pairs and -1 for negative sample pairs. Additionally, w_o and w_m denote the weights of each loss, while OE and ME refer to the original EEG signal set and mirror EEG set, respectively.

Classification Loss. This paper uses cross-entropy classification loss to train the temporal SWT EEG signal recognition model:

$$L_c = -\frac{1}{N} \sum_{i=1}^{N} y_i \ln(p_i) \tag{2}$$

Therefore, the final loss function of the model training is the sum of the MCL and the classification loss:

$$L = L_c + L_d \tag{3}$$

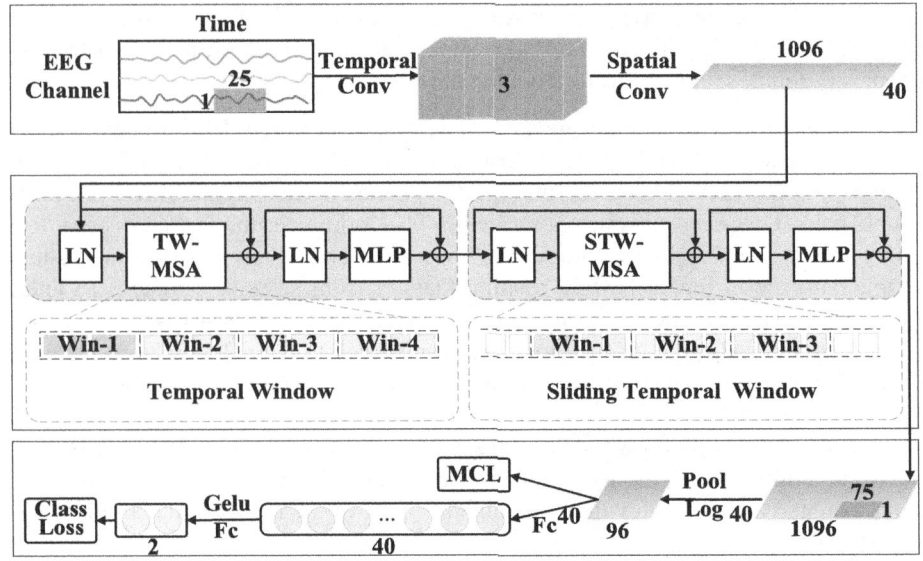

Fig. 3. Illustration of the overall architecture of the SWT model

3.3 SWT Model

The overall architecture of the SWT model consists of three parts: the CNN-based feature extraction module, the sliding temporal window based multi-head self-attention module and the classification module, as shown in Fig. 3.

CNN-Based Feature Extraction Block. Initially, drawing inspiration from the bandpass filter utilized in the FBCSP algorithm [36], a temporal convolution C_t with a kernel size of T × 1 is used processing temporal information. Second, a spatial convolution C_s with a kernel size of 1 × C is applied to fuse information from each EEG channel. Finally, a batch normalization layer aiming to normalize the distribution of data in a batch is added [37]. The CNN-based feature extraction module can learn features with spatiotemporal information which can be described as

$$F = BN(C_s(C_t(X))) \tag{3}$$

Temporal Sliding Window based Multi-head Self-attention Module. The temporal sliding window-based multi-head self-attention module is devised to establish a multi-head self-attention layer along the temporal dimension, aimed at capturing discriminative features of EEG signals with manageable computational load. This module comprises two stages, as depicted in Fig. 1. In the first stage, multiple self-attention layers are employed based on a temporal window, while the second stage facilitates information exchange between sequences in adjacent windows. Each stage is composed of two layer normalization (LN) layers, a temporal window multi-head self-attention layer, and a multi-layer perceptron (MLP) layer, integrated within a residual network structure. The sole discrepancy between these two stages lies in the division of temporal windows. A detailed description of each layer in this module is provided below:

(1) LN layer performs a layer normalization on the output feature F of the CNN-based feature extraction module, normalizing the values along the feature dimensions of each sample [38].

(2) Temporal window multi-head self-attention layer (TW-MSA) is used to extract local temporal dependencies of input vectors in the time dimension, computing attention scores within a single window. It takes the layer-normalized feature as input, then segments it into a set of non-overlapping windows of size M. Self-attention scores computation is performed within each local window to enhance the model's perception of local information, and improve computational efficiency, as shown in Fig. 3. The window size in TW-MSA is empirically set to 8. The specific calculation steps of TW-MSA are as follows:

$$Q_i = W_i^Q \cdot LN(F)$$
$$K_i = W_i^K \cdot LN(F) \qquad (5)$$
$$V_i = W_i^V \cdot LN(F)$$

$$h_i = Attention(Q_i, K_i, V_i) = \text{Softmax}\left(\frac{Q_i K_i^T}{\sqrt{d_k}}\right) V_i \qquad (6)$$

$$O = concat(h_1, \cdots, h_H) W^o + F \qquad (7)$$

where h_i represents the i-th attention head, H denotes the number of attention heads, $W_o \in R^{Hd_v \times d_{\text{mod } el}}$ represents the linear transformation matrix, $d_{\text{mod } el}$ is the dimension of the input embedding. $W_i^Q \in R^{d_{\text{mod } el} \times d_q}$, $W_i^K \in R^{d_{\text{mod } el} \times d_k}$, $W_i^V \in R^{d_{\text{mod } el} \times d_v}$ are the corresponding weight matrices. d_q, d_k, d_v represent the dimensions of the query, key, and value respectively.

(3) MLP layer consists of a fully connected layer, a GELU [39] activation function layer, and another fully connected layer. Finally, the results of the input and output features are summed to produce the final output of the residual network structure.

$$A = O + FC(GELU(FC(LN(O)))) \qquad (8)$$

(4) Sliding temporal window multi-head self-attention layer (STW-MSA) calculates the attention scores in a cyclically shifted local windows for information interaction. As illustrated in Fig. 4, STW-MSA shifts temporal windows to the right with a step size of M/2, and calculates self-attention scores within the new temporal window.

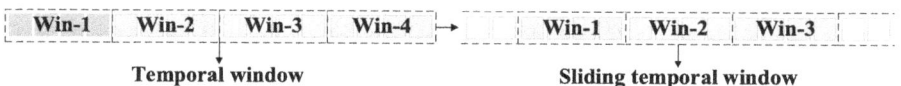

Fig. 4. Illustration of TW-MSA and STW-MSA

The sliding temporal window significantly reduces the computational complexity of the transformer model. While the computational complexity of the global multi-head self-attention mechanism is quadratic in relation to the length of the input sequence, as depicted in Eq. (8), the computational complexity of TW-MSA is linearly related to the length of the input vector, as illustrated in Eq. (9). Here, L represents the length of the input vector, and D denotes the input dimension. Consequently, TW-MSA exhibits a distinct advantage in terms of computational complexity.

$$\Omega(MSA) = 4LD^2 + 2L^2D \tag{9}$$

$$\Omega(TW-MSA) = 4LD^2 + 2M^2LD \tag{10}$$

Classification Module. In the classification module, a square non-linear function is first applied. An average pooling layer is used to reduce the temporal dimension of features, followed by a logarithmic activation function. The EEG features are then inputted into a fully connected layer for classification. Finally, the softmax function is applied to compute the prediction probabilities. Assuming the output of the temporal multi-head self-attention module is A, the output of the classification module is:

$$Y = Softmax(FC(Log(AvgPool(Square(A))))) \tag{11}$$

Finally, the predicted probabilities for the mirror EEG signal corresponding to left MI and right MI are interchanged and then added to the predicted probabilities of the original EEG trial to obtain the final predicted probabilities [5]:

$$[Y_l, Y_r] = [Y_l^o + Y_r^o] + [Y_r^m + Y_l^m] \tag{12}$$

4 Experimental Data and Setup

4.1 Experimental Data

The effectiveness of our proposed method was evaluated on BCI Competition IV dataset 2a and BCI Competition IV dataset 2b [42].

4.2 Experimental Setup

In the experiment, the window size M was set to 8. To divide the input feature vector of length L into a whole number of non-overlapping windows, a 4.48-s EEG signal segment with 1120 sampling point from 0.5 s before MI cue onset to 3.98 s after MI cue onset was used. The bandpass filtering with a range of 4–38 Hz or 0–38 Hz was applied to the EEG signal, followed by channel-wise logarithmic sliding normalization. The preprocessed EEG signal was then input into the MCL-SWT model. During model training, Adam optimizer [43] was used as the optimization method with a weight decay parameter set to 0.05. As the new subject experiment setup is applied, the two category and three channels consisting of C3, Cz, and C4 MI classification task as in dataset 2b is employed in this paper. Since the batch size is set to 100, the input data dimension would be $100 \times 1 \times 1120 \times 3$, and the output shapes and parameter quantities for each layer are shown in Table 1. The temporal convolution kernel T is set to 25, and the spatial convolution kernel C is set to 3. The weight coefficients w_o and w_m are empirically in (11) are set to 0.2 and 0.3, respectively.

Table 1. The output shape and parameters number of each layer

Block	Layer	Output Shape	Param
Input	Input	[100, 1, 1120, 3]	0
Feature Extraction Block	Temporal Conv	[100, 40, 1096, 3]	1040
	Spatial Filter	[100, 40, 1096, 1]	4840
	Feature Normalization	[100, 40, 1096, 1]	80
	Rearrange	[100, 1096, 40]	0
(S)TW-MSA	Layer Norm	[100, 1096, 40]	80
	Query Projection	[100, 1096, 40]	1640
	Key Projection	[100, 1096, 40]	1640
	Value Projection	[100, 1096, 40]	1640
	Attention Score	[100, 8, 137, 8, 8]	0
	Projection	[100, 1096, 40]	1640
	Concatenate	[100, 1096, 40]	0
	Residual Add	[100, 1096, 40]	0
	Layer Norm	[100, 1096, 40]	80
	Linear	[100, 1096, 160]	6560
	GELU	[100, 1096, 160]	0
	Linear	[100, 1096, 40]	6440
	Residual Add	[100, 1096, 40]	0
	Layer Norm	[100, 1096, 40]	80
	Query Projection	[100, 1096, 40]	1640
	Key Projection	[100, 1096, 40]	1640
	Value Projection	[100, 1096, 40]	1640
	Attention Score	[100, 8, 137, 8, 8]	0
	Projection	[100, 1096, 40]	1640
	Concatenate	[100, 1096, 40]	0
	Residual Add	[100, 1096, 40]	0
	Layer Norm	[100, 1096, 40]	80
	Linear	[100, 1096, 160]	6560
	GELU	[100, 1096, 160]	0
	Linear	[100, 1096, 40]	6440
	Residual Add	[100, 1096, 40]	0
Classification Block	Average Pool	[100, 40, 69]	0
	Log	[100, 40, 69]	0
	Linear1	[100, 40]	109640
	Gelu	[100, 40]	0
	Linear2	[100, 2]	82

4.3 Dataset Division

This paper aims to investigate the classification performance of subject-independent motor imagery EEG signals, so the dataset is partitioned in a new subject setup as follows: when utilizing the training sessions of all 9 subjects in dataset 2a as the training set, the testing sessions (session 4 and session 5) of all 9 subjects in dataset 2b are employed as the testing set; conversely, when using the training sessions (session 3) of all 9 subjects in dataset 2b as the training set, the testing sessions of all 9 subjects in dataset 2a are utilized as the testing set. 5. Consequently, the trial data used for testing and training originate from entirely different subjects.

5 Results

5.1 Performance Comparison of Subject-Independent MI Recogination

This section evaluates motor imagery recognition performance of the MCL-SWT model on new subjects setup. Since the subjects in dataset 2a are entirely different from those in dataset 2b, the subjects in the testing set are new. As training typically converges within 500 epochs, the maximum number of epochs for training was set to 500 to ensure model convergence. Due to significant randomness affecting the test accuracy at a specific epoch, the evaluation metrics used in the experiment include: (1) Maximum test accuracy over 500 epochs (Max Accuracy); (2) Average test accuracy from epochs 401 to 500 (Average Accuracy); (3) Test accuracy at the epoch with the lowest training loss (Accuracy). Table 2 presents the results in terms of "Accuracy/Kappa coefficient," with the maximum result highlighted in bold. The first row indicates the encoding format as "dataset - bandpass filter." For example, "2a-0 Hz" means the model was trained on dataset 2b and tested on dataset 2a using a 0–38 Hz bandpass filter for preprocessing. Five state-of-the-art models were compared in the experiment, including Shallow ConvNet, Deep ConvNet, EEGNet, FBCNet, and ATCNet. In additional, the ablation experiments are also conducted. The SWT and MCL-SWT indicate the SWT model without and with MCL, separately.

The experimental results in Table 2 lead to the following conclusions: (1) MCL-SWT achieved accuracies of 66.48% and 75.62%, surpassing the best state-of-the-art model by 2.82% and 2.17%, respectively. (2) The proposed temporal MCL-SWT model demonstrates superior performance across different datasets and bandpass filters. (3) MCL contributes to a further increase in the recognition accuracy and kappa value of SWT. (4) The proposed SWT model exhibits resistance to overfitting, as evidenced by the small gap between the maximum accuracy/Kappa and the average accuracy/Kappa of SWT. (5) The experimental results suggest that MCL-SWT possesses better inter-subject generalization ability.

5.2 Sensitivity Analysis of Parameters

The number of temporal multi-head self-attention blocks and the number of heads in each self-attention block are two primary hyperparameters of the SWT model. This section investigates the impact of these hyperparameters on the model's performance, with the experimental results presented in Table 3. Increasing the number of heads or blocks leads to larger attention scores, resulting in a more complex model.

The following conclusions can be drawn from the experimental results in Table 3: (1) In this experimental setup, the performance of the model deteriorates as the number of temporal multi-head self-attention blocks increases. This may be due to the limited number of training samples in motor imagery EEG, which is insufficient to fully train a model

Table 2. Classification performance (Accuracy and Kappa value) on the new subject

		2a-0 Hz	2a-4 Hz	2b-0 Hz	2b-4 Hz
Average Accuracy /Kappa	Shallow	63.72/0.29	60.76/0.25	72.7/0.46	67.06/0.34
	Deep	60.55/0.25	60.8/0.25	71.55/0.44	65.93/0.32
	EEGNet	60.58/0.25	59.52/0.23	71.21/0.44	65.91/0.32
	FBCNet	60.56/0.25	59.73/0.23	70.55/0.43	65.29/0.31
	ATCNet	58.87/0.22	57.45/0.20	68.47/0.36	64.69/0.30
	SWT	66.00/0.33	63.68/0.28	74.50/0.49	69.71/0.40
	MCL-SWT	**66.56/0.33**	**64.74/0.3**	**75.85/0.52**	**72.54/0.45**
Accuracy /Kappa	Shallow	63.66/0.28	61.18/0.26	73.45/0.47	68.7/0.36
	Deep	61.27/0.26	61.34/0.26	72.89/0.46	66.51/0.34
	EEGNet	61.19/0.26	58.72/0.22	72.24/0.45	66.2/0.33
	FBCNet	60.86/0.25	58.36/0.22	71.18/0.44	65.53/0.31
	ATCNet	59.34/0.23	57.78/0.20	69.13/0.39	63.78/0.27
	SWT	66.13/0.33	63.58/0.28	74.58/0.49	69.79/0.40
	MCL-SWT	**66.48/0.33**	**64.52/0.3**	**75.62/0.51**	**73.27/0.47**
Max Accuracy /Kappa	Shallow	67.28/0.34	65.36/0.32	75.18/0.51	71.69/0.45
	Deep	66.05/0.33	65.35/0.32	73.73/0.48	71.68/0.45
	EEGNet	65.51/0.32	64.51/0.3	73.31/0.47	71.16/0.44
	FBCNet	66.46/0.33	63.89/0.29	72.93/0.46	70.98/0.43
	ATCNet	64.11/0.29	62.18/0.28	70.84/0.43	68.22/0.36
	SWT	66.82/0.34	64.81/0.3	75.49/0.51	74.12/0.48
	MCL-SWT	**67.37/0.35**	**65.49/0.31**	**76.37/0.53**	**75.49/0.51**

Table 3. The performance evaluation (2b-0 Hz) on different hyperparameter: (1) the head number of self-attention; (2) the number of temporal multi-head self-attention block

	4 heads	8 heads	10 heads
1 block	74.71/0.49	74.5/0.49	74.73/0.49
2 block	74.3/0.48	74.2/0.48	74.45/0.49
3 block	73.16/0.46	73.67/0.47	73.26/0.46

with multiple blocks; (2) The number of self-attention heads has a minimal impact on the model's performance; (3) The model exhibits robustness to various hyperparameters.

5.3 Model Complexity Analysis

The number of parameters and the inference time of the model are important indicators for assessing complexity. Therefore, this section compares these two aspects. Inference time is measured as the average time taken for 1000 runs. The experiments were conducted on Intel i7 10700K and NVIDIA GeForce RTX 3090. The specific experimental results are shown in Table 4.

Table 4. Model complexity analysis

	Shallow	Deep	EEGNet	FBCNet	ATCNet	MCL-SWT
Parameter number/M	10	268	3	3	37	155
Inference time/ms	0.56	1.42	2.48	37.64	15.37	8.36

6 Conclusion

In this paper, we propose a Mirror Contrastive Loss (MCL) based on Sliding Window Transformation (SWT) model for subject-independent EEG signal recognition. By leveraging mirror EEG signals and MCL, the model aims to enhance sensitivity to the spatial location of ERD/ERS by contrasting the original EEG signals with their mirror counterparts. This modification is intended to improve the model's spatial awareness of ERD/ERS occurrences. Additionally, we introduce a temporal SWT that calculates self-attention scores in sliding windows, enhancing model performance with manageable computational complexity. The performance of MCL-SWT was evaluated on subject-independent motor imagery EEG signal recognition tasks. Experimental comparisons with state-of-the-art methods and ablation experiments demonstrate the superior performance of MCL-SWT. The proposed MCL serves as a general loss for MI-EEG recognition and can be integrated into various backbone networks. In future work, we aim to explore combining the proposed MCL with transfer learning in MI-EEG recognition.

Acknowledgments. This work is jointly supported by the National Natural Science Foundation of China (Grant Nos. 61906152, 61976177, 62076201, 62376213 and U21A20524) and the Scientific Research Program Founded by Shaanxi Provincial Education Department of China under Grant 23JK0556.

References

1. Benson, P.J.: Decoding brain-computer interfaces. Science **360**(6389), 615–616 (2018)
2. Shanechi, M.M.: Brain–machine interfaces from motor to mood. Nat. Neurosci. **22**(10), 1554–1564 (2019)
3. Tang, X., Yang, C., Sun, X., Zou, M., Wang, H.: Motor imagery EEG decoding based on multi-scale hybrid networks and feature enhancement. IEEE Trans. Neural Syst. Rehabil. Eng. **31**, 1208–1218 (2023)

4. Thomas, S.: Building a brain–computer interface to restore communication for people with paralysis. Nat. Electron. **6**(12), 924–925 (2023)
5. Luo, J., et al.: A shallow mirror transformer for subject-independent motor imagery BCI. Comput. Biol. Med. **164**, 107254 (2023)
6. Altaheri, H., et al.: Deep learning techniques for classification of electroencephalogram (EEG) motor imagery (MI) signals: a review. Neural Comput. Appl. **35**(20), 14681–14722 (2023)
7. Wei, F., Xu, X., Jia, T., Zhang, D., Wu, X.: A multi-source transfer joint matching method for inter-subject motor imagery decoding. IEEE Trans. Neural Syst. Rehabil. Eng. **31**, 1258–1267 (2023)
8. Kwon, O.-Y., Lee, M.-H., Guan, C., Lee, S.-W.: Subject-independent brain–computer interfaces based on deep convolutional neural networks. IEEE Trans. Neural Netw. Learn. Syst. **31**(10), 3839–3852 (2019)
9. Zhang, K., Robinson, N., Lee, S.-W., Guan, C.: Adaptive transfer learning for EEG motor imagery classification with deep Convolutional Neural Network. Neural Netw. **136**, 1–10 (2021)
10. Jia, Z., Lin, Y., Wang, J., Yang, K., Liu, T., Zhang, X.: MMCNN: a multi-branch multi-scale convolutional neural network for motor imagery classification. In: Machine Learning and Knowledge Discovery in Databases: European Conference, ECML PKDD 2020, Ghent, Belgium, 14–18 September 2020, Proceedings, Part III, pp. 736–751. Springer, Cham (2020)
11. Dai, G., Zhou, J., Huang, J., Wang, N.: HS-CNN: a CNN with hybrid convolution scale for EEG motor imagery classification. J. Neural Eng. **17**(1), 016025 (2020)
12. Yang, L., Song, Y., Jia, X., Ma, K., Xie, L.: Two-branch 3D convolutional neural network for motor imagery EEG decoding. J. Neural Eng. **18**(4), 0460c7 (2021)
13. Schirrmeister, R.T., et al.: Deep learning with convolutional neural networks for EEG decoding and visualization. Hum. Brain Mapp. **38**(11), 5391–5420 (2017)
14. Lawhern, V.J., Solon, A.J., Waytowich, N.R., Gordon, S.M., Hung, C.P., Lance, B.J.: EEGNet: a compact convolutional neural network for EEG-based brain–computer interfaces. J. Neural Eng. **15**(5), 056013 (2018)
15. Mane, R., et al.: FBCNet: a multi-view convolutional neural network for brain-computer interface. arXiv preprint arXiv:2104.01233 (2021)
16. Zhang, S., et al.: Learning EEG representations with weighted convolutional Siamese network: a large multi-session post-stroke rehabilitation study. IEEE Trans. Neural Syst. Rehabil. Eng. **30**, 2824–2833 (2022)
17. Hou, Y., et al.: GCNs-net: a graph convolutional neural network approach for decoding time-resolved EEG motor imagery signals. IEEE Trans. Neural Netw. Learn. Syst. (2022)
18. Luo, J., et al.: Improving the performance of multisubject motor imagery-based BCIs using twin cascaded softmax CNNs. J. Neural Eng. **18**(3), 036024 (2021)
19. Hermosilla, D.M., et al.: Shallow convolutional network excel for classifying motor imagery EEG in BCI applications. IEEE Access **9**, 98275–98286 (2021)
20. Larochelle, H., Hinton, G.E.: Learning to combine foveal glimpses with a third-order Boltzmann machine. In: Advances in Neural Information Processing Systems, vol. 23 (2010)
21. Vaswani, A., et al.: Attention is all you need. In: Advances in Neural Information Processing Systems, vol. 30 (2017)
22. Zhang, D., Yao, L., Chen, K., Monaghan, J.: A convolutional recurrent attention model for subject-independent EEG signal analysis. IEEE Signal Process. Lett. **26**(5), 715–719 (2019)
23. Altaheri, H., Muhammad, G., Alsulaiman, M.: Physics-informed attention temporal convolutional network for EEG-based motor imagery classification. IEEE Trans. Industr. Inf. **19**(2), 2249–2258 (2022)
24. Wen, Y., He, W., Zhang, Y.: A new attention-based 3D densely connected cross-stage-partial network for motor imagery classification in BCI. J. Neural Eng. **19**(5), 056026 (2022)

25. Amin, S.U., Altaheri, H., Muhammad, G., Abdul, W., Alsulaiman, M.: Attention-inception and long-short-term memory-based electroencephalography classification for motor imagery tasks in rehabilitation. IEEE Trans. Ind. Inf. **18**(8), 5412–5421 (2021)
26. Li, Y., Guo, L., Liu, Y., Liu, J., Meng, F.: A temporal-spectral-based squeeze-and-excitation feature fusion network for motor imagery EEG decoding. IEEE Trans. Neural Syst. Rehabil. Eng. **29**, 1534–1545 (2021)
27. Ko, D.-H., Shin, D.-H., Kam, T.-E.: Attention-based spatio-temporal-spectral feature learning for subject-specific EEG classification. In: 2021 9th International Winter Conference on Brain-Computer Interface (BCI), pp. 1–4. IEEE (2021)
28. Fan, C.-C., Yang, H., Hou, Z.-G., Ni, Z.-L., Chen, S., Fang, Z.: Bilinear neural network with 3-D attention for brain decoding of motor imagery movements from the human EEG. Cogn. Neurodyn. **15**, 181–189 (2021)
29. Tao, W., et al.: ADFCNN: attention-based dual-scale fusion convolutional neural network for motor imagery brain-computer interface. IEEE Trans. Neural Syst. Rehabil. Eng. **32**, 154–165 (2024)
30. Li, H., Zhang, X., Wan, Y., Zhang, X.: DSA-MFNet: deep-shallow attention based multi-frame fusion network for EEG motor imagery classification, pp. 2021–2026
31. Vaswani, A., et al.: Attention is all you need. Adv. Neural. Inf. Process. Syst. **30**, 5998–6008 (2017)
32. Dosovitskiy, A., Beyer, L., Kolesnikov, A., Weissenborn, D., Houlsby, N.: An Image is worth 16×16 words: transformers for image recognition at scale. In: International Conference on Learning Representations (2020)
33. Liu, Z., et al.: Swin transformer: hierarchical vision transformer using shifted windows. In: Proceedings of the IEEE/CVF International Conference on Computer Vision, pp. 10012–10022 (2021)
34. Zhang, H., Zhao, X., Wu, Z., Sun, B., Li, T.: Motor imagery recognition with automatic EEG channel selection and deep learning. J. Neural Eng. **18**(1), 016004 (2021)
35. Autthasan, P., et al.: MIN2Net: end-to-end multi-task learning for subject-independent motor imagery EEG classification. IEEE Trans. Biomed. Eng. **69**(6), 2105–2118 (2021)
36. Ang, K.K., Chin, Z.Y., Wang, C., Guan, C., Zhang, H.: Filter bank common spatial pattern algorithm on BCI competition IV datasets 2a and 2b. Front. Neurosci. **6**, 39 (2012)
37. Ioffe, S., Szegedy, C.: Batch normalization: accelerating deep network training by reducing internal covariate shift. In: International Conference on Machine Learning, pp. 448–456 (2015)
38. Ba, J.L., Kiros, J.R., Hinton, G.E.: Layer normalization. arXiv preprint arXiv:1607.06450 (2016)
39. Hendrycks, D., Gimpel, K.: Gaussian error linear units (gelus). arXiv preprint arXiv:1606.08415 (2016)

40. Hadsell, R., Chopra, S., LeCun, Y.: Dimensionality reduction by learning an invariant mapping. In: 2006 IEEE Computer Society Conference on Computer Vision and Pattern Recognition (CVPR 2006), vol. 2, pp. 1735–1742. IEEE (2006)
41. Wu, C.-Y., Manmatha, R., Smola, A.J., Krahenbuhl, P.: Sampling matters in deep embedding learning. In: Proceedings of the IEEE International Conference on Computer Vision, pp. 2840–2848
42. Tangermann, M., et al.: Review of the BCI competition IV. Front. Neurosci. **6**, 55 (2012)
43. Kingma, D.P., Ba, J.: Adam: a method for stochastic optimization. arXiv preprint arXiv:1412.6980 (2014)

Active Urination Detection Using EEG Based on FBCNet

Anan Gan[1], Banghua Yang[1]($\boxtimes$), Yonghuai Zhang[2], Xingye He[2], Fenqi Rong[1], Liang Chang[1], and Aolei Yang[1]

[1] School of Mechanical Engineering and Automation, Shanghai University, Shanghai 200444, China
yangbanghua@shu.edu.cn

[2] Shanghai Shaonao Sensing Technology Co., Ltd., Shanghai 201900, China

Abstract. Purpose: In the field of medical care, urination detection for bedridden patients is particularly challenging, and existing research has focused on monitoring bladder pressure, which does not reflect the patient's true intentions. Therefore, this study proposes a deep learning-based method for active urination detection using EEG to explore brain activities related to urination.

Methods: This study introduced the FBCNet deep learning model, combined with Integrated Gradients (IG) for interpretability analysis, to explore the activity patterns of the brain in states with and without sense of urination. FBCNet was used to decode the intention to urinate from EEG, while IG was used to assess the contribution of different frequency bands and channels to the classification.

Results: FBCNet achieved an accuracy of $65.83 \pm 0.14\%$ in the task of urination detection, with the highest accuracy for a single subject reaching $88.60 \pm 0.06\%$. The analysis showed that in the state of urination, brain activity was mainly concentrated in the sensorimotor cortex under the Delta band.

Conclusion: This study reveals that brain activities related to urination mainly occur in the sensorimotor cortex, a process that involves first perceiving the presence of the urination and then generating the control intention for urination motor. Additionally, it is proven that the degree of urination is determined by the feedback strength of motor intention for urination.

Significance: As the first study to use EEG to detect urination, this study provides a new perspective for clinical monitoring of urination and opens up new possibilities for future clinical application.

Keywords: Active Urination Detection · Sensorimotor Cortex · EEG · Integrated Gradients · FBCNet

1 Introduction

In the field of medical care, monitoring the urge to urinate for bedridden patients has always been a challenging issue. Many bedridden patients, due to various reasons such as postoperative recovery, neurological diseases, or geriatric diseases, may temporarily or permanently lose the ability to autonomously express the urge to urinate. This not only increases the physical discomfort of the patient but may also lead to complications such as urinary tract infections, which are detrimental to the patient's recovery. Therefore, the development of a technology that can actively monitor the urge to urinate is of great significance for improving the quality of life and the effectiveness of medical care for bedridden patients. Urination monitoring technology based on electroencephalogram (EEG) signals, which identifies the urge to urinate by analyzing brain activity patterns, provides a potential solution for patients who cannot express themselves autonomously. This method can actively monitor and alert caregivers for appropriate treatment before the patient feels significant discomfort, thereby reducing the patient's pain and the risk of complications.

Research on urination monitoring mainly focuses on bladder pressure detection [1]. The first type of ultrasonic technology: using ultrasonic imaging technology to provide information on urinary dysfunction. By placing ultrasonic wave transmitters and receivers on the skin, the bladder is scanned and its shape is determined [2–5]. The second type of optical technology: using near-infrared spectroscopy (NIRS) to measure light absorption, which is proportional to the bladder's fullness. The system is usually placed above the skin at the midline of the lower abdomen, consisting of an LED light source and a photoelectric detector that measures the absorption of emitted light [6,7]. The third type of electrical bioimpedance technology: by measuring the impedance changes of the bladder during activity to estimate the bladder urine volume. This technology is used to measure the electrical properties of biological tissues by applying a small high-frequency current to a specific body part and measuring the voltage generated by this section using traditional or dry electrodes [8–13]. However, these methods can intuitively reflect bladder pressure but cannot express an individual's subjective urination intention.

In response to the shortcomings of existing research, this study proposes a deep learning-based method for active urination detection. By collecting the electrical activity of the cerebral cortex and using advanced deep learning models and interpretability methods, we explore brain wave patterns related to the urge to urinate.

2 Methods

Experiment Data. Considering that subjects need to go to the bathroom to urinate during the experimental break, this study used the Neusen wireless EEG acquisition system at a sampling rate of 1000 Hz. To fully capture brain activity, a 64-channel EEG cap equipped with the international 10–20 system standard was used.

As shown in Fig. 1, the experimental paradigm design includes two main stages: before and after urination, each containing three sessions, each lasting about three minutes. The subjects were five healthy adult males, aged 22 to 27, with no past medical history and no bad habits. Before the experiment began, we assessed the degree of urge to urinate for each participant through a questionnaire to ensure that the EEG cap was worn when they felt significant pressure in the bladder. In addition, we informed the subjects to stay relaxed during the experiment and to avoid unnecessary body movements.

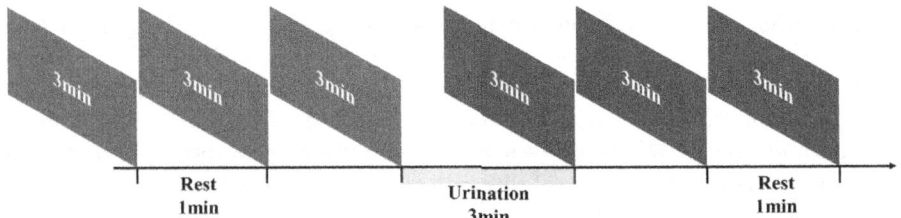

Fig. 1. Experimental Paradigm Schematic Diagram.

Preprocessing. The preprocessing of experimental data was implemented using the python-MNE programming [14]. First, the Cz channel was set as the reference electrode for signal re-referencing. Then, five external channels (ECG, HEOL, HEOR, VEOU, VEOL) were removed. Next, a band-pass filter was used to extract signals within the 1–100 Hz frequency band, which is considered to contain the main activities. Subsequently, Independent Component Analysis (ICA) was applied to identify and remove ocular and cardiac artifacts from the EEG channels based on the five external channels. Finally, continuous signal were divided into 4-s trails and labeled according to the presence or absence of the urge to urinate before and after urination (presence as 1, absence as 0).

FBCNet. The FBCNet used in this study is an architecture specifically designed to extract spatio-temporal-frequency features from EEG, fully capturing the complex characteristics of EEG in the state of urination [15]. As shown in Fig. 2, The core of the FBCNet architecture consists of the following four stages: (1) Multi-view data representation. (2) Spatial transformation learning. (3) Temporal feature extraction. (4) Classification. Multi-view data representation involves using multiple narrow-band filters to spectrally filter the raw EEG data, enabling the model to learn rich feature from different frequency bands. Spatial transformation learning utilizes deep convolutional layers to learn spatial discriminative patterns for each view. Temporal feature extraction employs a novel variance layer to extract temporal information after effective spatial transformation, which helps to capture dynamic temporal patterns associated with the urge to urinate. Finally, a fully connected (FC) layer integrates the features from the variance layer and classifies them into the given categories, i.e., the presence or absence of urination.

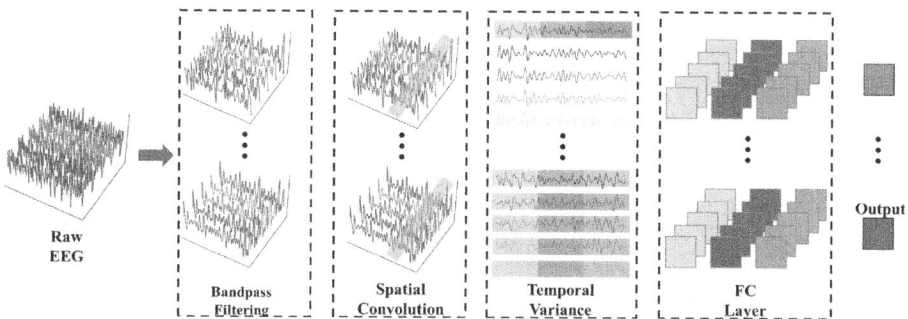

Fig. 2. FBCNet Architecture.

Integrated Gradients. Integrated Gradients (IG) has higher compatibility and performance than common interpolation methods and is plug-and-play, requiring no modification to the model structure [16], as shown in Fig. 3. Therefore, we introduced IG to enhance the interpretability of FBCNet and to examine the classification contribution of different bands and channels. Specifically, IG calculates the contribution of the baseline and samples to the model by linearly interpolating and summing the gradients between them. For input features, they can be either a single band feature or a single channel feature. The equation for calculating the classification contribution is as follows:

$$F = x' + \frac{m}{M}(x - x') \tag{1}$$

$$\phi_i^{IG}(\theta, x, x') = (x_i - x'_i)\frac{1}{M}\sum_{m=1}^{M}\frac{\partial \Theta(F)}{\partial x_i} \tag{2}$$

$m = 1, 2, \ldots, M$, M denotes the steps in Riemann path integration. x' is the baseline of x, F is the integration path between x' and x, θ is the model mapping(FBCNet).

Band and Channel Division. The IG method generates a classification contribution matrix the same size as the original input data, quantifying the impact of each input feature on the model output. Analyzing the performance of the trained model on the test set allows for the summation and averaging of categorized contributions from both the frequency band and channel dimensions. This process can help identify specific frequency bands and channels associated with brain activity.

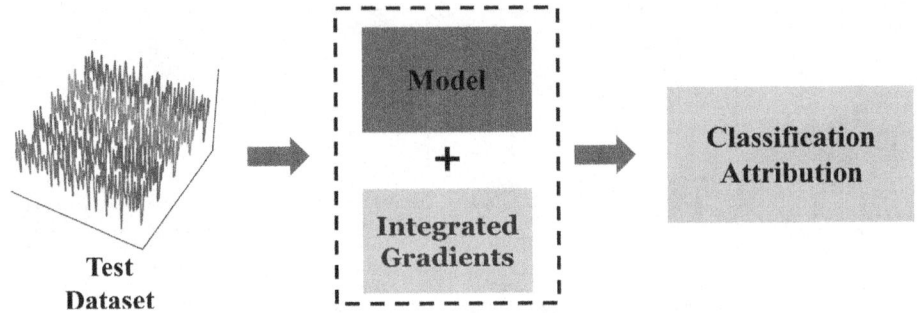

Fig. 3. Integrated Gradients Schematic Diagram.

Traditional EEG band division is used in this study to define seven different bands: Delta (0.5–4 Hz), Theta (4–8 Hz), Alpha (8–13 Hz), Beta1 (14–19 Hz), Beta2 (19–24 Hz), Beta3 (24–30 Hz), and Gamma (30–50 Hz). Given that the experimental data is in a resting state, most of the relevant information will be concentrated in the frequency bands below 30 Hz. In particular, the Beta band, due to its wide frequency range, is further divided into three sub-bands: Beta1, Beta2, and Beta3, to more finely explore brain activities within this band.

The cap used in this study adopted the international 10–20 system standard, with 59 EEG channels fully covering 12 major brain regions. These brain regions include the Frontal pole (Fp), Anterior frontal (AF), Frontal (F), Frontal central (FC), Frontal temporal (FT), Central/sensorimotor cortex (C), Temporal (T), Central parietal (CP), Temporal parietal (TP), Parietal (P), and Occipital pole (PO). This comprehensive channel layout provides a high-resolution view of brain activity.

3 Results

Urination Detection. The data of the five subjects were classified at the individual level, with the training set consisting of the first two sessions of data before and after the urge to urinate, totaling four sessions. The test set consisted of the third session of data before and after the urge to urinate, totaling two sessions. This division was made because the continuity of resting-state data results in a certain degree of time correlation between the sliced data, which can easily lead to data leakage affecting the validity of the results. A five-fold cross-validation was used to divide the training set for training.

In terms of model training parameter selection, the Adam optimizer was used with a learning rate of 1e−3, for 1500 epochs, with m set to 16, and the loss function was cross-entropy. The classification accuracy is shown in Table 1. Before using FBCNet for classification, Support Vector Machine (SVM) and Random Forest (RF) were also selected for classification. The classification results show that FBCNet performed the best, followed by RF, with SVM showing the lowest results. The classification accuracies for the three models were $65.83 \pm 0.14\%$,

Table 1. Classification Accuracies of Three Models.

Model	sub1	sub2	sub3	sub4	sub5	ACC
SVM	46.96 ± 0.05%	44.06 ± 0.02%	47.24 ± 0.09%	54.23 ± 0.04%	43.82 ± 0.04%	47.26 ± 0.04%
RF	53.80 ± 0.10%	47.68 ± 0.06%	51.38 ± 0.02%	62.05 ± 0.04%	73.56 ± 0.04%	57.69 ± 0.10%
FBCNet	61.94 ± 0.10%	66.15 ± 0.09%	50.89 ± 0.01%	61.57 ± 0/18%	88.60 ± 0.06%	65.83 ± 0.14%

57.69 ± 0.10%, and 47.26 ± 0.04%, respectively. On an individual subject basis, Subject 5 achieved the highest result of 88.60%, while the lowest was Subject 2 with only 47.68% (below the chance level of 50%). A box plot can reflect more information about the model's performance on the test set after five-fold cross-validation. The box plots of the classification accuracy for the three models across the five subjects are shown in Fig. 4.

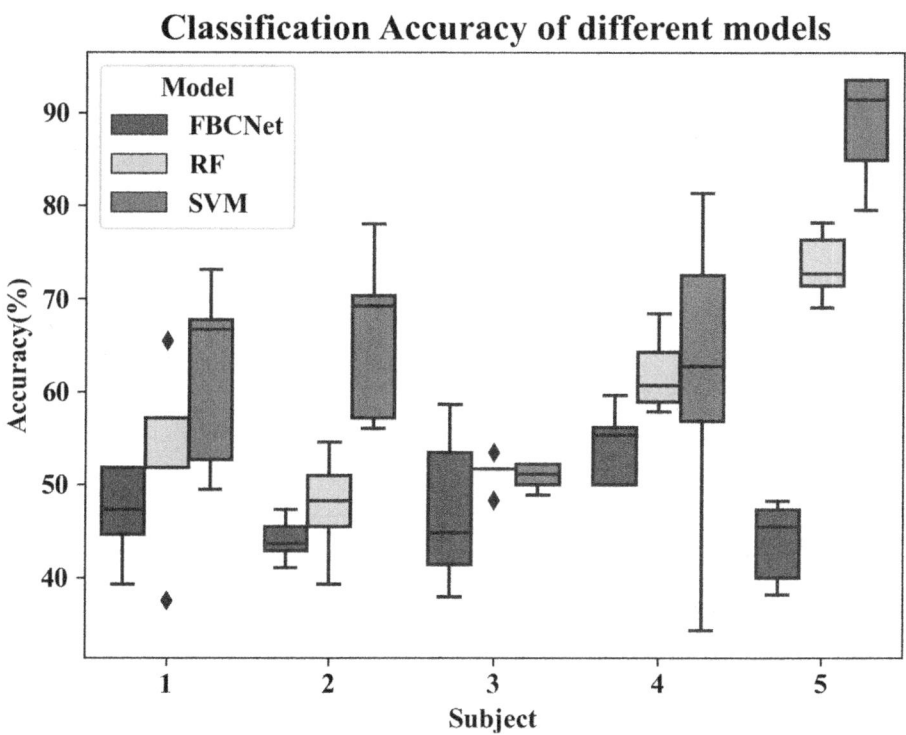

Fig. 4. Box Plots of Classification Accuracies for Three Models.

The raw EEG in the states of having or not having the urination are high-dimensional and not directly distinguishable. Principal Component Analysis (PCA) was used to reduce the dimensions to two dimensions for visualization (Fig. 5A)). There is an overlap between the data with and without the urination in the two principal component directions. FBCNet extracts features from the frequency domain to the spatial domain and then to the time domain to detect urination, and PCA is used to reduce the dimensions to two dimensions, where each point represents a trial. Figure 5(B) shows that the presence and absence of the urge to urinate are distinguishable in the two principal component directions. To more clearly display the differences, linear regression lines are used, as shown in Figs. 5(C) and 5(D).

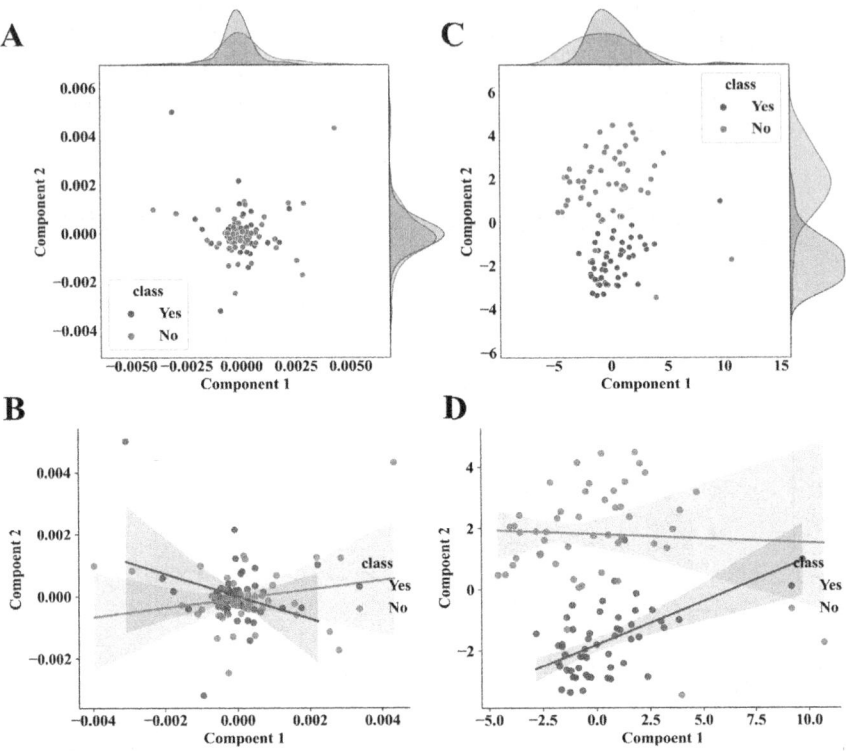

Fig. 5. Feature Distribution of Test Data Before and After Classification.

Urination-Related Band and Channel Analysis. Figure 6(A) shows the average time-frequency plot for all trials (trails) in a single session for Subject 5. Upon comprehensive analysis of the time-frequency plots of all trials, it was found that EEG activity tends to be significantly concentrated in the Delta band (0.5-4 Hz). Activity in this frequency band is typically associated with resting or sleep states, suggesting that similar neural mechanisms may be involved in the perception of the urination. Figure 6(B) shows the average classification contribution of all subjects calculated by IG for different frequency bands, further confirming the importance of the Delta band in the classification of urination detection. Figure 6(C) shows the classification contribution of these three brain regions in all subjects' trails, also indicating that they have a significant impact on the task of urination detection.

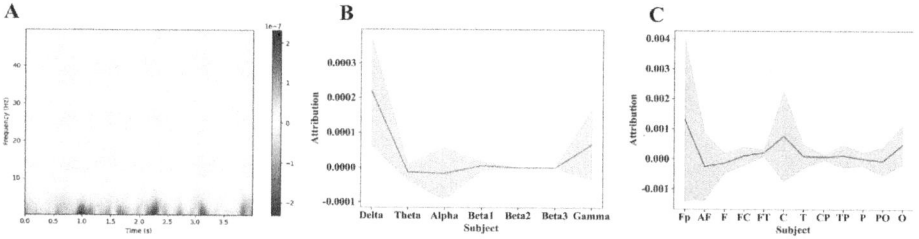

Fig. 6. (A) Time-Frequency Plot of Subject 4; (B) Brain Topography of Subject 4; (C) Average Classification Contribution of Different Bands; (D) Average Classification Contribution of Different Channels.

4 Discussion

This study fills the research gap in urination detection based on EEG. Given the lack of prior theoretical guidance, this study employed a deep learning model, combined with an interpretability method, to reveal the characteristics of EEG in the state of urination. The methods used not only enhance our understanding of the data but also provide a new analytical framework for future research.

In the experimental design, special attention was given to the issue of correlation in continuous resting-state data, and a time-independent training-test data division strategy was adopted to avoid potential data leakage problems.

From the perspective of spatio-temporal-frequency features, due to the lack of guiding cues and evoked paradigms, the temporal features of resting-state data are not prominent. However, through classification contribution analysis, it is possible to identify brain regions related to the control signals of the urination in terms of spatial features. In particular, the Filter Bank characteristics of the FBCNet model can effectively reveal EEG activities in specific frequency bands, which are mainly concentrated in the Delta band of the SMC, frontal pole and occipital region. From Fig. 6(D), it can be seen that the contribution of the frontal pole has a larger standard deviation, with outliers. Compared to SMC, it is more referenceable and shows a trend of spreading from the center to both ends, further proving that the activity involves a dynamic process in the SMC, that is, perceiving the existence of the urination and then feeding back the motor intention for urination. At the same time, the classification results show a large individual difference at the individual level, which depends on the strength of the motor intention of urination. Although this study has achieved innovative results, there are some limitations. The current study's sample is limited to males, with no female participants included, therefore, it has not been possible to explore the potential impact of gender differences on urination detection. Future research will expand the sample to include participants of different genders to achieve a more comprehensive understanding. Additionally, based on the conclusions of this study, further design and optimization of decoding algorithms will be undertaken to improve the accuracy and reliability of urination detection.

Acknowledgements. This study was supported by National Key Research and Development Program of China (No. 2023YFF1203500, No. 2023YFF1203503).

References

1. Hafid, A., et al.: State of the art of non-invasive technologies for bladder monitoring: a scoping review. Sensors (Basel). **23**(5), 2758 (2023). https://doi.org/10.3390/s23052758. PMID: 36904965; PMCID: PMC10007578
2. Nagle, A.S., Bernardo, R.J., Varghese, J., Carucci, L.R., Klausner, A.P., Speich, J.E.: Comparison of 2D and 3D ultrasound methods to measure serial bladder volumes during filling: steps toward development of non-invasive ultrasound urodynamics. Bladder **5**, e32 (2018). https://doi.org/10.14440/bladder.2018.565
3. Akkus, Z., Kim, B.H., Nayak, R., Gregory, A., Alizad, A., Fatemi, M.: Fully automated segmentation of bladder SAC and measurement of detrusor wall thickness from transabdominal ultrasound images. Sensors **20**, 4175 (2020). https://doi.org/10.3390/s20154175
4. Rix, A., et al.: Advanced ultrasound technologies for diagnosis and therapy. J. Nucl. Med. **59**, 740–746 (2018). https://doi.org/10.2967/jnumed.117.200030
5. Correas, J.-M., et al.: Advanced ultrasound in the diagnosis of prostate cancer. World J. Urol. **39**(3), 661–676 (2020). https://doi.org/10.1007/s00345-020-03193-0
6. Shadgan, B., Stothers, L., Molavi, B., Mutabazi, S., Mukisa, R., Macnab, A.: Near infrared spectroscopy evaluation of bladder function: the impact of skin pigmentation on detection of physiologic change during voiding. In: Proceedings of the Photonic Therapeutics and Diagnostics XI, San Francisco, CA, USA, 7–8 February 2015, pp. 106–116 (2015)
7. Koven, A., Herschorn, S.: NIRS: past, present, and future in functional urology. Curr. Bladder Dysfunct. Rep. **17**, 241–249 (2022). https://doi.org/10.1007/s11884-022-00665-4
8. Palla, A., Crema, C., Fanucci, L., Bellagente, P.: Kalman-based approach to bladder volume estimation for people with neurogenic dysfunction of the urinary bladder. In: Proceedings of the International Conference on Computers Helping People with Special Needs, Linz, Austria, 13–15 July 2016, pp. 521–528 (2016)
9. Li, Y., et al.: Analysis of measurement electrode location in bladder urine monitoring using electrical impedance. Biomed. Eng. Online **18**, 34 (2019). https://doi.org/10.1186/s12938-019-0651-4
10. Li, R., Gao, J., Li, Y., Wu, J., Zhao, Z., Liu, Y.: Preliminary study of assessing bladder urinary volume using electrical impedance tomography. J. Med. Biol. Eng. **36**(1), 71–79 (2016). https://doi.org/10.1007/s40846-016-0108-1

11. Leonhäuser, D., et al.: Evaluation of electrical impedance tomography for determination of urinary bladder volume: comparison with standard ultrasound methods in healthy volunteers. Biomed. Eng. Online **17**, 95 (2018). https://doi.org/10.1186/s12938-018-0526-0
12. Noyori, S.S., Nakagami, G., Noguchi, H., Mori, T., Sanada, H.: A small 8-electrode electrical impedance measurement device for urine volume estimation in the bladder. In: Proceedings of the 2021 43rd Annual International Conference of the IEEE Engineering in Medicine & Biology Society (EMBC), 1–5 November 2021, pp. 7174–7177 (2021)
13. Shin, S.-C., Moon, J., Kye, S., Lee, K., Lee, Y.S., Kang, H.-G.: Continuous bladder volume monitoring system for wearable applications. In: Proceedings of the 2017 39th Annual International Conference of the IEEE Engineering in Medicine and Biology Society (EMBC), Jeju Island, Republic of Korea, 1–15 July 2017, pp. 4435–4438 (2017)
14. Gramfort, A., et al.: MEG and EEG data analysis with MNE-Python. Front. Neurosci. **7**(267), 1–13 (2013). https://doi.org/10.3389/fnins.2013.00267
15. Mane, R., et al.: FBCNet: an efficient multi-view convolutional neural network for brain-computer interface. arXiv preprint arXiv:2104.01233 (2021)
16. Sanchez-Lengeling, B., et al.: Evaluating attribution for graph neural networks. Proc. Adv. Neural Inf. Process. Syst. **33**, 5898–5910 (2020)

D2CAN: Domain-Guided Contrastive Adversarial Network for EEG-Based Cross-Subject Cognitive Workload Decoding

Ruichao Zhan, Dongyang Li, Song Wang, and Quanying Liu[✉]

Department of Biomedical Engineering, Southern University of Science and Technology, Shenzhen, China
{12232591,lidy2023}@mail.sustech.edu.cn, liuqy@sustech.edu.cn

Abstract. Feature engineering and deep learning methods have allowed to decoding cognitive workload with neural data. However, the low efficiency and low cross-subject performance remain challenges. In this study, we propose a domain-guided cross-subject contrastive adversarial learning framework for cognitive workload decoding based on electroencephalogram (EEG). This framework employs contrastive adversarial learning to train an EEG encoder that projects individual EEG data from diverse sources into a low-dimensional invariant subspace, where subjects from both source and target domains share a common representation. The robustness of the model is enhanced by confusing the domain-specific representations of new subjects in the target domain and effectively interfering with the domain discriminator. These shared representations are subsequently used to classify cognitive workload. We conduct extensive experiments to investigate the influence of noise intensity from different sources and the contribution of different brain regions to decoding. Our method achieves state-of-the-art (SOTA) performance in EEG-based workload decoding, addressing the challenges of cross-subject variability and accuracy, and paving the way for more reliable cognitive workload assessment in diverse populations.

Keywords: EEG · Contrastive learning · Adversarial learning · Cognitive workload · Neural decoding

1 Introduction

Cognitive workload status describes the ratio of cognitive resource demands required to perform a task [22]. In real-world scenarios, both overworkload and underworkload conditions can affect the operator's state, subsequently affecting task execution efficiency. Maintaining an appropriate cognitive workload level is crucial for ensuring safe and efficient task execution [26]. The recorded physiological data provide a foundation for the objective and continuous assessment of cognitive workload [28]. EEG is considered the most sensitive physiological indicator of cognitive workload due to its high temporal resolution and non-invasiveness. However, EEG signals are characterized by randomness, non-stationarity, and noise, leading to individual differences in EEG data. This variability poses challenges to the reusability and generalization of EEG-based cognitive workload assessment models in practical scenarios [21, 25].

To address these challenges, we propose an EEG-based cognitive workload classification framework utilizing a **D**omain-guided **C**ross-subject **C**ontrastive **A**dversarial **N**etwork (**D2CAN**). Our contributions have three points:

- We propose a cross-subject learning framework D2CAN, allowing quick adaptation to new subjects. This framework effectively learns the temporal and spatial characteristics of EEG and achieves SOTA performance in downstream cognitive workload decoding tasks.
- We employ an adversarial learning strategy to enhance domain generalization across different subjects, which is crucial for practical applications of EEG-based cognitive workload decoding in diverse real-world scenarios.
- We conduct extensive experiments to investigate the biological interpretability of EEG-based cognitive workload decoding, including contributions of temporal information, spatial information, and frequency bands, as well as the robustness to noise.

2 Related Work

Feature engineering, deep learning, and transfer learning are extensively employed in the cross-subject decoding of cognitive workload. In traditional feature engineering methods, EEG is used to extract features for classifying cognitive workload from time domain [1], frequency domain [2,8], time-frequency domain [24], and functional connectivity network [5]. For instance, the extracted spectral features enabled the classification of cross-subject cognitive overworkload and underworkload in arithmetic tasks [15, 20].

However, these methods not only exhibit limited classification performance but are also unsuitable for practical applications.

Learning robust feature representations from cognitive overworkload and underworkload classification data is an advantage of deep learning methods. Jiao et al. [7] designed a multi-layer CNN model for cross-subject cognitive workload classification. Ni et al. [13] proposed a hierarchical recurrent network that combines temporal modeling and adversarial training for cognitive workload classification. Although deep learning outperforms traditional feature engineering methods in cross-subject tasks, these models still face the challenge of insufficient generalization due to individual differences. Furthermore, deep learning models typically require large amounts of annotated data, which are difficult to obtain in cognitive workload classification experiments.

To address the issue of inter-subject variability in cognitive workload recognition, transfer learning methods such as domain adaptation (DA) and domain generalization (DG) are proposed [18]. DA methods strive to minimize the discrepancy between data distributions of the source domain (i.e., the training subjects) and the target domain (i.e., the testing subjects). One typical method is the Domain Adversarial Neural Network (DANN) [3], which aligns feature representations of the source and target domains through adversarial training. Building on this, researchers have developed advanced models for cross-subject cognitive workload recognition [29]. DG methods, on the other hand, learn domain-invariant feature representations from source domains, primarily using contrastive learning techniques to align features across subjects. These contrastive learning methods have achieved excellent classification results in cross-subject emotion recognition [10,17,18].

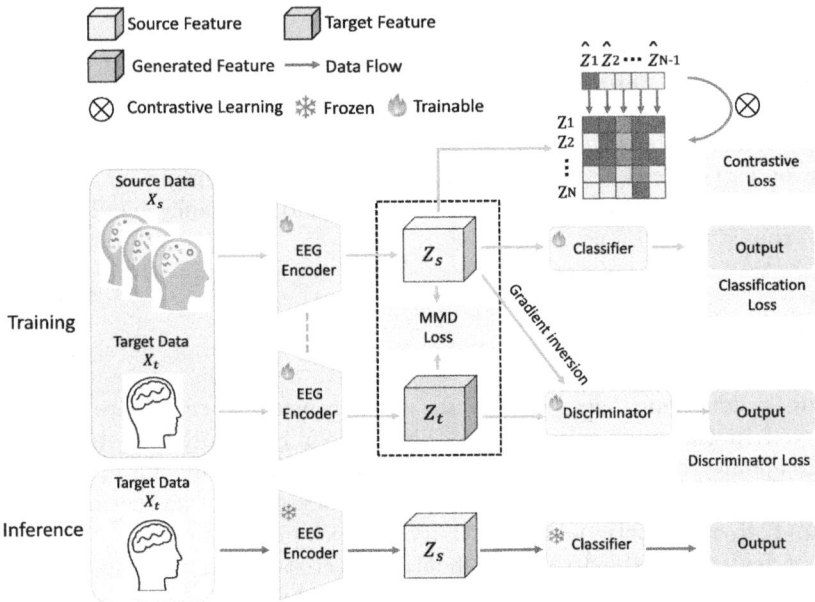

Fig. 1. Framework of D2CAN. During the training process, labeled data from the source subject domain and unlabeled data from the target subject domain are projected into a low-dimensional invariant subspace within the same high-dimensional space using the same EEG Encoder to obtain a shared representation. The data from the source domain imposes a gradient reversal perturbation to confuse the learning of the discriminator, while the classifier outputs a classification after contrastive learning. During the inference process, data from the target domain is directly classified using the trained EEG encoder and classifier.

3 Method

We propose a framework D2CAN, as illustrated in Fig. 1. The framework comprises five modules: feature extractor, feature alignment, source domain contrastive learning, classifier, and domain discriminator. D2CAN model input is raw EEG data collected under overworkload and underworkload tasks, and the output is the predicted cognitive workload labels.

3.1 Contrastive Adversarial Network (CAN)

D2CAN is trained in a supervised manner. Specifically, In the training stage, the train set is utilized as the source domain, and the test set is used as the target domain. The source domain comprises both classification labels and domain labels, and the target domain includes only domain labels. Both domains are jointly input into D2CAN for training.

In domain adversarial training, the objective is to minimize prediction loss by simultaneously training feature extractor and classifier while a domain classifier discriminates between features from domains. Through adversarial training and gradient reversal, the feature extractor learns to generate domain-invariant features, thereby confusing the domain classifier and reducing domain discrepancies. By optimizing the feature extractor to decrease the domain classifier's discriminative capability, feature consistency across domains is achieved [3,4,29]. In the testing stage, the trained feature extractor and classifier subsequently be used to predict cognitive workload label.

3.2 EEG Encoder

To extract the shared feature space representation from data across different subject domains, we employed the EEGNetV4 architecture by Lawhern et al. [9]. This architecture learns the projection from the EEG window $X_i \in \mathbb{R}^{C \times T}$ to the latent representation $z_i \in \mathbb{R}^F$. Compared to other types of projection layers and time series models, EEGNetV4 was selected as our encoder for its small size of parameters and robustness.

3.3 Feature Alignment

To minimize the distribution discrepancy between the source domain and the target domain, Maximum Mean Discrepancy (MMD) is employed as the distance discrepancy measure. We employed MMD to reduce the statistical distribution differences between domains in the feature space, thereby improving D2CAN's effectiveness in transferring knowledge from the source domain to the target domain [10,29].

3.4 Classifier

We designed a classifier comprising two fully connected layers to decode and classify feature representations extracted by the feature extractor. Specifically, the first fully connected layer employs the Rectified Linear Unit (ReLU) activation function to introduce nonlinearity, while the second layer utilizes the sigmoid activation function to produce outputs interpretable as probabilities.

3.5 Contrastive Learning

To enhance the classifier's ability to distinguish between different cognitive loads, we employ a contrastive learning approach using positive and negative representation samples before the final classification step. During training, each representation from the source domain subjects is compared with other positive and negative samples within the same batch. Samples of the same class are considered positive, while those of different classes are considered negative. This approach aims to refine the model's ability to discern inter-class differences while minimizing intra-class variability, ultimately improving overall classification performance.

3.6 Domain Discriminator

To enhance D2CAN ability to distinguish feature projections from different sources, a domain discriminator utilizing fully connected layers is employed. The domain discriminator identifies whether the feature comes from the source domain or the target domain. It employs gradient reversal to modulate the gradients propagated to the EEG encoder in training stage. By introducing features from the source domain and applying gradient reversal operations, the discriminator is deliberately confused. Consequently, the adversarial training strategy enhances EEG encoder capacity to accurately project shared representations across domains, thereby improving the network's robustness and generalization across different subjects.

3.7 Loss Function

In the training stage, we calculate the MMD of the source domain features and the target domain features [10,29], the formula is as follows:

$$\mathcal{L}_{\mathrm{mmd}} = \left\| \frac{1}{N_s} \sum_{i=1}^{N_s} \phi\left(X_i\right) - \frac{1}{N_t} \sum_{j=1}^{N_t} \phi\left(X_j\right) \right\|^2, \tag{1}$$

where $\phi(X)$ is the kernel function. Here we use Radial Basis Function (RBF) as the kernel.

$$\mathcal{L}_{contrast}(\theta) = -\frac{1}{B} \sum_{i=1}^{B} \log \frac{\exp\left(s\left(\hat{z}_i, z_i\right)\right)}{\sum_{j=1}^{B} \exp(s\left(\hat{z}_i, z_j\right))}, \tag{2}$$

where s is the cosine similarity; $\hat{z}_i$ and $\hat{z}_i = f_\theta(X_i)$ are the latent representation and the corresponding EEG-based prediction, respectively. B is the batch size, and z_j are other samples in the same batch for comparison.

$$\mathcal{L}_{\text{cls}} = -\sum_{c=1}^{C} y_c \log y_c^{\text{predict}}, \tag{3}$$

where y_c is the ground-truth label of class c, y_c predict is the model prediction label, and C is the number of cognitive levels to recognize.

$$\mathcal{L}_{\text{dis}} = -\sum_{j=0}^{J} d_j \log d_j^{\text{predict}}, \tag{4}$$

where $J = \{0, 1\}$ is the domain labels; d_j is the predicted label output. The total loss function of the proposed model is as follows.

$$\mathcal{L}_{total}(\theta) = \alpha \mathcal{L}_{cls} + \theta \mathcal{L}_{mmd} + \lambda \mathcal{L}_{contrastive} + \gamma \mathcal{L}_{dis}, \tag{5}$$

where α, θ, λ and γ are the hyperparameters for controlling classification loss, the weights of feature alignment loss, contrastive loss, and domain loss, respectively.

4 Experiments

4.1 EEG Datasets and Preprocessing

The EEG data under different cognitive workload is from a public competition dataset [14], which was collected under NASA's Multi-Attribute Task Battery (MATB) task. This dataset comprises EEG signals from 15 subjects (average age 25 years). Three independent repeated experiments were conducted, with a one-week interval between each sesson. Each subject completed the MATB-II task, which consisted of three 5-minute blocks. These blocks presented different difficulty levels in a pseudo-random manner, resulting in three levels of cognitive workload. The MATB-II task, developed by NASA, evaluates mental capacity through multiple concurrent subtasks, including tracking, system monitoring, communication, and resource management. Cognitive workload is quantified based on the subject's performance across these subtasks. Statistical analysis of the dataset was conducted to derive and validate three distinct levels of cognitive workload, specifically low, medium, and high.

Data preprocessing involved seven steps to ensure the quality and consistency of the EEG data. First, the data was segmented into 2-s non-overlapping epochs and referenced to the right mastoid electrode. A high-pass FIR filter at 1 Hz was applied to remove low-frequency noise. Channels with average amplitude exceeding 2 standard deviations were rejected, and spherical interpolation was used to estimate the removed channels. Second-order blind identification (SOBI) was performed for source separation, followed by automated independent component labeling to remove muscle, heart, and eye artifacts with a 95%

threshold. A low-pass FIR filter at 40 Hz was then applied. Finally, the data was re-referenced using the common average reference (CAR) and down-sampled to 250 Hz.

In this study, we utilized the low and high cognitive workload data from Session 1 to conduct binary classification for cross-subject cognitive workload classification. Each subject provided 149 samples for both low and high cognitive workloads, resulting in a total of 4,470 samples. For feature extraction, we employed Power Spectral Density (PSD). Specifically, the PSD was extracted for five frequency bands using the Short-Time Fourier Transform (STFT) [30]. Consequently, this process yielded a total of 1,220 dimensions of PSD features.

4.2 D2CAN Performance

Table 1 presents the performance of D2CAN compared to other methods on the dataset. We implemented methods combining PSD features with widely used traditional classifiers, including SVM, LDA, and KNN. We replicated results with existing deep learning models as baselines using raw EEG data as input, including EEGNetV4 [9], ShallowCNN [16], EEGConformer [19] and DANN [3], as well as DDA model [29]. Compared to these baseline models, D2CAN achieved the superior performance in terms of accuracy, sensitivity, F1 score, specificity and Matthews Correlation Coefficient (MCC), with an average accuracy increase of 9.71%.

To optimize D2CAN performance and identify the most effective settings, the hyperparameter configurations presented in Fig. 2 systematically explore the impact of various parameters on D2CAN accuracy. Figure 2A shows the impact of different loss function coefficients on model accuracy with a batch size set to 32. We set the coefficients for classification loss and domain classification loss to 0.5 and 1, respectively, followed by coefficients for the MMD loss and contrastive loss. The results indicate that optimal classification performance is achieved when the loss function coefficients are set to (1, 1, 1, 0.5).

Figure 2B demonstrated the impact of learning rate variations with different batch sizes on model accuracy. The findings suggest that setting the learning rate to 3e−4 and the batch size to either 32 or 64 yields the best model performance.

Table 1. The comparison of different baselines.

Model	Accuracy (%)	Sensitivity (%)	F1 Score (%)	Specificity (%)	MCC (%)
SVM	67.90 ± 7.34	81.83 ± 12.17	71.72 ± 6.07	53.96 ± 17.89	38.58 ± 13.18
LDA	70.07 ± 7.78	76.33 ± 14.22	71.73 ± 6.27	63.80 ± 23.04	42.76 ± 14.82
KNN	65.75 ± 4.70	80.13 ± 9.15	69.97 ± 4.03	51.36 ± 12.76	33.58 ± 8.69
EEGNetV4	72.21 ± 12.13	74.41 ± 26.80	69.91 ± 21.01	69.75 ± 25.08	47.82 ± 20.55
ShllowCNN	71.99 ± 11.84	78.70 ± 23.26	72.42 ± 16.23	65.28 ± 31.36	48.14 ± 21.88
EEGConformer	71.57 ± 12.76	78.43 ± 24.23	72.18 ± 15.93	64.70 ± 34.82	48.00 ± 23.41
DANN	62.77 ± 9.71	62.37 ± 28.38	58.35 ± 22.87	63.18 ± 22.34	27.67 ± 19.39
DDA	73.47 ± 6.91	73.47 ± 7.15	73.11 ± 7.34	**82.51 ± 7.88**	48.02 ± 14.11
D2CAN (Ours)	**81.79 ± 5.89**	**87.02 ± 8.11**	**82.75 ± 5.34**	76.55 ± 11.63	**64.68 ± 12.15**

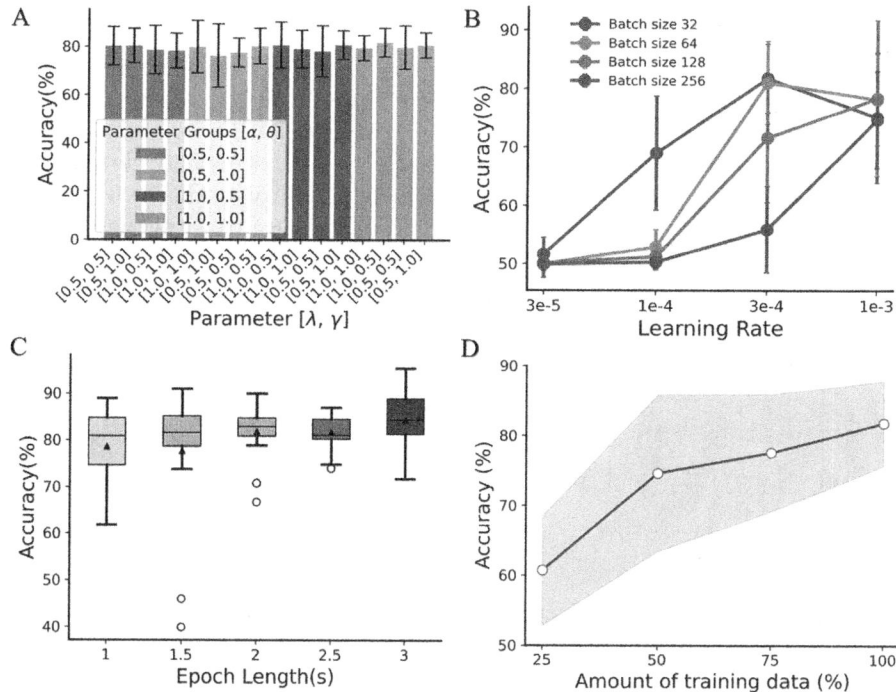

Fig. 2. Impact of Hyperparameters and Training Data on Accuracy. (A) Accuracy across different loss function coefficient settings, with $\alpha, \theta, \lambda, \gamma$ set to 0.5 or 1. (B) Influence of batch size and learning rate on model classification accuracy, with batch sizes of 32, 64, 128, and 256, and learning rates of 1e−3, 1e−4, 3e−4, and 3e−5. (C) Model classification accuracy at different epoch lengths, set to 1 s, 1.5 s, 2 s, 2.5 s, and 3 s. (D) Impact of varying training data volumes on classification accuracy, with training data set to 25%, 50%, and 75% of the original data volume.

4.3 Temporal Analysis

To evaluate D2CAN capability in classifying cognitive workload across various time scales, we resegmented the original EEG data into multiple epoch lengths (Fig. 2C). Specifically, we segmented the EEG data into epochs of 1 s, 1.5 s, 2 s, 2.5 s, and 3 s to examine the model's reliability in cognitive workload classification under different task durations.

Figure 2C showed the distribution of D2CAN accuracy across different epoch lengths. The accuracy generally remained stable, indicating robustness over various time scales. Notably, epoch lengths of 2 and 2.5 s yielded higher accuracy and reduced variability. These findings emphasized the model's adaptability and consistent performance across varying input data lengths. This suggests its potential applicability in practical tasks that involve different cognitive workloads.

4.4 Spatial Analysis

To explore the impacts of different brain regions on the cognitive workload decoding task. Inspired by [11,27], we divided the scalp EEG into five brain lobes (i.e., frontal, temporal, central, parietal and occiptial lobe). The experiment eliminated data from some brain region electrodes to simulate electrode failure or data loss. In this way, we can assess the model's ability to deal with missing information from specific lobe [12]. The experimental results are shown in Fig. 3.

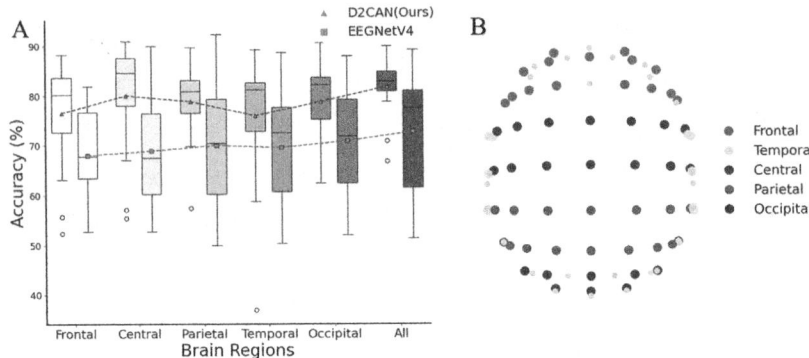

Fig. 3. Accuracy after the removal of electrodes from different brain regions. (A) The classification accuracy of D2CAN and EEGNetV4 in different brain regions. (B) The basis of electrode distribution removal for the brain area, color-coded as red for Frontal, blue for Temporal, green for Central, purple for Parietal, and yellow for Occipital lobe. (Color figure online)

As depicted in Fig. 3, this study precisely delineated and excluded data from specific brain regions. Despite these exclusions, the D2CAN model consistently exhibited higher accuracy and reduced variability relative to EEGNetV4, illustrating its capacity to adeptly leverage signals from adjacent regions to offset information deficits, thereby markedly augmenting its robustness and dependability. These results not only confirm the model's superior adaptability and performance in handling complex EEG signals but also underscore its viability for future research applications.

It is imperative to recognize that while theoretically each EEG electrode is capable of capturing signals from the entire brain, electrodes distal to active regions are likely to register only attenuated signals that are susceptible to noise interference. Consequently, for the purposes of this experiment, it is posited that electrodes situated within specific brain regions are predominantly associated with activity localized to those areas.

4.5 Frequency Band Analysis

To investigate the impact of EEG frequency bands on decoding performance and to uncover the most relevant neural oscillations to cognitive workload, we filter EEG signals into four main frequency bands: δ (0.1–4 Hz), θ (4–8 Hz), α (8–12 Hz), and β (12–30 Hz). By evaluating the classification accuracy using EEG data in these difference frequency bands, we aim to identify the frequency band associated with high cognitive workload.

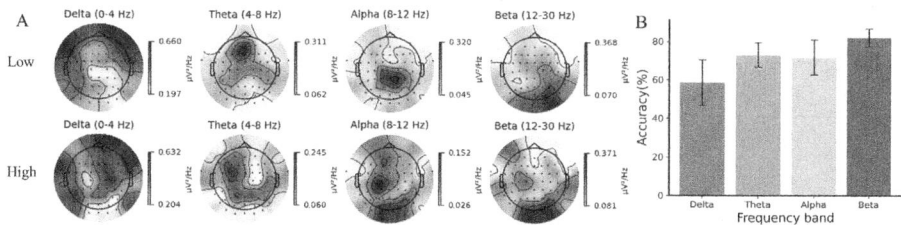

Fig. 4. Impact of Different EEG Frequency Bands on Model Classification Accuracy. (A) The PSD distribution in low and high cognitive workload. The color bar represents the mean PSD for each frequency band. (B) The model's classification accuracy in each of the four frequency bands. (Color figure online)

In Fig. 4A, we extracted and visualized EEG features from different frequency bands for low and high difficulty cognitive load tasks. The visualization results align with previous studies [30], indicating that for higher β frequencies, the strongest activity is concentrated in the occipital lobe, while the temporal lobe shows relatively low activation or inhibition compared to the occipital lobe.

As shown in Fig. 4B, our model achieved the highest accuracy (82.03%) in the β band, which is consistent with the overall performance of the model. This finding corroborates previous research suggesting that high cognitive load tasks are primarily associated with high-frequency bands, demonstrating that our model effectively captures the EEG characteristics of these task. Conversely, the lowest accuracy was observed in the δ band (58.57%), likely due to noise interference from eye movements, heartbeats, and other artifacts within this frequency range.

4.6 Noise Analysis

To evaluate the robustness of the model under various signal-to-noise ratio (SNR) conditions for EEG signal noise, this study introduced Gaussian noise ranging from −10 dB to +10 dB across all channels, simulating noisy environments commonly encountered in EEG data. The objective was to assess the model's sensitivity to noise and its reliability in practical applications. Additionally, by incorporating varying degrees of label noise into the training data, the study further explored the model's performance under conditions of label interference, aiming to understand its controllability and effectiveness when faced with such disturbances.

As illustrated in Fig. 5A, the model exhibits high performance in low noise environments (5, 10 dB), confirming its robustness. Despite a decline in accuracy

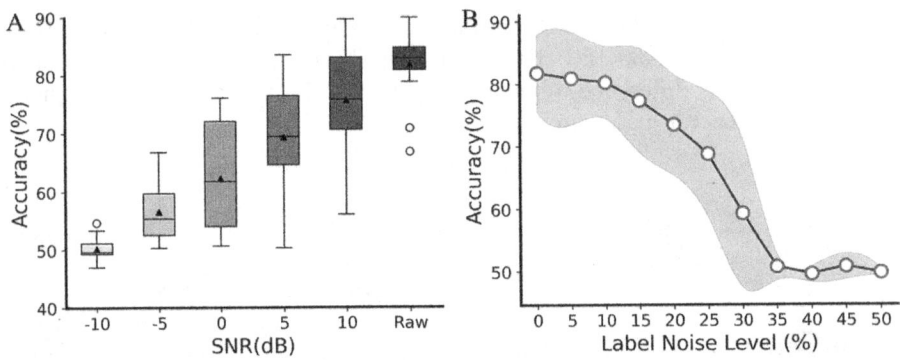

Fig. 5. Model Robustness in SNR or Label Noise. (A) The resilience of the model to varying SNR from -10 dB to 10 dB. (B) The resilience of the model to varying Label noise levels. The label noise level represents the proportion of classification labels that are flipped in the training dataset.

in high noise conditions (0, -5 dB), the model continues to demonstrate effective noise suppression capabilities. At extremely high noise levels (-10 dB), accuracy reaches 50%, indicating that the model's performance is primarily determined by the inherent characteristics of EEG data. Figure 5B showed that the model maintains high accuracy and low variance under low label noise, with a linear decrease in accuracy to 50% under moderate to high label noise, further substantiating that the model's performance relies on the intrinsic features of EEG data.

4.7 Features Distribution

To assess the effectiveness of our method in extracting invariant features across subjects, we employed the t-SNE algorithm to visualize the features of subject 15. Figure 6A illustrated the distribution of data points from both the source and target domains after dimensionality reduction via t-SNE. Figure 6B showed the distribution of data points within the target domain, highlighting the samples under different cognitive workload conditions.

4.8 Ablation Study

To validate the roles and effectiveness of various components within our model, we conducted a series of ablation experiments. We systematically removed each module one at a time, while keeping other variables constant, and assessed D2CAN performance on the test set by comparing accuracy and standard deviation. In our experiments, '-DD' denotes the removal of the domain discriminator module, '-CON' indicates the removal of the contrastive learning module, and '-MMD' represents the removal of the feature alignment module. These configurations were tested to determine their impact on model performance.

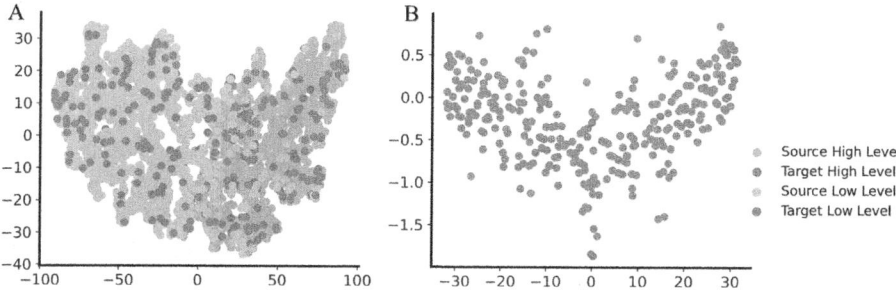

Fig. 6. Features visualization. Blue and pink dots represent the high and low cognitive workload levels, respectively. (A) The comparison of features between the source and target domains under different cognitive workload levels. (B) The feature distribution within the target domain at these levels. (Color figure online)

The results of the ablation experiments, as presented in Table 2, demonstrate that the removal of any training module from the original model adversely affects performance. Specifically, the exclusion of the MMD module resulted in the most substantial decline in model performance. This finding indicates that the MMD module is essential for aligning feature distributions and improving model generalization in EEG data analysis. Furthermore, these ablation study results underscore that incorporating the MMD module markedly enhances the accuracy and robustness of the model when processing EEG data with complex distributions.

Table 2. Results summary for different configurations.

Model	Accuracies (%)	Sensitivity (%)	F1 Score (%)	Specificity (%)	MCC (%)
D2CAN (Ours)	**81.79 ± 5.89**	**87.02 ± 8.11**	**82.75 ± 5.34**	**76.55 ± 11.63**	64.68 ± 12.15
-MMD	76.35 ± 12.16	84.74 ± 14.42	78.32 ± 10.86	67.96 ± 19.85	54.91 ± 25.82
-DD	75.21 ± 12.15	73.19 ± 15.35	62.01 ± 27.41	53.42 ± 22.81	**75.21 ± 11.73**
-CON	75.15 ± 11.36	75.15 ± 11.76	74.43 ± 13.03	74.09 ± 15.42	51.38 ± 24.56
-DD,CON	76.82 ± 8.99	76.82 ± 9.30	76.59 ± 9.46	76.02 ± 15.41	54.56 ± 18.51
-DD,MMD	73.22 ± 12.07	73.22 ± 12.50	71.71 ± 14.41	65.15 ± 24.00	48.89 ± 24.18
-CON,MMD	73.38 ± 9.99	73.38 ± 10.34	72.62 ± 10.92	70.78 ± 21.88	48.87 ± 20.46
EEGNetV4	72.08 ± 10.90	74.41 ± 26.80	69.91 ± 21.01	69.75 ± 25.08	47.82 ± 20.55

5 Discussion

We developed the D2CAN framework, which achieved SOTA performance in cross-subject cognitive workload decoding. The primary innovation of D2CAN is its domain-guided contrastive adversarial learning strategies, which have proven effective in various neural signal processing tasks [17,18,29,30]. For instance, in applications such as emotion recognition and attention monitoring, these strategies have effectively addressed the generalization challenges posed by individual differences [10,18]. By leveraging domain-guided contrastive adversarial learning, D2CAN not only improved cross-subject performance but also demonstrated advantages in limited data and model generalization.

Our research explores the intricate relationship between brain activity and cognitive workload, describing the temporal and spatial dimensions of how workload information is processed. By examining EEG data across multiple time windows, we identified that cognitive information predominantly manifests within the β frequency band, corroborating previous research findings [30]. Furthermore, our spatial experiments revealed that cognitive workload information is primarily encoded in the occipital and central regions of the brain. These findings not only confirm the involvement of posterior regions in visual and spatial processing but also underscore the critical role of central regions in the integration of complex cognitive functions. Additionally, we observed performance variations between interdisciplinary and intradisciplinary settings, which can be attributed to factors such as individual variability in EEG signals, heterogeneity in brain structure and function, and differential noise distribution during recording.

There are several limitations associated with D2CAN. First, we utilized some target data without classification labels during the training stage, which may not facilitate real-time decoding in practical applications. Second, D2CAN was evaluated on a single dataset, thereby its robustness and generalizability have not been verified on other datasets. Third, D2CAN is only evaluated on a single dataset, its robustness and generalizability have not been validated on other datasets. Future research should focus on zero-shot EEG decoding tasks, which may involve more adaptable neural network architectures or training on larger EEG datasets. Meta-learning also present promising avenues for exploration [6, 23].

6 Conclusion

In this study, we developed an EEG-based cognitive workload decoding framework (D2CAN). By employing a contrastive adversarial network, our method enables the model to achieve SOTA compared to other methods and better generalization performance in different subjects.

Acknowledgement. The authors would like to thank Youzhi Qu, Chen Wei, and Xinke Shen for their invaluable comments and suggestions that greatly improved this manuscript. This work was funded in part by the National Key R&D Program of China (2021YFF1200804), Shenzhen Science and Technology Innovation Committee (2022410129, KCXFZ20201221173400001, KJZD20230923115221044).

References

1. Causse, M., et al.: EEG/ERP as a measure of mental workload in a simple piloting task. Procedia Manuf. **3**, 5230–5236 (2015)

2. Choi, M.K., et al.: Development of an EEG-based workload measurement method in nuclear power plants. Ann. Nuclear Energy **111**, 595–607 (2018)
3. Ganin, Y., Lempitsky, V.: Unsupervised domain adaptation by backpropagation. In: International Conference on Machine Learning, pp. 1180–1189. PMLR (2015)
4. Ganin, Y., et al.: Domain-adversarial training of neural networks. J. Mach. Learn. Res. **17**(59), 1–35 (2016)
5. Guan, K., et al.: EEG based dynamic functional connectivity analysis in mental workload tasks with different types of information. IEEE Trans. Neural Syst. Rehabil. Eng. **30**, 632–642 (2022)
6. Han, J.-W., et al.: META-EEG: meta-learning-based class-relevant EEG representation learning for zero-calibration brain-computer interfaces. Expert Syst. Appl. **238**, 121986 (2024)
7. Jiao, Z., et al.: Deep convolutional neural networks for mental load classification based on EEG data. Pattern Recogn. **76**, 582–595 (2018)
8. Kumar, N., Kumar, J.: Measurement of cognitive load in HCI systems using EEG power spectrum: an experimental study. Procedia Comput. Sci. **84**, 70–78 (2016)
9. Lawhern, V.J., et al.: EEGNet: a compact convolutional neural network for EEG-based brain-computer interfaces. J. Neural Eng. **15**(5), 056013 (2018)
10. Lee, P., et al.: Inter-subject contrastive learning for subject adaptive EEG-based visual recognition. In: 2022 10th International Winter Conference on Brain-Computer Interface (BCI), pp. 1–6. IEEE (2022)
11. Li, D., et al.: Visual decoding and reconstruction via EEG embeddings with guided diffusion. arXiv preprint arXiv:2403.07721 (2024)
12. Luo, Z., et al.: Mapping effective connectivity by virtually perturbing a surrogate brain (2024). arXiv:2301.00148 [q-bio.NC]
13. Ni, Z., et al.: Improving cross-state and cross-subject visual ERP-based BCI with temporal modeling and adversarial training. IEEE Trans. Neural Syst. Rehabil. Eng. **30**, 369–379 (2022)
14. Passive BCI-Hackathon—Neuroergonomics conference 2021 (2021). https://www.neuroergonomicsconference.um.ifi.lmu.de/pbci/
15. Plechawska-Wójcik, M., et al.: A three-class classification of cognitive workload based on EEG spectral data. Appl. Sci. **9**(24), 5340 (2019)
16. Schirrmeister, R.T., et al.: Deep learning with convolutional neural networks for EEG decoding and visualization. Hum. Brain Map. **38**(11), 5391–5420 (2017)
17. Shen, X., et al.: Contrastive learning of shared spatiotemporal EEG representations across individuals for naturalistic neuroscience. arXiv preprint arXiv:2402.14213 (2024)
18. Shen, X., et al.: Contrastive learning of subject-invariant EEG representations for cross-subject emotion recognition. IEEE Trans. Affect. Comput. (2022)
19. Song, Y., et al.: EEG conformer: convolutional transformer for EEG decoding and visualization. IEEE Trans. Neural Syst. Rehabil. Eng. **31**, 710–719 (2022)
20. Walter, C., et al.: Online EEG-based workload adaptation of an arithmetic learning environment. Front. Hum. Neurosci. **11**, 286 (2017)
21. Wan, Z., et al.: A review on transfer learning in EEG signal analysis. Neurocomputing **421**, 1–14 (2021)
22. Wickens, C.D.: Multiple resources and performance prediction. Theor. Issues Ergon. Sci. **3**(2), 159–177 (2002)
23. Xie, Y., et al.: Cross-dataset transfer learning for motor imagery signal classification via multi-task learning and pre-training. J. Neural Eng. **20**(5), 056037 (2023)

24. Zarjam, P., Epps, J., Lovell, N.H.: Beyond subjective selfrating: EEG signal classification of cognitive workload. IEEE Trans. Auton. Mental Dev. **7**(4), 301–310 (2015)
25. Zhang, J., Wang, Y., Li, S.: Cross-subject mental workload classification using kernel spectral regression and transfer learning techniques. Cogn. Technol. Work **19**(4), 587–605 (2017). https://doi.org/10.1007/s10111-017-0425-3
26. Zhang, L., et al.: Cognitive load measurement in a virtual reality-based driving system for autism intervention. IEEE Trans. Affect. Comput. **8**(2), 176–189 (2017)
27. Zhang, S., et al.: The neural correlates of ambiguity and risk in human decision-making under an active inference framework (2023)
28. Zhou, Y., et al.: Cognitive workload recognition using EEG signals and machine learning: a review. IEEE Trans. Cogn. Dev. Syst. **14**(3), 799–818 (2021)
29. Zhou, Y., et al.: Cross-subject cognitive workload recognition based on EEG and deep domain adaptation. IEEE Trans. Instrum. Measur. (2023)
30. Zhou, Y., et al.: Cross-task cognitive workload recognition based on EEG and domain adaptation. IEEE Trans. Neural Syst. Rehabil. Eng. **30**, 50–60 (2022)

Group-Specific Fusion Model and Its Application in Identifying Multimodal Co-varying Diagnostic Patterns for Psychiatric Disorders

Siyuan Cao[1,2,3], Chuang Liang[1,2,3], Qi Zhu[1,2,3], Rongtao Jiang[5], Daoqiang Zhang[1,2,3], Vince D. Calhoun[4(✉)], and Shile Qi[1,2,3(✉)]

[1] Department of Computer Science and Technology, Nanjing University of Aeronautics and Astronautics, Nanjing, China
shile.qi@nuaa.edu.cn
[2] Key Laboratory of Brain-Machine Intelligence Technology, Ministry of Education, Nanjing University of Aeronautics and Astronautics, Nanjing, China
[3] MIIT Key Laboratory of Pattern Analysis and Machine Intelligence, Nanjing University of Aeronautics and Astronautics, Nanjing, China
[4] Tri-Institutional Center for Translational Research in Neuroimaging and Data Science (TReNDS) Georgia State University, Georgia Institute of Technology, Emory University, Atlanta, GA, USA
vcalhoun@gsu.edu
[5] Department of Radiology and Biomedical Imaging, Yale University, New Haven, CT, USA

Abstract. Multimodal fusion offers a complementary perspective for understanding brain function and structure. Nevertheless, most existing multimodal fusion methods are unsupervised and thus ignore the diagnostic information. Previous supervised fusion models stack subjects from different diagnostic groups into a single feature matrix, resulting in potential interference between diagnostic groups. In this study, we propose a group-specific multimodal fusion method, called Group-specific Multiset Canonical Correlation Analysis with Logistic Regression (GMCCALR) that can effectively leverage group specific information by incorporating a logistic regression term into mCCA and utilizing a group-specific fusion strategy. Results in human neuroimaging data show that GMCCALR provide higher diagnostic accuracy across several psychiatric disorders, including schizophrenia, schizoaffective disorder, bipolar disorder and autism spectrum disorder. We also show the identified disorder-specific multimodal patterns are replicable across multiple cohorts. Overall, results demonstrate the reliability and generalizability of the proposed GMCCALR approach and its efficacy for identifying multimodal co-varying diagnostic patterns for psychiatric disorders.

Keywords: Multimodal fusion · group-specific fusion · diagnostic information · psychiatric disorders · classification

---

S. Cao and C. Liang—These authors contributed equally to this work.

1 Introduction

Existing neuroimaging techniques, such as functional magnetic resonance imaging (fMRI) and structural MRI, have provided remarkable new insights into human brain function and structure. However, any neuroimaging modality alone provides only a limited view into brain function or structure, which may neglect the hidden relationships between different modalities. Multimodal fusion can utilize the complementary information captured from multiple modalities to reveal the potential functional-structural covariations [1–5], especially supervised fusion that can reflect the relationship between the brain and the interested clinical measures [6].

Existing multimodal fusion models can be classified into two categories: unsupervised fusion methods and supervised fusion methods. Multiset Canonical Correlation Analysis + Joint Independent Component Analysis (mCCA + jICA) combined the advantages of mCCA [7] and jICA [8] by maximizing inter-modality correlation with mCCA and then ensuring independence among components with jICA [9]. MUlti-modal Structured Embedding (MUSE) encourages the maintenance of each modality's structure in the fusion model by structured self-supervision and then uses a common self-reconstruction loss and a structured self-supervision loss to optimize the obtained joint representation [10]. Multi-view Factorization AutoEncoder (MAE) incorporated biological interaction networks as an "external" domain knowledge source into the models through network regularization [11]. However, these methods are unsupervised and thus do not leverage existing diagnostic information. Discriminative Sparse Canonical Correlation Analysis (DSCCA) incorporates diagnostic information by maximizing intra-group correlation and minimizing inter-group correlation [12]. Outcome-relevant Sparse Canonical Correlation Analysis (OSCCA) regularizes the similarity matrices between subjects to leverage diagnostic information [13]. Multi-Task Sparse Canonical Correlation Analysis with Logistic Regression (MTSCCALR) assigns one-vs-all (OVA) tasks and introduces a logistic regression terms into SCCA to utilize diagnostic information [14]. However, the aforementioned SCCA-based method identifies a single component rather than multiple components as identified in mCCA, hindering the identification of multiple multimodal co-varying components [15]. Furthermore, these multimodal fusion methods typically stack subjects from different diagnostic groups into a single feature matrix during fusion, resulting in information mutual interference between diagnostic groups, which may diminish diagnostic accuracy [14].

To address the limitations of these methods, in this study, we proposed a method called Group-specific Multiset Canonical Correlation Analysis with Logistic Regression (GMCCALR). We incorporated a logistic regression term into mCCA to effectively leverage diagnostic information and ensure the variability of the number of components. Additionally, we utilize a group-specific fusion strategy to optimize the model for diagnostic specificity. To evaluate the effectiveness, GMCCALR was compared with six alternative methods in terms of classification accuracy for several psychiatric disorders. Results demonstrate that GMCCALR outperforms other fusion methods in identifying multimodal co-varying diagnostic patterns for psychiatric disorders.

2 Methods and Materials

The main ideas of the proposed group-specific multimodal fusion method include: leveraging diagnostic information by incorporating a logistic regression term $z_d^T A_{i_d}^{(m)} - \ln(1 + e^{A_{i_d}^{(m)}})$ into the objective function of mCCA; performing fusion separately on each diagnostic group to remove information mutual interference between diagnostic groups.

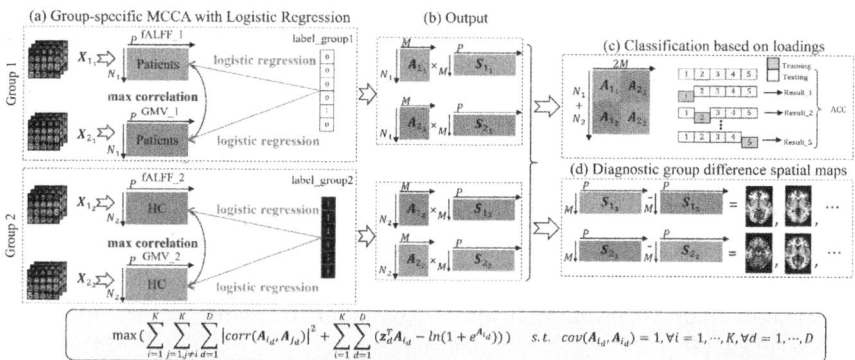

Fig. 1. Flowchart of the proposed GMCCALR. (a) Input feature matrices and GMCCALR. (b) Output of GMCCALR, including loadings (mixing matrices) and spatial maps. (c) Classification based on loadings. (d) Diagnostic group difference spatial maps.

2.1 GMCCALR

We assume that $X_{i_d} = \left[x_{i_d}^{(1)}, x_{i_d}^{(2)}, \cdots, x_{i_d}^{(P)}\right]$ is the feature matrix of d-th diagnostic group of modality i (subjects-by-voxels, $N_d \times P$), where i denotes the modality index. First, principal component analysis (PCA) was employed on X_{i_d} to reduce the dimensionality into a feature subspace matrix $Y_{i_d} = \left[y_{i_d}^{(1)}, y_{i_d}^{(2)}, \cdots, y_{i_d}^{(M)}\right]$ (subjects-by-components, N_d by M). Additionally, assume a set of weights $W_{i_d} = \left[w_{i_d}^{(1)}, w_{i_d}^{(2)}, \cdots, w_{i_d}^{(M)}\right]$, which are orthogonal in the Euclidean space, and each of them is an M by one vector. Linear combinations (loadings) A_{i_d} is obtained using the following formula:

$$A_{i_d} = Y_{i_d} \cdot W_{i_d} \qquad (1)$$

To effectively leverage diagnostic information while removing information mutual interference between diagnostic groups, as shown in Fig. 1(a), we incorporate a logistic regression term into mCCA objective function and utilize group-specific fusion strategy, resulting in the objective function of GMCCALR:

$$m = 1$$

$$\left\{w_{1_d}^{(1)}, w_{2_d}^{(1)}, \cdots, w_{K_d}^{(1)}\right\} = \underset{w_{i_d}^{(1)}}{\arg\max} \left\{ \begin{array}{l} \sum_{i,j=1}^{K} \sum_{d=1}^{D} \left|corr\left(A_{i_d}^{(1)}, A_{j_d}^{(1)}\right)\right|^2 \\ +\sum_{i=1}^{K} \sum_{d=1}^{D} \left(z_d^T A_{i_d}^{(1)} - \ln(1 + e^{A_{i_d}^{(1)}})\right) \end{array} \right\} \quad (2)$$

$$\forall m = 2 : M$$

$$\left\{w_{1_d}^{(m)}, w_{2_d}^{(m)}, \cdots, w_{K_d}^{(m)}\right\} = \underset{w_{i_d}^{(m)}}{\arg\max} \left\{ \begin{array}{l} \sum_{i,j=1}^{K} \sum_{d=1}^{D} \left|corr\left(A_{i_d}^{(m)}, A_{j_d}^{(m)}\right)\right|^2 \\ +\sum_{i=1}^{K} \sum_{d=1}^{D} \left(z_d^T A_{i_d}^{(m)} - \ln(1 + e^{A_{i_d}^{(m)}})\right) \end{array} \right\} \quad (3)$$

$$s.t. \begin{cases} w_{i_d}^{(m)} \perp \left\{w_{i_d}^{(1)}, w_{i_d}^{(2)}, \cdots, w_{i_d}^{(m-1)}\right\} \\ cov\left(A_{i_d}^{(m)}, A_{i_d}^{(m)}\right) = 1, \forall i = 1, 2, \cdots, K, \forall d = 1, 2, \cdots D \end{cases}$$

Here, i, j denotes modality index, d denotes diagnostic group index, z_d is the label vector of the d-th diagnostic group, K represents the number of modalities, D represents the number of diagnostic groups, and $w_{i_d}^{(m)}$ represents the m-th column vector of W_{i_d}. The number of components M was estimated based on the minimum description length (MDL) criterion [16].

2.2 The Optimization and Convergence

Firstly, we construct the Lagrangian function in Stage 1 as follows:

$$L_1 = \sum_{i,j=1}^{K} \sum_{d=1}^{D} \left|corr\left(A_{i_d}^{(1)}, A_{j_d}^{(1)}\right)\right|^2 + \sum_{i=1}^{K} \sum_{d=1}^{D} \left(z_d^T A_{i_d}^{(1)} - \ln\left(1 + e^{A_{i_d}^{(1)}}\right)\right) + (1 + \alpha_{i_d})\left(cov\left(A_{i_d}^{(1)}, A_{i_d}^{(1)}\right) - 1\right)$$

(4)

Utilizing the gradient of the penalty term, we obtain the partial derivative of $w_{i_d}^{(1)}$ as follows:

$$\frac{\partial L_1}{\partial w_{i_d}^{(1)}} = -2 \sum_{j=1}^{K} cov(Y_{i_d}, Y_{j_d}) w_{j_d}^{(1)} \left|corr\left(A_{i_d}^{(1)}, A_{j_d}^{(1)}\right)\right|^2 + \frac{1}{n_d} Y_{i_d}^T (P(z_d = 1|Y_{i_d}) - z_d)$$
$$+ 2(1 + \alpha_{i_d}) cov(Y_{i_d}, Y_{i_d}) w_{i_d}^{(1)}$$

(5)

where $P(z_d = 1|Y_{i_d}) = \frac{1}{1 + e^{-A_{i_d}^{(1)}}}$ is the class posterior probability [17].

For convenience, assume $g_1 = \frac{\partial L_1}{\partial w_{i_d}^{(1)}}$. Since g_1 is nontrivial, it is challenging to find a closed-form solution for $w_{i_d}^{(1)}$. Therefore, we take the partial derivative of g_1 with respect to $w_{i_d}^{(1)}$ to obtain the Hessian matrix H_1.

$$H_1 = -4 \sum_{j=1}^{K} \left(cov(Y_{i_d}, Y_{j_d}) w_{j_d}^{(1)}\right)^2 \left|corr\left(A_{i_d}^{(1)}, A_{j_d}^{(1)}\right)\right|^2 + \frac{1}{n_d} Y_{i_d}^T$$
$$(P(z_d = 1|Y_{i_d}).*(1 - P(z_d = 1|Y_{i_d})))Y_{i_d} + 2(1 + \alpha_{i_d}) cov(Y_{i_d}, Y_{i_d})$$

(6)

We then employ the Newton's method to obtain an approximate closed-form solution for $w_{i_d}^{(1)}$. The solution process of Newton's method can be formalized as follows:

$$w_{i_d}^{(1)} = w_{i_d}^{(1)} - H_1^{-1} g_1 \tag{7}$$

Secondly, we construct the Lagrangian function from Stage 2 to Stage M as follows:

$$L_m = \sum_{i,j=1}^{K} \sum_{d=1}^{D} \left| corr\left(A_{id}^{(m)}, A_{jd}^{(m)}\right) \right|^2 + \sum_{i=1}^{K} \sum_{d=1}^{D} \left(z_d^T A_{id}^{(m)} - \ln\left(1 + e^{A_{id}^{(m)}}\right) \right)$$
$$+ (1 + \alpha_{id}) \left(cov\left(A_{id}^{(m)}, A_{id}^{(m)}\right) - 1 \right) + (1 + \beta_{id}) \sum_{l=1}^{m-1} w_{id}^{(m)T} w_{id}^{(l)} \tag{8}$$

Similarly, we obtain the partial derivative of L_m with respect to $w_{i_d}^{(m)}$ as g_m and calculate the partial derivative of g_m with respect to $w_{i_d}^{(m)}$ to obtain the Hessian matrix H_m. We then employ the Newton's method to solve $w_{i_d}^{(m)}$. The process can be expressed as follows:

$$g_m = -2 \sum_{j=1}^{K} cov(Y_{id}, Y_{jd}) w_{jd}^{(m)} \left| corr\left(A_{id}^{(m)}, A_{jd}^{(m)}\right) \right|^2 + \frac{1}{n_d} Y_{id}^T (P(z_d = 1|Y_{id}) - z_d)$$
$$+ 2(1 + \alpha_{id}) cov(Y_{id}, Y_{id}) w_{id}^{(m)} + (1 + \beta_{id}) \sum_{l=1}^{m-1} w_{id}^{(l)} \tag{9}$$

$$H_m = -4 \sum_{j=1}^{K} \left(cov(Y_{id}, Y_{jd}) w_{jd}^{(m)} \right)^2 \left| corr\left(A_{id}^{(m)}, A_{jd}^{(m)}\right) \right|^2$$
$$+ \frac{1}{n_d} Y_{id}^T (P(z_d = 1|Y_{id}). *(1 - P(z_d = 1|Y_{id}))) Y_{id} + 2(1 + \alpha_{id}) cov(Y_{id}, Y_{id}) \tag{10}$$

$$w_{i_d}^{(m)} = w_{i_d}^{(m)} - H_m^{-1} g_m \tag{11}$$

Note that the parameters α_{i_d} and β_{i_d} will be fine-tuned to their optimal values during the five-fold cross-validation process.

Ultimately, we obtain the weight matrix W_{i_d}, as shown in Fig. 1(b), which allows us to derive the loadings A_{i_d} based on Eq. (1) and the derivation of corresponding spatial map S_{i_d} can be formulated as follows:

$$S_{i_d} = A_{i_d}^{-1} \cdot X_{i_d} \tag{12}$$

For classification and visualization of differences between diagnostic groups, we obtain the multi-diagnostic group loadings A_i and diagnostic group difference spatial maps $S_{i_{d_1,d_2}}$ of the corresponding components between d_1-th diagnostic group and d_2-th diagnostic group as shown in Fig. 1(d), based on Eq. (13–14).

$$A_i = [A_{i_1}; A_{i_2}; \cdots; A_{i_D}] \tag{13}$$

$$S_{i_{d_1,d_2}} = S_{i_{d_1}} - S_{i_{d_2}} \tag{14}$$

2.3 Human Neuroimaging Data

42 schizophrenia (SZ) (39.6 ± 13.1, 35M) and 47 age-gender matched healthy controls (HCs, 37.0 ± 11.8, 32M) from Center for Biomedical Research Excellence (COBRE); 147 SZs (37.4 ± 11.1, 102 M) and 147 age-gender matched HCs (39.5 ± 11.8, 112 M) from Function Biomedical Informatics Research Network (FBIRN); 178 SZs (34.5 ± 13.2, 124M), 134 schizoaffective disorder (SAD) (36.3 ± 12.3, 58M), 143 bipolar disorder (BP) (36.2 ± 13.2, 47M) and 219 age-gender matched HCs (38.8 ± 12.6, 129M) from the Bipolar and Schizophrenia Network of Intermediate Phenotypes (BSNIP) study; 421 autism spectrum disorder (ASD) (13.5 ± 5.5, 421M) and 389 matched HCs (13.7 ± 6.2, 389M) from the Autism Brain Imaging Data Exchange (ABIDE) study were used as the discovery and the validation cohorts in our study. Fractional low-frequency fluctuation amplitude (fALFF) from fMRI and gray matter volume (GMV) from sMRI were calculated as features for the fusion input.

2.4 Classification Between SZ and HC

As is shown in Fig. 1(c), with loadings as features, we perform five-fold cross-validation within COBRE, FBIRN, and BSNIP, individually, to compare GMCCALR with OSCCA, MTSCCALR, DSCCA, jICA, mCCA + jICA and mCCA in terms of classification accuracy between SZ and HC. Furthermore, to validate the robustness of our method, we also compare loadings-based classification accuracy between SZ and HC across the COBRE, FBIRN, and BSNIP cohorts.

2.5 Classification on Other Psychiatric Disorders

To evaluate the generalizability of the proposed GMCCALR, we also compared the classification accuracy between ASD/BP/SAD and HC, using the loadings identified from GMCCALR, OSCCA, MTSCCALR, DSCCA, jICA, mCCA + jICA and mCCA.

3 Results

Figure 2(a–c) and Fig. 3(a–f) shows classification accuracy comparison in diagnosing SZ and HC within and across three cohorts. Compared to OSCCA, MTSCCALR, DSCCA, jICA, mCCA + jICA and mCCA, when multimodal loadings were used as features, GMCCALR achieved the highest classification accuracy, whether considering all IC or single IC with the highest contribution to classification. Furthermore, we visualized the diagnostic spatial maps of the components obtained by GMCCALR that contribute the most to classification. Overlap was observed in spatial maps across different testing sets, such as decreased fALFF in calcarine and bilateral lingual gyrus, increased fALFF in dorsolateral prefrontal cortex, and decreased GMV in anterior cingulate gyrus and bilateral hippocampus in SZ. Additionally, partial overlaps were identified in spatial maps across different training sets, such as decreased fALFF in bilateral lingual gyrus and decreased GMV in bilateral hippocampus in SZ. These results demonstrate the interpretability and robustness of the proposed GMCCALR in identifying multimodal co-varying diagnostic patterns for SZ.

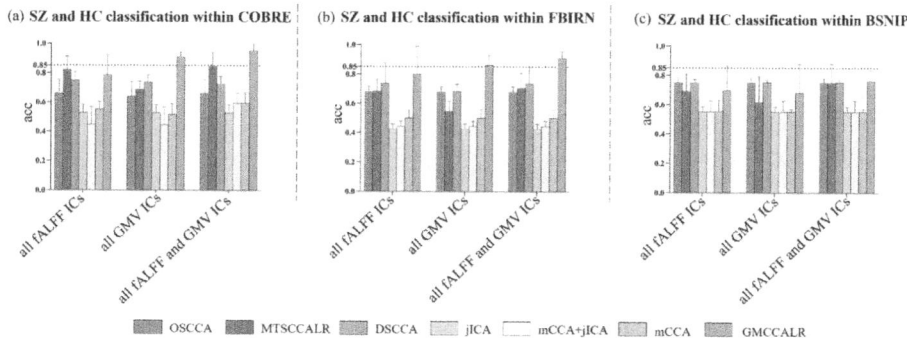

Fig. 2. Classification accuracy comparison for seven fusion methods in differentiating SZ and HC within (a) COBRE, (b) FBIRN, and (c) BSNIP. The black dashed line in (a-c) represents a threshold of 0.85.

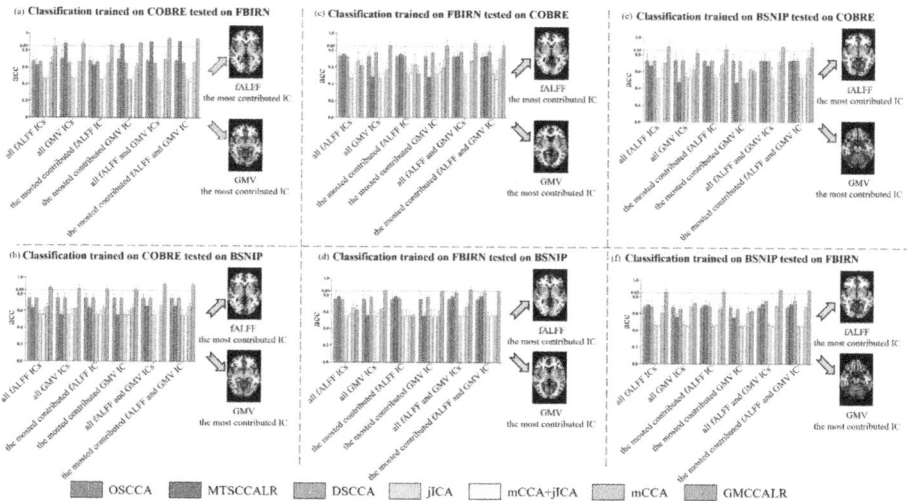

Fig. 3. Classification accuracy comparison for seven fusion methods in differentiating SZ and HC across COBRE, FBIRN, and BSNIP cohorts and the identified diagnostic spatial maps of the most contributed IC obtained by GMCCALR. (a–b) Comparisons of classification accuracy trained on COBRE and tested on FBIRN and BSNIP. (c–d) Comparisons of classification accuracy trained on FBIRN and tested on COBRE and BSNIP. (e–f) Comparisons of classification accuracy trained on BSNIP and tested on COBRE and FBIRN. The black dashed line in (a-f) represents a threshold of 0.85 and the Z-scored spatial maps are visualized at $|Z|>2$.

One of the strengths of the proposed GMCCALR lies in its ability to maintain high accuracy for the diagnosis across several psychiatric disorders. Results shown in Fig. 4(a–c) indicate that GMCCALR outperforms OSCCA, MTSCCALR, DSCCA, jICA, mCCA + jICA and mCCA in terms of classifying ASD/BP/SAD and HC. Similarly, multimodal features yield the best classification accuracy compared to single modality.

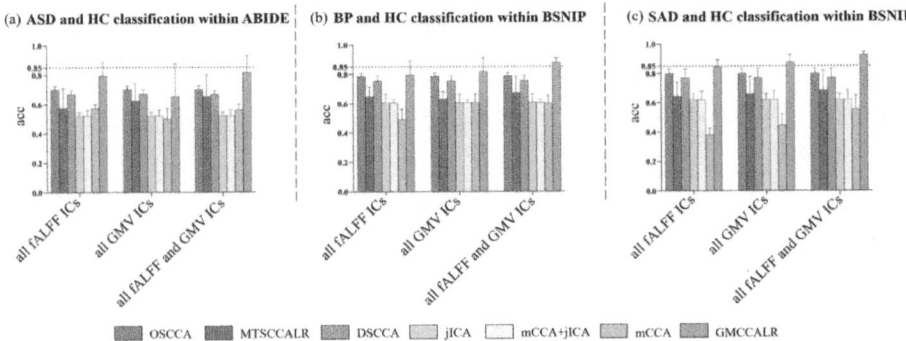

Fig. 4. Comparison of seven fusion methods in differentiating (a) ASD/(b) BP/(c) SAD and HC.

4 Discussion and Conclusion

In this study, we proposed a novel group-specific multimodal fusion method, called GMCCALR, to effectively identify multimodal co-varying diagnostic patterns for psychiatric disorders. The advantages of the proposed method include: 1) a logistic regression term was incorporated into mCCA to fully utilize the diagnostic information; 2) a group-specific fusion strategy was utilized to avoid mutual interference between diagnostic groups which may diminish diagnostic accuracy; 3) generalizability across different cohorts; 4) improved performance in classifying across several psychiatric disorders and HC. Overall, results demonstrate the superior capability of GMCCALR in identifying multimodal co-varying diagnostic patterns for psychiatric disorders.

Results of classification within COBRE, FBIRN and BSNIP indicated the superiority of GMCCALR in diagnosis of SZ and the highest diagnostic accuracy of GMCCALR can be maintained across different cohorts. Compared with OSCCA, MTSCCALR, DSCCA, jICA, mCCA + jICA and mCCA, GMCCALR consistently achieved the highest diagnostic accuracy for SZ and did not exhibit significant fluctuations for different training or testing sets across three cohorts. Among spatial maps of the most contributed components in classification obtained by GMCCALR, replicable abnormal brain regions in SZ were identified across different training sets and testing sets. Specifically, the functional abnormalities in the bilateral lingual gyrus [18] and the structural abnormalities in the bilateral hippocampus [19] in SZ, have been confirmed in previous studies. These results demonstrate the robustness and stability of GMCCALR to identify multimodal co-varying diagnostic patterns for SZ. Notably, comparing with single modality, the superior classification performance of multimodal features implies that even after fusion, the joint analysis of multiple modalities still shows stronger robustness in diagnosing psychiatric disorders.

In the application of GMCCALR and its comparative methods to other psychiatric disorders, such as ASD, BP, and SAD, the multimodal loadings obtained by GMCCALR still achieved the highest diagnostic accuracy. This provides evidence of the feasibility of GMCCALR in identifying multimodal neuroimaging co-varying diagnostic patterns for different psychiatric disorders. All the above results demonstrate the reliability and generalizability of the proposed GMCCALR to identify multimodal co-varying diagnostic patterns for psychiatric disorders.

In summary, this study proposed a novel group-specific multimodal fusion method, called GMCCALR, and validated its effectiveness in distinguishing between several psychiatric disorders and HC. To the best of our knowledge, this is the first proposed group-specific multimodal fusion method that utilizes a group-specific fusion strategy and incorporates a logistic regression term into mCCA to fully leverage the diagnostic information, aiming at identifying multimodal neuroimaging co-varying diagnostic patterns. Based on GMCCALR, the highest diagnostic accuracy was achieved for several different psychiatric disorders across multiple cohorts. The identified replicable disorder specific multimodal patterns highlight the reliability and generalizability of the proposed GMCCALR in identifying multimodal co-varying diagnostic patterns for psychiatric disorders.

Acknowledgments. This work was supported by the National Natural Science Foundation of China (62376124), the Natural Science Foundation of Jiangsu Province, China (BK20220889) the Key Research and Development Plan of Jiangsu Province, China (BE2023668), and the National Institutes of Health (R01MH118695). The authors declare no conflict of interests.

Compliance with Ethical Standards. Written informed consent was obtained for COBRE, FBIRN, BSNIP and ABIDE cohorts under protocols approved by the Institutional Review Boards.

References

1. Qi, S., et al.: Derivation and utility of schizophrenia polygenic risk associated multimodal MRI frontotemporal network. Nat. Commun. **13**(1), 4929 (2022)
2. Liang, C., et al.: Psychotic symptom, mood, and cognition-associated multimodal MRI reveal shared links to the salience network within the psychosis spectrum disorders. Schizophrenia Bull. **49**(1), 172–184 (2023)
3. Qi, S., et al.: Reward processing in novelty seekers: a transdiagnostic psychiatric imaging biomarker. Biol. Psychiatry **90**(8), 529–539 (2021)
4. Qi, S., et al.: Three-way parallel group independent component analysis: fusion of spatial and spatiotemporal magnetic resonance imaging data. Hum. Brain Mapp. **43**(4), 1280–1294 (2022)
5. Qi, S., et al.: Parallel group ICA+ICA: joint estimation of linked functional network variability and structural covariation with application to schizophrenia. Hum. Brain Mapp. **40**(13), 3795–3809 (2019)
6. Qi, S., et al.: Multimodal fusion with reference: searching for joint neuromarkers of working memory deficits in schizophrenia. IEEE Trans. Med. Imaging **37**(1), 93–105 (2017)
7. Kettenring, J.R.: Canonical analysis of several sets of variables. Biometrika **58**(3), 433–451 (1971)
8. Calhoun, V.D., Adalı, T., Kiehl, K.A., Astur, R., Pekar, J.J., Pearlson, G.D.: A method for multitask fMRI data fusion applied to schizophrenia. Hum. Brain Mapp. **27**(7), 598–610 (2006)
9. Sui, J., et al.: Three-way (N-way) fusion of brain imaging data based on mCCA+ jICA and its application to discriminating schizophrenia. Neuroimage **66**, 119–132 (2013)

10. Bao, F., et al.: Integrative spatial analysis of cell morphologies and transcriptional states with MUSE. Nat. Biotechnol. **40**(8), 1200–1209 (2022)
11. Ma, T., Zhang, A.: Integrate multi-omics data with biological interaction networks using Multi-view Factorization AutoEncoder (MAE). BMC Genomics **20**(Suppl 11), 944 (2019)
12. Yan, J., Risacher, S.L., Nho, K., Saykin, A.J., Shen, L.: Identification of discriminative imaging proteomics associations in Alzheimer's disease via a novel sparse correlation model. In: PSB, pp. 94–104. World Scientific (2017)
13. Yan, J., et al.: Joint exploration and mining of memory-relevant brain anatomic and connectomic patterns via a three-way association model. In: 2018 IEEE 15th International Symposium on Biomedical Imaging (ISBI 2018), pp. 6–9: IEEE (2018)
14. Du, L., et al.: Identifying diagnosis-specific genotype–phenotype associations via joint multi-task sparse canonical correlation analysis and classification. Bioinformatics **36**(1), i371–i379 (2020)
15. Mihalik, A., et al.: Canonical correlation analysis and partial least squares for identifying brain–behavior associations: a tutorial and a comparative study. Biol. Psychiatry: Cogn. Neurosci. Neuroimaging **7**(11), 1055–1067 (2022)
16. Li, Y.O., Adalı, T., Calhoun, V.D.: Estimating the number of independent components for functional magnetic resonance imaging data. Hum. Brain Mapp. **28**(11), 1251–1266 (2007)
17. Zaidi, N.A., Webb, G. I.: A fast trust-region newton method for softmax logistic regression. In: Proceedings of the 2017 SIAM International Conference on Data Mining, pp. 705–713. SIAM (2017)
18. Hoptman, M.J., et al.: Amplitude of low-frequency oscillations in schizophrenia: a resting state fMRI study. Schizophrenia Res. **117**(1), 13–20 (2010)
19. Brosch, K., et al.: Reduced hippocampal gray matter volume is a common feature of patients with major depression, bipolar disorder, and schizophrenia spectrum disorders. Mol. Psychiatry **27**(10), 4234–4243 (2022)

Multi-category Brain Tumor Segmentation via Multi-scale and Cross-category Relation Modeling

Dongzhe Li, Baoyao Yang(✉), Yuebin Xie, Weide Zhan, and Jingsong Lin

Guangdong University of Technology, Guangzhou, China
ybaoyao@gdut.edu.cn

Abstract. Accurate multi-category brain tumor image segmentation is crucial for early diagnosis. In recent years, deep learning-based methods have boosted the performance of brain tumor segmentation. However, existing standard models struggle to capture sufficient global context information, which may lead to the neglect of important background knowledge of brain tumors or local features during the segmentation process. Moreover, due to the presence of various sub-regions with significant morphological heterogeneity, size, and positional variations across multiple tumor categories, traditional feature extraction methods fail to adequately capture the complex spatial semantic relations in multi-category tumors. This results in insufficient accuracy and robustness of segmentation results. In this paper, we propose a Multi-scale and Cross-category Relation Modeling method to capture enough global context and handle morphological heterogeneity relations across tumor categories. The network consists of two primary modules: the Adaptive Context Attention Multi-scale Feature Learning Module (ACML) and the Multi-category Feature Enhancement Module (MFEM). ACML dynamically focuses on the relation between local features and global dependencies by incorporating attention mechanisms, effectively extracting multi-scale information. MFEM models the structural relations among multi-categories of brain tumors at various levels, highlighting the most crucial features within each category and integrating them into the overall network structure. Comprehensive experiments conducted on the BraTS2019 and BraTS2020 benchmark datasets demonstrate that our method achieves significant improvements over existing state-of-the-art methods.

Keywords: Brain Tumor Segmentation · Attention Mechanism · Cross-Category Relation

1 Introduction

The delineation of brain tumor regions is typically manually performed by clinical doctors based on their experience. However, this process is not only time-consuming but also prone to subjectivity. Thus, the development of automatic and accurate brain tumor segmentation methods is crucial. This can significantly enhance the efficiency of clinical workflows while reducing the workload of radiologists and other medical professionals. Recently, models such as U-Net [1], Attention U-Net [2], SegResNet [3], Cascaded U-Net [4], and nnUNet [5] have become dominant in the field of brain tumor segmentation. Above mentioned methods typically use element-wise addition or concatenation to directly integrate features from different levels of the encoder. Such operations fail to adequately consider differential information between different levels, potentially introducing redundant information. This weakens the network's ability to precisely locate and refine subtle boundaries. Moreover, there are typically four categories of sub-regions in brain tumor images: Background, Enhancing Tumor (ET), Edema (ED), and the Necrotic Tumor Core (NCR). According to clinical experience from some experts [28], these sub-regions are closely interconnected spatially. For example, ET is typically surrounded by ED and ED cannot be surrounded by NCR only. But due to the limitation of receptive fields, it is challenging to distinguish multiple categories of sub-regions, often resulting in the mixing of different tissue areas and consequently leading to a loss of detail and segmentation ambiguity.

To delve into information at different levels, many methods try to aggregate detailed information from different scales. Zhao [6] et al. proposed a pyramid network, which aggregates feature maps obtained from multiple extended convolutional blocks. The integration of contextual information on multiple scales can also be realized by means of pooling operations [7]. While these techniques may assist in capturing objects across various scales, they assume that the contextual dependencies across all image regions are homogeneous and lack adaptive capabilities, thereby neglecting the relationships between local representations of different classes and their structural interconnections. Generally, features extracted from different levels possess different characteristics. Deep features carry richer semantic information, while shallow features contain more low-level visual information. Therefore, exploring contextual dependencies and category relationships can comprehensively learn the distinct characteristics of each layer and enhance the perception of organs or lesion areas.

In this paper, we propose we propose a Multi-scale and Cross-category Relation Modeling Network. Our network incorporates Adaptive Context Attention(ACML) and Multi-Category Feature Enhancement Modules(MFEM) to aggregate multi-scale information and enhance features for multiple categories of brain tumors. Specifically, ACML dynamically combines local features with global dependencies, while introducing spatial attention to capture positional information and category relations for localization and classification. And MFEM captures spatial relations between categories through spatial attention, and then extracts specific feature maps to enhance feature representations for each tumor sub-region. This enables the model to better capture subtle features of each region, leading to enhanced segmentation precision across multiple categories. Moreover, we integrate Adaptive Context Attention with Graph Convolutional Network (GCN) to focus precisely on the relations between adjacent pixels of lesions, thus facilitating the extraction of more tumor details.

This paper makes the following key contributions:

1. This paper proposes a Multi-category Brain Tumor Segmentation Network via Multi-scale and Cross-category Relation Modeling, which effectively aggregates multi-scale information across different levels while preserving multi-category feature information at each scale.
2. Two fundamental modules ACML and MFEM are introduced in our network. ACML is designed to acquire richer contextual and local feature representations, which allows the model better to understand the relations between different scales of the image. MFEM focuses on learning the spatial relations of different categories and enhancing their feature representations.
3. A comprehensive experiments was conducted on BraTS2019, BraTS2020 datasets, which demonstrated that the proposed method exhibits competitive performance.

2 Related Work

2.1 Medical Image Segmentation

Most medical image segmentation methods are inspired by Fully Convolutional Networks (FCNs) [16] or UNet [1]. FCN achieves pixel-wise dense prediction by replacing fully connected layers with convolutional layers and restores the resolution of input images through single-step upsampling. UNet adopts a symmetric encoder-decoder structure and skips connections, effectively improving the classification accuracy of small-sized objects. For volumetric medical image segmentation, researchers have proposed various extensions and variants, such as applying 3D operations to UNet to avoid spatial information loss and enhancing feature representation capability through integrated attention mechanisms. In terms of medical image segmentation methods, they can be categorized into two types: medical general-purpose and medical-specific, based on organ or lesion features. For medical general-purpose methods, U-Net and its variants, despite their widespread use in medical image segmentation, struggle to capture crucial global information due to the localized nature of convolutional operations. Accurate segmentation often relies on understanding relations between distant parts of the image, which these models might miss. While some emerging methods introduce the Transformer architecture to achieve better global contextual dependencies. Models like UTNet [18] and TransUNet [17] successfully capture long-range dependencies at different scales by applying self-attention modules and hybrid transformer architectures, demonstrating robust segmentation capabilities. However, the computational cost associated with these techniques becomes a major limitation when dealing with large medical images, which can be computationally expensive to process.

2.2 Multi-scale Feature Aggregation

Scale cues play a crucial role in capturing contextual information of the target. Multi-scale features demonstrate advantages over single-scale features when dealing with natural scale variations. Various methods for encoding multi-scale contextual information have been extensively studied. In addition to encoder-decoder architectures [22], constructing image pyramids is also a commonly used method to capture multi-scale contextual information. For example, DeepLabV3 [23] integrates multi-scale contextual information by using parallel atrous spatial pyramid pooling (ASPP) blocks with different dilation rates. Dense-ASPP [24] adopts a more compact way to stack ASPP modules. CPFNet [25] combines multi-scale and global contextual information to enhance feature representation. FCP-Net combines channel attention and spatial attention blocks, as well as multiple receptive fields, for multi-scale contextual modeling. Additionally, some studies design two auxiliary branches in each module to generate and merge context information from coarse to fine, thereby improving overall performance. Recently, CMD-Net [26] utilizes NAS technology to explore the efficiency of multi-scale connections and constructs a restricted multi-scale dense connection network. While this approach achieves high performance while reducing computational costs, NAS technology itself faces challenges such as large search space and high computational costs. CMM-Net [27] has developed an end-to-end neural network that integrates global contextual features and multi-scale features at each level of U-Net. However, effectively integrating these features remains a challenge.

2.3 Deep Attention Mechanism

In recent years, attention mechanisms have been widely applied in the field of computer vision, especially playing a significant role in tasks such as object detection [19] and image segmentation [20]. By introducing attention mechanisms, models can automatically learn and focus on the most relevant regions or features in the image, thereby improving the performance and generalization ability of the model. This design intends to highlight important local features while suppressing noise and irrelevant information in global features, thus better capturing key information and structures in the image. In image segmentation tasks, this mechanism can help the model better understand the correlation between different regions in the image, thereby improving the modeling of long-range dependencies and enhancing the accuracy and robustness of segmentation. Therefore, attention mechanisms have become an indispensable and important component in models that need to capture global dependencies [21].

3 Method

3.1 Overview

The proposed network is demonstrated in Fig. 1(a). The model is based on the U-Net architecture [1], which initially concatenates the four modalities of MRI images and then obtains feature maps $X \in \mathbb{R}^{C \times H \times W}$ at different scales through successive downsampling operations, where C, W, H represent the channel, width and height dimensions. At the bottleneck layer, the feature maps of each scale

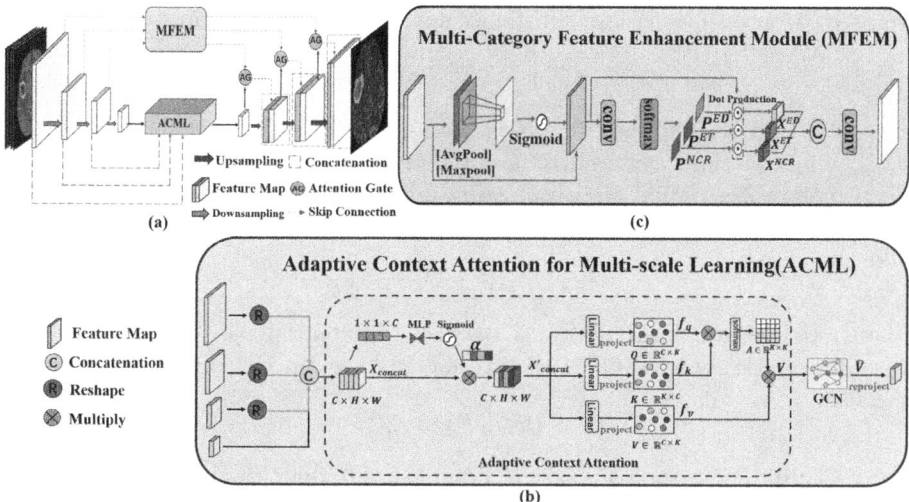

Fig. 1. Overview of Multi-category Brain Tumor Segmentation Network via Multi-scale and Cross-category Relation Modeling.

are fed into the ACML to adaptively capture multi-scale global dependencies and local details of the images. In each layer's skip connection, we incorporate the MFEM to learn spatial relations between brain tumor regions and highlight features of each class. This aids in restoring positional information and spatial details of tumors at various scales during the upsampling process.

3.2 Adaptive Context Attention for Multi-scale Feature Learning

The features at different levels in the encoder have distinct characteristics. Specifically, high-level features contain richer semantic information, aiding in object localization, while low-level features contain more information in detail, supporting the identification of intricate object boundaries. Combining deep-level features with shallow-level features can achieve a more comprehensive and rich feature representation, which helps to improve the accuracy of segmentation. Figure 1(b) illustrates the structure of the Adaptive Context Attention for Multi-scale Learning Module. Firstly, we integrate feature maps from all scales, the feature maps at each layer are adjusted to align with the size of the bottom-layer feature map and then concatenated to obtain $X_{concat} \in \mathbb{R}^{C \times H \times W}$. To adaptively capture the relations between different scales and gain global context, weights are assigned to each layer through generating scale attention $\alpha \in \mathbb{R}^{1 \times 1 \times W}$. The updated feature map can be formulated as: $X'_{concat} \in \mathbb{R}^{C \times H \times W}$. To further explore the dependencies between different scales, we projected [13] features from X'_{concat} into the interaction coordinate space. X'_{concat} is transformed into query, key, value vectors $X'_q, X'_v \in \mathbb{R}^{C \times (H \times W)}$, $X'_k \in \mathbb{R}^{(H \times W) \times C}$. And the representation of interaction coordinate spaces can be formulated as:

$$f'_{q,v,k} = (X'_{q,v,k})_{project} \tag{1}$$

where $f'_{q,v} \in \mathbb{R}^{C \times K}$, $f'_k \in \mathbb{R}^{K \times C}$, K is the number of feature vectors.

To infer relations between different nodes, we construct an adaptive adjacency matrix $A \in \mathbb{R}^{K \times K}$ to obtain a graph $V \in \mathbb{R}^{C \times K}$, which can selectively focus on information from different positions. $\otimes$ acts as a matrix product in the following equation:

$$V = (f'_q \otimes f'_k) \otimes f'_v = A \otimes f'_v \tag{2}$$

Finally, to capture local relations between different pixels, we utilize Graph Convolutional Networks (GCN) [14] to model long-range dependencies in image regions.

$$\hat{V} = g(A^T V W) \tag{3}$$

where the updated graph $\hat{V} \in \mathbb{R}^{C \times K}$, $g(\cdot)$ is a non-linear activation function, W is the weight matrix$\in \mathbb{R}^{K \times K}$. By constructing graphs and adaptively building long-range dependency relations between regions through node interactions, we simultaneously overcome the inefficiency of convolutional operations and reduce computational complexity when reasoning about relations.

3.3 Multi-category Feature Enhancement Module

In brain tumor images, multiple categories of information are contained, and the relations between these categories are crucial for segmentation, localization, and classification. Therefore, we propose the Multi-category Feature Enhancement Module to capture the spatial relations between brain tumor sub-regions, as shown in Fig. 1(c). To understand the relations between different regions and better perceive the spatial distribution of targets in the image, for the feature maps X at each layer, we first concentrate attention on specific regions of multi-category brain tumor images through spatial attention, enabling the model to focus on more tumor-related features. Then, to accurately distinguish sub-regions of multi-categories, we generate probability distributions for different categories, resulting in specific probability map P^{ED}, P^{ET}, P^{NCR} for each category:

$$P^{NCR,ED,ET} = \text{split}(\text{softmax}(\text{conv}(\tilde{X}))) \tag{4}$$

With probability maps, we generate feature maps X^{NCR}, X^{ED}, X^{ET} for each category to highlight the features of sub-regions. Finally, these feature maps are concatenated and convolved to obtain the final feature map $\tilde{X}_r$.

$$\tilde{X}_r = conv(concat[\tilde{X} \odot P^{NCR}, \tilde{X} \odot P^{ED}, \tilde{X} \odot P^{ET}]) \tag{5}$$

It is worth noting that we connect the upsampled feature maps and $\tilde{X}_r$ through attention gates [15] to eliminate ambiguities in irrelevant noise features, highlighting significant feature information. This mechanism addresses the U-Net model's accuracy issues in tumor segmentation while aiding in restoring tumor positional information during the upsampling process.

Table 1. Comparison on the BraTS2019 Dataset

Model	Dice (%) ↑				HD95 (mm) ↓			
	ET	TC	WT	Ave	ET	TC	WT	Ave
TransBTS [8]	80.86	81.19	89.35	83.80	5.642	6.048	4.332	5.460
Nestedformer [9]	82.11	86.42	**91.18**	86.57	5.534	5.906	5.317	5.585
SF-Net [10]	80.08	82.33	88.61	83.67	4.787	7.440	7.288	6.505
ACM-Net [11]	80.63	87.15	88.08	85.28	4.564	7.774	**3.862**	5.400
Eoformer [12]	**82.94**	86.83	90.39	86.72	**4.053**	5.843	5.822	5.239
Ours	82.74	**87.67**	90.59	**87.00**	5.296	**5.694**	4.472	**5.154**

4 Experiment

4.1 Dataset

For evaluation, public brain tumor segmentation datasets BraTS 2019 and BraTS 2020 [8] were selected as benchmark datasets. These datasets consist of 335 and 369 annotated brain tumor samples, respectively. Each case contains four preprocessed modalities (T1, T1Gd, T2, T2-FLAIR) that have completed standardized preprocessing procedures. The segmentation labels have the following numerical values: label 1 represents non-enhancing tumor core (NCR&NET), label 2 represents peritumoral edema (ED), label 4 represents enhancing tumor (ET), and label 0 represents the background region. The three sub-regions defined for online evaluation are as follows: 1) Enhancing tumor (ET) region, including the ET label; 2) Tumor core (TC) region, including ET and NCR & NET labels; 3) Whole tumor (WT) region, including ET, NCR & NET, and ED labels.

4.2 Implementation Details and Evaluation Metrics

The MRI volumes of each subject are sized $155 \times 240 \times 240$. Each modality scan is sliced, and the size of each cropped slice is 160×160. The datasets are split into 80% for training our model and 20% for testing. The model was trained for a total of 400 epochs and the batch size is set to 18. All the programs were implemented under the PyTorch framework. The training process is conducted on V100-SXM2-32 GB GPUs. For optimization, we utilize the Adam optimizer with a learning rate set to 1e−3.

The performance of the model is evaluated using two widely accepted metrics: Dice score and 95% Hausdorff Distance (HD). The Hausdorff95 metric quantifies the distance between the model's predictions and the ground-truth segmentation. For each metric, three distinct regions are assessed separately: the enhancing tumor (ET, label 1), whole tumor (WT, labels 1, 2, and 4), and the tumor core (TC, labels 1 and 4). The Loss function L_{loss} is formulated as a combination of soft Dice loss and cross-entropy loss, as referenced in [7].

Table 2. Comparison on the BraTS2020 Dataset

Model	Dice (%) ↑				HD95 (mm)↓			
	ET	TC	WT	Ave	ET	TC	WT	Ave
TransBTS [8]	80.89	83.25	90.10	84.08	5.873	6.875	4.876	5.824
Nestedformer [9]	82.85	86.48	**91.20**	86.84	5.721	6.115	4.598	5.528
SF-Net [10]	81.10	83.84	89.01	84.65	**4.305**	7.661	7.720	6.562
ACM-Net [11]	82.42	87.75	90.08	86.75	4.492	7.624	3.956	5.375
Eoformer [12]	**83.54**	87.12	90.87	87.17	5.911	6.041	**3.852**	5.268
Ours	83.20	**87.89**	90.93	**87.34**	5.134	**5.863**	4.529	**5.175**

4.3 Comparison with SOTA Methods

Our network is evaluated in comparison to five advanced segmentation methods. This comparison includes two techniques focused on multi-modal feature fusion (SF-Net [10] and ACM-Net [11]) and three contemporary Transformer-based methods (Nestedformer [9], TransBTS [8], and EoFormer [12]), the last of which specifically focuses on edge-oriented segmentation. As shown in Table 1, proposed network achieves Dice scores of 82.74%, 87.67%, and 90.59% for the enhancing tumor (ET), tumor core (TC), and whole tumor (WT) regions, respectively, on the BraTS2019 dataset. By integrating multi-category interaction alongside Region-guided Graph Reasoning Modules, our network achieves the top Dice scores for the TC region, highlighting its ability to model the relationships of tumor sub-regions =effectively. Likewise, as demonstrated in Table 2, our network maintains its superior performance on the BraTS2020 dataset, obtaining the top Dice scores in the TC region. While the Dice score for the WT region is slightly lower than that of Nestedformer and the ET region is marginally behind EoFormer, our method significantly outperforms the remaining approaches by a considerable margin.

Notably, our method attains the highest mean score compared to all the methods evaluated. Figures 2 and 3 illustrate the detailed results of our approach on the BraTS2019 and BraTS2020 datasets visually, respectively. It is evident that our method accurately captures the shapes and edges of brain tumors. Additionally, it successfully identifies scattered edema regions (ED) surrounding the tumors. Within the tumor areas, even small necrotic (NCR) regions and overlapping enhanced tumor regions (ET) are precisely distinguished. These results demonstrate the model's exceptional capability in modeling dependencies of multi-category and accurately delineating brain tumor boundaries. This success can be attributed to the effective integration of region-based and contour-based constraints, underscoring the method's robustness and accuracy in tumor segmentation tasks.

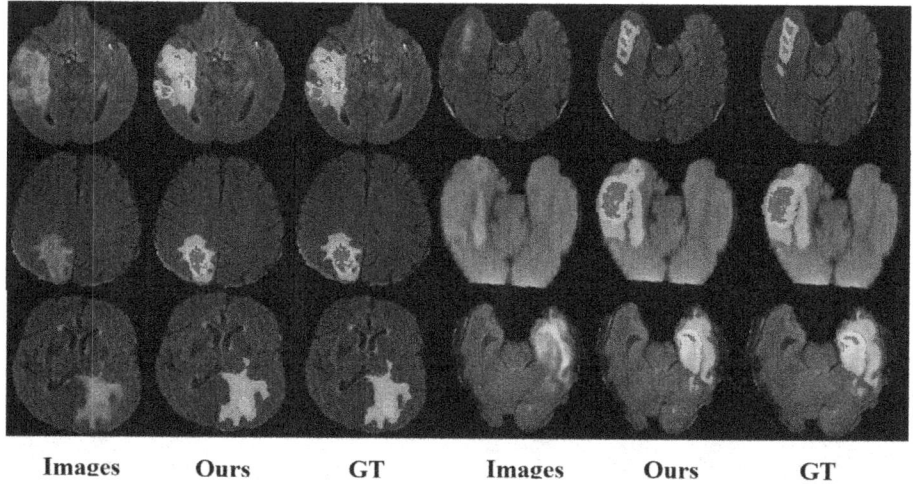

Fig. 2. The visual results on BraTS2019. (Green: ED, Yellow: ET, Red: NCR) (Color figure online)

4.4 Ablation Study

We conduct ablation study on the BraTS2019 dataset. Tge standard Unet [1] is used as the baseline model. Specifically, we compare the performance of four models: (1) the baseline Unet, (2) Unet augmented with MFEM, (3) Unet augmented with ACML, and (4) Unet augmented with both ACML and MFEM. Through the analysis of the comparison results of each component, we can see that each component has contributed to the improvement of the segmentation results. Comparing the results of the Unet+MFEM model with (1), we found that the Unet+MFEM model improved the segmentation accuracy for all categories, especially showing significant enhancement in the segmentation of WT and TC. This validates that the multi-category spatial information learned through MFEM can effectively enhance the discrimination between categories and the accuracy of tumor localization. Comparing the Unet+ACML model with (1), we found that the Unet+ACML model is slightly superior to the Unet model overall. There is a significant improvement in the segmentation of all categories, confirming that the global contextual information learned through multi-scale aggregation can enhance segmentation effectiveness. It is noteworthy that compared to (2), ET has shown a greater improvement in (3) over (1), which strongly indicates that MFEM can simultaneously focus on local features. By combining Unet+MFEM and Unet+ACML, the segmentation performance of brain tumors can be further improved, with an average Dice score reaching the highest at 87.00%.

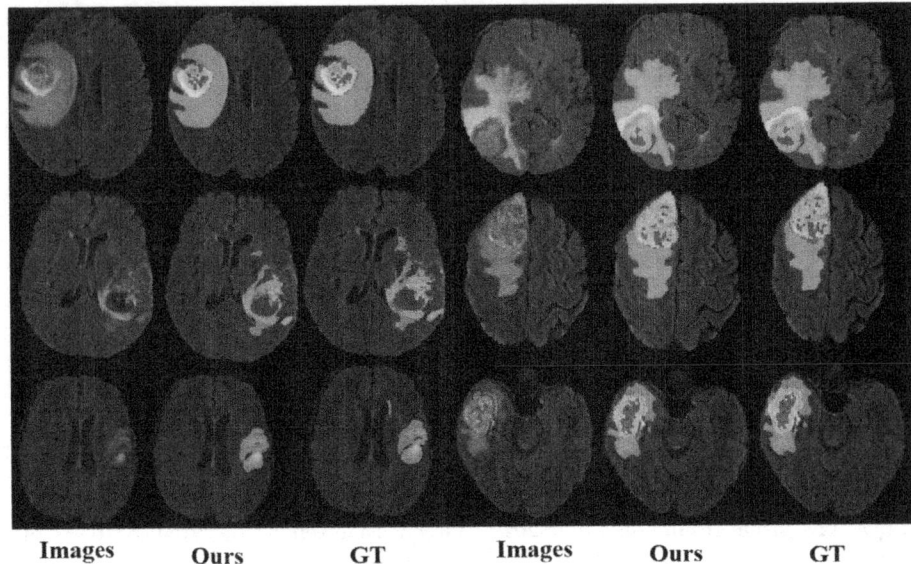

Fig. 3. The visual results on BraTS2020. (Green: ED, Yellow: ET, Red: NCR) (Color figure online)

Table 3. Ablation Study

Model	Dice(%)			
	ET	TC	WT	Ave
(1) Unet	79.10	81.00	87.93	82.67
(2) Unet+MFEM	80.80	85.34	90.17	85.43
(3) Unet+ACML	82.55	86.28	90.06	86.69
(4) Unet+ACML+MFEM	**82.74**	**87.67**	**90.59**	**87.00**

5 Conclusion

In this work, we present a novel approach for adaptively capturing multi-scale as well as multi-category semantic relations in brain glioma images. We introduce a ACML module, which integrates semantic information from different levels through attention mechanisms, gradually aggregating relevant contextual features. Additionally, we devise an MFEM module to assist the network in focusing on class-specific regions relevant to brain tumor images at each scale. We conducted experiments on the public datasets BraTS 2019 and BraTS 2020, and the results indicate that our proposed method outperforms other state-of-the-art methods in terms of overall performance (Table 3).

Acknowledgement. This research was partially funded by the National Natural Science Foundation of China (NSFC) under Grant 62102098, the Natural Science Foundation of Guangdong Province under Grant 2024A1515010186, and the Regional Joint Fund Project of the Basic and Applied Basic Research Foundation of Guangdong Province under Grant 2022A1515140096.

References

1. Ronneberger, O., Fischer, P., Brox, T.: U-net: convolutional networks for biomedical image segmentation. In: International Conference on Medical Image Computing and Computer-Assisted Intervention, pp. 234–241. Springer (2015)
2. Oktay, O., et al.: Attention U-net: learning where to look for the pancreas. arXiv, preprint arXiv:1804.03999 (2018)
3. Myronenko, A., et al.: 3D MRI brain tumor segmentation using autoencoder regularization. In: MICCAI Brainlesion Workshop, pp. 311–320. Springer (2019)
4. Jiang, Z., Ding, C., Liu, M., Tao, D.: Two-stage cascaded U-net: 1st place solution to brats challenge 2019 segmentation task. In: MICCAI Brainlesion Workshop, pp. 231–241. Springer (2020)
5. Isensee, F., Jager, P.F., Full, P.M., Vollmuth, P., Maier-Hein, K.H.: nnU-net for brain tumor segmentation. In: MICCAI Brainlesion Workshop, pp. 118–132. Springer (2021)
6. Zhao, H., Shi, J., Qi, X., Wang, X., Jia, J.: Pyramid scene parsing network. In: Proceedings of the IEEE Conference on Computer Vision and Pattern Recognition, pp. 2881–2890 (2017)
7. Liu, W., Rabinovich, A., Berg, A.C.: Parsenet: Looking wider to see better. arXiv, preprint arXiv:1506.04579 (2015)
8. Wang, W., Chen, C., Ding, M., Yu, H., Zha, S., Li, J.: Transbts: multi-modal brain tumor segmentation using transformer. In: International Conference on Medical Image Computing and Computer-Assisted Intervention, pp. 109–119 (2021)
9. Xing, Z., Yu, L., Wan, L., Han, T., Zhu, L.: NestedFormer: nested modality-aware transformer for brain tumor segmentation. In: International Conference on Medical Image Computing and Computer-Assisted Intervention, pp. 140–150 (2022)
10. Liu, Y., Mu, F., Shi, Y., Chen, X., Li, J.: Sf-net: a multi-task model for brain tumor segmentation in multi-modal MRI via image fusion. Signal Process. Lett. 1799–1803 (2022)
11. Yuzhou, Z., Hong, L., Enmin, S., Chih-Cheng, H.: Sf-net: a 3D cross-modality feature interaction network with volumetric feature alignment for brain tumor and tissue segmentation. J. Biomed. Health Inf. 1799–1803 (2022)
12. She, D., Zhang, Y., Zhang, Z., Li, H., Yan, Z., Sun, X.:EoFormer: edge-oriented transformer for brain tumor segmentation. In: International Conference on Medical Image Computing and Computer-Assisted Intervention, pp. 333–343. Springer (2023)
13. Huang, Z., Li, Y.: Interpretable and accurate fine-grained recognition via region grouping. In: Proceedings of the IEEE/CVF Conference on Computer Vision and Pattern Recognition (CVPR) (2020)
14. Yao, Y., et al.: Non-salient region object mining for weakly supervised semantic segmentation. In: Proceedings of the IEEE/CVF Conference on Computer Vision and Pattern Recognition (CVPR), pp. 2623–2632 (2021)

15. Oktay, O., Schlemper, J., Folgoc, L.L., Lee, M.: Attention U-net: learning where to look for the pancreas. In: MIDL, pp. 2626–2637. IEEE (2018)
16. Milletari, F., Navab, N., Ahmadi, S.A.:V-net: fully convolutional neural networks for volumetric medical image segmentation. In: 4th International Conference on 3D Vision (3DV), pp. 565–571. IEEE (2016)
17. Chen, J., Lu, Y., Yu, Q.:TransUNet: transformers make strong encoders for medical image segmentation. arXiv:2102.04306 (2021)
18. Gao, Y., Zhou, M., Metaxas, D.N.: UTNet: a hybrid transformer architecture for medical image segmentation. In: International Conference on Medical Image Computing and Computer-Assisted Intervention, pp. 61–71. IEEE (2021)
19. Yu, H., Tian, Y., Ye, Q., Liu, Y.: Spatial transform decoupling for oriented object detection. In: Proceedings of the AAAI Conference on Artificial Intelligence, aaai.v38i7.28502 (2023)
20. Chen, B., Liu, Y., Zhang, Z.: TransAttUnet: multi-level attention-guided U-Net with transformer for medical image segmentation. In: IEEE Transactions on Emerging Topics in Computational Intelligence, pp. 55–68. IEEE (2024)
21. Zhu, Q., Li, P., Li, Q.: Attention retractable frequency fusion transformer for image super resolution. In: Proceedings of the IEEE/CVF Conference on Computer Vision and Pattern Recognition (CVPR), pp. 1756–1763(2023)
22. Badrinarayanan, V., Kendall, A., Cipolla, R.: SegNet: a deep convolutional encoder-decoder architecture for image segmentation. IEEE Trans. Pattern Anal. Machine Intell. 2481–2495 (2017)
23. Chen, L.-C., Zhu, Y., Papandreou, G., Schroff, F., Adam, H.: Encoder-decoder with atrous separable convolution for semantic image segmentation. In: Ferrari, V., Hebert, M., Sminchisescu, C., Weiss, Y. (eds.) ECCV 2018. LNCS, vol. 11211, pp. 833–851. Springer, Cham (2018). https://doi.org/10.1007/978-3-030-01234-2_49
24. Yang, M., Yu, K., Zhang, C., Li, Z., Yang, K.: Denseaspp for semantic segmentation in street scenes. In: Proceedings of the IEEE/CVF Conference on Computer Vision and Pattern Recognition (CVPR), pp. 3684–3692 (2018)
25. Feng, S., Zhao, H., Shi, F., Cheng, X., Wang, M.: CPFNet: context pyramid fusion network for medical image segmentation. IEEE Trans Med. Imaging 3008–3018 (2020)
26. Sahin, C.: CMD-Net: self-supervised category-level 3D shape denoising through canonicalization. In: MDPI (2022)
27. Sahin, C.: CMD-Net: 3D CMM-net with deeper encoder for semantic segmentation of brain tumors in BraTS2021 challenge. In: International MICCAI Brainlesion Workshop, pp. 333–343 (2021)
28. Kotowski, K., Adamski, S., Machura, B., Zarudzki, L., Nalepa, J.: Coupling nnU-nets with expert knowledge for accurate brain tumor segmentation from MRI. In: International MICCAI Brainlesion Workshop, Springer (2022)

Consistent Brain Age Difference in Childhood Autism Spectrum Disorder and its Subtypes

Fangling Sun[1,2,3], Chuang Liang[1,2,3], Wei Shao[1,2,3], Zening Fu[4], Daoqiang Zhang[1,2,3], Rongtao Jiang[5(✉)], Shile Qi[1,2,3(✉)], and Vince D. Calhoun[4]

[1] Department of Computer Science and Technology, Nanjing University of Aeronautics and Astronautics, Nanjing, China
shile.qi@nuaa.edu.cn
[2] Key Laboratory of Brain-Machine Intelligence Technology, Ministry of Education, Nanjing University of Aeronautics and Astronautics, Nanjing, China
[3] MIIT Key Laboratory of Pattern Analysis and Machine Intelligence, Nanjing University of Aeronautics and Astronautics, Nanjing, China
[4] Tri-institutional Center for Translational Research in Neuroimaging and Data Science (TReNDS) Georgia State University, Institute of Technology, Emory University, Atlanta, GA, USA
[5] Department of Radiology and Biomedical Imaging, Yale University, New Haven, CT, USA
rongtao.jiang@yale.edu

Abstract. Autism spectrum disorder (ASD) is a neurodevelopmental condition characterized by impairments in social interaction and behavior as well as structural abnormalities regarding brain development. The difference between neuroimaging-predicted brain age and the chronological age (predicted age difference; PAD) is a potential biomarker reflecting individual differences in brain developmental trajectories. However, current findings of PAD in ASD are inconsistent due to the biases introduced by different prediction models. Here, 6 classic 3-dimensional convolutional neural network (3D-CNN) models were used to assess brain age to identify consistent and reliable PAD and its interpretable features for ASD. No significant PAD differences between ASD and typically developing controls (TDC) were observed, but PAD differences in younger age and subtypes were found. Occlusion sensitivity analysis showed that default mode network (DMN) and salience network (SAN) drove the atypical brain developmental trajectories in ASD. Our findings were replicable across 6 CNN models, showing the promise of neuroimaging targets for exploring atypical developmental patterns in ASD.

Keywords: Brain Age Prediction · Autism Spectrum Disorder · Convolutional Neural Network · Interpretability

---

F. Sun and C. Liang—These authors contributed equally to this work.

1 Introduction

Autism spectrum disorder (ASD) is a neurodevelopmental condition characterized by deficits in social communication and interaction, and restricted, repetitive patterns of behavior, interests, or activities [1–3]. Existing neuroimaging studies have reported structural brain abnormalities in ASD, such as increased total brain volume [4], disrupted organization of cortical topology [5], and enlargement of gray/white matter (GM/WM) in the frontal, temporal, and parietal lobes [6, 7]. Moreover, ASD specific brain deviations have been found across different development stages [8] and subtypes [9]. Some of these alterations have been linked to atypical brain development trajectories in ASD [10]. Accumulating neuroimaging studies have been conducted to assess atypical brain development patterns [11], which reflect the neurodevelopmental mechanisms underlying ASD. The underlying hypothesis is that the brain age of healthy individuals should match their chronological age. Brain age can be estimated using information extracted from neuroimaging features, especially the structural magnetic resonance imaging (sMRI) [12]. The gap between the predicted brain age and the chronological age (predicted age difference or PAD) can provide an important early indicator of deviation from the normal developmental trajectory, which may promote understanding of neurodevelopmental disorders, such as ASD.

Deep learning techniques, which are capable of integrating the feature extraction, feature reduction and prediction phases of complex data into a unified computational framework, have been widely applied in brain age prediction community. Convolutional neural networks (CNN) have attracted researchers for brain age estimation and pretrained networks become increasingly popular because of their automatic feature extraction capability and high performance. A 3-dimensional visual geometry group (3D-VGG) model was constructed using the T1-weighted UK Biobank dataset [13], uncovering a mild correlation between the age differences and the image-derived phenotype of multimodalities. 3D residual neural network (ResNet) was employed for multimodal brain age estimation model, which takes T1-weighted, WM, GM and Jacobian map as the input to each CNN network [14]. 3D dense convolutional network (DenseNet) was also modified for brain age prediction in normal aging and dementia [15].

Previous work has reported positive PAD in ASD, but case-control differences were not observed [16]. Neuroimaging studies found that brain maturation patterns were similar between typically developing controls (TDC) and ASD [17]. However, PAD in ASD was significantly lower than TDC in both the Autism Brain Imaging Data Exchange (ABIDE) dataset and the Healthy Brain Network (HBN) dataset [18], suggesting that ASD exhibited the characteristics of delayed development. While another study found that ASD showed accelerated brain development in youth followed by a delay after preadolescence [19]. Thus, the findings from brain age prediction of ASD are often inconsistent, possibly due to the differences in sample size, preprocessing strategies, and experimental setup including feature selection methods and prediction models.

In this paper, we aim to: 1) identify the consistent and reliable PAD for ASD by comparing 6 classic CNN models, including VGG, ResNet, ResNeXt, DenseNet, ShuffleNet, MobileNet, that worked directly on the raw MRI data without feature extraction (Fig. 1a); 2) calculate the PAD differences between ASD and TDC in different age ranges; 3) estimate the PAD differences between ASD subtypes (Fig. 1b) and 4) identify the interpretable regions for ASD brain age prediction (Fig. 1c).

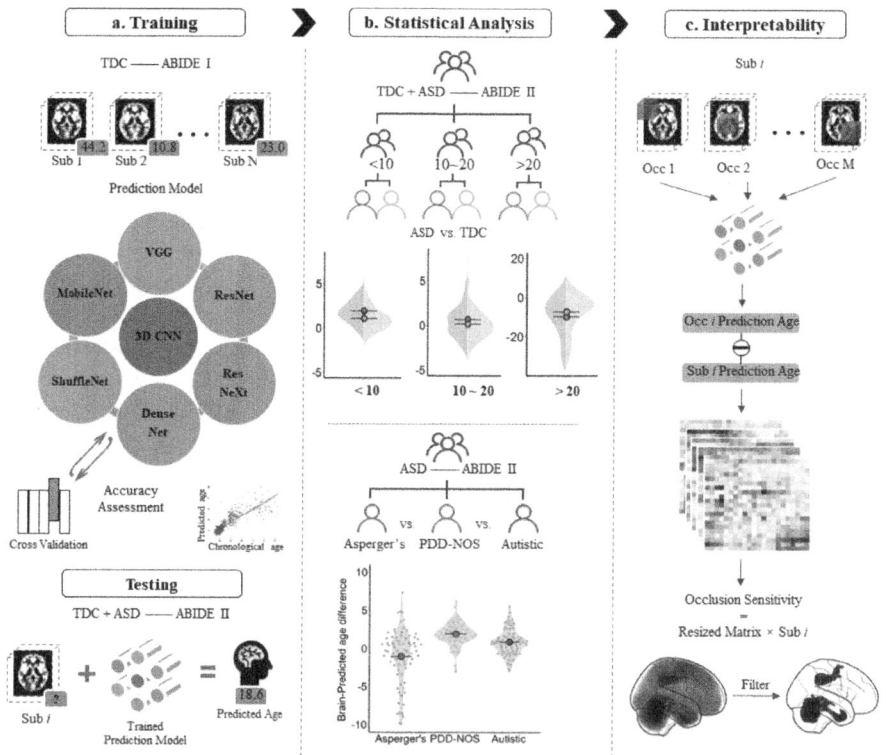

Fig. 1. Flowchart of the study design. a. (Training & Testing) training CNN models separately under the same settings and applying it on the validation and the testing sets. b. (Statistical Analysis) PAD differences between ASD and TDC in different age ranges and ASD subtypes. c. (Interpretability) using occlusion sensitivity analysis to identify the interpretable regions for ASD brain age prediction.

2 Materials and Methods

2.1 Participants

563 TDCs (mean ± standard deviation age: 16.9 ± 7.6; gender: 465M/98F) from ABIDE I, 358 ASDs (age: 14.8 ± 10.2; gender: 303M/55F) and 429 demographically-matched TDCs (age: 12.7 ± 6.6; gender: 284M/145F) from ABIDE II (https://fcon_1000.projects.nitrc.org/indi/abide/) were used as the training and testing dataset, respectively. Ethical approval was granted by each relevant Ethics Committees, and informed consent was obtained from each subject prior to scanning according to each site's Institutional Review Boards.

2.2 Brain Age Prediction Models

Brain age prediction models were based on 6 classic CNN architectures, including VGG-11, ResNet-18, ResNeXt-101, DenseNet-169, ShuffleNet-v2 and MobileNet-v2, in which we replaced the 2D convolutional and pooling layer to 3D and added a batch normalization layer before the pooling layer to build the model directly with the raw imaging data and reduce the reliance on data pre-processing.

The VGG network was proposed with a smaller convolution kernel (3x3) with a deeper architecture that showed better performance [20]. ResNet is regarded as a breakthrough in deep learning which is accomplished by shortcut connections in the feedforward process of residual learning to address the degradation problem in model training [21]. ResNeXt is modified from the ResNet with a strategy of splitting and merging the building block [22]. While preserving similar computational complexity, the ResNeXt divides the residual block into multiple substructures with the same topology and then aggregates them, serving as a more effective way of gaining accuracy than going deeper or wider. To improve information flow between the layers, DenseNet is designed to connect the output of the previous layer not just to the next layer but to all subsequent layers by forming a so-called dense block that has several advantages over ResNet, such as strong feature propagation, feature reuse, and a smaller number of network parameters [23]. ShuffleNet is built by applying the shuffled channels and pointwise group convolutions to reduce the computational cost for implementation in portable devices [24]. ShuffleNet-V2 is an improved version with equalizing the channel width, selecting a reasonable group number, avoiding network fragmentation and reducing element-wise operations. MobileNet is another lightweight CNN that uses the depth-wise separable convolution for portable and mobile applications [25]. The upgraded version of MobileNet-v2 applies two additional modifications, including the linearization of the bottlenecks and the use of inverted residuals to improve performance and allow to separate the network expressiveness from its capacity.

2.3 Occlusion Sensitivity Analysis

Occlusion sensitivity analysis [26] was performed to identify the interpretability of ASD brain age predictive features. Occlusion sensitivity analysis involves 'masking out' a (e.g., cubic) region in an image and transferring the 'occluded' image to a trained model. The model's response is expected to vary when applied to an image with masked regions, especially if the masked portion encompasses crucial features influencing the prediction of brain age. To visualize the significant areas within the image that strongly impact the model's predictions, a heatmap can be generated by repeatedly masking different regions of the image and observing the resulting changes in prediction. This technique provides insights into the specific regions that contribute to the model's output in the context of brain age estimation.

3 Experiments and Results

3.1 Experimental Setup

During training, we set the batch size to 4, the learning rate to 0.0001 with a constant decay of 0.1 after 20 epochs, and the number of epochs to 100. The optimal number of model training iterations was determined by assessing the prediction error and implementing early stopping [27] if the validation error did not improve in 25 epochs. Five-fold cross validation was used to ensure an objective evaluation by avoiding biased results. We optimized the parameters of the prediction model using stochastic gradient descent (SGD) optimization and backpropagation with the mean squared error (MSE) function as the loss function during the training process. After training, the mean absolute error (MAE) and Pearson correlation coefficients between the model-predicted and chronological age were calculated in the testing set to evaluate the prediction performance. PAD (predicted brain age - chronological age) was calculated for each individual.

3.2 The Performance of 6 Brain Age Prediction Models

Table 1. Performance in the training & testing set.

Model	ABIDE I (TDC-Training)	ABIDE II (TDC + ASD—Testing)		TDC vs. ASD
	MAE (years)	MAE (years)	Correlation (r)	p-value
VGG	2.30 ± 0.24	3.12 ± 0.02	0.85 ± 0.01	0.8575
ResNet	2.57 ± 0.27	3.32 ± 0.17	0.85 ± 0.02	0.5932
ResNeXt	2.79 ± 0.50	3.78 ± 0.88	0.71 ± 0.23	0.4405
DenseNet	2.55 ± 0.34	3.13 ± 0.10	0.86 ± 0.01	0.8272
ShuffleNet	2.60 ± 0.21	3.44 ± 0.12	0.82 ± 0.02	0.9403
MobileNet	2.44 ± 0.31	3.29 ± 0.13	0.85 ± 0.01	0.9145

As shown in Table. 1, most prediction models achieved high predictive accuracy, with the MAE being around 2 years in the training set, and around 3 years in the testing set. The group differences of PAD between ASD and TDC were not significant for all 6 prediction models ($p > 0.05$).

3.3 Age Specific PAD Differences Between ASD and TDC

The testing set was divided into childhood (<10 years old), adolescence (10–20 years old) and adulthood (> 20 years old) groups. The PAD differences between ASD and TDC in different age groups under 6 prediction models was calculated. Results showed that the PAD in ASD was significantly higher than TDC only in childhood, and marginally higher in adolescence, but not significant in adulthood ($p > 0.05$, Fig. 2).

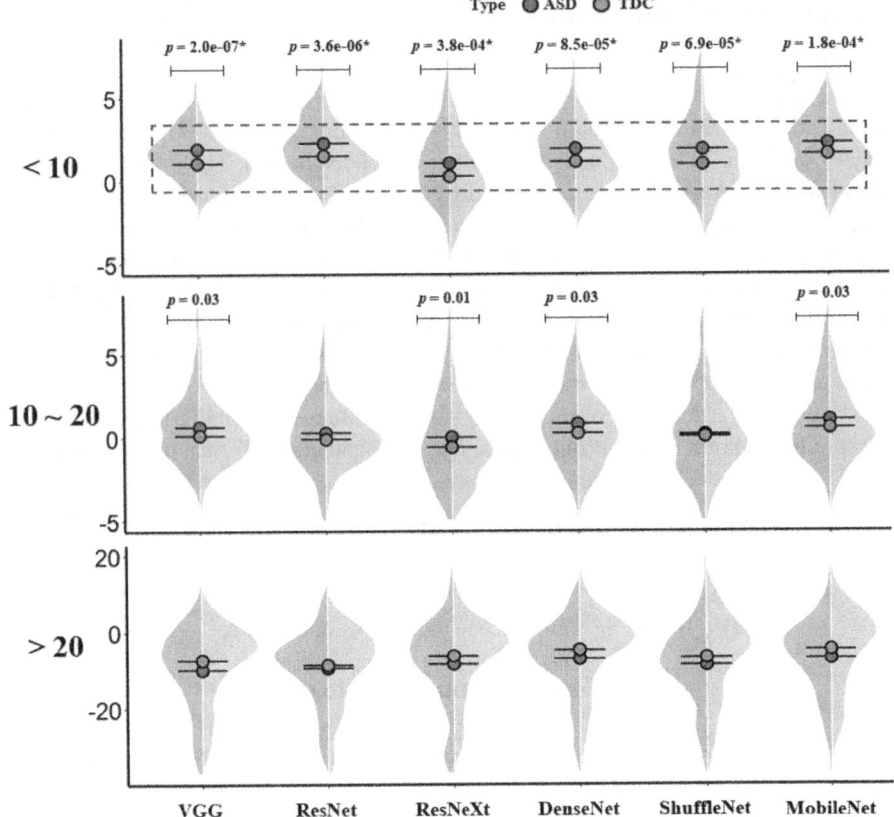

Fig. 2. The PAD differences between ASD and TDC in childhood (<10 years), adolescence (10–20 years) and adulthood (> 20 years) under 6 prediction models.

3.4 PAD Differences Among ASD Subtypes

The PAD differences among ASD subtypes (Asperger's, PDD-NOS and Autistic) were also calculated. As shown in Fig. 3, The PAD in Asperger's was significantly lower than PDD-NOS and Autistic for all 6 models. Furthermore, all the models showed the negative PAD for Asperger's and the positive PAD for the other two subtypes. The Autistic group achieved the highest correlation between the predicted age and the chronological age across the 6 models.

3.5 Consistent Features in ASD Brain Age Prediction

Saliency maps were estimated through occlusion sensitivity analysis to identify the interpretable features in ASD brain age estimation. Results showed insula, anterior cingulate cortex (ACC), posterior cingulate cortex (PCC), middle temporal cortex (MTC), inferior temporal cortex (ITC) and fusiform gyrus (FG) were consistently identified across the 6 prediction models (Fig. 4), which are key brain regions in the default mode network (DMN) and salience network (SAN).

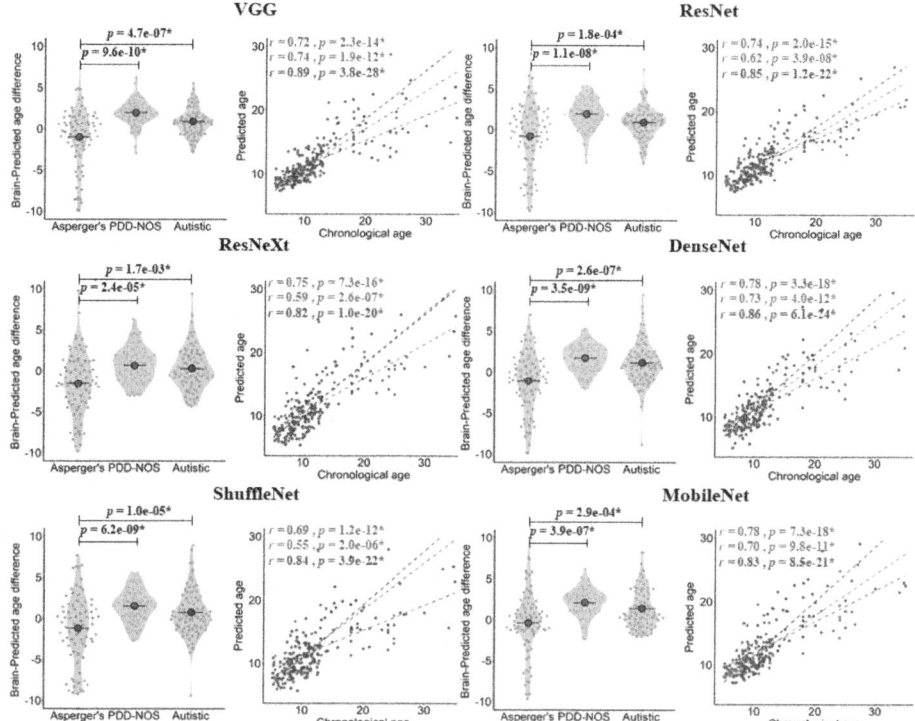

Fig. 3. The PAD differences and the correlation between the predicted brain age and the chronological age among ASD subtypes (Asperger's: blue; PDD-NOS subjects: green; Autistic: red) based on 6 prediction models.

4 Discussion

This study contributed to identifying the consistent and reliable brain patterns in ASD brain age estimation by comparing 6 3D-CNN models. Results showed that (1) most prediction models achieved a good prediction accuracy (around 2 years MAE in the training set and 3 years MAE in the testing set); (2) ASD showed no PAD differences in comparison to TDC; (3) Significant PAD differences were observed in childhood (<10 years old) and among ASD subtypes. (4) DMN and SAN were consistently identified as having high predictive power for brain age across 6 models.

A major finding was that there were no PAD differences between ASD and TDC under 6 3D-CNN prediction models. Consistent with our study, previous research showed that the brain maturation patterns were similar between ASD and TDC on average [17]. The lack of PAD differences between ASD and TDC suggested preserved brain development in ASD. Another interesting finding was that the PAD differences in brain development for ASD were only observed in childhood, but not in the following adolescence and adulthood. This is consistent with previously reported age-related differences in ASD [19]. Brain development in ASD seemed to be accelerated in younger individuals but

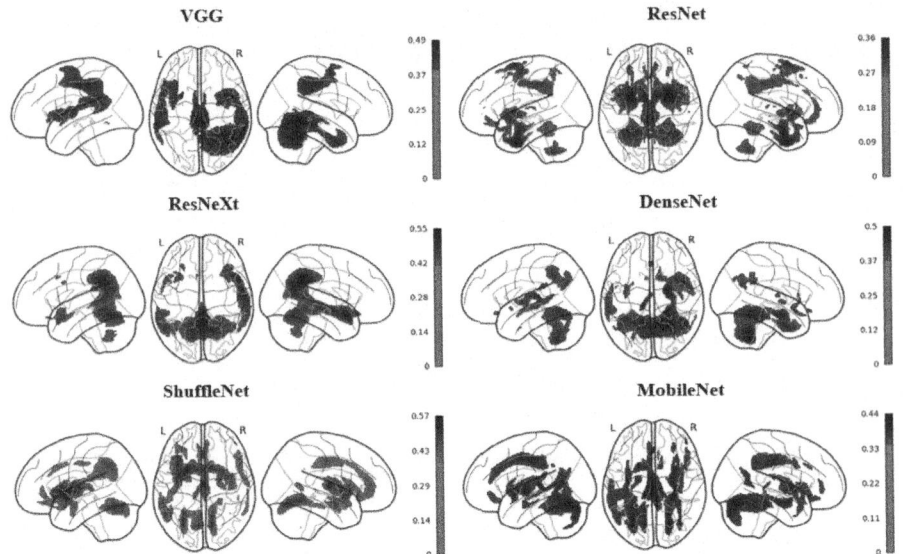

Fig. 4. The saliency maps of prediction models. Higher saliency represents the regional importance in brain age estimation.

delayed in older ones. Moreover, the Asperger's PAD was lower than the other accelerated growth subtypes due to the incidence of developmental delays. Although previous studies have revealed a connection between brain age and disease severity in ASD [17], differences in PAD among subtypes have not been clearly reported. Our result suggested that there was significant heterogeneity of PAD within ASD that presented with a wide range of symptoms and varying levels of impairments.

With respect to the features, DMN and SAN were consistently identified as interpretable brain networks in ASD brain age estimation over 6 models. The MTC and cingulate played important roles during brain development and showed obvious gender differences in brain age estimation of ASD [18]. The insula was a core area of SAN, which is reported to be involved in emotional regulation and decision-making and associated with brain developmental trajectories in ASD [28]. Moreover, a study using only temporal, cingulate or insula as features to predict ASD brain age also achieved high model performance [16]. Our results suggest that the atypical brain developmental trajectories in ASD may be driven by DMN and SAN. It is worth noting that the above results were reproducible across multiple predictive models, which overcame the limitation of bias due to model selection and further enhanced our confidence in the reliability and generalizability of the findings.

In conclusion, this study identified the consistent and reliable atypical brain development patterns and interpretable regions within ASD by comparing 6 classic 3D-CNN models. There were no PAD differences between ASD and TDC, and further analyses revealed distinct brain development patterns in younger ASD and its subgroups, highlighting the heterogeneity of brain development patterns among different development statuses and subtypes of ASD. The contributed regions across the predictions from our

6 models indicate that DMN and SAN drove deviations in brain growth trajectories in ASD, which may underlie the neurodevelopmental differences in ASD. The identified predictive signatures may serve as potential neuroimaging targets for the clinical exploration of atypical neurodevelopmental patterns in ASD.

Acknowledgments. This work was supported by the Natural Science Foundation of Jiangsu Province, China (BK20220889), the National Natural Science Foundation of China (62376124, 82172061 and 81771444), the Key Research and Development Plan of Jiangsu Province, China (BE2023668, BE2022677), and the Fundamental Research Funds for the Central Universities, (NJ2023032).

References

1. Hodges, H., Fealko, C., Soares, N.: Autism spectrum disorder: definition, epidemiology, causes, and clinical evaluation. Transl. Pediatr. **9**(Suppl 1), S55 (2020)
2. Qi, S., et al.: Reward processing in novelty seekers: a transdiagnostic psychiatric imaging biomarker. Biol. Psychiatry **90**(8), 529–539 (2021)
3. Qi, S., et al.: The relevance of transdiagnostic shared networks to the severity of symptoms and cognitive deficits in schizophrenia: a multimodal brain imaging fusion study. Transl. Psychiatry **10**(1), 149 (2020)
4. Courchesne, E., et al.: Unusual brain growth patterns in early life in patients with autistic disorder: an MRI study. Neurology **57**(2), 245–254 (2001)
5. Zheng, W., et al.: Multi-feature based network revealing the structural abnormalities in autism spectrum disorder. IEEE Trans. Affect. Comput. **12**(3), 732–742 (2021)
6. Carper, R.A., Moses, P., Tigue, Z.D., Courchesne, E.: Cerebral lobes in autism: early hyperplasia and abnormal age effects. Neuroimage **16**(4), 1038–1051 (2002)
7. Palmen, S.J., et al.: Increased gray-matter volume in medication-naive high-functioning children with autism spectrum disorder. Psychol. Med. **35**(4), 561–570 (2005)
8. Hazlett, H.C., Poe, M.D., Gerig, G., Smith, R.G., Piven, J.: Cortical gray and white brain tissue volume in adolescents and adults with autism. Biol. Psychiatry **59**(1), 1–6 (2006)
9. Qi, S., et al.: Common and unique multimodal covarying patterns in autism spectrum disorder subtypes. Mol. Autism **11**(1), 90 (2020)
10. Guo, X., Chen, H., Long, Z., Duan, X., Zhang, Y., Chen, H.: Atypical developmental trajectory of local spontaneous brain activity in autism spectrum disorder. Sci. Rep. **7**(1), 39822 (2017)
11. Ecker, C., Bookheimer, S.Y., Murphy, D.G.M.: Neuroimaging in autism spectrum disorder: brain structure and function across the lifespan. Lancet Neurol. **14**(11), 1121–1134 (2015)
12. Mishra, S., Beheshti, I., Khanna, P.: A Review of neuroimaging-driven brain age estimation for identification of brain disorders and health conditions. IEEE Rev. Biomed. Eng. **16**, 371–385 (2023)
13. Dinsdale, N.K., et al.: Learning patterns of the ageing brain in MRI using deep convolutional networks. Neuroimage **224**, 117401 (2021)
14. Jonsson, B.A., et al.: Brain age prediction using deep learning uncovers associated sequence variants. Nat. Commun. **10**(1), 5409 (2019)
15. Lee, J., et al.: Deep learning-based brain age prediction in normal aging and dementia. Nat. Aging **2**(5), 412–424 (2022)
16. Kaufmann, T., et al.: Common brain disorders are associated with heritable patterns of apparent aging of the brain. Nat. Neurosci. **22**(10), 1617–1623 (2019)

17. Tunc, B., et al.: Deviation from normative brain development is associated with symptom severity in autism spectrum disorder. Mol. Autism **10**, 46 (2019)
18. Wang, Q., et al.: Predicting brain age during typical and atypical development based on structural and functional neuroimaging. Hum. Brain Mapp. **42**(18), 5943–5955 (2021)
19. He, C., et al.: Structure–function connectomics reveals aberrant developmental trajectory occurring at preadolescence in the autistic brain. Cereb. Cortex **30**(9), 5028–5037 (2020)
20. Simonyan, K., Zisserman, A.: Very Deep Convolutional Networks for Large-Scale Image Recognition (2015). 2015-04-10T16:25:04
21. He, K., Zhang, X., Ren, S., Sun, J.: Deep residual learning for image recognition (2015). 2015-12-10T19:51:55
22. Xie, S., Girshick, R., Doll\'ar, P., Tu, Z., He, K.: Aggregated residual transformations for deep neural networks (2017). 2017-04-11T01:53:41
23. Huang, G., Liu, Z., Maaten, L.V.D., Weinberger, K.Q.: Densely connected convolutional networks (2018). 2018-01-28T17:12:02
24. Zhang, X., Zhou, X., Lin, M., Sun, J.: ShuffleNet: an extremely efficient convolutional neural network for mobile devices (2017). 2017-12-07T18:06:34
25. Howard, A.G., et al.: Efficient convolutional neural networks for mobile vision applications (2017). 2017-04-17T03:57:34
26. Zeiler, M.D., Fergus, R.: Visualizing and understanding convolutional networks. In: Computer Vision–ECCV 2014, pp. 818–833. Springer International Publishing (2014)
27. Morgan, N., Bourlard, H.: Generalization and parameter estimation in feedforward nets: some experiments. In: Advances in Neural Information Processing Systems, vol. 2, pp. 630–637. Morgan Kaufmann Publishers Inc. (1990)
28. Corps, J., Rekik, I.: Morphological brain age prediction using multi-view brain networks derived from cortical morphology in healthy and disordered participants. Sci. Rep. **9**(1), 9676 (2019)

Brain-Aware Readout Layers in GNNs: Advancing Alzheimer's Early Detection and Neuroimaging

Jiwon Youn[1], Dong Woo Kang[2], Hyun Kook Lim[3], and Mansu Kim[1(✉)]

[1] AI Graduate School, Gwangju Institute of Science and Technology, Gwangju, Republic of Korea
mansu.kim@gist.ac.kr
[2] Department of Psychiatry, Seoul St. Mary's Hospital, College of Medicine, The Catholic University of Korea, Seoul, Republic of Korea
[3] Department of Psychiatry, Yeouido St. Mary's Hospital, College of Medicine, The Catholic University of Korea, Seoul, Republic of Korea

Abstract. Alzheimer's disease (AD) is a neurodegenerative disorder characterized by progressive memory and cognitive decline, affecting millions worldwide. Diagnosing AD is challenging due to its heterogeneous nature and variable progression. This study introduces a novel brain-aware readout layer (BA readout layer) for Graph Neural Networks (GNNs), designed to improve interpretability and predictive accuracy in neuroimaging for early AD diagnosis. By clustering brain regions based on functional connectivity and node embedding, this layer improves the GNN's capability to capture complex brain network characteristics. We analyzed neuroimaging data from 383 participants, including both cognitively normal and preclinical AD individuals, using T1-weighted MRI, resting-state fMRI, and FBB-PET to construct brain graphs. Our results show that GNNs with the BA readout layer significantly outperform traditional models in predicting the Preclinical Alzheimer's Cognitive Composite (PACC) score, demonstrating higher robustness and stability. The adaptive BA readout layer also offers enhanced interpretability by highlighting task-specific brain regions critical to cognitive functions impacted by AD. These findings suggest that our approach provides a valuable tool for the early diagnosis and analysis of Alzheimer's disease.

Keywords: Graph Neural Network · Alzheimer's disease · brain-aware readout layer · early diagnosis

1 Introduction

Alzheimer's disease (AD) is characterized by progressive impairment of memory and cognitive functions, affecting approximately 47 million people worldwide. A major challenge in diagnosing AD is its heterogeneity among individuals, as it typically manifests as a decline in cognitive abilities. Although the onset of AD correlates with the accumulation of amyloid proteins beyond a certain threshold, these proteins are not the direct cause of AD, and the future progression patterns also vary among individuals [1–3].

Early diagnosis of AD is critical due to no effective treatments for dementia associated with AD [4]. Recently, it has been reported that a preclinical stage of AD is asymptomatic and characterized by extremely minor cognitive declines, which can be captured by the Preclinical Alzheimer's Cognitive Composite (PACC) score [5]. The PACC score provides a measure of cognitive abilities that may decline even before noticeable symptoms appear by capturing subtle changes in episodic memory, executive function, and attention [6].

For decades, the ATN framework, β-amyloid deposition (A), tau pathology (T), and neurodegeneration (N) play important roles in understanding AD, has been focused on understanding AD pathology and identifying related biomarkers [7]. Recent studies have successfully demonstrated to separate AD and cognitive normal(CN) based on T1-weighted magnetic resonance imaging (MRI), while having a limited performance on early diagnosis separating CN with mild cognitive impairment (MCI) [8]. Another study based on positron emission tomography (PET) reported that reported significant reductions in metabolic activity in key brain regions associated with Alzheimer's disease, while encountering challenges related to resolution issues [9,10]. Recent studies have highlighted disruptions in the default mode network through functional MRI, showing connectivity reductions that correlate with cognitive declines and predict progression from mild cognitive impairment to Alzheimer's disease [11].

Recent Graph Neural Networks (GNNs) studies have demonstrated high effectiveness across various tasks, including recommendation systems, computer vision, and neuroimaging [12]. Specifically, recent bioinformatics studies have demonstrated that GNNs achieve higher performance than traditional machine learning techniques in tasks such as disease prediction, drug discovery, and patient classification [13–16]. Moreover, neuroimaging studies have demonstrated that GNNs are highly effective at capturing the intricate connectivity and patterns within brain networks outperforming traditional analysis techniques [17]. These studies suggest that GNN might have the potential to understate complex brain networks due to its ability to capture complex relationships within data.

In this paper, we propose a novel brain-aware readout layer (BA readout layer) for GNNs to improve their interpretability. By applying the proposed readout layer, the GNN model can not only capture the complex patterns inherent in brain data but also identify groups of regions whose functionalities are closely related to each other. Our major contribution is as follows: 1) The proposed BA readout layer is more interpretable than the conventional readout layer by leveraging brain functional network. 2) We demonstrated the flexibility of the proposed BA readout layer by implementing it across different GNN frameworks. 3) The empirical experiment on real data has been performed to demonstrate the effectiveness of the proposed model compared with competing methods.

2 Related Works

GNNs are a powerful framework for learning from graph-structured data. Unlike traditional neural networks that excel at processing grid-like data such as images or sequences, GNNs are designed to handle data represented in the form of graphs. In a graph $G = (\mathcal{V}, \mathbf{E})$, the set $\mathcal{V}$ represents nodes (or vertices) and $\mathbf{E}$ represents edges (or relationships) between the nodes.

The fundamental concept of GNN is the iterative updating of node representations by aggregating information from neighboring nodes in the graph using an aggregation layer. This process enables GNN to capture complex relational dependencies and structural properties inherent in the data. Finally, the readout layer has been used to aggregate node representations to produce a single representation for the entire graph.

2.1 Aggregation Layer

Graph Convolutional Networks (GCN). GCN is an aggregation layer based on convolution operation for graphs by using a localized first-order approximation of spectral graph convolutions [18]. The fundamental concept of the GCN is a layer-wise propagation rule, defined as:

$$H^{(l+1)} = \sigma\left(\tilde{D}^{-\frac{1}{2}} \tilde{A} \tilde{D}^{-\frac{1}{2}} H^{(l)} W^{(l)}\right)$$

where $\tilde{A} = A + I_N$ is the adjacency matrix with added self-loops, $\tilde{D}$ is the degree matrix of $\tilde{A}$, $H^{(l)}$ is the matrix of activation in the l-th layer, $W^{(l)}$ is a layer-specific weight matrix, and σ is an activation function like ReLU. The final layer uses a softmax activation for classification:

$$Z = \text{softmax}\left(\tilde{A} \text{ReLU}\left(\tilde{A} X W^{(0)}\right) W^{(1)}\right)$$

This method efficiently integrates both node features and graph structure, and the adjacency matrix is normalized to prevent numerical instabilities and ensure effective gradient flow during training.

GraphSAGE. GraphSAGE is also an aggregation layer based on an inductive framework for generating node embeddings in large graphs [19]. GraphSAGE learns embeddings by sampling and aggregating features from a node's local neighborhood, allowing it to generalize to previously unseen nodes. The three steps of GraphSAGE are described as follows: 1) sampling a fixed-size set of neighbors, 2) aggregating feature information using various functions (mean, LSTM, pooling), and 3) updating the node's representation. Here are the typical aggregation functions used in GraphSAGE:

1. **Mean Aggregator:** This function computes the mean of the feature vectors of the sampled neighbors.

$$h_{\mathcal{N}(v)}^k = \frac{1}{|\mathcal{N}(v)|} \sum_{u \in \mathcal{N}(v)} h_u^{k-1}$$

2. **LSTM Aggregator:** An LSTM is employed to aggregate the features from the neighborhood. This approach is order-sensitive.

$$h_{\mathcal{N}(v)}^k = \text{LSTM}(\{h_u^{k-1} : u \in \mathcal{N}(v)\})$$

Note: The LSTM aggregator processes the embeddings of the neighbors sequentially, and the output is the final hidden state of the LSTM.

3. **Pooling Aggregator:** Applies a neural network followed by a max-pooling operation to the features of each neighbor.

$$h_{\mathcal{N}(v)}^k = \max(\{\text{ReLU}(W \cdot h_u^{k-1} + b) : u \in \mathcal{N}(v)\})$$

Graph Attention Networks (GAT). Graph Attention Networks (GATs) is an aggregation layer with masked self-attention [20]. These self-attentional layers enable the GNN model to selectively focus on specific parts of a node's neighborhood, assigning different weights to neighboring nodes. This approach eliminates the need for computationally expensive matrix operations and does not require prior knowledge of the entire graph structure. The key components and equations of GATs are as follows:

1. **Attention Coefficients:** This equation computes the attention coefficients that indicate the importance of each node's features to the central node. The attention coefficients are normalized using the softmax function to ensure they sum to one and can be interpreted as probabilities.

$$\alpha_{ij} = \frac{\exp\left(\text{LeakyReLU}\left(a^T[Wh_i \| Wh_j]\right)\right)}{\sum_{k \in N(i)} \exp\left(\text{LeakyReLU}\left(a^T[Wh_i \| Wh_k]\right)\right)}$$

2. **Aggregation:** The node features are aggregated using the weighted sum of the features of its neighbors, with weights given by the attention coefficients.

$$\mathbf{h}_i' = \sigma\left(\sum_{j \in \mathcal{N}(i)} \alpha_{ij} \mathbf{W} \mathbf{h}_j\right)$$

2.2 Readout Layer

The readout layer, also known as the graph-level pooling layer, plays a crucial role in the GNN model by transforming node-level embeddings into a graph-level representation. This representation is essential for tasks such as graph classification, graph regression, and entire graph embeddings. Conventionally, GNNs utilize simple pooling operations, such as averaging or summing, as described as follows:

Mean Pooling Readout Layer. Mean pooling aggregates node features by computing the average. This process effectively captures the central tendency of feature distributions across the graph, which can be beneficial for representing the overall structure. The mean pooling operation can be mathematically described by:

$$h_G = \frac{1}{|V|} \sum_{v \in V} h_v$$

where h_v represents the node features, V is the set of all nodes in the graph, and h_G is the resulting graph-level representation.

Add Readout Layer. Add pooling, or sum pooling, aggregates node features by computing the sum of all node embeddings. This method preserves the scale of node features, making it suitable for tasks where the magnitude of features impacts the output. The add pooling operation is expressed as:

$$h_G = \sum_{v \in V} h_v$$

where h_v and V are defined as above, and h_G is the graph-level representation.

While computationally efficient, these methods often limit capturing the intricate topology and heterogeneous relationships within the graph data.

3 Methods

3.1 Data Description

We collected a total of 383 Alzheimer's Disease patients, including 243 CN and 140 preclinical AD participants, from the Catholic Aging Brain Imaging(CABI) database. All participants, matched for age, gender, and education level, underwent T1-weighted MRI, resting-state functional MRI(rs-fMRI), and 18F-Florbetaben PET(FBB-PET). This database contains brain scans of patients who visited the outpatient clinic at the Catholic Brain Health Center, Yeouido St. Mary's Hospital, The Catholic University of Korea, between 2017 and 2022. The study was conducted under ethical and safety guidelines set forth by the Institutional Review Board of Yeouido St. Mary's Hospital, College of Medicine, The Catholic University of Korea (IRB number: SC22RIDI0153). The informed consent was waived by the IRB because we only used retrospective data. The detailed demographic information is described in Table 1.

3.2 Data Preprocessing

T1w MRI preprocessing was performed based on Micapipe to align normalized T1w MRI and parcellated brain into 374 regions, consisting of 360 cortical regions from Glasser atlas and 14 subcortical regions from aseg subcortical atlas [21]. The rs-fMRI had been pre-processed using Micapipe [22], including

Table 1. Demographic Information

	Mean ± standard deviation
Age (years)	70.96 ± 8.09
Education (years)	11.56 ± 4.92
Gender (Male/female)	114/270
CDR (0/0.5/1.0)	294/89/1
PACC5	0.24 ± 0.59

the following steps: slice timing and head motion correction, skull stripping, intensity normalization, and band-pass filtering. The pre-processed data were registered onto the MNI152 standard space. The FBB-PET preprocessing was performed based on several steps: registration of the PET image to the subject's T1-weighted MRI, partial volume correction using the PETPVC toolbox [23], inter-subject spatial normalization into MNI space. Finally, the standardized uptake value ratio (SUVR) was calculated for all parcellation by dividing the update of reference regions (i.e., cerebellumPons).

3.3 Brain-Aware Graph Neural Network

Motivation. Unlike social networks or chemical molecular networks, brain networks have the unique characteristic that nodes (i.e., brain regions) are organized into networks based on similar roles or functions. Therefore, simple readout layers that average or sum node features are insufficient for capturing the complex and heterogeneous characteristics of brain networks. To overcome limitations, we propose a novel readout layer clustering brain regions regarding similar tasks and allowing the GNN model to represent details brain characteristics. By leveraging these clusters, we aim the proposed GNN model to learn the intricate and heterogeneous features of brain networks, leading to more accurate and meaningful graph-level representations.

Brain-Aware Readout Layer (BA Readout Layer). The BA readout layer is designed to effectively capture the complex and heterogeneous characteristics of brain networks by clustering brain regions based on functional connectivity and node embedding. As shown in Fig. 1(D), the BA readout layer is expressed as:

$$h_{BA} = [f(h_{c_1}), f(h_{c_2}), ..., f(h_{c_i})]^T, \text{where } h_{c_i} = \frac{1}{|V_{c_i}|} \sum_{v \in V_{c_i}} h_v$$

where h_{BA} represents group-level representation, $f(.)$ denotes a linear transformation function used to aggregate embeddings, V_{c_i} represents a set of nodes in c_i cluster, and h_{c_i} denotes a cluster-level embedding in the c_i cluster.

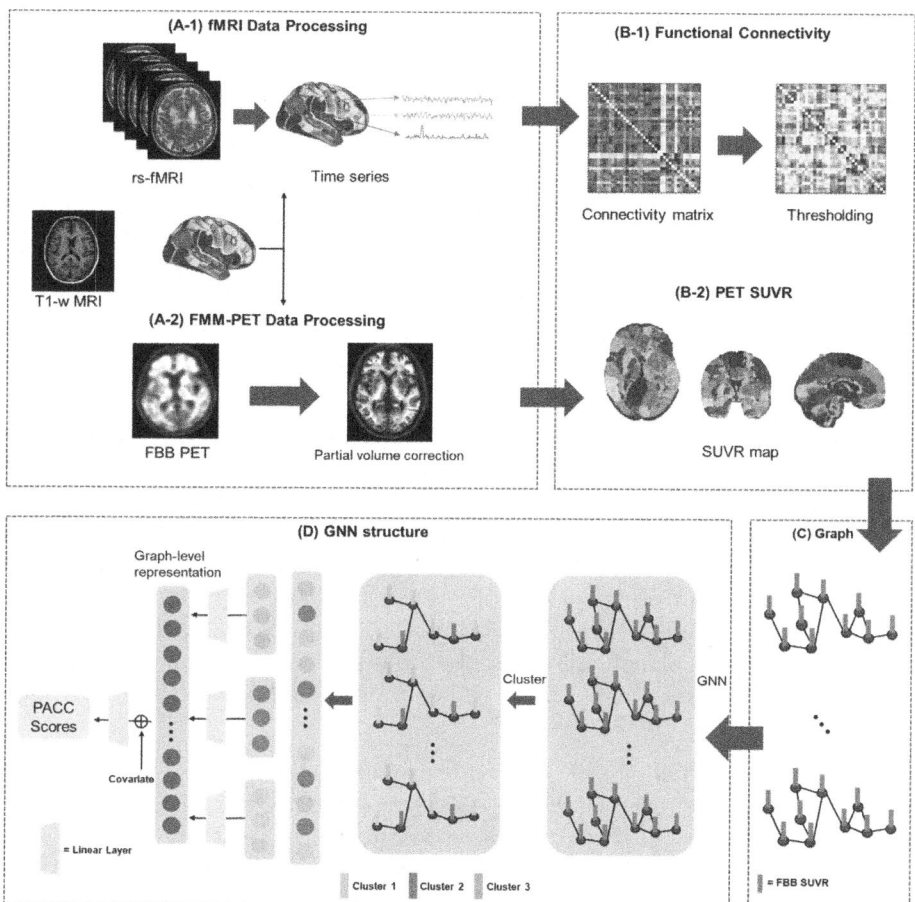

Fig. 1. Overview of data processing and analysis pipeline for neuroimaging and graph neural network application. (A-1) fMRI data processing involves acquiring fMRI timeseries, followed by brain parcellation and extraction of time courses. (A-2) For FBB-PET data, The T1-weighted images are processed by partial volume correction to produce PET SUVR maps. (B-1) Functional connectivity is calculated from fMRI data, resulting in connectivity matrices that are further refined through thresholding. (B-2) PET SUVR maps are obtained from processed FBB-PET images. (C) The processed data are then represented as graphs, which are input into a GNN. (D) The GNN structure processes these graphs to yield graph-level representations and predicts clinical scores such as PACC.

Briefly, the brain-aware clusters and their members have been selected by using a simple auxiliary neural network. An auxiliary network consisted of a single-layer artificial neural network with softmax function feed position embedding of each node, $\mathbf{C} = \text{softmax}(f(\mathcal{P}))$, and compute its probability of belonging to each cluster. The position embedding, $\mathcal{P} = [p_1, p_2, ..., p_i] \in \mathbb{R}^{N \times N}$, is a onehot vector, where the i-th entry of p_i is 1 and all other entries are 0, which make embedding invariant to the order of node.

$$\mathbf{C} = [c_1, c_2, ..., c_i] \in \mathbb{R}^{N \times p}$$

where N denotes number of nodes in the graph, p denotes number of brain-aware cluster, and c_i denotes selection probability vector of the i-th cluster for all nodes.

The cluster-level embedding, $f(h_{c_i})$, is computed by aggregating node embedding, h_{c_i}, based on c_i, selected members of the c_i cluster. Finally, the group-level representation is obtained by concatenating all cluster-level embeddings. This representation is then used in further analysis.

Architecture. The pre-processed fMRI and FBB-PET are utilized to construct the input graph, as described in Fig. 1(A)–(C). The proposed network model encodes node-level representation based on the existing GNN encoder and then graph-level representation based on the proposed BA readout layer. Using the BA readout layer, nodes are clustered into groups based on their similarities, indicated by different node colors in Fig. 1(D), and are aggregated to generate a comprehensive graph-level representation that captures the intricate structure of the brain network. The final representation is subsequently concatenated with covariate variables (e.g., age, gender, and education) and passed through a linear layer to predict the PACC score.

Step 1: Graph Construction. For given neuroimage, the input graph g is represented as a set of triplet $\{(\mathcal{V}, \mathbf{E}, \mathcal{P})\}$. $\mathcal{V}$ represents the set of nodes, expressed as $\mathcal{V} = \{v_1, v_2, \ldots, v_N\} \in \mathbb{R}^{N \times d_{in}}$, where each v_i can be neuroimaging measures with d_{in} features, such as brain region volume, SUVR of PET image, or thickness. $\mathbf{E}$ is the set of edges, consisting of pairs (e_i, e_j). Each (e_i, e_j) denotes the connectivity information between nodes (i.e., brain regions), and represents a non-negative undirected connection (either functional or structural connectivity). $\mathcal{P}$ represents the positional information of the nodes, where the i-th entry of p_i is 1 and all other entries are 0. The node position is assigned based on the brain atlas so that the nodes of all graph instances are ordered in a consistent sequence. Additionally, $y = \{y_1, y_2, \ldots, y_N\}$ represents the label for graph classification or the value for regression tasks.

Step 2: Node Embedding. In this step, we perform node embedding using the GNN models mentioned in Sect. 2.1. The GNN model, such as GCN, GraphSAGE, or GAT, aggregates node attributes to generate node-wise embedding, h_v.

$$h_v^{(k)} = \text{COMBINE}^{(k)}\left(\text{AGGREGATE}^{(k)}\left(\{h_u^{(k-1)} : u \in \mathcal{N}(v)\}\right), h_v^{(k)}\right) \quad (1)$$

Here, $h_v^{(k)} \in \mathbb{R}^{N \times d}$ represents the embedding of node v at layer k, where d is the hidden dimension. AGGREGATE$^{(k)}$ denotes the aggregation function at layer k, and COMBINE$^{(k)}$ represents the combination function at layer k. The function AGGREGATE processes local feature information from the node's neighbors to capture the local structural context within the graph. The COMBINE function integrates the aggregated neighbor information with the node's existing features to update and refine its embedding.

Step 3: Graph-Level Representation. Finally, we applied the BA readout layer to generate the graph-level representation for each input graph. Briefly, the brain-aware clusters and their members have been selected by using a simple auxiliary neural network, and the graph-level representation is obtained by concatenating cluster-level embeddings.

4 Experiments and Results

4.1 Experimental Setups

We have applied the proposed BA readout layer on various GNN models and assessed their predictive performance compared with the model with conventional readout layers. Many researchers have successfully adopted the GNN model for brain connectivity analysis. We have carefully chosen three GNN models: 1) GCN, 2) GraphSAGE, and 3) GAT, and two readout layers: 1) mean-pooling layer, and 2) add pooling layer. For all benchmark algorithms, we have constructed a graph where a cortico-cortical functional connectivity is based on Pearson's correlation with a predefined atlas as the edges of the graph and the pre-calculated SUVR for each brain region as node attributes of the graph.

In this study, we focus on the early detection of AD at the preclinical stage, which predicting the PACC score. We have applied a five-fold cross-validation strategy to examine the performance of the model, in terms of R^2 score. The model hyperparameters, such as hidden dimensions and learning rates, optimizing the number of clusters for each model, are tuned by Bayesian search on the training set implemented in Sweep methodology from Weights and Biases (WandB) [24]. Bayesian search strategy efficiently searches the most effective combination of hyperparameters using Bayesian optimization. Once optimal hyper-parameters are determined, the trained model is applied to the test set to generate the final performance.

4.2 Performance Comparison Across Readout Layer on Various GNN Model

To assess the effectiveness of the proposed method(i.e., BA readout layer), we have compared the predictive power of various GNN models with conventional readout layer, such as mean readout and add readout layer, against those using BA readout layer. The predictive power of the model is evaluated by predicting the PACC score of CN and preclinical AD subjects.

Table 2. Comparison of Readout Methods Across Different GNN Architectures with Performance Measured by R^2 Score

	Mean	Add	Yeo BA readout	adaptive BA readout
GCN	0.1539 ± 0.04	0.0877 ± 0.07	$\mathbf{0.1734 \pm 0.05}$	0.1517 ± 0.02
GraphSAGE	0.1917 ± 0.02	0.0441 ± 0.05	0.1921 ± 0.03	$\mathbf{0.1969 \pm 0.02}$
GAT	0.2019 ± 0.03	0.0758 ± 0.03	$\mathbf{0.2141 \pm 0.08}$	0.2044 ± 0.03

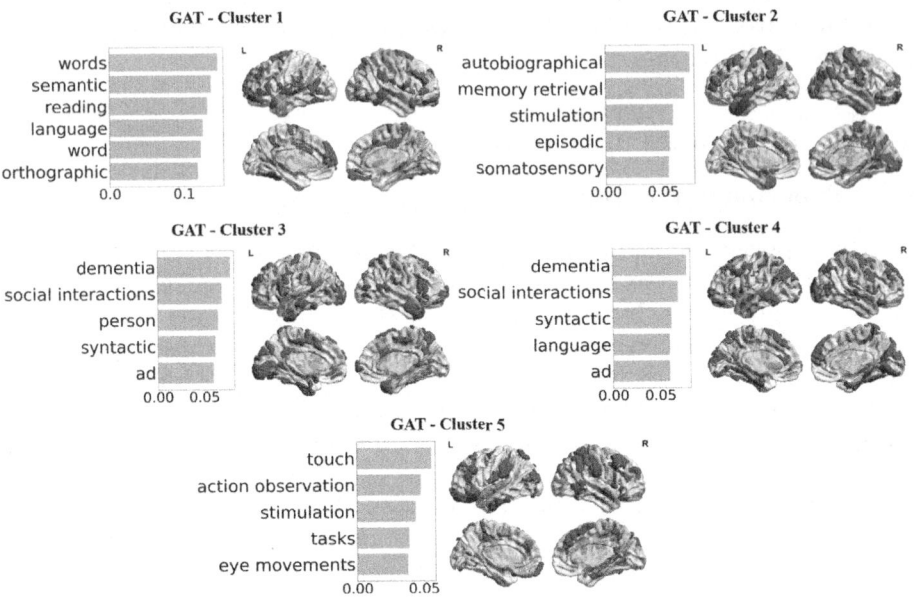

Fig. 2. Visualization of brain function clusters derived from GAT

Table 2 indicated that our model outperforms to predict PACC for all GNN models compared with those with mean and add readout layer. Specifically, the GCN and GAT with BA readout layer using Yeo network yield the best performance, 0.1734 ± 0.05, and 0.2141 ± 0.08, of R^2 scores, respectively, and the GraphSAGE with adaptive BA readout layer yield the best performance, 0.1969 ± 0.02, of R^2 scores.

4.3 Effectiveness of Prior Knowledge in BA Readout Layer

Additionally, we examine the effectiveness of incorporating prior knowledge into the BA readout layer. In detail, we apply the well-known Yeo 7 functional network as a predefined cluster of BA readout layer, called as Yeo BA readout layer, and compared with the proposed BA readout layer, called adaptive BA readout layer [25]. Overall, the integration of prior knowledge with the proposed layer (i.e., Yeo BA readout layer) demonstrates enhanced performance compared to layers without such knowledge (i.e., adaptive BA readout layer). However, we noted that the adaptive BA readout layer exhibited a lower standard deviation than the Yeo BA readout layer. This suggests that the adaptive BA readout layer offers greater robustness and stability in its performance.

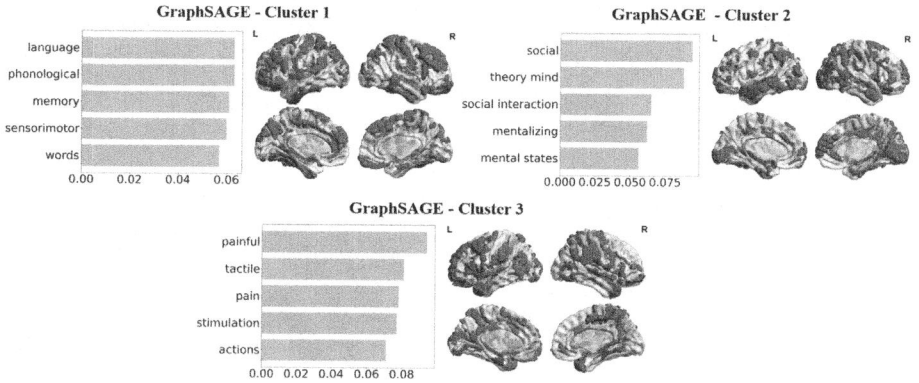

Fig. 3. Visualization of brain function clusters derived from GraphSAGE

4.4 Interpretability of the GNN Model with BA Readout Layer

The advantage of the adaptive BA readout layer is that be able to select a set of task-specific brain regions of the clusters, regardless of the GNN model. Particularly, the members of clusters are AD-specific brain regions, which play roles in memory, cognitive, or sensory function, that improve the model's interpretability. We utilized Neurosynth [26], a platform that synthesizes large amounts of human brain imaging data to decode patterns of neural activity associated with psychological terms and cognitive states to interpret selected brain regions of our model. Top 5 Neurosynth topics were investigated and the topics related to anatomical terminology were excluded.

Figure 2 illustrates five distinct clusters and their top five Neurosynth topics identified by the GAT with the BA readout layer. The brain regions within Cluster 1 are associated with language-related tasks, those within Cluster 2 are linked to memory functions, those within Clusters 3 and 4 are associated with dementia and social interactions, and those within Cluster 5 are related to sensory-motor functions. We found that the model identified the anterior cingulate, temporal pole, anterior insula, orbitofrontal cortex, and prefrontal cortex as members of clusters 3 and 4. These regions play important roles in cognitive functions and emotional responses or memory decline which is associated with AD [27–29].

Figure 3 illustrates three distinct clusters and their top five Neurosynth topics identified by the GraphSAGE with the BA readout layer. The brain regions within cluster 1 are associated with language and memory processing, which is crucial for detecting early signs of cognitive decline related to linguistic abilities and memory retention [30,31]. The brain regions within cluster 2 are linked to social cognition, essential for understanding and interacting with others, which can be impaired in Alzheimer's disease [32,33]. Those within Cluster 3 are associated with sensory processing and pain.

Figure 4 illustrates five distinct clusters and their top five Neurosynth topics identified by the GCN with the BA readout layer. The brain regions within each cluster are associated with specific brain functions relevant to cognitive disorders, particularly Alzheimer's disease. For example, the brain regions within clusters 1 and 2 are associated with memory [34], those within clusters 4 and 5 are linked to language tasks, essential for communication, and which deteriorate over the course of Alzheimer's disease [35,36].

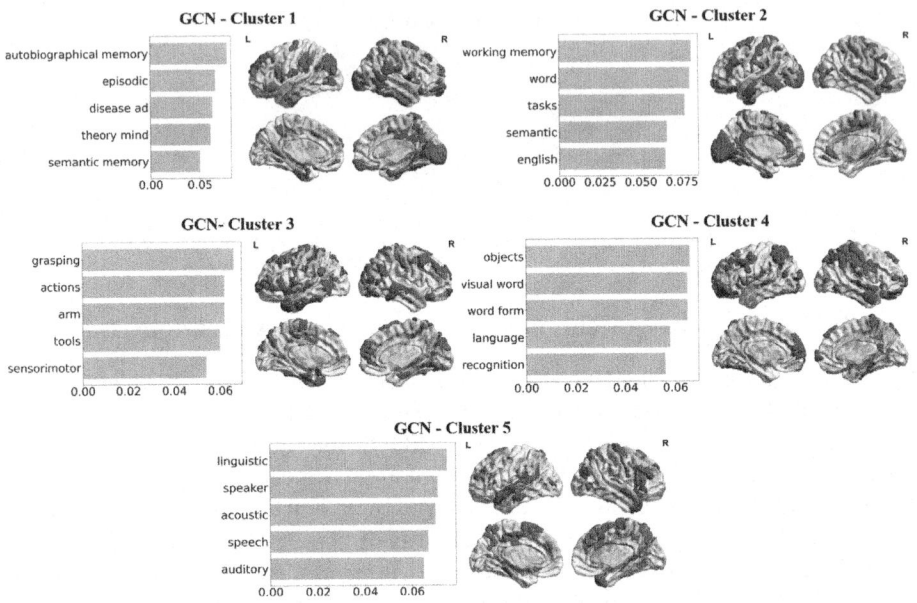

Fig. 4. Visualization of brain function clusters derived from GCN

5 Conclusion

AD presents significant challenges in both diagnosis and treatment, particularly due to its heterogeneous manifestation and the varying progression among individuals. This study introduces a novel approach to improve the interpretability and predictive power of GNNs in analyzing brain networks for early diagnosis of AD. We propose a brain-aware readout layer (BA readout layer) that clusters brain regions based on functional connectivity and node embedding, allowing the GNN model to capture intricate and heterogeneous characteristics of brain networks more effectively.

Our experiments demonstrate that GNN models incorporating the BA readout layer outperform those using conventional readout layers in predicting the

PACC score. Specifically, the GCN and GAT models with the Yeo BA readout layer, as well as the GraphSAGE model with the adaptive BA readout layer, achieve the highest R^2 scores, indicating superior predictive performance.

Moreover, the adaptive BA readout layer provides robustness and stability, exhibiting lower standard deviation compared to the Yeo BA readout layer. Furthermore, unlike the Yeo BA readout, which focuses solely on functional connectivity, the adaptive BA readout considers functional connectivity as well as node embedding, allowing for a richer and more meaningful graph-level representation. This suggests that the adaptive approach is more reliable across different datasets and scenarios. The interpretability of our model is further enhanced by the ability to identify task-specific brain regions within the clusters, which are associated with key cognitive functions and areas known to be affected by AD.

In summary, our proposed brain-aware readout layer for GNNs offers a promising tool for improving the diagnosis of Alzheimer's disease by leveraging the complex patterns inherent in brain data. By effectively clustering brain regions based on functional connectivity and node embedding, our approach enhances both the interpretability and performance of GNN models in neuroimaging analysis. Future research can further explore the potential of this method in other neurological disorders and refine the clustering techniques to capture even more nuanced brain network characteristics.

Acknowledgments. This work was partly supported by Institute of Information & communications Technology Planning & Evaluation (IITP) grant funded by the Korea government(MSIT) (No. RS-2021-II212068, Artificial Intelligence Innovation Hub, No.2019-0-01842, Artificial Intelligence Graduate School Program (GIST)). This work was also partly supported by the National Research Foundation of Korea (NRF) grant funded by the Korean government (Ministry of Science and ICT) (NRF-2022R1F1A106 8529, 2022R1I1A1A01053710). We also appreciate the high-performance GPU computing support of HPC-AI Open Infrastructure via GIST SCENT.

Disclosure of Interests. The authors have no competing interests to declare that are relevant to the content of this article.

References

1. Silverman, D.H., Small, G.W., Phelps, M.E.: Clinical value of neuroimaging in the diagnosis of dementia: sensitivity and specificity of regional cerebral metabolic and other parameters for early identification of Alzheimer's disease. Clin. Positron Imaging **2**(3), 119–130 (1999)
2. Serrano-Pozo, A., Das, S., Hyman, B.T.: APOE and Alzheimer's disease: advances in genetics, pathophysiology, and therapeutic approaches. Lancet Neurol. **20**(1), 68–80 (2021)
3. Selkoe, D.J., Hardy, J.: The amyloid hypothesis of Alzheimer's disease at 25 years. EMBO Mol. Med. **8**(6), 595–608 (2016)
4. Aël Chetelat, G., Baron, J.C.: Early diagnosis of Alzheimer's disease: contribution of structural neuroimaging. Neuroimage **18**(2), 525–541 (2003)

5. Hanseeuw, B.J., et al.: Association of amyloid and tau with cognition in preclinical Alzheimer disease: a longitudinal study. JAMA Neurol. **76**(8), 915–924 (2019)
6. Donohue, M.C., et al.: The preclinical Alzheimer cognitive composite: measuring amyloid-related decline. JAMA Neurol. **71**(8), 961–970 (2014)
7. Jack, C.R., Jr., et al.: NIA-AA research framework: toward a biological definition of Alzheimer's disease. Alzheimer's Dement. **14**(4), 535–562 (2018)
8. Salvatore, C., et al.: Magnetic resonance imaging biomarkers for the early diagnosis of Alzheimer's disease: a machine learning approach. Front. Neurosci. **9**, 307 (2015)
9. Høilund-Carlsen, P.F., et al.: FDG-PET versus amyloid-PET imaging for diagnosis and response evaluation in Alzheimer's disease: benefits and pitfalls. Diagnostics **13**(13), 2254 (2023)
10. Chapleau, M., Iaccarino, L., Soleimani-Meigooni, D., Rabinovici, G.D.: The role of amyloid pet in imaging neurodegenerative disorders: a review. J. Nucl. Med. **63**(Supplement 1), 13S-19S (2022)
11. Sheline, Y.I., Raichle, M.E.: Resting state functional connectivity in preclinical Alzheimer's disease. Biol. Psychiatry **74**(5), 340–347 (2013)
12. Wu, Z., Pan, S., Chen, F., Long, G., Zhang, C., Philip, S.Y.: A comprehensive survey on graph neural networks. IEEE Trans. Neural Netw. Learn. Syst. **32**(1), 4–24 (2020)
13. Boll, H.O., et al.: Graph neural networks for clinical risk prediction based on electronic health records: a survey. J. Biomed. Inform. 104616 (2024)
14. Zhang, L., Zhao, Y., Che, T., Li, S., Wang, X.: Graph neural networks for image-guided disease diagnosis: a review. iRADIOLOGY **1**(2), 151–166 (2023)
15. Zhou, J., et al.: Graph neural networks: a review of methods and applications. AI Open **1**, 57–81 (2020)
16. Jiang, D., et al.: Could graph neural networks learn better molecular representation for drug discovery? A comparison study of descriptor-based and graph-based models. J. Cheminform. **13**, 1–23 (2021)
17. Li, X., et al.: BrainGNN: interpretable brain graph neural network for fMRI analysis. Med. Image Anal. **74**, 102233 (2021)
18. Kipf, T.N., Welling, M.: Semi-supervised classification with graph convolutional networks. arXiv preprint arXiv:1609.02907 (2016)
19. Hamilton, W., Ying, Z., Leskovec, J.: Inductive representation learning on large graphs. Adv. Neural Inf. Process. Syst. **30** (2017)
20. Velickovic, P., Cucurull, G., Casanova, A., Romero, A., Lio, P., Bengio, Y., et al.: Graph attention networks. Stat **1050**(20), 10–48550 (2017)
21. Glasser, M.F., et al.: A multi-modal parcellation of human cerebral cortex. Nature **536**(7615), 171–178 (2016)
22. Cruces, R.R., et al.: Micapipe: a pipeline for multimodal neuroimaging and connectome analysis. Neuroimage **263**, 119612 (2022)
23. Thomas, B.A., et al.: PETPVC: a toolbox for performing partial volume correction techniques in positron emission tomography. Phys. Med. Biol. **61**(22), 7975 (2016)
24. Ferrell, B.J.: Fine-tuning strategies for classifying community-engaged research studies using transformer-based models: algorithm development and improvement study. JMIR Formative Res. **7**, e41137 (2023)
25. Yeo, B.T., et al.: The organization of the human cerebral cortex estimated by intrinsic functional connectivity. J. Neurophysiol. (2011)
26. Yarkoni, T., Poldrack, R.A., Nichols, T.E., Van Essen, D.C., Wager, T.D.: Large-scale automated synthesis of human functional neuroimaging data. Nat. Methods **8**(8), 665–670 (2011)

27. Molnar-Szakacs, I., Uddin, L.Q.: Anterior insula as a gatekeeper of executive control. Neuros. Biobehav. Rev. **139**, 104736 (2022)
28. Carter, C.S., Botvinick, M.M., Cohen, J.D.: The contribution of the anterior cingulate cortex to executive processes in cognition. Rev. Neurosci. **10**(1), 49–58 (1999)
29. Arnold, S.E., Hyman, B.T., Van Hoesen, G.W.: Neuropathologic changes of the temporal pole in Alzheimer's disease and pick's disease. Arch. Neurol. **51**(2), 145–150 (1994)
30. Clarke, N., Barrick, T.R., Garrard, P.: A comparison of connected speech tasks for detecting early Alzheimer's disease and mild cognitive impairment using natural language processing and machine learning. Front. Comput. Sci. **3**, 634360 (2021)
31. Beltrami, D., Gagliardi, G., Rossini Favretti, R., Ghidoni, E., Tamburini, F., Calzà, L.: Speech analysis by natural language processing techniques: a possible tool for very early detection of cognitive decline? Front. Aging Neurosci. **10**, 369 (2018)
32. Demichelis, O.P., Coundouris, S.P., Grainger, S.A., Henry, J.D.: Empathy and theory of mind in Alzheimer's disease: a meta-analysis. J. Int. Neuropsychol. Soc. **26**(10), 963–977 (2020)
33. Wilson, N.A., Ahmed, R., Piguet, O., Irish, M.: Disrupted social perception in frontotemporal dementia and Alzheimer's disease-associated cognitive processes and clinical implications. J. Neurol. Sci. **458**, 122902 (2024)
34. Verma, M., Howard, R.J.: Semantic memory and language dysfunction in early Alzheimer's disease: a review. Int. J. Geriatr. Psychiatry **27**(12), 1209–1217 (2012)
35. Swords, G.M., Nguyen, L.T., Mudar, R.A., Llano, D.A.: Auditory system dysfunction in Alzheimer disease and its prodromal states: a review. Ageing Res. Rev. **44**, 49–59 (2018)
36. Massoud, F., Chertkow, H., Whitehead, V., Overbury, O., Bergman, H.: Word-reading thresholds in Alzheimer disease and mild memory loss: a pilot study. Alzheimer Dis. Assoc. Disord. **16**(1), 31–39 (2002)

TSICNet: Importance of Connectome Information for Epilepsy Classification

Zonghan Du[1(✉)] and Zhongyuan Lai[2]

[1] School of Computer Science, Sun Yat-sen University, Guangzhou 510275, China
duzh9@mail2.sysu.edu.cn
[2] DeepVerse Inc., Shanghai 200135, China

Abstract. Epilepsy is a chronic brain disease characterized by recurrent seizures. These episodes are usually triggered by abnormal firing of neurons in the brain and appear as brief, recurring episodes. From the point of view of mitigation and treatment, a major problem is timely and accurate classification of epileptic seizure episodes, which is important for rapid intervention and possible prevention of future episodes. In recent years, with the rapid development of artificial intelligence technology, its application in the field of epilepsy diagnosis is increasingly extensive. In particular, electroencephalography (EEG), as a non-invasive neurophysiological examination method, plays an important role in the diagnosis of epilepsy. In particular, it allows the learning of *connectomic* information from the relative positions of EEG electrodes on the scalp. In this paper, a deep learning model, **T**emporal-**S**patio-**I**mportance **C**orrelation **N**etwork (TSICNet), is proposed to classify the onset of epileptic seizures. Our model design combines a variety of neural network components and training methods, such as graph, spatio-temporal and separable convolutional networks, as well as sliding windows data extraction techniques and ATCNet's time-sensitive attention mechanism, to realize effective recognition and extraction of epilepsy data features. In particular, TSICNet makes heavy use of learned connectomic information to enhance classification accuracy. Experimental results show excellent performance on the CHB-MIT Scalp EEG dataset and HUH neonatal epilpsy dataset, and has good results in accuracy and True (TPR) and False Positive Rates (FPR). Moreover, ablation analysis shows that each part of the model contributes significantly, thus confirming the validity our design principles.

Keywords: Electroencephalogram · Connectome · Epilepsy Classification · Graph Neural Networks

1 Introduction

Epilepsy is one of the most common neurological diseases [32]. The most recognizable symptom of epilepsy are frequent and often severe *seizures*. The effects of an episode can range from minor lapses in attention [27] to abnormal movements (convulsions) of patients' body, in the sense that they lost partial or complete control. Such episodes could also entail loss of consciousness, bowel control, cognitive functions and sensations [4]. Patients' actions while within these abnormal periods will potentially endanger their physical well-being, and a particularly severe episode could lead to injury or even death. In the twentieth century, epilepsy has become one of the most prevalent neurological ailments; its sociological and economic costs are substantial, since it affects 50 million people globally, and out of the three million afflicted in the United States, around 450, 000 are under the age of 17. In this sense, epilepsy incurs a substantial social cost as well [7]. In addition, there is also social stigma associated to people suffering from epilepsy, in the sense that they are often perceived to be "uncontrollable" or uninhibited [4], a phenomena which is directly related to the strong link between epilepsy and the frequency of seizures. If untreated, it is also often fatal: patients are usually unaware of their predicament until too late. However, if discovered early, epileptic fits can be addressed sufficiently via medications [19]; however, the efficacy of such interventions depends critically on early diagnosis of epilepsy. Conventional diagnostic techniques are mainly based on after-the-event curative techniques, which focusses on medical intervention *after* an epileptic episode. However, such measures have been widely acknowledged to be inadequate for cases where the first episode might be the most severe, and which severity will gravely influence the frequency and intensity of all subsequent epileptic episodes. Hence, the *diagnosis* of epileptic seizures has become a crucial but still unsolved problem in the treatment of epileptic seizures.

From a neurobiological point of view, epilepsy (and, by association, epileptic seizures), can be traced to well-defined neurological sources [12]. It is known that seizures are strongly associated with abnormally excessive electrical discharges in well-defined regions of the brain [23], and, hence, can be essentially divided into three distinct types: generalized, focal and epileptic spasms [27,32]. The first type originate in networks spanning part of one brain hemisphere, while generalized ones begin in bilateral distributed neuronal networks. These point to the overt fact that the origins of epilepsy are *strongly* spatially-dependent on brain region; in other words, the point of origin of epileptic seizures influences their later dynamics. We take this into account in the design of our model, where we enable the learning of the *connectome*, in the form of a connectivity graph extracted from the EEG signals. Having connectomic information as prior greatly increases the accuracy of our model, as we demonstrate via ablation studies.

An EEG is a trace of electrical activity of the brain, which manifests as a time series of voltage values [18]. The detection and measurement of this signal can be easily done via electrodes affixed to the scalp in a regular pattern, which has been chosen to ensure comprehensive coverage of all important regions of the brain. The voltage difference between electrodes are then measured. EEG signals have excellent temporal resolution, and hence is very suitable for diagnostic tasks, where the requirement for *timeliness* is of great importance. An EEG trace containing an epileptic episode (*ictus*) can generally be divided into four distinct states, namely *interictal, preictal, ictal* and *postictal*. Pre- and postictal are the periods prior and subsequent to an epileptic episode, respectively, while the ictal and interictal periods refer to the interval during which an episode occurs and in-between episodes, respectively. In summary, analysis of signals collected from this set of electrodes form the clinical gold standard for diagnosis of neurological episodes.

In recent years, along with the increasing prevalence of machine and deep learning algorithms, the use these techniques to analyze EEG signals towards the aim of accurate epilepsy diagnosis has become a concrete trend. Due to the inherent complexity of the task (involving the analysis of time series data containing multiple time scales and data sources), the design of deep models which are robust and accurate, on the one hand, and interpretable, on the other, remains a challenging task. There already exist a large selection of deep models which have been optimized for the task of classifying and/or predicting epileptic seizures; some of the more recent advances in deep learning architecture (such as transformers [3]) have been deployed for this purpose as well.

Despite this wide selection of models, it is widely acknowledged that a significant barrier to the more widespread adoption of deep learning models for accurate epilepsy prediction is the lack of medical interpretability and transparent way for input of neuroscientific prior knowledge to the prediction model. In most works on the use of deep learning for epilepsy diagnosis and prediction, models are inevitably designed to be end-to-end, with no way of providing human oversight in the computational process. Apart from lack of interpretability, this black-box state of such models also make them susceptible to inaccuracies in data preprocessing and handling [16]. Such setbacks not only hamper the widespread use of deep models for fast and robust epilepsy diagnosis, it also prevents the building-up of trust in such approaches in medicine. By incorporating medical prior information in the form of a learnable connectomic graph, as well modular design of components, we aim to alleviate some of these problems in our proposed model. Our main contributions are the following:

- We present a model which is accurate and robust for epilepsy classification. We base our design principles on the importance of *connectomic* information in ensuring accuracy and performance.
- We performed extensive numerical experimentation which demonstrates the superiority of our method as compared to state-of-the-art models performing a similar task. To ensure a wide scope of data source and quality (reflecting the generalizability of our model), we evaluated our model on two different datasets covering a wide range of patient demographics;
- We show, via ablation studies, that the integration of connectomic information is crucial to achieving the SOTA performances which we display in this work. This insight can be utilized in the design of future EEG analysis frameworks, by emphasizing the importance of such information.

2 Related Work

Due to the long history and understanding of epilepsy as the most common neurological disease, there is a multitude of works relating to the traditional diagnosis and treatment, as well as more modern diagnostic approaches based on deep and machine learning, which is then directly relevant to our work. In this section we review some of these.

2.1 The Role of EEG in Epilepsy Diagnostics and Treatment

As a neurological disease, epilepsy has had a long history of diagnosis and treatment development. Medically, it is defined as a disease characterized by one or more seizures with a relatively high recurrence risk [7]. Diagnosis is based on a description of seizure behavior and EEG manifestations, further aided by neuroimaging and other genetic investigations [14]. In general, the most commonly method for diagnosis remains the recording and analysis of EEG signals [18,30]. In this respect, clinical applications of EEG include diagnosis, selection of antiepileptic drugs (AED) therapy, evaluation of response to treatment, determination of candidacy for drug withdrawal, and surgical localization [4]. In a clinical setting, EEGs are valuable tools for epilepsy diagnosis, since they are relatively inexpensive and easy to obtain [22,36], and contains the clearest signature of seizure onset. However, EEG signals alone are also inadequate to produce a robust and accurate diagnosis; their efficacy is affected by various clinical factors: age, seizure type, presence of AED therapy, and proximity of the EEG recording to seizure activity [12]. Hence the accelerating shift to deep learning methods to substantially increase accuracy and robustness of diagnosis.

2.2 Deep Learning in Epilepsy Diagnosis

Since the EEG waveform consists of a time series of voltage fluctuations, it is natural to consider whether deep learning methods for time series analysis are useful for EEG analysis. In this context deep learning offers several advantages: automated detection of epileptic seizures (removing the need for manual feature engineering, which is complex for a signal as heterogeneous as a scalp EEG) [18]; the availability of large datasets with which to train such models to high levels of accuracy [14] and robustness; as well as the flexibility of designing deep learning models to take biases and priors particular to the disease into account [29], which enables clinicians to personalize models towards specific patients. In the past, a multitude of different deep models have been deployed to the task of epilepsy seizure detection. One of the earliest work by [34] used an artificial neural network for 5-class classifications of EEG segments, while [21] performs places emphasis on feature engineering by first doing clustering of the wavelet-decomposed coefficients of EEG signals before classification using a multilayer perceptron. [35] takes advantage of the expressive power of pyramidal one-dimensional convolutional neural networks for classification of the Bonn University dataset; additional architectures based on the convolutional layer are [1,13,24]. There are also models based on the analysis of time series, e.g., the models of [6,11] which are based on the LSTM and stacked and Bi-LSTMs, respectively.

2.3 Relevance of the Connectome to Disease Diagnostics

The connectome is an important topological input information which is increasingly receiving attention from neuroscientists as well as computer scientists. The connection between spatial information and dynamics of epileptic seizure outbreak has been studied previously and has increasingly received attention from physicians and neuoscientists [5,20,23]. In recent years, along with the advance in deep learning, in particular the increasing expressiveness of graph neural networks, connectomic information has also been taken into account in several early works [17,23]. However, to our knowledge, there has not been a previous design which is able to comprehensively integrate connectomic information with pure EEG-based signals.

3 Methods

3.1 Datasets

The datasets used in this work consists of:

- the CHB-MIT Scalp EEG database [9]: The dataset consists of EEG recordings from 22 pediatric subjects with intractable epilepsy collected by the Boston Children's Hospital and detailed in [8]. The recordings were obtained by monitoring the subjects for up to several days after discontinuation of anti-epileptic drugs to characterize their seizures and assess their suitability for surgical intervention. A total of 22 subjects (5 males, ages 3–22 years; 17 females, ages 1.5–19). In this experiment, five subjects were randomly selected for numerical experimentation and model validation; distribution of the different genders and ages to test the generalization of the model is shown in Table 1. For each data set in the table, we used 80% for training and the remaining 20% for testing.

Table 1. Information on five subjects

Subject	Age	Gender	EEG seconds	Number of seizures	Seizure seconds	Proportion
chb01	11	female	23925	7	442	1.85%
chb02	11	male	8159	3	172	2.11%
chb03	14	female	25200	7	402	1.60%
chb05	7	female	18000	5	558	3.1%
chb08	3.5	male	18000	5	919	5.11%
chb_5_mix	-	-	93284	27	2493	2.67%

– the Helsinki University Hospital (HUH) neonatal epilepsy dataset [33]: this dataset consists of EEG recordings from 79 infants admitted to the NICU at the HUH between 2010 and 2014. These recordings were separately annotated by three experts and hence contains discrepancies between the labeled EEG traces. On average, a total of 460 seizures were annotated per expert; out of this, 39 neonates had seizures and 22 were seizure free, by consensus. This situation with the dataset labels enables us to compare the labels from different experts with our model classification results; for this work, we randomly sampled 10 datapoints, from both male and female neonates, respectively, so as to ensure fullest possible coverage of the dataset.

3.2 Model Training

The model was trained and tested on a single GPU (RTX 4090 24 GB) using the Pytorch and braindecode [28] frameworks. For all experiments, we used the following training configuration: the training was done using the Adam optimizer and the cross-entropy loss function with a learning rate of 0.0001, a batch size of 64, and training for 20 epochs. We also tried several different configurations of hyperparameters, the training results from which is detailed in the Results section. In addition, the pre-processing of all EEG data only uses a bandpass filter for 8–30 Hz filtering (Fig. 1).

3.3 Model Overview

TSICNet. Here we describe the TSICNet in detail. The primary submodule in TSICNet is a temporal convolutional layer with the kernel size set to $(1, 64)$; it extracts time domain features and outputs N feature maps. Since the convolution kernel size is $(1, 64)$ and the sampling rate is 256, information above 4Hz can be extracted. The extracted features are input into the spatial GCNN. In the GCNN, given the graph $G = (V, E)$, where V represents the set of nodes, and E represents the set of edges between nodes in E, V data can be used on the matrix $X \in \mathbb{R}^{n \times d}$, where $n = |V|$ and d the input feature dimension. The edge set E can be used to construct the weighted adjacency matrix $A \in \mathbb{R}^{n \times n}$, where $A_{ii} = 1$, $i = 1, 2, \ldots, n$. The data in other positions of the matrix are Pearson correlation coefficients. On the other hand, the GCNN learns elements of the matrix $I \in \mathbb{R}^{n \times n}$, namely the importance matrix. I can be understood as a measure of the importance of different input channels for the connection of different nodes (or edges) in the graph. Between adjacent layers of GNN, the feature transformation can be written as:

$$Z^{l+1} = Z^l (A \odot I) \qquad (1)$$

where $l = 0, 1, \ldots, l-1$, l and $Z^0 = X$, $H^L = Z$, I to learn the importance of the matrix. The Pearson correlation coefficient is a number between -1 and 1 that measures the strength of the linear relationship between two variables. When this coefficient is positive, it means that there is a positive correlation between

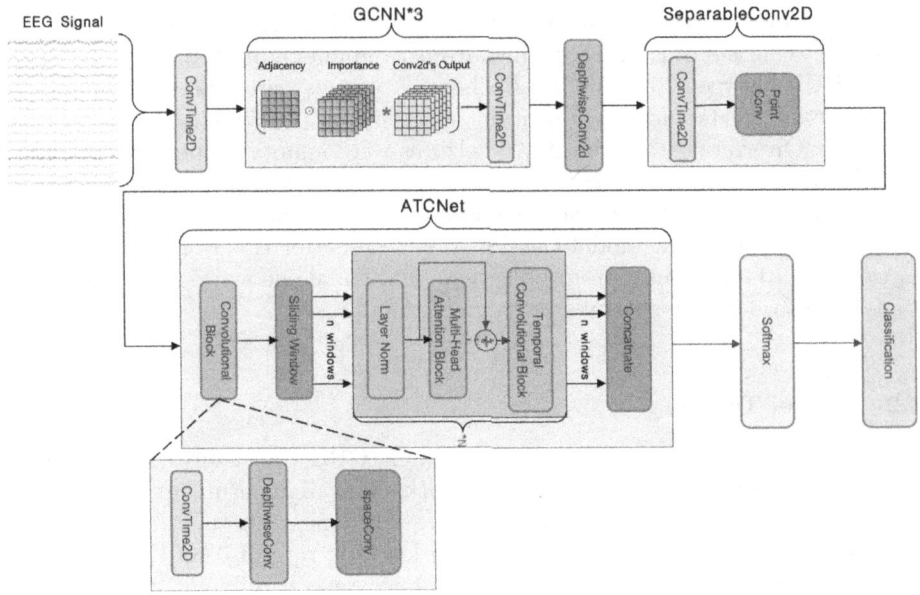

Fig. 1. TSICNet structure diagram

the two variables, and conversely. If the value is 0, there is no linear relationship between the two variables. The mathematical formula for Pearson's correlation coefficient is:

$$r = \frac{\sum_{i=1}^{n}(x_i - \bar{x})(y_i - \bar{y})}{\sqrt{\sum_{i=1}^{n}(x_i - \bar{x})^2}\sqrt{\sum_{i=1}^{n}(y_i - \bar{y})^2}} \quad (2)$$

n is the number of samples, x_i and y_i are the first I observations of the two variables under consideration, and $\bar{x}$, $\bar{y}$ the respective sample means. From this, we establish an correlation adjacency matrix with a size of [23, 23] as shown in Fig. 2, where the value 23 represents the number of sensor channels in the CHB-MIT Scalp EEG setup. Each value of r_{ij} in the matrix represents the correlation between the first i and j EEG channels. Since we regard the EEG channels as an undirected graph, the adjacency matrix is symmetric with a unit diagonal. As shown in Eq. 1, I initializes a square coefficient matrix in the network that can be updated as the network learns, with its initial values randomly drawn from the standard normal distribution. The A is a learnable prefactor that adjusts the importance between different nodes so that the adjacency matrix becomes fully adaptive according to the input data. It is elementwise-multiplied by the importance matrix I; this operation weights the adjacency matrix, with the weights provided by I. This means that different input channels will have different effects on the connections in the diagram, and this effect is learned. Finally, the result is multiplied by the EEG dataset matrix after convolution and batch normalization.

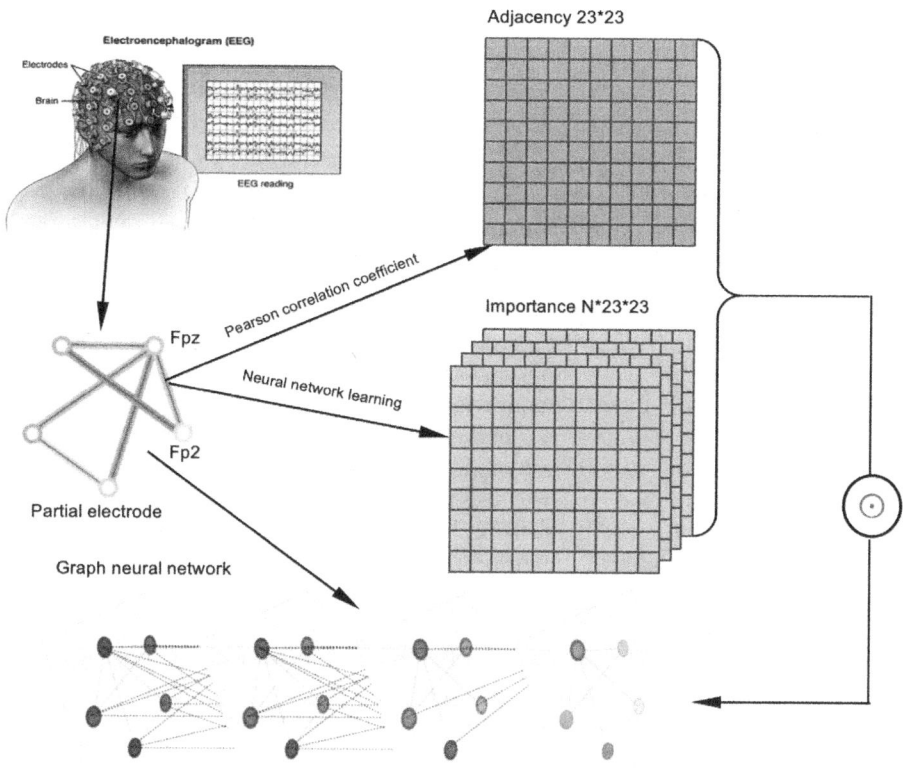

Fig. 2. The relationship between connectomic information from EEG traces and the adjacency matrix in TSICNet.

By making I a learnable parameter, the model adaptively adjusts the structure of the graph during training. This allows the model to learn which input channels are more important for a particular graph connection, thereby better capturing the relationship between the graph structure and the input data. Multiplying the weighted adjacency matrix with the input data **x** actually fuses the structure information of the graph with the input data. This fusion allows the model to consider both the structure of the graph and the characteristics of the input data, allowing for more complex classification tasks.

The final layer in TSICNet is a separable convolution that explicitly decouples relationships within and between feature maps by first learning a kernel that individually summarizes each feature map and then merges the results to produce an output.

4 Results

4.1 Baseline Models

We perform extensive experiments comparing the performance of TSICNet with a number of SOTA baseline models. All of these baselines are models that were optimized for tasks on EEG data. Apart from this set of conventional deep EEG models, we also compared TSICNet with a group of GCNN-based deep models. These models are briefly described in the following:

- ATCNet [2]: ATCNet is an attention-based framework for motor imagery classification. The framework consists of self-attention, temporal convolution submodules for feature extraction, and includes a convolutional-layer-based sliding window for data augmentation;
- EEGNet [15]: the EEGNet is a classic deep model for EEG tasks. The backbone of this model are convolutional layers which are organized serially in such a way to enhance generalizability across different BCI paradigms;
- EEGInception [26]: the EEG-Inception model fuses conventional convolutional layers in a chain of Inception modules, and optimized for event-related potential (ERP) tasks. The Inception modules are chosen for their multiscale feature resolution;
- EEG-ITNet [25]: this model is also based on the Inception framework, augmented by causal convolutions with dilations. The authors of this paper have also developed a visualization method for more effective EEG analysis result interpretation;
- EEGConformer [31]: the EEGConformer is a novel architecture based on convolutional transformer submodules, in which the convolutional layers learn the low-level local features, while the self-attention modules extracts the global correlations within the local features;

Apart from the deep EEG models listed above, we also compared TSICNet to several GCNN-based models.

- AST-GCN [38]: the AST-GCN is a graph-convolutional network for traffic forecasting. It consists of static and dynamic attribute augmentation module which integrates external factors for input to a conventional spatio-temporal GCN;
- AS-GCN [37]: the AS-GCN is a text dependency-tree based model which learns syntactical and word-level relations in sentences, which is then applied to a sentiment-classification task;
- ST-GCN [10]: the ST-GCN can be considered to be the original graph-convolutional based model which takes both spatial and temporal relationships between nodes into account. The original task, as reported in this publication, was for point-of-interest (POI) recommendation.

4.2 Comparison with Baseline Models

We compare the TSICNet with several baseline models on the CHB-MIT dataset. In addition to the accuracy, we also compute the True $TPR = \frac{TP}{TP+FN}$ and False Positive Rates $FPR = \frac{FP}{FP+TN}$ indicators. These are commonly used in binary classification tasks to complement ROC curves. Here TP (True Positive) indicates the number of samples that are positive and returned by the model, and FN (False Negative) indicates the number of samples that are actually positive but computed by the model to be negative. The other two classes, FP (False Positive) and TN (True Negative) can be obtained by taking the complement of the normalized values of these factors. Table 2 shows a comparison of our model to several baseline, SOTA models, evaluated on the CHB-MIT dataset. Results obtained in this table are averaged values over the full sampled patient cohort. We see that our model achieves superior accuracy compared to the baselines; this is true as well in terms of the TPR and FPR factors. We consider the evaluation of TPR and FPR in order to mitigate the problem of our model overfitting on the majority label, given that the labels in our dataset is significantly unbalanced. In Table 4 we report on results on individual members of the sampled cohort; the column labeled chb_5_mix is a special instance constructed by mixing sampled data from the individual members of the cohort; results of model evaluation on this constructed cohort is meant to be indicative of the generalizability of our model. From Table 4 it is clear that the performance of our model is transferable across patients, indicating a high level of model generalizability on this cohort. Figure 3 show the six confusion matrices corresponding to the models under evaluation. We see that, in terms of absolute numbers, the FPs and TNs obtained from our model are an order of magnitude smaller than other baseline models, indicating the robustness of TSICNet under real use cases. Table 3 shows model evaluation and baseline comparison on the HUH dataset. For this dataset, we randomly sampled ten individuals (ten male and ten female), in order to show that our model generalizability systematically transfers across gender as well as samples. We see that again, for this dataset we are able to achieve full SOTA accuracies and TPR/FPR numbers (performance of our model is shown in the first column).

Table 2. Performance comparison of baseline models with TSICNet on the CHB-MIT dataset

-	TSICNet	ATCNet	EEGNet	EEGInception	EEGITNet	EEGConformer
Accuracy	**99.90%**	98.10%	98.08%	97.85%	97.32%	98.51%
TPR	**99.72%**	56.25%	70.31%	70.17%	32.10%	78.55%
FPR	0.09%	0.22%	0.81%	1.04%	**0.07%**	0.69%

Table 3. Performance comparison of baseline models with TSICNet on the HUH dataset

	Subject	TSICNet			ATCNet			EEGNet			EEGInception			EEGITNet			EEGConformer		
		ACC	TPR	FPR	ACC	TPR	FPR	ACC	TPR	FPR	ACC	TPR	FPR	ACC	TPR	FPR	ACC	TPR	FPR
Female	1	99.60%	99.07%	0.08%	98.20%	97.88%	1.60%	95.10%	92.43%	3.29%	89.80%	72.90%	0.06%	91.40%	87.12%	6.01%	99.10%	98.80%	0.72%
	5	99.95%	99.92%	0.06%	99.95%	99.95%	0.00%	99.90%	99.89%	0.00%	99.95%	99.92%	0.06%	79.80%	78.44%	0.00%	99.95%	99.92%	0.00%
	9	99.95%	99.42%	0.00%	99.95%	99.42%	0.00%	99.80%	98.28%	0.05%	96.45%	59.20%	0.00%	97.65%	72.99%	0.00%	99.85%	98.85%	0.05%
	14	99.50%	99.09%	0.10%	99.50%	99.29%	0.30%	99.00%	98.68%	0.69%	98.80%	97.87%	0.30%	88.25%	76.21%	0.00%	99.60%	99.60%	0.40%
	17	99.70%	90.62%	0.00%	99.85%	99.90%	0.15%	99.80%	98.43%	0.21%	99.65%	89.06%	0.00%	97.35%	17.19%	0.00%	99.65%	95.31%	0.21%
	19	99.90%	95.83%	0.00%	99.95%	99.92%	0.00%	99.95%	99.92%	0.05%	99.80%	91.67%	0.00%	99.80%	91.67%	0.00%	99.95%	99.92%	0.00%
	22	99.65%	99.81%	0.41%	99.40%	99.92%	0.82%	99.60%	99.81%	0.48%	98.90%	96.10%	0.07%	90.90%	99.63%	12.31%	99.65%	99.92%	0.48%
	31	99.95%	99.92%	0.00%	99.95%	99.92%	0.00%	99.80%	98.35%	0.06%	99.00%	98.01%	0.00%	97.30%	70.33%	0.00%	99.95%	99.45%	0.00%
	41	99.80%	99.78%	0.00%	99.25%	99.78%	5.37%	99.15%	99.39%	2.99%	95.50%	99.96%	44.78%	77.55%	75.04%	0.00%	99.80%	99.89%	1.00%
	44	99.20%	86.55%	0.00%	97.55%	58.82%	0.00%	94.40%	5.88%	0.00%	85.80%	99.65%	15.10%	94.05%	0.00%	0.00%	99.75%	99.90%	0.27%
	F_Avg	**99.72%**	**97.00%**	**0.06%**	99.36%	95.49%	0.82%	98.65%	89.13%	0.78%	96.37%	89.58%	6.03%	91.41%	66.86%	1.83%	**99.72%**	**99.17%**	0.31%
Male	4	99.95%	99.92%	0.07%	99.95%	99.92%	0.00%	99.90%	99.92%	0.14%	98.55%	94.56%	0.00%	92.20%	70.73%	0.00%	99.95%	99.61%	0.00%
	7	99.95%	99.78%	0.00%	99.85%	99.78%	0.13%	99.80%	99.56%	0.13%	98.60%	93.89%	0.00%	99.60%	98.25%	0.00%	99.90%	99.78%	0.06%
	15	99.75%	99.20%	0.12%	99.70%	99.47%	0.25%	99.35%	97.86%	0.31%	88.65%	100%	13.96%	95.35%	75.13%	0.00%	99.65%	98.40%	0.06%
	16	99.15%	94.76%	0.17%	98.15%	98.88%	1.96%	98.10%	92.13%	0.98%	98.80%	97.75%	1.04%	89.60%	22.10%	0.00%	98.90%	96.63%	0.75%
	20	99.35%	96.67%	0.18%	99.45%	98.67%	0.41%	98.40%	91.00%	0.29%	99.25%	95.33%	0.06%	97.00%	80.67%	0.12%	99.90%	98.00%	0.24%
	21	99.95%	99.92%	0.00%	99.95%	99.92%	0.00%	99.95%	99.92%	0.00%	99.95%	99.92%	0.05%	99.85%	93.02%	0.00%	99.95%	99.92%	0.00%
	38	99.85%	99.87%	0.21%	99.75%	99.92%	0.31%	99.35%	99.56%	0.58%	0.00%	94.45%	70.79%	0.00%	85.20%	22.11%	0.00%	99.30%	96.84%	0.12%
	39	99.40%	96.75%	0.06%	99.15%	95.56%	0.12%	96.4%	78.70%	0.00%	99.10%	94.67%	0.00%	86.20%	18.34%	0.00%	98.75%	94.38%	0.36%
	40	99.85%	99.35%	0.00%	99.10%	96.56%	0.13%	93.35%	71.83%	0.13%	86.20%	40.65%	0.00%	97.80%	91.61%	0.46%	99.45%	99.57%	0.59%
	52	99.90%	98.32%	0.00%	99.95%	99.92%	0.00%	99.95%	100%	99.40%	89.92%	0.00%	99.15%	85.71%	0.00%	99.95%	99.92%	0.00%
	M_Avg	**99.71%**	**98.46%**	**0.08%**	99.50%	**98.87%**	0.33%	98.46%	92.77%	0.20%	93.30%	87.76%	1.51%	94.20%	71.80%	0.06%	99.57%	98.30%	0.22%
	Avg	**99.72%**	97.73%	**0.07%**	99.43%	97.18%	0.58%	98.56%	90.95%	0.49%	94.84%	88.67%	3.77%	92.80%	69.33%	0.95%	99.66%	**98.74%**	0.27%

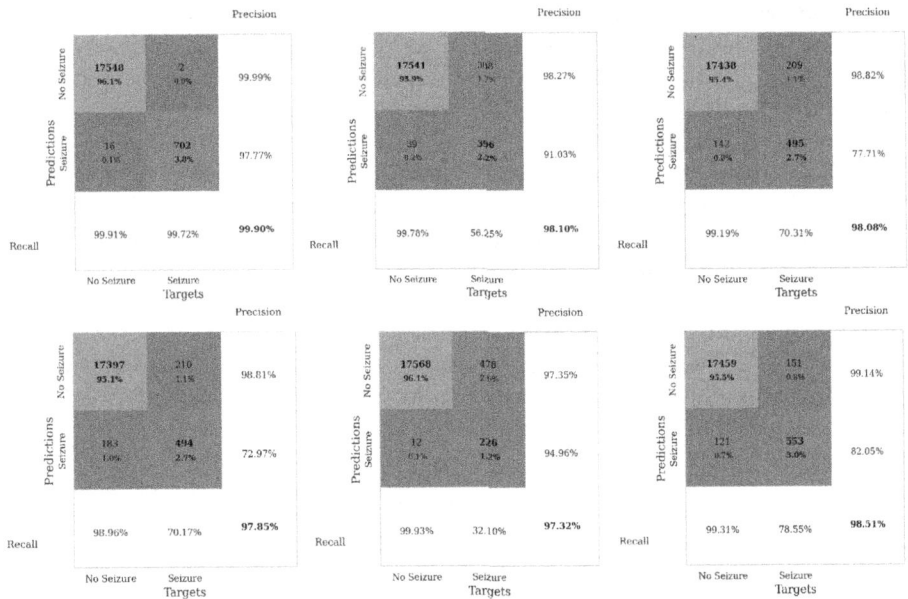

Fig. 3. The confusion matrix of 6 different models, from left to right and from top to bottom, is TSICNet, ATCNet, EEGNet, EEGInception, EEGITNet, and EEGConformer.

4.3 Comparison with Graph-Convolution-Based Models

In addition to conventional deep models for EEG analysis, we also compared our model design with three existing combination GCNN models; this was done in

Table 4. Data set results for a single subject and a mix of five subjects

-	chb_5_mix	chb01	chb02	chb03	chb05	chb08
Accuracy	99.90%	99.48%	99.63%	99.69%	99.72%	99.98%
TPR	99.72%	98.02%	97.55%	95.70%	96.37%	99.96%
FPR	0.09%	0.45%	0.30%	0.24%	0.17%	0.05%

order to verify the validity of our design principles (as mentioned in the model description section), in particular with respect to the inclusion and training of the GCNN. The results are reported in Table 5 and confirms the superiority of our proposed combined GCNN architecture. We note that some of the GCNN-based models examined under this study have structures similar to ours; e.g., the ASTGCNN [38] is also a fusion of graph convolution with the attention mechanism, while the STGCNN [10] learns relationships between both spatial and temporal points, similarly to the way we integrate this information via spatial and temporal convolutional layers in the TSICNet.

Table 5. Performance comparison of four models combined with GCNN

-	TSICNet	ASTGCNN	ASGCNN	STGCNN
Accuracy	**99.55%**	97.89%	96.77%	98.90%
TPR	**96.73%**	72.34%	68.55%	75.99%
FPR	0.34%	0.41%	0.65%	**0.18%**

4.4 Ablation Studies

We performed ablation studies to highlight the roles of the different modules in achieving our reported classification results. Overall ablation results, reported in Table 6, show the decrease in accuracy and TPR/FPR when we remove different submodules from our framework; this indicates the efficacy of our model design principles in translation to actual results. We note that a larger decrease in performance was observed when the GCNN was removed as compared to the case when the ATCNet submodule was excised; this indicates the importance of the connectome learning module (implemented in the adjacency matrix of the GCNN) in our model's results, and validates the hypothesis forming the major basis of our model design. Removal of additional submodules led to larger decrease in accuracy, which is to be expected.

Decoding Methods. Data extraction for model training was conducted using various sliding windows in the CHB-MIT Scalp dataset, with each window representing a time unit of 1 s. Each time unit comes with its own label (seizure

Table 6. TSICNet Ablation Studies

Removed	Accuracy	TPR	FPR
None	99.90%	99.72%	0.18%
ATCNet	99.55%	96.73%	0.34%
GCNN	99.06%	83.52%	0.31%
ATCNet+GCNN	98.08%	70.31%	0.81%
Conv+GCNN	98.10%	56.25%	0.22%

or no seizure). We ablate with respect to three types of sliding Windows: in the first case, each window includes 1 s of data, the step length is 1 (full length of the window), and the label of this second is fixed by this data. In the second method, each window contains 6 s of data and the step size is 1 s, while in the final case we add 4 s of data on the basis of the second window, and the step length is 1 s, that is, a total of 10 s of data to judge the 5 to 6 s of labels in the window. The schematic diagram of the three sliding windows is shown in Fig. 4.

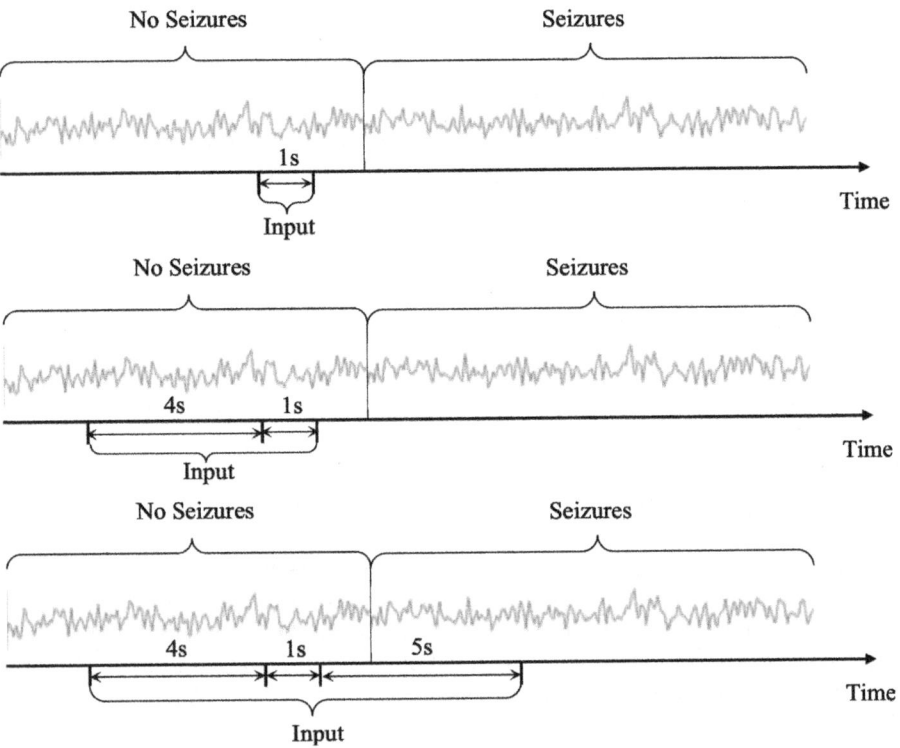

Fig. 4. Schematic comparison between our three sliding windows information extraction schemes.

From this set of ablation results we observe an expected increase in model accuracy with amount of training data, although the variance between these sets are not substantial. However, we do observe a sharp drop in the TPR when reducing the length of the learning window, which is indicative of increasing overfitting. The experimental results are shown in Table 7.

Table 7. Performance comparison of different sliding windows

-	10 s	6 s	1 s
Accuracy	**99.75%**	99.35%	99.17%
TPR	**99.71%**	96.16%	90.63%
FPR	**0.20%**	0.52%	0.48%

Feature Extraction and Learning from ATCNet and TSICNet. As our TSICNet design incorporates parts of the existing ATCNet, the question of how to integrate both these modules is of interest. We studied four different combination methodologies, the first two being combination in series, but in various different orders; the latter two involves combinations in parallel and applying either addition or concatenation to the resulting learned feature vectors for feature fusion. The experimental comparison results of the four combination methods are shown in Table 8. It can be seen that a series combination method with ATCNet at the end is the best combination method.

Table 8. Comparison of different combinations of ATCNet

-	ATCNet (back)	ATCNet (front)	Concatenation	Addition
Accuracy	**99.90%**	99.06%	99.64%	99.75%
TPR	**99.72%**	78.55%	96.78%	99.71%
FPR	**0.09%**	0.36%	0.24%	0.20%

5 Conclusion

Epilepsy is one of the most prevalent neurological diseases, affecting millions of people worldwide and imposing substantial social and economic burdens. In recent years, deep learning methods have become widely applied in numerous areas of scientific research, due, in part, to its ability to automatically extract the most relevant features and the availability of extensive, multimodal datasets for almost all areas. In this work, we have endeavoured to take advantage of this feature learning capability of deep learning and incorporate, as far as possible, important prior knowledge and information regarding epilepsy. Based on these considerations, we design TSICNet, a novel framework for epilepsy classification which is able to: first, capture the main indicators of an epileptic seizures, and second, extract the most relevant features according to these indicators. The main design principles involved combines key features from various elemental modules to maximize the accuracy of model classification; for this task, our model combines CSP capabilities with connectomic and multi-level attention in order to maximize the amount of relevant extracted information. Comparison with baseline models show SOTA performance in comparison with five current models, each based on various different learning structures. Comparison with GCNN-based deep models also substantiate our reasoning regarding the use of graph-based layers in our model. Finally, ablation studies show the efficacy of our design principles from various perspectives. We regard this work as a hint towards the possibility of robust and accurate epilepsy diagnosis models, by basing model design principles on knowledge and understanding of prior medical information and data.

Disclosure of Interests. There are no conflicts of interests.

References

1. Acharya, U.R., Hagiwara, Y., Adeli, H.: Automated seizure prediction. Epilepsy Behav. **88**, 251–261 (2018). https://doi.org/10.1016/j.yebeh.2018.09.030
2. Altaheri, H., Muhammad, G., Alsulaiman, M.: Physics-informed attention temporal convolutional network for EEG-based motor imagery classification. IEEE Trans. Industr. Inf. **19**(2), 2249–2258 (2022)
3. Bhattacharya, A., Baweja, T., Karri, S.: Epileptic seizure prediction using deep transformer model. In: Int. J. Neural Syst., p. 2150058. (2022). https://doi.org/10.1142/S0129065721500581 pMID: 34720065
4. Blume, W.T.: Diagnosis and management of epilepsy. CMAJ **168**(4), 441–448 (2003)
5. Bonilha, L., et al.: The brain connectome as a personalized biomarker of seizure outcomes after temporal lobectomy. Neurology **84**(18), 1846–1853 (2015). https://doi.org/10.1212/WNL.0000000000001548, https://www.neurology.org/doi/abs/10.1212/WNL.0000000000001548
6. Thara, D.K., PremaSudha, B.G., Xiong, F.: Epileptic seizure detection and prediction using stacked bidirectional long short term memory. Pattern Recogn. Lett. **128**, 529–535 (2019). https://doi.org/10.1016/j.patrec.2019.10.034, https://www.sciencedirect.com/science/article/pii/S0167865519303125
7. Fisher, R.S., et al.: ILAE official report: a practical clinical definition of epilepsy. Epilepsia **55**(4), 475–482 (2014). https://doi.org/10.1111/epi.12550
8. Goldberger, A.L., et al.: PhysioBank, PhysioToolkit, and PhysioNet: components of a new research resource for complex physiologic signals. Circulation [Online] **101**(23), e215–e220 (2000)
9. Guttag, J.: CHB-MIT scalp EEG database (version 1.0.0). PhysioNet (2010). https://doi.org/10.13026/C2K01R

10. Han, H., et al.: STGCN: a spatial-temporal aware graph learning method for POI recommendation. In: 2020 IEEE International Conference on Data Mining (ICDM), pp. 1052–1057. IEEE (2020)
11. Hussein, R., Palangi, H., Ward, R.K., Wang, Z.J.: Optimized deep neural network architecture for robust detection of epileptic seizures using EEG signals. Clin. Neurophysiol. **130**(1), 25–37 (2019). https://doi.org/10.1016/j.clinph.2018.10.010, https://www.sciencedirect.com/science/article/pii/S1388245718313464
12. Iasemidis, L., et al.: SEIZURES | a new look into epilepsy as a dynamical disorder : seizure prediction, resetting and control. In: Schwartzkroin, P.A. (ed.) Encyclopedia of Basic Epilepsy Research, pp. 1295–1302. Academic Press, Oxford (2009). https://doi.org/10.1016/B978-012373961-2.00267-8, https://www.sciencedirect.com/science/article/pii/B9780123739612002678
13. Johansen, A.R., Jin, J., Maszczyk, T., Dauwels, J., Cash, S., Westover, M.: Epileptiform spike detection via convolutional neural networks. In: Proceedings of the IEEE International Conference Acoustics Speech Signal Process, vol. 2016, pp. 754–758 (2016). https://doi.org/10.1109/ICASSP.2016.7471776
14. Kiral-Kornek, I., et al.: Epileptic seizure prediction using big data and deep learning: toward a mobile system. EBioMedicine **27**, 103–111 (2018). https://doi.org/10.1016/j.ebiom.2017.11.032, https://www.sciencedirect.com/science/article/pii/S235239641730470X
15. Lawhern, V.J., Solon, A.J., Waytowich, N.R., Gordon, S.M., Hung, C.P., Lance, B.J.: EEGNET: a compact convolutional neural network for EEG-based brain-computer interfaces. J. Neural Eng. **15**(5), 056013 (2018)
16. Li, X., et al.: Interpretable deep learning: interpretation, interpretability, trustworthiness, and beyond. Knowl. Inf. Syst. **64**(12), 3197–3234 (2022). https://doi.org/10.1007/s10115-022-01756-8
17. Maher, C., et al.: Deep learning distinguishes connectomes from focal epilepsy patients and controls: feasibility and clinical implications. Brain Commun. **5**(6), fcad294 (2023)
18. Mallick, S., Baths, V.: Novel deep learning framework for detection of epileptic seizures using EEG signals. Front. Comput. Neurosci. **18**, 1340251 (2024). https://doi.org/10.3389/fncom.2024.1340251, https://www.frontiersin.org/articles/10.3389/fncom.2024.1340251
19. Manole, A.M., et al.: State of the art and challenges in epilepsy–a narrative review. J. Pers. Med. **13**(4), 623 (2023). https://doi.org/10.3390/jpm13040623
20. Morgan, V.L., et al.: Presurgical temporal lobe epilepsy connectome fingerprint for seizure outcome prediction. Brain Commun. **4**(3), fcac128 (2022). https://doi.org/10.1093/braincomms/fcac128
21. Orhan, U., Hekim, M., Ozer, M.: EEG signals classification using the K-means clustering and a multilayer perceptron neural network model. Expert Syst. Appl. **38**(10), 13475–13481 (2011). https://doi.org/10.1016/j.eswa.2011.04.149, https://www.sciencedirect.com/science/article/pii/S0957417411006762
22. Ouichka, O., Echtioui, A., Hamam, H.: Deep learning models for predicting epileptic seizures using IEEG signals. Electronics **11**(4), 605 (2022). https://doi.org/10.3390/electronics11040605
23. Proix, T., Jirsa, V.K.: Using the connectome to predict epileptic seizure propagation in the human brain. BMC Neurosci. **16**(Suppl 1), P110 (2015). https://doi.org/10.1186/1471-2202-16-S1-P110
24. Roy, S., Kiral-Kornek, I., Harrer, S.: ChronoNet: a deep recurrent neural network for abnormal EEG identification. In: Riaño, D., Wilk, S., ten Teije, A. (eds.) AIME

2019. LNCS (LNAI), vol. 11526, pp. 47–56. Springer, Cham (2019). https://doi.org/10.1007/978-3-030-21642-9_8
25. Salami, A., Andreu-Perez, J., Gillmeister, H.: EEG-ITNet: an explainable inception temporal convolutional network for motor imagery classification. IEEE Access **10**, 36672–36685 (2022)
26. Santamaria-Vazquez, E., Martinez-Cagigal, V., Vaquerizo-Villar, F., Hornero, R.: EEG-inception: a novel deep convolutional neural network for assistive ERP-based brain-computer interfaces. IEEE Trans. Neural Syst. Rehabil. Eng. **28**(12), 2773–2782 (2020)
27. Scharfman, H.E.: The neurobiology of epilepsy. Curr. Neurol. Neurosci. Rep. **7**(4), 348–354 (2007). https://doi.org/10.1007/s11910-007-0053-z
28. Schirrmeister, R.T., et al.: Deep learning with convolutional neural networks for EEG decoding and visualization. Hum. Brain Mapp. **38**(11), 5391–5420 (2017). https://doi.org/10.1002/hbm.23730
29. Shoeibi, A., et al.: An overview of deep learning techniques for epileptic seizures detection and prediction based on neuroimaging modalities: methods, challenges, and future works. Comput. Biol. Med. **149**(C), 106053 (2022). https://doi.org/10.1016/j.compbiomed.2022.106053
30. Singh, A., Velagala, V.R., Kumar, T.: The application of deep learning to electroencephalograms, magnetic resonance imaging, and implants for the detection of epileptic seizures: a narrative review. Cureus **15**(7), e42460 (2023). https://doi.org/10.7759/cureus.42460
31. Song, Y., Zheng, Q., Liu, B., Gao, X.: EEG conformer: convolutional transformer for EEG decoding and visualization. IEEE Trans. Neural Syst. Rehabil. Eng. **31**, 710–719 (2022)
32. Stafstrom, C.E., Carmant, L.: Seizures and epilepsy: an overview for neuroscientists. Cold Spring Harb. Perspect. Med. **5**(6), a022426 (2015). https://doi.org/10.1101/cshperspect.a022426
33. Stevenson, N.J., Tapani, K., Lauronen, L., Vanhatalo, S.: A dataset of neonatal EEG recordings with seizure annotations. Sci. Data **6**, 190039 (2019). https://doi.org/10.1038/sdata.2019.39
34. Tzallas, A.T., Tsipouras, M.G., Fotiadis, D.I.: Automatic seizure detection based on time-frequency analysis and artificial neural networks. Comput. Intell. Neurosci. **2007**, 80510 (2007). https://doi.org/10.1155/2007/80510
35. Ullah, I., Hussain, M., ul Haq Qazi, E., Aboalsamh, H.: An automated system for epilepsy detection using EEG brain signals based on deep learning approach. Expert Systems with Applications **107**, 61–71 (2018). https://doi.org/10.1016/j.eswa.2018.04.021, https://www.sciencedirect.com/science/article/pii/S0957417418302513
36. West, J., Bozorgi, Z.D., Herron, J., Chizeck, H.J., Chambers, J.D., Li, L.: Machine learning seizure prediction: one problematic but accepted practice. J. Neural Eng. **20**(1), 016008 (2023). https://doi.org/10.1088/1741-2552/acae09
37. Zhang, C., Li, Q., Song, D.: Aspect-based sentiment classification with aspect-specific graph convolutional networks. arXiv preprint (2019). arXiv:1909.03477
38. Zhu, J., Wang, Q., Tao, C., Deng, H., Zhao, L., Li, H.: AST-GCN: attribute-augmented spatiotemporal graph convolutional network for traffic forecasting. IEEE Access **9**, 35973–35983 (2021)

Brain-Inspired AI

A Brain-Inspired Distributed Long-Term Memory Guided Online Continual Learning Method

Yuyang Han[1], Xiuxing Li[2], Qixin Wang[1], Tianyuan Jia[1], and Xia Wu[2](✉)

[1] School of Artificial Intelligence, Beijing Normal University, No. 19 Xinjiekouwai Street Haidian District, Beijing, People's Republic of China
[2] School of Computer Science and Technology, Beijing Institute of Technology, No. 5, Zhongguancun South Street, Haidian District, Beijing, People's Republic of China
wuxia@bit.edu.cn

Abstract. Online Continual Learning (CL) is a challenging scenario that focuses on enabling models to incrementally learn new knowledge from sequential, non-i.i.d. data with only a single pass through the data. Replay-based methods have shown great potential in this scenario, but they still suffer from catastrophic forgetting. In contrast, the human brain exhibits great advantages in online CL scenario, with memories swiftly encoded as short-term memories in the hippocampus and subsequently consolidated into distributed long-term memories within the neocortex. Inspired by this neural mechanism, we introduce the **Distributed Long-Term Memory Guided CL(DLTMG-CL)** method, which guides the network to retain old knowledge while learning new information by consolidating and preserving distributed long-term memories. This approach facilitates efficient learning in online CL scenario. Our algorithm has achieved state-of-the-art (SOTA) performance among various CL methods.

Keywords: Continual Learning · Catastrophic Forgetting · Online Continual Learning · Brain-inspired Method

1 Introduction

Online continual learning (CL) aims to train a model capable of acquiring new knowledge incrementally while retaining previously learned knowledge [21]. Different from offline CL that permits multiple accesses to the data of the same task during learning, online CL only allows a single access. In the context of online CL, tasks with distinct data distributions are presented sequentially, necessitating the model to perform well on both current and past tasks with only one shot for each data (i.e., one epoch). However, connectionism-based neural network architectures, such as CNNs and Transformers, inevitably undergo parameter change as the training data distribution shifts, thus impacting their performance on previous tasks. This phenomenon is known as *catastrophic orgetting* [15].

The past decade has seen the rapid development of CL in many methods, and most of them can be used both in offline CL and online CL scenarios. There are four main directions to address *catastrophic forgetting* in CL. **Regularization-based** methods [8,12,23] aim to determine task-specific weights and maintain these weights as invariant as possible when learning new tasks. **Parameter isolation** methods [18] allocate separate networks or neurons to each task to prevent interference between tasks. **Replay-based** methods [3,5,16,17] store samples from previous tasks and combine them with current data for training. **LLM-based** methods [13] leverage advancements in Large Language Models (LLMs) and MultiModal Large Language Models (MM-LLMs) to enhance model generalization by incorporating text information encoded by pre-trained LLMs as additional supervisory signals in image classification tasks. In these directions, replay-based approaches are currently regarded as the most effective ones, but they still perform poorly in online CL scenario.

In contrast, the human brain possesses an inherent advantage in real world which is very similar to online CL scenario. People don't need to repeat the same visual stimuli again and again to remember. Existing research in neuroscience has explored and found that the neocortex exhibits a slow learning rate, employing overlapping **distributed representations** to extraction the inherent statistical pattern in the environment as long-term memory. Meanwhile, the hippocampus can learn non-overlapping representation rapidly, which is called short-term memory [14]. Many researchers, inspired by this phenomenon, have designed a series of brain-inspired CL algorithms [3,19]. However, their modeling approach does not fully consider the overlapping distributed storage mechanism of long-term memory encoding in the neocortex, as well as the interaction between the hippocampus and the neocortex, leaving room for improvement.

Inspired by the neural mechanism in human brain, we proposed a novel brain-inspired replay-based online CL scenario method named **Distributed Long-Term Memory Guided Continual Learning(DLTMG-CL)**. In our study, we aim to develop a model that exhibits robust performance on both new and previous tasks. To achieve this, we designed a hippocampus network that incorporates replay of old samples while learning new information. To better balance the integration of new and old knowledge, we employ a series of neocortex networks to simulate distributed long-term memory storage. Periodically, we selected the neocortex network that demonstrates optimal preservation of long-term memory to guide the learning process of the hippocampus network. Specifically, this is accomplished by constraining the hippocampus network to have a similar predictive distribution on old samples as that of the best performing neocortex network. To simulate the interaction between the hippocampus and neocortex, the hippocampus network parameters are periodically utilized to update the neocortex networks, enabling the integration of new and old memories. By simulating distributed long-term memory storage and the interactive dynamics between hippocampus and neocortex information, our approach achieves state-of-the-art (SOTA) performance in online CL scenarios.

Our main contributions can be summarized as follows:

- We proposed a novel replay-based online CL scenario method named **Distributed Long-Term Memory Guided Continual Learning(DLTMG-CL)** inspired by the distributed long-term memory in neocortex and the interaction between neocortex and hippocampus in human brain. We designed a hippocampus network to learn short-term memory while maintaining a constraint with the neocortex network, which learns long-term memory.
- We simulate the distributed storage of long-term memories by periodically generating neocortex networks, in which memories from distinct phases are stored across different networks. Additionally, we enable interaction between the hippocampus and neocortex by dynamically updating the parameters within the neocortex networks. Compared to existing approaches, our method achieves state-of-the-art (SOTA) performance.

2 Related Work

In this section, we first summarize some mainstream CL methods in CL scenario. Secondly, we focus on the approaches in online CL scenario.

2.1 Mainstream CL Methods

As mentioned in Sect. 1, currently mainstream CL methods can be broadly categorized into four main types.

Firstly, replay-based methods reinforce memory consolidation by revisiting a subset of past data stored in a buffer. Prominent examples include LiDER [4], ER [17], A-GEM [6], iCaRL [16], DER [5] and CLSER [3]. While acknowledged as the most effective strategy currently available, replay-based methods face the issue of high storage resource consumption and sometimes also pose privacy concerns.

Secondly, regularization-based methods try to find important parameter associated to old tasks and penalize them change when adapting to new ones, as seen in techniques such as EWC [8] and SI [23]. For instance, EWC employs the diagonal Fisher information matrix, while SI relies on path integrals of gradient vector fields. While these methods operate within a smaller parameter space, their efficacy is somewhat diminished in scenarios involving long sequences of tasks due to multiple tasks sharing the same network.

Thirdly, parameter isolation methods allocate independent parameters to each task, as exemplified by approaches like HAT [18]. These techniques is a kind of modular approaches. These methods offer the advantage of not requiring access to previous data. However, due to the lack of appropriate selection of independent parameters, they are often more suited to task-IL scenarios. Additionally, they are frequently criticized for the excessively large number of parameters, which can lead to computational inefficiencies.

Lastly, prompt-based methods leverage LLMs to generate additional supervision or replay information. These methods typically incorporate image category labels or more intricate textual descriptions of images like [13]. While these methods demonstrate significant potential for generalization and the ability to mitigate catastrophic forgetting, they may encounter issues related to pre-training leakage.

Recent research endeavors have begun integrating elements from these four streams to further augment model performance.

2.2 Online CL Methods

Recently, a significant number of researchers have focused on online CL scenario due to its greater challenges and better adaptability to real-world applications. Most of these online CL scenario methods use sample replay as a CL strategy. Some previous works use gradients information to select good samples for replay in online CL such as GSS [1], others use inference information to choose samples like MIR [2] and OCS [22]. CBA [20] employs meta-learning techniques to introduce a non-linear mapping, aligning the label space of new tasks with that of old tasks and achieve online CL in cases where labels are meaningless. In addition to methods specifically designed and applied to online CL scenario, mainstream CL methods can also be utilized within online CL scenario. However, both general and specialized approaches demonstrate relatively weak performance in online CL scenario. In conclusion, as a more challenging paradigm of CL, online CL scenario necessitates further in-depth exploration.

3 Methodology

In this section, we first introduce the problem setting in online CL scenario and then give a detailed description of our method **DLTMG-CL**.

3.1 Problem Setting

For a sequence of tasks $\mathcal{T} = \{\mathcal{T}_1, \mathcal{T}_2, ..., \mathcal{T}_N\}$, each task $\mathcal{T}_t$ contains a dataset $\mathcal{D}_t = \{(x_t, y_t) \mid x_t \in X_t, y_t \in Y_t\}$. X_t represents the data in $\mathcal{T}_t$ and Y_t represents the labels of X_t that belong to $\mathcal{C}_t$. In CL setting, the classes in different task $\mathcal{C}_t$ are disjoint, as $\mathcal{C}_i \cap \mathcal{C}_j = \emptyset, i, j = 1, ..., t, i \neq j$.

CL algorithms aim to find a model $f_\theta : \mathcal{X} \to \mathbb{R}^d, d = |\mathcal{C}_t|, t = 1, 2, ..., N$ that performs well on both current task and old tasks after learning each task sequentially. However, the catastrophic forgetting happens. For example, after learning $\mathcal{T}_t$, we get f_{θ_t}, and when $\mathcal{T}_{t+1}$ comes, as the data distribution shifts, some of the weights in θ_t will be overwritten. Thus the performance in $\mathcal{T}_t$ will decrease after learning $\mathcal{T}_{t+1}$.

In online CL method, although sample storage is permitted, the model can only access (x_t, y_t) once during task t, which poses a greater challenge to our algorithm. Replay-based method solves this problem by storing some of the samples in each task into a data buffer $\mathcal{M}$ and combine them with next task's data together when training next task. The loss function can be described as:

$$\ell_{RE} = \alpha \ell(f_\theta(X_t), Y_t) + \beta \ell(f_\theta(X_\mathcal{M}), Y_\mathcal{M}) \tag{1}$$

where α and β are weight coefficients. The loss function ℓ is typically the cross-entropy loss. However, this approach remains ineffective in online CL scenario.

To address this, we propose a brain-inspired replay-based method for online CL scenario **DLTMG-CL**, which will be detailed in the following section.

3.2 Distributed Long-Term Memory Guided Continual Learning: Framework Formulation

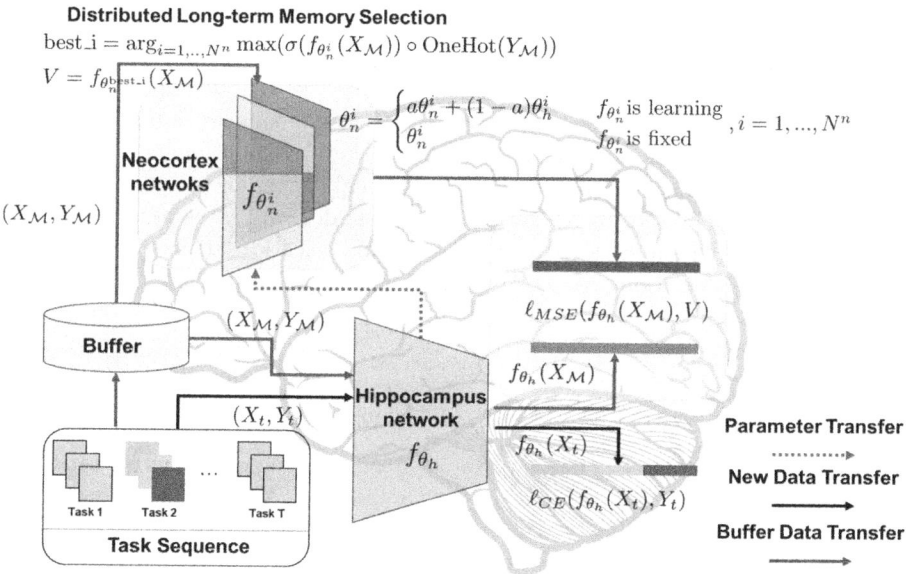

Fig. 1. Distributed Long-Term Memory Guided Continual Learning(DLTMG-CL) Framework. DLTMG-CL has two main components, a hippocampus network and several neocortex networks. The hippocampus network is trained using data derived from both current task and buffer. During training, it is crucial to select effective long-term memory features by the neocortex network acting as a guidance or a constraint. Once completion of training, the hippocampus network updates the parameters of the neocortex network. To prevent memory forgetting induced by these updates, the neocortex network will be fixed at regular training intervals as long-term memory. This fixed version represents the consolidation of memories acquired during that phase. Ultimately, this process results in distributed long-term memory in different neocortex networks.

Initially, we design the hippocampus network f_{θ_h}. For task $\mathcal{T}_t$, $\mathcal{D}_t$ sampled from a non-i.i.d distribution comes combined with the samples in data buffer $\mathcal{M}$ as

$(X_t, Y_t) = \{(X_{\mathcal{D}_t}, Y_{\mathcal{D}_t}), (X_\mathcal{M}, Y_\mathcal{M})\}$. The first goal of f_{θ_h} is to predict the right label of those samples as:

$$\ell = \ell_{CE}(f_{\theta_h}(X_t), Y_t) = \ell_{CE}(f_{\theta_h}(X_{\mathcal{D}_t}), Y_{\mathcal{D}_t}) + \ell_{CE}(f_{\theta_h}(X_\mathcal{M}), Y_\mathcal{M}) \qquad (2)$$

where ℓ_{CE} represents the cross-entropy loss function.

To enhance the hippocampus network's capacity to recognize old memories, we use constraints within the predictive space of old memories between the hippocampus and neocortex networks $f_{\theta_n^i}$. For the neocortex networks are specifically designed to consolidate old memories as long-term memories, they possess superior recognition capabilities for old memories. Concurrently, the neocortex networks must continuously update based on the hippocampus network to adapt to new tasks effectively. In order to optimize old memory recognition, it is important to select a neocortex network that yields the best performance in this regard as a guidance. To achieve this, the current parameters of the neocortex network are periodically fixed and used as a consolidation of long-term memories, with a saving frequency of r_d. Subsequently, a new neocortex network is initialized with the parameters of the previously fixed network to undertake subsequent tasks. This process facilitates the distributed storage of long-term memories. The details are shown in Fig. 2. When constraining the predictive space of old memories between the hippocampus and neocortex networks, we propose a strategy to choose the distributed long-term memory that demonstrates optimal efficacy in recognizing old memories as a guidance. This selection strategy is what we term as 'Distributed Long-Term Memory Guided'. The specific mathematical formulation is expressed as follows:

$$\text{best\_i} = \arg_{i=1,...,N^n} \max(\sigma(f_{\theta_n^i}(X_\mathcal{M})) \circ \text{OneHot}(Y_\mathcal{M})) \qquad (3)$$

$$V = f_{\theta_n^{\text{best\_i}}}(X_\mathcal{M}) \qquad (4)$$

where best_i represents the index of the best-performing old memory neocortex network and V represents the feature extracted from this network. $\sigma(\cdot)$ is the softmax function, N^n indicates the total number of neocortex networks.

After getting V, the constraint between neocortex network and hippocampus network can be expressed as according to [3]:

$$\ell = \ell_{CE}(f_{\theta_h}(X_t), Y_t) + \beta \ell_{MSE}(f_{\theta_h}(X_\mathcal{M}), V) \qquad (5)$$

where β is a weight coefficient and ℓ_{MSE} represents the mean squared error (MSE) loss function.

To simulate the interaction between the hippocampus and the neocortex, facilitating the generation of distributed long-term memories for new knowledge within the neocortex, we design a mechanism that the hippocampus parameters update the current active parameters of the neocortex network. Instead of updating the parameters of neocortex network every time like [3], the interaction between hippocampus and neocortex is defined as:

$$\theta_n^i = \begin{cases} a\theta_n^i + (1-a)\theta_h^i & f_{\theta_n^i} \text{ is learning} \\ \theta_n^i & f_{\theta_n^i} \text{ is fixed} \end{cases}, i = 1, ..., N^n \qquad (6)$$

The overall details of our method are outlined in pseudocode 1.

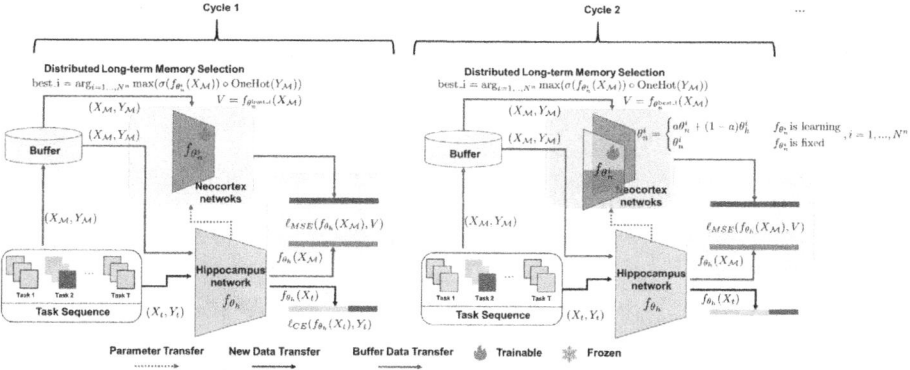

Fig. 2. This figure illustrates the process of updating the distributed long-term memory networks. Specifically, during training, the neocortex model undergoes parameter updates from the hippocampus model to acquire new knowledge. After each cycle, the neocortex model is frozen to preserve all knowledge learned during that period as long-term memory. A new neocortex model is then created, initialized with the parameters of the previous model, to learn new knowledge in the subsequent phase.

4 Experiment

4.1 Experimental Settings

Experimental Datasets. We choose many common CL datasets to evaluate our method.

- Split-MNIST [11]. Split-MNIST has 5 tasks, each task contains 2 classes.
- Split-CIFAR10 [9]. Split-CIFAR10 has 5 tasks, each task contains 2 classes.
- Split-CIFAR100 [9]. There are 10 tasks for each task contains 10 classes.
- Split-TinyImageNet [10]. There are 10 tasks for each task contains 20 classes.

Evaluation Metrics

- **Average Accuracy(AA)**: $AA_k = \frac{1}{k}\sum_{j=1}^{k} a_{k,j}$ which $a_{k,j} \in [0,1]$ denote the classification accuracy evaluated on the test set of the j-th after learning task k.
- **Forgetting Measure(FM)**: the average of the difference between its maximum performance obtained in the past and its current performance. $FM_k = \frac{1}{k-1}\sum_{j=1}^{k-1} f_{j,k}$ and $f_{j,k} = max_{i \in [1,...,k-1]}(a_{i,j} - a_{k,j}), \forall j < k$.

Implementation Details. We use MLP and ResNet [7] as our backbone for both neocortex network and hippocampus network. We adopted a two-layer fully connected MLP with 100 ReLU units per hidden layer for Split-MNIST and ResNet-18 for the other datasets. All backbones are optimized using SGD optimizer and the learning rate is 0.03. For all those datasets, we set 1 epoch for each task. For Split-MNIST and Split-CIFAR10, we set $r_d = 50$ and for others we set $r_d = 100$.

Algorithm 1. DLTMG-CL Framework

Input: Sequence of datasets: $D = \{D_1, \ldots, D_N\}$. weight coefficient β, update rates r, distribute step r_d, decay parameters α, data buffer $\mathcal{M}$
Initialization: $\theta_p = \theta_n$, global step $g = 0$, $\mathcal{M} \leftarrow \emptyset$, number of noecortex networks $N^n = 1$

1: **for** $t = 1 \ldots N$ **do**
2:    **for** $batch = 1, \ldots, N_t$ **do**
3:       $g = g + 1$
4:       **if** $g//r_d > N^n$ **then**
5:          fixed the $f_{\theta_n^{N^n}}$ and create a new neocortex network initialized by $f_{\theta_n^{N^n}}$
6:          $N^n = N^n + 1$
7:       **end if**
8:       **if** $\mathcal{M}$ is not empty **then**
9:          $(X_t, Y_t) = \{(X_{\mathcal{D}_t}, Y_{\mathcal{D}_t}), (X_\mathcal{M}, Y_\mathcal{M})\}$
10:         Find the long-term memory V that demonstrates the highest prediction accuracy on the data within the buffer as follows:

$$\text{best\_i} = \arg_{i=1,\ldots,N^n} \max(\sigma(f_{\theta_n^i}(X_\mathcal{M})) \circ \text{OneHot}(Y_\mathcal{M}))$$

$$V = f_{\theta_n^{\text{best\_i}}}(X_\mathcal{M})$$

11:         Update the hippocampus network using :

$$\ell = \ell_{CE}(f_{\theta_h}(X_t), Y_t) + \beta \ell_{MSE}(f_{\theta_h}(X_\mathcal{M}), V)$$

12:       **else**
13:          $(X_t, Y_t) = \{(X_{\mathcal{D}_t}, Y_{\mathcal{D}_t})\}$
14:         Update the hippocampus network using :

$$\ell = \ell_{CE}(f_{\theta_h}(X_{\mathcal{D}_t}), Y_{\mathcal{D}_t})$$

15:       **end if**
16:       update $= \mathcal{U}(0, 1)$
17:       Update neocortex parameters from 0 to N^n if update $< r$

$$\theta_n^i = \begin{cases} a\theta_n^i + (1-a)\theta_h^i & f_{\theta_n^i} \text{ is learning} \\ \theta_n^i & f_{\theta_n^i} \text{ is fixed} \end{cases}, i = 1, \ldots, N^n$$

18:       $\mathcal{M} \leftarrow \text{Reservoir}(\mathcal{M}, \mathcal{D}_t)$
19:    **end for**
20: **end for**

4.2 Results in Online Class Incremental Learning(Class-IL) Setting

We first compare our methods with other replay-based online CL methods. Table 1 provides the comparison with some replay-based online CL methods in class incremental learning(Class-IL) setting. Class-IL setting is a CL setting that task-id is not provided during inference phase. DLTMG-CL shows the highest performance for most of the datasets. In simple datasets such as Split-MNIST and Split-CIFAR10, our method has great advantages especially both in small buffer size and large buffer size. In particular, our performance is higher than CLSER in most cases for our methods is inspired by this work. For instance, when employing a small buffer size of 500, DLTMG-CL exhibits performance improvements of 19.61%, 16.53%, and 13.80% over the baseline methods ER, DERPP, and one of the current state-of-the-art methods CLSER in the Split-CIFAR10 dataset.

We further examine and analyze the performance of DLTMG-CL using the FM metric. The FM metric quantifies the average difference between the highest accuracy achieved for each task and the accuracy at the end of the learning process for all tasks, providing an indirect measure of catastrophic forgetting. Our experimental results indicate that with smaller buffer size, DLTMG-CL maintains a high level of AA and exhibits high FM performance. This suggests that while our method can achieve high accuracy, it may experience catastrophic forgetting for some tasks. However, with larger buffer size, our method consistently shows low FM values, indicating that it not only maintains high AA but also effectively mitigates catastrophic forgetting.

Table 1. Comparison with other replay-based works in online class-IL scenario

Buffer	Method	Split-MNIST AA↑	FM↓	Split-CIFAR10 AA↑	FM↓	Split-CIFAR100 AA↑	FM↓	Split-TinyImageNet AA↑	FM↓
200	ER [17]	81.46	19.98	35.21	50.28	9.87	**46.10**	6.57	**39.67**
	DERPP [5]	84.37	17.75	40.17	**41.84**	8.94	51.01	**7.15**	42.81
	CLSER [3]	87.97	7.81	42.01	42.33	9.68	51.82	6.23	46.7
	DLTMG-CL(ours)	**88.86**	**4.86**	**42.43**	51.91	**11.24**	52.49	6.87	46.87
500	ER [17]	86.64	15.33	42.32	40.80	13.98	43.08	7.78	**41.57**
	DERPP [5]	90.67	9.36	43.44	40.63	15.07	45.66	**8.66**	42.54
	CLSER [3]	92.05	8.02	44.48	**36.83**	15.85	**43.15**	6.9	44.02
	DLTMG-CL(ours)	**92.26**	**5.65**	**50.62**	44.46	**16.13**	46.26	8.42	42.86
5120	ER [17]	92.83	7.25	54.87	34.64	22.73	33.92	17.02	31.02
	DERPP [5]	95.7	3.78	58.51	**26.56**	17.61	44.53	12.31	39.93
	CLSER [3]	95.73	3.28	52.34	34.4	23.18	34.35	17.28	31.79
	DLTMG-CL(ours)	**96.22**	**0.99**	**58.55**	32.82	**23.96**	**33.78**	**17.73**	**30.82**

Results of other methods referenced from [3,20].

In addition, we investigate the extent of forgetting for each task across various datasets. Our analysis, depicted in Table 2, reveals that our method outperforms the baseline method ER across most tasks in the task sequence, particularly demonstrating higher accuracy in earlier tasks. This suggests that our approach maintains good memory retention for more distant knowledge, reflecting the efficiency of our distributed long-term memory. Furthermore, the results demonstrate that our method not only performs well on short task sequences but also maintains a high level of performance on long task sequences.

Table 2. Accuracy for each task after learning the final task of ER and DLTMG-CL.

Method	$a_{1,5}$	$a_{2,5}$	$a_{3,5}$	$a_{4,5}$	$a_{5,5}$	ACC
ER	18.2	**19.6**	18.7	**56.3**	77.45	38.05
DLTMG-CL	**61.65**	5.65	**37.1**	34.4	**79.6**	**43.6**

Method	$a_{1,10}$	$a_{2,10}$	$a_{3,10}$	$a_{4,10}$	$a_{5,10}$	$a_{6,10}$	$a_{7,10}$	$a_{8,10}$	$a_{9,10}$	$a_{10,10}$	ACC
ER	14.2	13.2	18.3	**12.6**	**25.1**	**19.6**	20.6	12.9	11.7	60.9	20.91
DLTMG-CL	**20.4**	**22.4**	**28.3**	12.4	15.4	13.4	**22.6**	**18.3**	**17.6**	**68.8**	**23.96**

Method	$a_{1,10}$	$a_{2,10}$	$a_{3,10}$	$a_{4,10}$	$a_{5,10}$	$a_{6,10}$	$a_{7,10}$	$a_{8,10}$	$a_{9,10}$	$a_{10,10}$	ACC
ER	13.3	7.3	12.8	**14.5**	13.4	9.1	**9.0**	**10.6**	**7.0**	39.9	13.69
DLTMG-CL	12.9	**13.8**	**15.9**	14.3	**14.9**	**13.8**	7.1	8.7	5.5	**52.3**	**15.92**

Note: Results of other methods referenced from [3, 20].

We also investigate the impact of different distributed storage intervals r_d on the results. The distributed storage interval determines the number of steps after which the current memory is updated and consolidated into distributed long-term memory. We test intervals of 10, 50, 100, 150, and 200 on Split-CIFAR10. The results indicate that performance is optimal when consolidation occurs at approximately one-third of the total batch size for each task. Detailed results are presented in the Fig. 3. In Fig. 3, DLTMG-CL-X means DLTMG-CL methods with the $r_d = X$.

4.3 Ablation Study

To evaluate the effectiveness of our proposed distributed long-term memory storage and the interaction mechanism between the hippocampus and neocortex, we conduct a series of ablation experiments. We perform these experiments on the Split-CIFAR100 dataset, comparing our method against the baseline ER method with buffer sizes of 200, 500, and 5120. The results show that our method consistently outperforms the baseline, especially with smaller buffer sizes. Specifically, the accuracy improvements for buffer size 200 are 0.12%, 1.12%, 1.37%, and for buffer size 500 are 1.78%, 0.28%, 2.15%. However, for a buffer size of 5120, improvements are observed only when both mechanisms are applied simultaneously. Additionally, we find that the distributed long-term memory storage mechanism contributed more significantly to the performance gains compared to the interaction mechanism, further validating the effectiveness of our proposed DLTMG-CL. Detailed results are presented in Table 3.

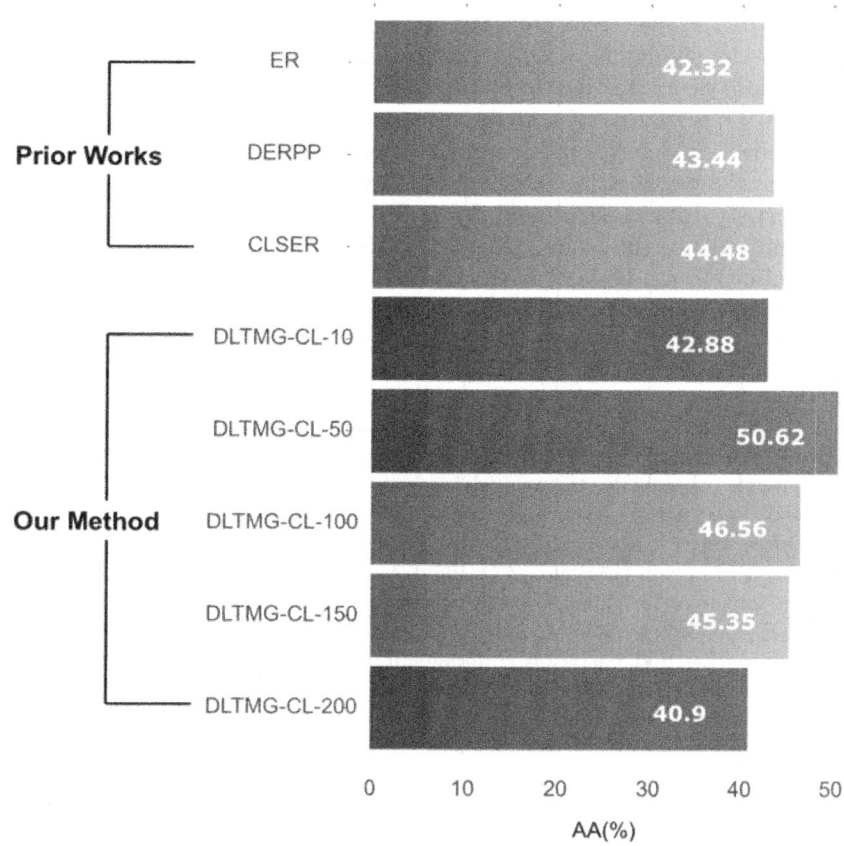

Fig. 3. The impact of different distributed storage intervals r_d in Split-CIFAR10 with 500 buffer size.

Table 3. The results of ablation study.

Method	Details	Split-CIFAR100					
		$\mathcal{M}=200$		$\mathcal{M}=500$		$\mathcal{M}=5120$	
		AA↑	FM↓	AA↑	FM↓	AA↑	FM↓
ER	-	9.87	**46.10**	13.98	**43.08**	22.73	33.92
DLTMG-CL (ours)	no distributed memory	9.99	52.13	15.76	45.20	20.08	44.26
DLTMG-CL (ours)	no interaction	10.99	54.31	14.26	46.47	23.09	39.07
DLTMG-CL (ours)	full	**11.24**	52.49	**16.13**	46.26	**23.96**	**33.78**

5 Conclusion

In this work, we propose a novel brain-inspired replay-based online CL methods called DLTMG-CL. Our approach simulates the distributed long-term memory storage in the neocortex and the interaction between the neocortex and hippocampus by employing a hippocampus network to learn short-term memories and various neocortex networks to generate and store distributed long-term memories. We design a novel distributed long-term memory selection algorithm to choose the best long-term memory to maintain a constraint between hippocampus and neocortex, and periodically update neocortex network to make sure memories can be consolidated. We evaluate our method in online Class-IL scenario and most of the results show SOTA performance compared with other classical online CL methods. Our work provides inspiration for developing novel CL algorithms based on brain mechanisms.

References

1. Aljundi, R., et al.: Online continual learning with maximal interfered retrieval. In: Wallach, H., Larochelle, H., Beygelzimer, A., d' Alché-Buc, F., Fox, E., Garnett, R. (eds.) Advances in Neural Information Processing Systems, vol. 32. Curran Associates, Inc. (2019)
2. Aljundi, R., Lin, M., Goujaud, B., Bengio, Y.: Gradient Based Sample Selection for Online Continual Learning. Curran Associates Inc., Red Hook, NY, USA (2019)
3. Arani, E., Sarfraz, F., Zonooz, B.: Learning fast, learning slow: a general continual learning method based on complementary learning system. In: International Conference on Learning Representations (2022)
4. Bonicelli, L., Boschini, M., Porrello, A., Spampinato, C., Calderara, S.: On the effectiveness of lipschitz-driven rehearsal in continual learning. In: Conference on Neural Information Processing Systems (2022)
5. Buzzega, P., Boschini, M., Porrello, A., Calderara, S.: Rethinking experience replay: a bag of tricks for continual learning. In: International Conference on Pattern Recognition (2020)
6. Chaudhry, A., Ranzato, M., Rohrbach, M., Elhoseiny, M.: Efficient lifelong learning with a-gem. In: International Conference on Learning Representations (2019)
7. He, K., Zhang, X., Ren, S., Sun, J.: Deep residual learning for image recognition. In: 2016 IEEE Conference on Computer Vision and Pattern Recognition (CVPR), pp. 770–778 (2015). https://api.semanticscholar.org/CorpusID:206594692
8. Kirkpatrick, J., et al.: Overcoming catastrophic forgetting in neural networks. Proc. Natl. Acad. Sci. **114**(13), 3521–3526 (2017)
9. Krizhevsky, A.: Learning multiple layers of features from tiny images (2009). https://api.semanticscholar.org/CorpusID:18268744
10. Le, Y., Yang, X.S.: Tiny imagenet visual recognition challenge (2015). https://api.semanticscholar.org/CorpusID:16664790
11. Lecun, Y., Bottou, L., Bengio, Y., Haffner, P.: Gradient-based learning applied to document recognition. Proc. IEEE **86**(11), 2278–2324 (1998). https://doi.org/10.1109/5.726791
12. Li, Z., Hoiem, D.: Learning without forgetting. In: European Conference on Computer Vision, pp. 614–629 (2016)
13. Ni, B., et al.: Enhancing visual continual learning with language-guided supervision. ArXiv **abs/2403.16124** (2024). https://api.semanticscholar.org/CorpusID:268681357

14. O'Reilly, R.C., Norman, K.A.: Hippocampal and neocortical contributions to memory: advances in the complementary learning systems framework. Trends Cogn. Sci. **6**(12), 505–510 (2002). https://doi.org/10.1016/S1364-6613(02)02005-3, https://www.sciencedirect.com/science/article/pii/S1364661302020053
15. Parisi, G.I., Kemker, R., Part, J.L., Kanan, C., Wermter, S.: Continual lifelong learning with neural networks: a review. Neural Netw. **113**, 54–71 (2019)
16. Rebuffi, S.A., Kolesnikov, A., Sperl, G., Lampert, C.H.: ICARL: incremental classifier and representation learning. In: The IEEE/CVF Computer Vision and Pattern Recognition Conference (2017)
17. Rolnick, D., Ahuja, A., Schwarz, J., Lillicrap, T.P., Wayne, G.: Experience replay for continual learning. In: Conference on Neural Information Processing Systems (2019)
18. Serra, J., Suris, D., Miron, M., Karatzoglou, A.: Overcoming catastrophic forgetting with hard attention to the task. In: International Conference on Machine Learning (2018)
19. van de Ven, G.M., Siegelmann, H.T., Tolias, A.S.: Brain-inspired replay for continual learning with artificial neural networks. Nat. Commun. (2020)
20. Wang, Q., Wang, R., Wu, Y., Jia, X., Meng, D.: Cba: improving online continual learning via continual bias adaptor. In: 2023 IEEE/CVF International Conference on Computer Vision (ICCV), pp. 19036–19046 (2023). https://doi.org/10.1109/ICCV51070.2023.01749
21. Wang, Z., Liu, L., Duan, Y., Tao, D.: Continual learning through retrieval and imagination. In: AAAI Conference on Artificial Intelligence (2022)
22. Yoon, J., Madaan, D., Yang, E., Hwang, S.J.: Online coreset selection for rehearsal-based continual learning. ArXiv **abs/2106.01085** (2021). https://api.semanticscholar.org/CorpusID:235293695
23. Zenke, F., Poole, B., Ganguli, S.: Continual learning through synaptic intelligenc. In: International Conference on Machine Learning (2017)

Memory Sequence Length of Data Sampling Impacts the Adaptation of Meta-reinforcement Learning Agents

Menglong Zhang, Fuyuan Qian, and Quanying Liu(✉)

Department of Biomedical Engineering, Southern University of Science and Technology, Shenzhen 518055, China
liuqy@sustech.edu.cn

Abstract. Fast adaptation to new tasks is extremely important for embodied agents in the real world. Meta-reinforcement learning (meta-RL) has emerged as an effective method to enable fast adaptation in unknown environments. Compared to on-policy meta-RL algorithms, off-policy algorithms rely heavily on efficient data sampling strategies to extract and represent the historical trajectories. However, little is known about how different data sampling methods impact the ability of meta-RL agents to represent unknown environments. Here, we investigate the impact of data sampling strategies on the exploration and adaptability of meta-RL agents. Specifically, we conducted experiments with two types of off-policy meta-RL algorithms based on Thompson sampling and Bayes-optimality theories in continuous control tasks within the MuJoCo environment and sparse reward navigation tasks. Our analysis revealed the long-memory and short-memory sequence sampling strategies affect the representation and adaptive capabilities of meta-RL agents. We found that the algorithm based on Bayes-optimality theory exhibited more robust and better adaptability than the algorithm based on Thompson sampling, highlighting the importance of appropriate data sampling strategies for the agent's representation of an unknown environment, especially in the case of sparse rewards.

Keywords: Meta-Reinforcement Learning · Embodied Agent · Data Sampling Strategy · Task Adaptation · Task Representation

1 Introduction

The realization of embodied intelligence relies on an agent's ability to fast adapt and generalize to unfamiliar environments. The core of adaptation and generalization is to transfer the knowledge learned during training to new task scenarios. Meta-RL is considered as one of the most effective approaches to facilitate fast adaptation to new tasks for embodied intelligence. The goal of meta-RL is to learn a policy within a given task distribution, which can efficiently adapt to a new task distribution with minimal data acquisition [3,5,23]. This goal of meta-RL, i.e., learning for fast adaptation, is well aligned with the learning strategy

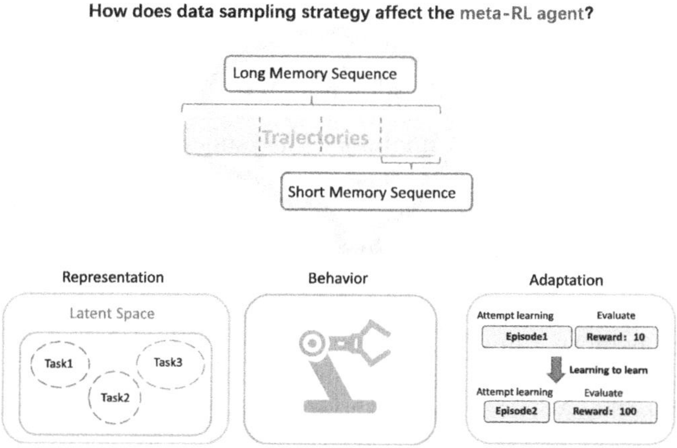

Fig. 1. Motivations of our work.

of humans and embodied AI. Therefore, understanding the mechanism of fast adaptation in meta-RL agents will shed light on the human and embodied AI.

Task representation, as a critical component of meta-RL, impacts the agent's generalization capabilities, particularly in complex control and navigation tasks. Agents are influenced by different task representations during training [22,28,30,31] and need to effectively learn representations to abstract the shared structures in the task distribution. The ability of task representation of an agent mainly depends on data sampling, such as utilizing data relevant to task generalization. Therefore, how to effectively sample task-relevant data during training is crucial, especially for training off-policy meta-RL agents within an off-policy framework. Existing meta-RL algorithms can be categorized into the policy gradient methods [5,18,33] and context-based methods [3,20,27,34]. These meta-RL methods can be categorized into on-policy and off-policy based on the relationship between the policy employed during the learning process and the policy used to generate the data. Compared to on-policy meta-RL algorithms, the off-policy algorithms are more sample-efficient and also rely more on suitable data sampling strategies [2,20,21]. However, how the data sampling strategy affects the off-policy meta-RL agents is still unclear.

In this study, we aim to understand how the data sampling strategy affects the online off-policy meta-RL algorithms, in terms of task representation, agent behaviors and adaptation ability (Fig. 1). To this end, we analyze the exploration capabilities of two types of off-policy meta-RL algorithms based on different data sampling strategies, specifically those based on Thompson sampling [24] in PEARL [20] and Bayes-optimal policy [4] in VariBAD [34]. By conducting experiments with two meta-RL algorithms based on Bayes-optimal policy and Thompson sampling using different data sampling strategies, we examine the influence of two data sampling strategies, i.e., long and short memory sequence, on task representation, agent behavior, and adaptability of off-policy meta-RL agents through experiments in continuous control tasks in MuJoCo [25] and complex navigation tasks.

Our findings are summarized as follows.

- Meta-RL based on Bayes-optimal policy has superior robustness to data sampling distributions compared to Thompson sampling-based Meta-RL method in sparse reward tasks. This robustness originates from the better representation of the unknown environment's dynamics and reward models (Fig. 5, 6, and 7).
- Experiments on complex robotic navigation tasks demonstrate that although the short memory sampling strategy enables PEARL to converge faster, it does not improve the agent's adaptability; in contrast, the relatively robust off-policy VariBAD algorithm exhibits stronger adaptability (Fig. 8, 9 and 10).
- The robustness of algorithms to short memory sequence or long memory sequence sampling strategy is associated with their adaptability capabilities.

2 Background and Related Work

In this section, we primarily introduce the foundational concepts of POMDPs and meta-RL, and related work on task representation in reinforcement learning.

2.1 POMDP and Meta-RL

A partially observable Markov decision process (POMDP) [9] framework offers a robust mathematical model for decision-making where agents must act under conditions of uncertainty and partial information. A POMDP is defined as a tuple $(S, A, O, T, Z, R, \gamma)$, where S is the state space, containing all possible states of the environment; A is the action space, containing all actions that the agent can perform; $T : S \times A \to \mathcal{P}(S)$ is the state transition function; $Z : S \times A \to \mathcal{P}(O)$ is the observation function, defining the probability distribution of generating observations given the next state and action; $R : S \times A \times S \to \mathbb{R}$ is the reward function, which computes the immediate reward based on the current state, chosen action, and resulting state, and γ is the discount factor. The optimization objective of a POMDP is to find a policy $\pi : \mathcal{H} \to A$, where $\mathcal{H}$ represents all possible sequences of historical information. In off-policy meta-RL methods, sufficient task representation from historical trajectories influences the online performance.

The adaptation process of meta-reinforcement learning agents in unknown environments can be seen as a generalization process within POMDPs with a similar distribution. In conventional reinforcement learning algorithms, policy π aims to maximize the expected discounted cumulative reward, expressed as:

$$\mathcal{J}^\pi = \mathbb{E}\left[\sum_{t=0}^{\infty} \gamma^t R(s_t, a_t, s_{t+1}) \mid \pi\right]. \qquad (1)$$

Meta-RL extends the foundational concepts of reinforcement learning by enabling agents to learn how to learn across a variety of tasks, rather than optimizing for a single task. This approach leverages past experience to rapidly adapt to new environments or tasks with minimal additional data. The framework of meta-RL is built around the idea that the skills acquired in previous tasks can inform the agent's policy on unseen tasks, thus reducing the time and data required for learning new tasks. The objective is to train a learning algorithm that can quickly adapt to new tasks using only a few interactions:

$$\max_{\theta} \mathbb{E}_{\tau \sim p(\tau|\theta,T)} \left[\sum_{t=0}^{T} \gamma^t r_t \right], \quad (2)$$

where θ represents the meta-parameters of the policy, and T represents a task sampled from a distribution of tasks. In meta-RL, we aim to perform well across a variety of such tasks.

2.2 Task Representation in Reinforcement Learning

Effective task representation involves capturing the essential features of different tasks in a manner that accentuates their commonalities and differences, thus enabling the agent to adapt learned strategies to new, yet similar, scenarios. An expressive representation that captures task variations is vital for reducing the number of interactions needed to adapt to new tasks [8]. Previous work has employed reconstruction loss to train auto-encoders to generate low-dimensional representations of tasks, which are then used to assist in policy learning [11,17,29]. This method is also applicable to the extraction of common features in multi-task settings, often utilized in multi-task reinforcement learning [31,32]. Additionally, some works have used contrastive learning in the latent representation space to obtain robust representations across multiple tasks [12,28,30].

In this paper, we utilize two of the most fundamental and effective meta-RL models. PEARL [20] employs an RNN encoder as either the task representation module or the inference module. VariBAD [2,34] uses a Variational Autoencoder (VAE) [10] as both the representation and prediction module, which not only extracts task representations but also predicts the environmental model during training.

3 Models and Data Sampling Strategy

In this section we introduce models used and show how to use long-term memory replay and short-term memory reply in two context-based meta-RL methods.

3.1 Thompson Sampling and PEARL

Thompson sampling [24] is a Bayesian approach to addressing the exploration-exploitation dilemma and has been effectively applied in the context of meta-RL [16]. In sequential decision-making, meta-learning can be divided into two phases: the first phase involves abstracting a representation of the distribution of training tasks, while the second phase allows the policy to leverage the prior knowledge of the task distribution acquired in the first phase to predict unknown task distributions, achieving rapid adaptation with minimal interactions with the environment. These models are typically comprised of two components: a task inference module and a policy module [1,19,20,34]. Thompson sampling can be described as the process of sampling actions from the policy based on posterior predictions [15]. According to Bayes' theorem, given the historical trajectory τ, we can update the posterior distribution of θ: $P(\theta \mid \tau) \propto P(\tau \mid \theta)P(\theta)$, where

$$P(\tau \mid \theta) = \prod_{t=0}^{T-1} P(s_{t+1} \mid s_t, a_t, \theta) P(r_t \mid s_t, a_t, \theta). \tag{3}$$

We employ PEARL as a meta-RL model based on Thompson sampling on historical information. The inference network (task encoder) q_ϕ encodes the agent's historical information to capture task-relevant sufficient statistics, and $\mathbf{z}$ is the latent embedding of the encoder. During the meta-training phase, the parameters within q_ϕ are optimized by modeling the Q-value function $Q_\theta(\mathbf{s}, \mathbf{a}, \mathbf{z})$ within the Soft Actor-Critic (SAC) [7] algorithm and constrained by the variational lower bound to enhance the inference module's ability to learn information pertinent to the current task being performed, the inference module and policy module use different experience.

In the meta-testing phase, as the agent tackles unknown tasks, it updates the posterior of task history in a manner similar to Thompson sampling through the inference network, enhancing its capability to explore unknown tasks. The structure of PEARL is depicted in the upper part of Fig. 2.

3.2 Bayes-Optimality and VariBAD

Bayes-optimality is a principle within the broader Bayesian decision theory that dictates selecting actions based on maximizing expected utility, considering all possible outcomes weighted by their probabilities. In the context of meta-RL, a Bayes-optimal policy aims to maximize the expected reward across a distribution of tasks by leveraging a posterior distribution over tasks [4]. This approach inherently balances exploration and exploitation by considering the uncertainty in the environment's dynamics and the reward function. By integrating prior knowledge with observations gathered during interactions with the environment, the Bayes-optimal approach adapts its strategy to better respond to unseen dynamics. In BAMDP, we aim to maximize the expected reward over T time steps:

$$\mathcal{J}(\pi) = \mathbb{E}_{b_0, \pi} \left[\sum_{t=0}^{T-1} \mathbb{E}_{R \sim b_t} [R(s_t, a_t)] \right], \tag{4}$$

where $b_t = p(r, p \mid \tau_{:t})$ represents the belief about the current environment dynamics and reward function based on the historical trajectory.

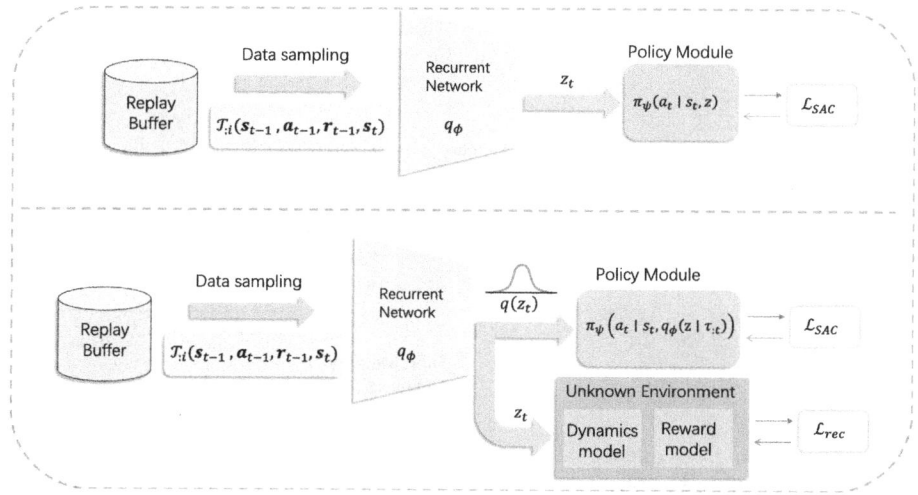

Fig. 2. PEARL and off-policy VariBAD famework.

VariBAD implements Bayesian optimal policies under the framework of Bayes-Adaptive MDP (BAMDP) [4,6] and it is designed to learn a latent representation of the environment's dynamics, rewards, and task-specific parameters through a variational autoencoder architecture (lower part of Fig. 2). During training, the agent learns a universal policy that is conditioned on both the current state and a latent variable that encodes task-specific information. This latent variable is updated as new data is collected, effectively allowing the agent to infer the underlying task dynamics and adapt its policy accordingly. VariBAD thus enables an agent to perform robustly across a variety of tasks by efficiently learning and updating its understanding of the task environment.

3.3 Long and Short Memory Sequence Sampling

In this paper, we investigate the impact of data sampling strategies on the prior information extracted by the representation module in off-policy meta-RL, leading to varying exploration outcomes by the algorithm. In our experiments, we have set up long and short memory sequence sampling strategies (Fig. 3). The long memory strategy involves storing all historical information from interactions with the environment in the replay buffer, from which the agent and inference network sample during training. In contrast, the short memory sequence sampling setting requires clearing the replay buffer at the start of each training iteration, ensuring that the agent and inference network can only sample the most recent historical data. Typically, Thompson sampling updates the posterior based on the most recent historical information. In our experiments described in Sect. 4, we observe that the long memory sequence can significantly disrupt the exploration effectiveness of PEARL, while the off-policy VariBAD remains robust under both settings. The exploration and exploitation trade-off is also reflected in the varying performance of the different representation modules.

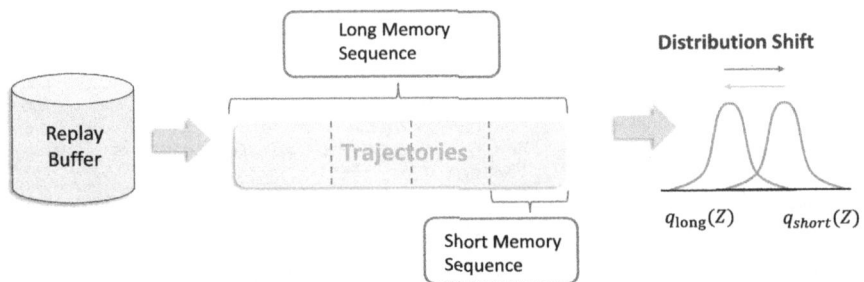

Fig. 3. Long memory sequence sampling and short memory sequence sampling. Different context sampling strategies can lead to shifts in the distribution of task representations, thereby affecting the agent's exploration and adaptation capabilities.

4 Experiments

In our study, we conduct comparisons of the performance of PEARL and off-policy VariBAD under various data sampling strategies in the continuous control task Sparse Half-Cheetah-Vel within the MuJoCo environment and two challenging navigation tasks, Ant-Semi-Circle and Sparse-Point-Robot. We analyze the robustness of the algorithms and their task adaptation capabilities based on their performance.

4.1 Task Setting

Sparse Half-Cheetah-Vel. We have modified the original Half-Cheetah environment in MuJoCo to adopt a sparse reward function, where the cheetah receives a reward of +1 only if it moves a distance greater than a specified threshold in each time step; otherwise, the reward is 0.

$$r_t^{sparse} = \begin{cases} 1, & \|x_t - x_{\text{goal}}\|_2 \leq r \\ 0, & \text{otherwise.} \end{cases}$$

Subsequently, we utilize the commonly used Half-Cheetah-Vel task in meta-RL, where the objective for the agent is to reach a target velocity as quickly as possible. The target velocity is randomly sampled from a range of 0.0 to 3.0. Consequently, the reward in this environment is structured as follows:

$$r_t = -|v_t - v_{\text{goal}}| - 0.05 \cdot \|a_t\|_2^2 + r_t^{sparse},$$

We set up 100 training tasks and 20 testing tasks.

Sparse-Point-Robot. We established the Sparse-Point-Robot environment to evaluate the performance of algorithms in a sparse reward navigation setting. Different goals are set on a semi-circle, with their locations unknown. At the start of each episode, the robot's initial position is randomly placed outside of the semi-circle. The objective is for the robot to locate the target within a single episode. The reward structure is configured as follows:

$$r = \begin{cases} 1, & \text{if } r \geq -\text{goal\_radius} \\ 0, & \text{otherwise.} \end{cases}$$

Ant-Semi-Circle. In the Ant-Semi-Circle task [2], an ant robot is required to navigate toward a goal that is randomly positioned on a semi-circle. Unlike the point robot scenario, this task employs the Ant model from the MuJoCo simulation, which introduces increased control complexity. The reward structure is set as follows:

$$r_t = -\|x_t - x_{\text{goal}}\|_1 - 0.1 \cdot \|a_t\|_2^2.$$

4.2 Convergence of Algorithms

We conducted experiments in the aforementioned three environments, initially testing the convergence of the algorithms on different tasks under default parameters (PEARL using short sequence, off-policy VariBAD using long sequence) (Fig. 4).

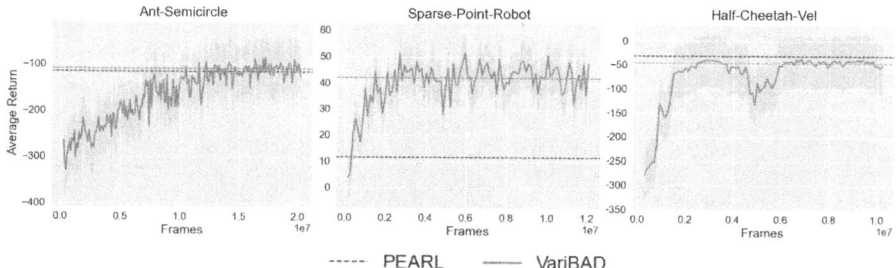

Fig. 4. Tasks training. Dashed lines correspond to the maximum return achieved by PEARL after 1e6 steps. Solid lines correspond to average return achieved by VariBAD. In the Ant-Semicircle and Half-Cheetah-Vel tasks, PEARL and VariBAD converge to similar average returns. However, in the Sparse-Point-Robot task, VariBAD significantly outperforms PEARL.

4.3 Robustness of Algorithms

For algorithm robustness, the analysis mainly focuses on the task representations during the meta-training and meta-testing phases. During the meta-training phase, for tasks involving the control of simulated robots such as Ant-SemiCircle and Half-Cheetah-Vel, PEARL is more susceptible to the influence of memory sequence sampling. Here, short memory sequence sampling proves more advantageous for PEARL's adaptation to new environments, while off-policy VariBAD remains relatively stable in comparison (Fig. 5).

We randomly generate 20 goals from each environment, and for each goal, the agent performs 40 runs, each containing 5 episodes. We utilize t-SNE [14] to visualize the latent embeddings of the task inference module at the last time step of the fifth episode during the meta-testing phase for PEARL (Fig. 6) and off-policy VariBAD (Fig. 7). For navigation tasks, especially the Sparse-Point-Robot task, PEARL does not sufficiently learn the task-relevant information of the environment, whereas off-policy VariBAD exhibits better performance, forming clusters for similar positions on the semicircle. In the case of Ant-Semi-Circle, off-policy VariBAD accurately represents each target's position on the semicircle in the latent space, whereas the clusters formed by PEARL are more dispersed. In the Half-Cheetah-Vel environment, the representation of off-policy VariBAD is more sensitive to the memory sequence sampling strategy because, inherently, based on Bayes-optimality, it requires sampling of the whole history, and the representation in continuous control tasks is significantly influenced by the history.

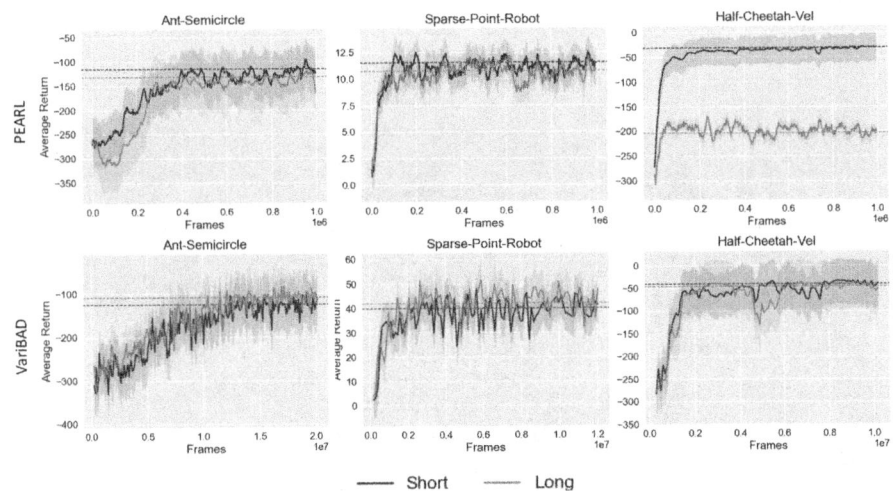

Fig. 5. The average return during the meta-training phase of PEARL and off-policy VariBAD after using short and long memory sampling strategies.

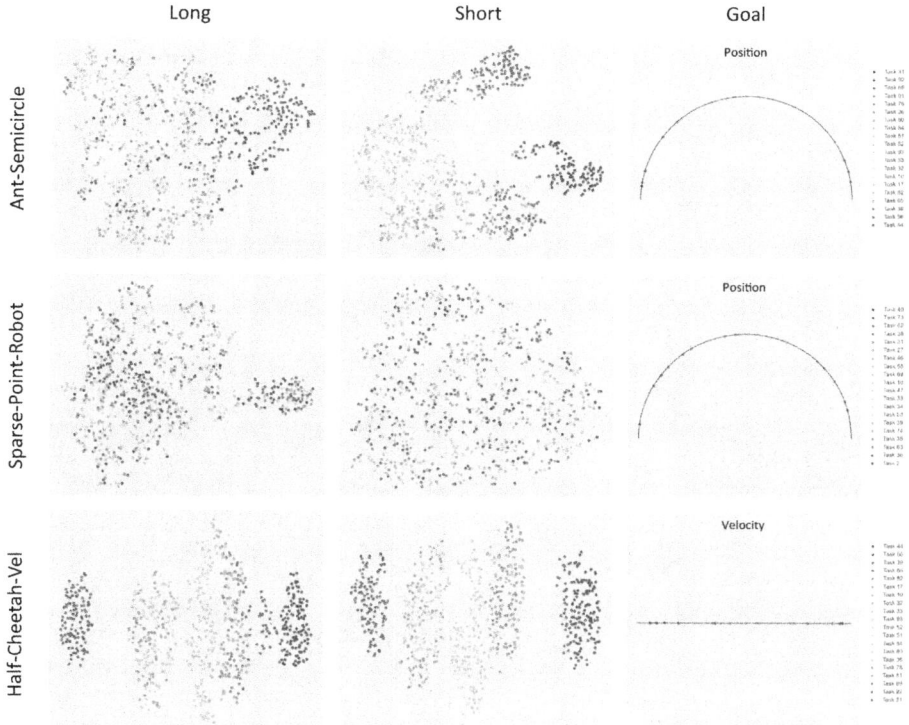

Fig. 6. The t-SNE visualization of latent embedding during PEARL adaptation to the environment.

4.4 Tasks Adaptation Performance

To compare the adaptability of PEARL and VariBAD using different sampling strategies in unknown environments, we conducted rollouts of 5 episodes in each environment (Fig. 8). The experiments revealed that the long memory sequence sampling strategy prevented PEARL from adapting effectively to navigation tasks and from achieving satisfactory results in the Sparse-Point-Robot task, consistent with its performance during the meta-training phase. In contrast, off-policy VariBAD demonstrated stable adaptability, successfully adjusting to tasks and achieving high average returns from the first episode across all three environments. Moreover, VariBAD showed less sensitivity to memory sequence sampling strategy, with agents trained using short memory sequence in the Sparse-Point-Robot task exhibiting enhanced exploratory capabilities.

Furthermore, to intuitively assess the impact of sampling strategies on the exploratory capabilities of agents during the meta-testing process, we visualized the exploration trajectories of agents in the Ant-Semi-Circle environment.

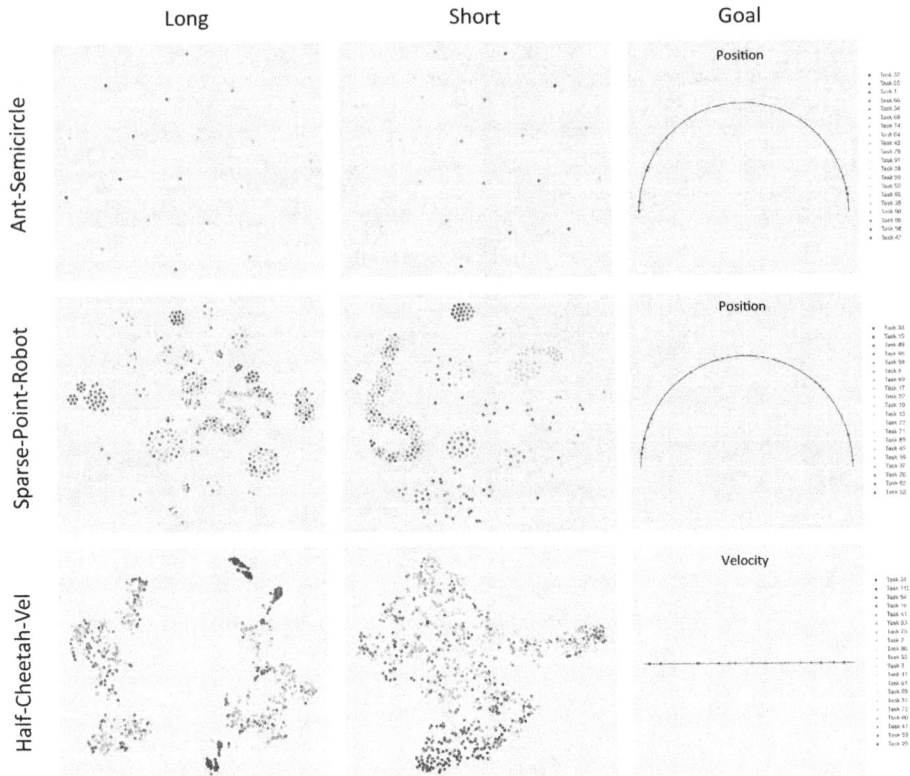

Fig. 7. The t-SNE visualization of latent embedding during off-policy VariBAD adaptation to the environment.

The length of the memory sequence affects PEARL's exploratory capabilities, primarily because Thompson sampling focuses on updating the encoder's posterior based on recent context, hence performing better with a short memory sequence length. Figure 9 shows that PEARL, when using short memory, can successfully reach the target in all five episodes; however, when switched to long memory, the agent fails to accurately locate the target for the same test task. Figure 10 demonstrates that off-policy VariBAD can successfully find the target in two episodes and is unaffected by the memory sequence length. This indicates that applying Bayes-optimality to off-policy meta-reinforcement learning results in stronger online adaptability to complex environments and more robust task representation compared to algorithms based on Thompson sampling.

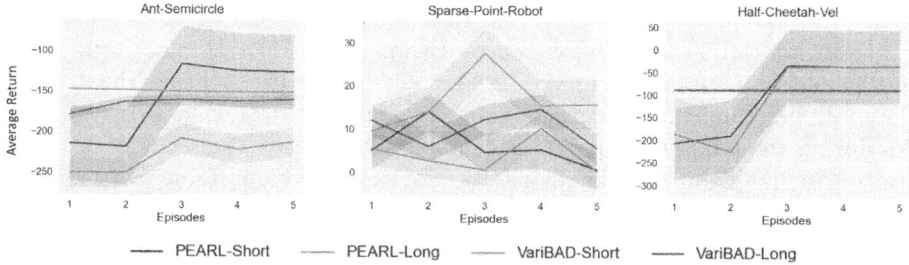

Fig. 8. Adaptation performance of PEARL and off-policy VariBAD using short and long memory sequence sampling strategy.

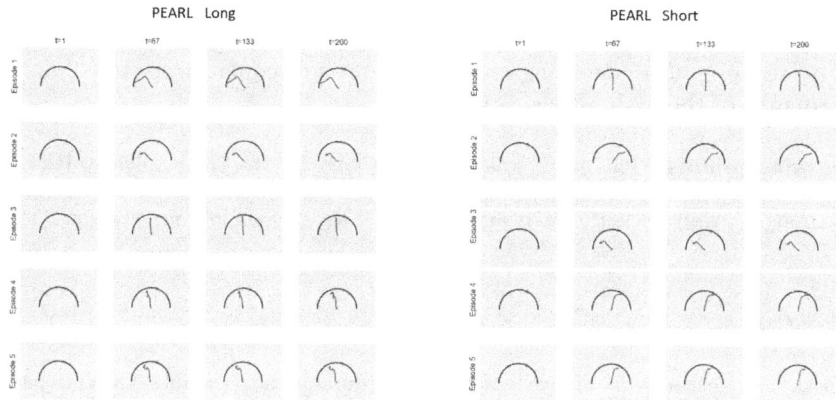

Fig. 9. Behavior visualization of PEARL during adaptation in Ant-Semi-Circle.

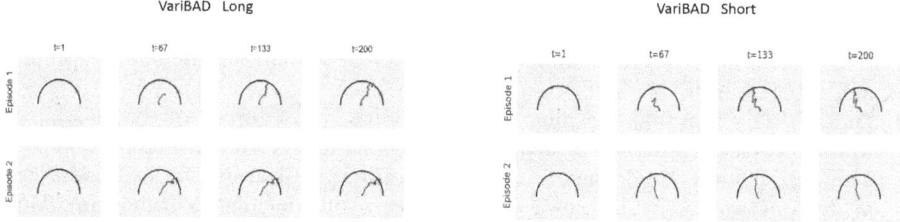

Fig. 10. Behavior visualization of off-policy VariBAD during adaptation in Ant-Semi-Circle.

5 Conclusion

Discussion. In this work, we comprehensively compared the impact of memory sequence length sampling on task representation, agent behavior, and the exploration and adaptation capabilities of two types of context-based off-policy meta-RL algorithms. The off-policy VariBAD algorithm, based on Bayes-optimality, demonstrated stronger robustness in sparse reward environments and its adaptability to unknown environments and task representations were less influenced by the training-time memory sequence. Although the PEARL algorithm, based on Thompson sampling, achieved similar average returns during the training phase, different memory lengths significantly affected its exploratory capabilities during the adaptation phase, fundamentally because the memory sequence length can cause shifts in the distribution of task representations.

Future Work. Task representation extraction, especially in multi-task scenarios, is crucial for an agent's ability to adapt to unknown tasks [13,22,26,31]. This paper explores the impact of memory sequence length on shifts in task feature distributions, providing insights into how to maximize the extraction of information relevant to the current task from limited historical data and generate a robust representation of the task distribution. This prompts us to employ effective representation learning methods in our subsequent work to enhance the performance of the task inference module.

Acknowledgments. This work was funded in part by the National Key R&D Program of China (2021YFF1200804), Shenzhen Science and Technology Innovation Committee (2022410129, KCXFZ20201221173400001, KJZD20230923115221044).

References

1. Beck, J., et al.: A survey of meta-reinforcement learning. arXiv preprint arXiv:2301.08028 (2023)
2. Dorfman, R., Shenfeld, I., Tamar, A.: Offline meta reinforcement learning-identifiability challenges and effective data collection strategies. Adv. Neural Inf. Process. Syst. **34**, 4607–4618 (2021)
3. Duan, Y., Schulman, J., Chen, X., Bartlett, P.L., Sutskever, I., Abbeel, P.: RL2: fast reinforcement learning via slow reinforcement learning. arXiv preprint arXiv:1611.02779 (2016)
4. Duff, M.O.: Optimal learning: computational procedures for bayes-adaptive Markov decision processes. University of Massachusetts Amherst (2002)
5. Finn, C., Abbeel, P., Levine, S.: Model-agnostic meta-learning for fast adaptation of deep networks. In: International Conference on Machine Learning, pp. 1126–1135. PMLR (2017)
6. Ghavamzadeh, M., Mannor, S., Pineau, J., Tamar, A., et al.: Bayesian reinforcement learning: a survey. Found. Trends® Mach. Learn. **8**(5-6), 359–483 (2015)
7. Haarnoja, T., Zhou, A., Abbeel, P., Levine, S.: Soft actor-critic: off-policy maximum entropy deep reinforcement learning with a stochastic actor. In: International Conference on Machine Learning, pp. 1861–1870. PMLR (2018)
8. Humplik, J., Galashov, A., Hasenclever, L., Ortega, P.A., Teh, Y.W., Heess, N.: Meta reinforcement learning as task inference. arXiv preprint arXiv:1905.06424 (2019)
9. Kaelbling, L.P., Littman, M.L., Cassandra, A.R.: Planning and acting in partially observable stochastic domains. Artif. Intell. **101**(1–2), 99–134 (1998)
10. Kingma, D.P., Welling, M.: Auto-encoding variational bayes. arXiv preprint arXiv:1312.6114 (2013)

11. Lange, S., Riedmiller, M.: Deep auto-encoder neural networks in reinforcement learning. In: The 2010 International Joint Conference on Neural Networks (IJCNN), pp. 1–8. IEEE (2010)
12. Laskin, M., Srinivas, A., Abbeel, P.: Curl: Contrastive unsupervised representations for reinforcement learning. In: International Conference on Machine Learning, pp. 5639–5650. PMLR (2020)
13. Lee, S., Chung, S.Y.: Improving generalization in meta-rl with imaginary tasks from latent dynamics mixture. Adv. Neural Inf. Process. Syst. **34**, 27222–27235 (2021)
14. Van der Maaten, L., Hinton, G.: Visualizing data using t-sne. J. Mach. Learn. Res. **9**(11) (2008)
15. Ortega, P.A., Braun, D.A.: A minimum relative entropy principle for learning and acting. J. Artif. Intell. Res. **38**, 475–511 (2010)
16. Ortega, P.A., et al.: Meta-learning of sequential strategies. arXiv preprint arXiv:1905.03030 (2019)
17. Péré, A., Forestier, S., Sigaud, O., Oudeyer, P.Y.: Unsupervised learning of goal spaces for intrinsically motivated goal exploration. arXiv preprint arXiv:1803.00781 (2018)
18. Raghu, A., Raghu, M., Bengio, S., Vinyals, O.: Rapid learning or feature reuse? Towards understanding the effectiveness of maml. arXiv preprint arXiv:1909.09157 (2019)
19. Raileanu, R., Goldstein, M., Szlam, A., Fergus, R.: Fast adaptation to new environments via policy-dynamics value functions. In: Proceedings of the 37th International Conference on Machine Learning, pp. 7920–7931 (2020)
20. Rakelly, K., Zhou, A., Finn, C., Levine, S., Quillen, D.: Efficient off-policy meta-reinforcement learning via probabilistic context variables. In: International Conference on Machine Learning, pp. 5331–5340. PMLR (2019)
21. Rusu, A.A., et al.: Meta-learning with latent embedding optimization. arXiv preprint arXiv:1807.05960 (2018)
22. Sodhani, S., Zhang, A., Pineau, J.: Multi-task reinforcement learning with context-based representations. In: International Conference on Machine Learning, pp. 9767–9779. PMLR (2021)
23. Stadie, B.C., et al.: Some considerations on learning to explore via meta-reinforcement learning. arXiv preprint arXiv:1803.01118 (2018)
24. Thompson, W.R.: On the likelihood that one unknown probability exceeds another in view of the evidence of two samples. Biometrika **25**(3–4), 285–294 (1933)
25. Todorov, E., Erez, T., Tassa, Y.: Mujoco: a physics engine for model-based control. In: 2012 IEEE/RSJ International Conference on Intelligent Robots and Systems, pp. 5026–5033. IEEE (2012)
26. Wang, B., Xu, S., Keutzer, K., Gao, Y., Wu, B.: Improving context-based meta-reinforcement learning with self-supervised trajectory contrastive learning. arXiv preprint arXiv:2103.06386 (2021)
27. Wang, J.X., et al.: Learning to reinforcement learn. arXiv preprint arXiv:1611.05763 (2016)
28. Wang, M., et al: Meta-reinforcement learning based on self-supervised task representation learning. In: Proceedings of the AAAI Conference on Artificial Intelligence, vol. 37, pp. 10157–10165 (2023)
29. Watter, M., Springenberg, J., Boedecker, J., Riedmiller, M.: Embed to control: a locally linear latent dynamics model for control from raw images. Adv. Neural Inf. Process. Syst. **28** (2015)

30. Yuan, H., Lu, Z.: Robust task representations for offline meta-reinforcement learning via contrastive learning. In: International Conference on Machine Learning, pp. 25747–25759. PMLR (2022)
31. Zhang, A., Sodhani, S., Khetarpal, K., Pineau, J.: Learning robust state abstractions for hidden-parameter block mdps. arXiv preprint arXiv:2007.07206 (2020)
32. Zhang, Y., Yang, Q.: A survey on multi-task learning. IEEE Trans. Knowl. Data Eng. **34**(12), 5586–5609 (2021)
33. Zintgraf, L., Shiarli, K., Kurin, V., Hofmann, K., Whiteson, S.: Fast context adaptation via meta-learning. In: International Conference on Machine Learning, pp. 7693–7702. PMLR (2019)
34. Zintgraf, L., et al.: Varibad: a very good method for bayes-adaptive deep rl via meta-learning. arXiv preprint arXiv:1910.08348 (2019)

Parameter-Efficient Fine-Tuning of ChatGLM to Mitigate Hallucinations in Chinese Abstractive Summarization

Yongjian Huang[1,2]($\boxtimes$) and Simin Wu[1]

[1] Guangdong University of Science and Technology, Guangdong, China
huangyongjian@gdust.edu.cn
[2] Guangzhou Xuanyuan Research Institute Co., Ltd., Guangdong, China

Abstract. Large language models, such as ChatGPT, have shown great promise in various natural language processing tasks, including abstractive summarization. However, these models are prone to generating hallucinated content—information not found in the original source. To tackle this issue, we introduce a mathematical framework for quantifying hallucinations and assess the effectiveness of different fine-tuning methods for mitigating this problem. We focus on fine-tuning ChatGLM due to its proficiency in both English and Chinese, as well as its ability to run on consumer-grade graphics cards. Our experiments, conducted on the XL-Sum datasets, utilize BLEU and ROUGE metrics to evaluate performance. The results reveal that P-Tuning is the most effective method for reducing hallucinations while maintaining high-quality summaries. This study not only enhances the reliability of abstractive summarization but also suggests the potential for reducing hallucinations in other NLP tasks through fine-tuning methods.

Keywords: Large Language Models · Natural Language Generation · Hallucination · Fine-tuning · P-tuning · ChatGLM

1 Introduction

Abstractive summarization models aim to extract essential information from long documents and to generate short, concise and readable text [1]. In recent years, large language models have revolutionized the field of Natural Language Processing (NLP), demonstrating unparalleled performance across a wide array of tasks. However, their practical application is often limited by a significant drawback: the generation of hallucinated content, or information that is not present in the source material. This issue also appears in abstractive summarization, where accurate and reliable summaries are generally desired. A further drawback of large language models such as GPT is their primary training in English, narrowing their capacity to understand and interact in other languages. This poses a challenge for those who are not native English speakers. To mitigate hallucinations in abstractive summarization, various strategies can be employed that fall into two main categories [2]: Data-Related Methods and Modeling and Inference Methods. Data-Related Methods encompass the curation of a reliable dataset annotated by experts, automated techniques for data cleaning to remove inconsistencies, and information augmentation to enhance the size and diversity of the training set. On the other hand, Modeling and Inference Methods involve architectural choices that can reduce hallucinations, such as the incorporation of attention mechanisms focused on factual consistency. During the training phase, specialized loss functions can be designed to penalize hallucinations, and these can be complemented by techniques like teacher-forcing or scheduled sampling. Post-processing steps can also be applied after the summary generation to further refine the output, including fact-checking against the source document or employing additional classifiers to weed out hallucinated information. By integrating these methods, the model's propensity for hallucinations can be substantially reduced, leading to more reliable and accurate summaries.

This paper aims to address this challenge by leveraging efficient fine-tuning technology to adapt Large Language Models for the specific task of mitigating hallucinations in abstractive summarization.

Our primary contributions are threefold:

- We introduce a mathematical framework for quantifying hallucinations and the effectiveness of fine-tuning, providing a rigorous foundation for evaluating the performance of our approach.
- We employ a fine-tuning approach that incorporates four distinct methods: LoRA, Prefix Tuning, Prompt Tuning, and P-tuning. Each method is reevaluated to understand its effectiveness in reducing hallucinatory outputs while maintaining the model's performance. Furthermore, we provide an in-depth analysis of the advantages and disadvantages associated with each of these fine-tuning techniques.
- We train a large language model that is not limited to English, specifically focusing on understanding Chinese abstractive summaries. This effort aims to bridge the language gap and extend the capabilities of large language models, thereby making them more accessible and useful to non-English speaking communities.

2 Preliminaries and Notations

In this section, we propose a mathematical framework to address these issues through fine-tuning and adaptive methods.

We define hallucination as the generation of a token x_{i+1} with a low conditional probability given the preceding tokens $x_1, x_2, \ldots, x_i$. Mathematically, this can be represented as:

$$H(x_{i+1}|x_1, x_2, \ldots, x_i; \theta) = \begin{cases} 1, & \text{if } P(x_{i+1}|x_1, x_2, \ldots, x_i; \theta) < \epsilon \\ 0, & \text{otherwise} \end{cases}$$

Here, ϵ is a threshold value below which a token is considered hallucinatory.

As the efficacy of LLMs across various tasks hinges on the delicate equilibrium between hallucination and creativity [3]. We define an objective function $J(\theta, \alpha)$ that encapsulates the trade-off between hallucination and creativity. The function is parameterized by α, a trade-off parameter that ranges from 0 to 1, and θ, the model parameters.

$$J(\theta, \alpha) = (1 - \alpha) \cdot E_{(x_1, \ldots, x_n) \sim P_{true}}$$

$$[DKL(P_{true}(x_{i+1}|x_1, x_2, \ldots, x_i) \| P_{model}(x_{i+1}|x_1, x_2, \ldots, x_i; \theta))]$$
$$- \alpha \cdot E_{(x_1,\ldots,x_n) \sim P_{true}} [C(x_{i+1}|x_1, x_2, \ldots, x_i; \theta)]$$

Here, DKL denotes the Kullback-Leibler divergence, and C represents the creativity measure.

To dynamically adjust the trade-off parameter α based on task requirements and input data, we introduce an adaptive method. The method employs gradient-based optimization to iteratively update α based on the model's performance on a validation set.

$$\alpha^{(t+1)} = \alpha^{(t)} - \eta \nabla_\alpha J(\theta, \alpha)$$

Here, η is the learning rate, and t represents the iteration number.

Hallucination mitigation aims to reduce the likelihood of generating hallucinatory tokens. We propose a regularization term $R(\theta)$ that penalizes the model for generating low-probability tokens that lead to hallucinations.

$$J_{new}(\theta, \alpha) = J(\theta, \alpha) + \lambda R(\theta)$$

Here, λ is a hyperparameter that controls the strength of the regularization.

The proposed mathematical framework offers a structured approach to fine-tuning large language models, with a focus on mitigating hallucinations while balancing creativity.

3 Fine-Tuning a Pretrained Language Model

We will use ChatGLM as the base model and fine-tune it for specific tasks to mitigate hallucination. ChatGLM is a large language model co-created by Tsinghua University and Zhipu.AI [4]. Designed with a focus on question-answering tasks, ChatGLM-6B is built on the General Language Model (GLM) framework and supports both English and Chinese languages. The model boasts 6 billion parameters and is open-source, making it an attractive choice for various applications. Additionally, its compatibility with model quantization methods allows for local deployment on standard GPUs.

For the fine-tuning process, we will employ four commonly used techniques for LLMs, namely LoRA, Prefix Tuning, Prompt Tuning and P-tuning. These methods have been proven effective in adapting pre-trained language models to domain-specific tasks and improving their performance.

LoRA: Lora, which stands for low-rank adaptation, is a technique that freezes the pre-trained model weights and injects trainable rank decomposition matrices into each layer of the Transformer architecture, greatly reducing the number of trainable parameters for downstream tasks [5]. A fully connected layer is used to reduce the dimensionality of the trainable layers from 'd' to 'r', and then it's mapped back to the original 'd' dimension through another fully connected layer. Here, 'r' is the rank of the matrix and is much smaller than 'd'. This results in a substantial reduction in computational complexity, changing it from 'd × d' to 'd × r + r × d'. In Lora, the parameters A and B are initialized using a random Gaussian distribution, as shown in Fig. 1.

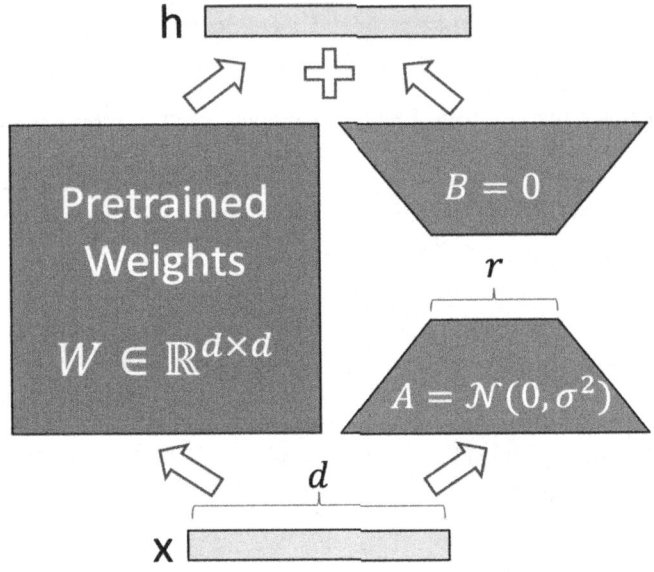

Fig. 1. LoRA (Fine-tuning is only required for the matrices B and A) [5]

Prefix Tuning: Prefix-tuning keeps language model parameters frozen and optimizes a sequence of continuous task-specific vectors, known as the prefix [6]. These prefix vectors are prepended to the input sequence and processed through the same transformer architecture as the original input. The idea is that these prefix vectors can encapsulate task-specific information, effectively guiding the model to generate more appropriate outputs for the task at hand. Because only the prefix vectors are being optimized, and not the entire set of model parameters, prefix-tuning is often more computationally efficient than full-model fine-tuning (Fig. 2).

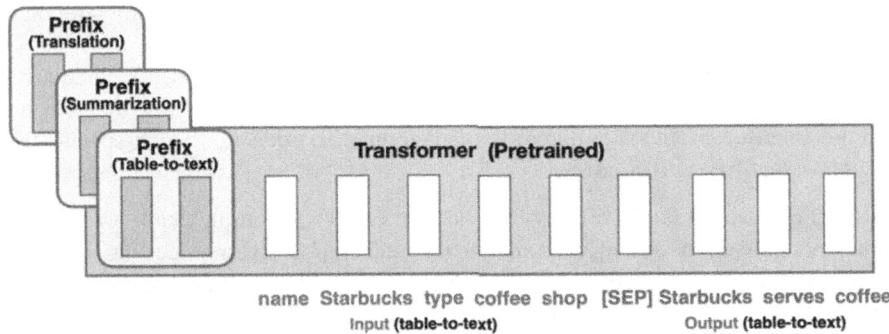

Fig. 2. Prefix- tuning freezes the LM parameters and only optimizes the prefix (the red prefix blocks) [6]

Prompt Tuning: Prompt tuning is a simple yet effective mechanism for learning "soft prompts" to condition frozen language models to perform specific downstream tasks [7]. It learned through backpropagation and can be tuned to incorporate signals from any number of labelled examples. In this approach, a fixed prompt is prepended to each input sequence during training. The tokens in this prompt are treated as trainable parameters, meaning they are optimized along with the rest of the model. The idea is that the prompt can help guide the model's behaviour for specific tasks (Fig. 3).

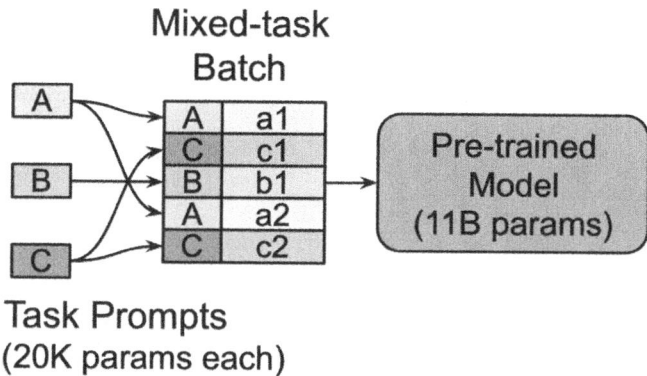

Fig. 3. Prompt tuning only requires storing a small task-specific prompt for each task, and enables mixed-task inference using the original pretrained model [7]

P-Tuning: P-tuning is a technique that autonomously generates templates to enable language models to handle specific tasks [8]. It employs [unused]* tokens as continuous prompts and fine-tunes them using annotated data. This approach often matches or even surpasses the performance of traditional fine-tuning methods but with fewer parameters to train. P-tuning v2 is an enhanced variant that extends the use of continuous prompts to every layer of the pre-existing model, making it more adaptable for natural language understanding tasks. In our research, we utilized P-tuning v2 for the fine-tuning process, as illustrated in Fig. 4.

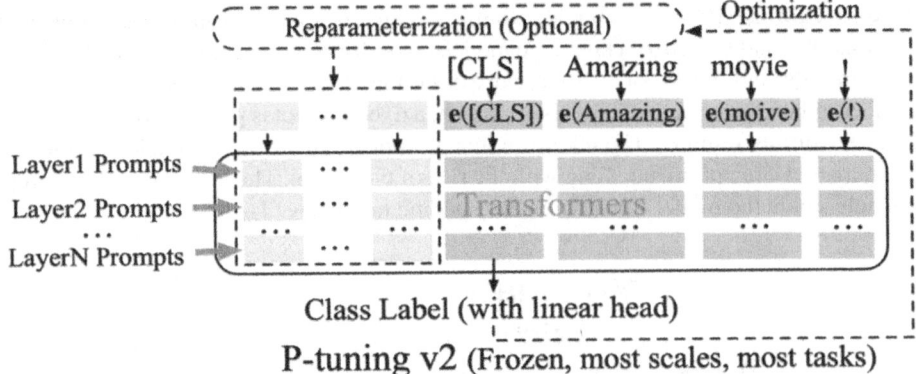

Fig. 4. P-tuning (Each layer of the transformer requires fine-tuning of the embedding) [8]

4 Experiment

4.1 Dataset

To evaluate the efficacy of our fine-tuning approach in mitigating hallucinations, we employ the XL-Sum dataset [9], which is a large-scale professionally annotated article summary pairs from BBC. It comprises 16,369 texts in Chinese, each annotated with a multi-sentence summary by language experts, XL-Sum serves as an ideal testing ground.

4.2 Evaluation Metrics

To evaluate our fine-tuning methods, we employ two metrics:

BLEU: BLEU, short for Bilingual Evaluation Understudy, serves as a metric for assessing a model's accuracy by quantifying the n-gram similarity between machine-generated text and a reference standard [10]. In the context of reducing hallucinations in large language models, a higher BLEU score indicates that the generated content closely aligns with the reference material, thereby suggesting fewer instances of hallucinated or fabricated information.

ROUGE: The ROUGE score, standing for "Recall-Oriented Understudy for Gisting Evaluation," serves as a widely-used metric for evaluating the quality of summarization models [11]. By measuring the N-gram overlap between reference and generated summaries, ROUGE offers a nuanced view of model performance through its variants—ROUGE-1, ROUGE-2, and ROUGE-L—which focus on unigram, bigram, and longest common subsequence overlaps, respectively. Importantly, this metric also provides insights into the model's ability to reduce hallucinations. A high ROUGE score not only indicates effective summarization but also suggests that the model's generated text closely aligns with the reference, thereby minimizing the likelihood of generating hallucinated information.

4.3 Results

Based on the data presented in Table 1, it's evident that different fine-tuning methods have varying impacts on the performance metrics of the model. P-Tuning emerges as the most effective method, consistently outperforming all other techniques across BLEU-4, Rouge-1, Rouge-2, and Rouge-L. For instance, P-Tuning achieved a BLEU-4 score of 6.4064, a significant leap from the original model's 2.73, and also led in Rouge metrics with scores like 20.4967 in Rouge-1 and 16.5956 in Rouge-L. Prefix Tuning and LoRA also showed notable improvements in these metrics, while Prompt Tuning lagged slightly but still performed better than the original model.

The results suggest that P-Tuning could be the most effective approach for enhancing summarization quality and potentially mitigating hallucinations. Prefix Tuning and LoRA follow closely, offering promising avenues for further research and development. Prompt Tuning, although generally scoring lower than the other methods, still shows an improvement over the original model, indicating that each of these fine-tuning methods has its merits.

Table 1. Performance analysis of fine-tuning methods based on BLEU and Rouge scores

Metrics	Original	LoRA	Prefix Tuning	Prompt Tuning	P-Tuning
BLEU-4	2.73	4.4266	4.51824	4.21824	6.4064
Rouge-1	15.9026	19.0612	19.7743	18.7743	20.4967
Rouge-2	2.0035	2.9178	2.9495	2.7495	3.6532
Rouge-L	12.0588	14.1009	14.9884	13.9884	16.5956

4.4 Discussion

The choice of ChatGLM as the base model for our experiments was motivated by several factors. Primarily, ChatGLM has demonstrated effectiveness in abstractive summarization tasks. Additionally, its bilingual capabilities make it versatile for handling both Chinese and English languages, broadening its applicability. Importantly, the model's architecture is optimized for deployment on consumer-grade graphics cards, making it accessible for a wider range of research and practical applications.

Previous research has indicated that methods aimed at reducing hallucinations in language models generally fall into two main categories: those that are data-related and those that are training-related [2]. Data-related methods often involve building a more faithful dataset, cleaning data automatically, or augmenting information. On the other hand, training-related methods focus on the model's architecture, training techniques, and post-processing steps to mitigate hallucinations. Fine-tuning, the method we have employed in this study, belongs to the training-related category. Fine-tuning was selected as the method of adaptation for several pragmatic reasons. In enterprise settings, it's common to have access to data and computational resources, making fine-tuning a feasible approach. The method allows for customization of the model to specific tasks without requiring extensive computational power, making it widely deployable in practical situations.

Our experiments indicate that P-Tuning is the most effective method for improving summarization quality and potentially reducing hallucinations. However, it's crucial to consider the intrinsic structures of the fine-tuning methods when interpreting these results. P-Tuning's effectiveness could be attributed to its ability to optimize a small set of parameters effectively, thereby retaining the model's generalization capabilities while adapting to the specific task. On the other hand, methods like Prefix Tuning and LoRA, while effective, may not offer the same level of task-specific adaptability due to their structural limitations. Prefix Tuning optimizes a sequence of task-specific vectors, which may not capture the nuances of the task as effectively as P-Tuning's more flexible approach. LoRA focuses on low-rank adaptations, which, although computationally efficient, might not be as effective for complex tasks that require a richer parameter space.

In summary, while P-Tuning stands out as the most effective method in our experiments, the choice of fine-tuning method should be context-dependent, taking into account both the specific requirements of the task and the intrinsic advantages and limitations of each method.

5 Conclusion

Our research demonstrates that among the various fine-tuning methods examined, P-Tuning stands out as the most effective in reducing hallucinations in abstractive summarization tasks. This suggests that P-Tuning could be a valuable approach for enhancing the reliability and accuracy of large language models. However, it's important to note that our study is focused on a specific set of tasks and models. As such, future research should extend this investigation to other fine-tuning methods and a broader range of NLP tasks. This will provide a more comprehensive understanding of how to mitigate hallucinations effectively across different applications and model architectures.

Acknowledgments. This research was financially supported by the Integration of Culture and Technology Key Special Project for Key Research and Development Plan of Guangdong Province (No. 2022B0101010005).

References

1. Yu, T., et al.: AdaptSum: towards low-resource domain adaptation for abstractive summarization. In: Conference of the North-American-Chapter of the Association-for-Computational-Linguistics - Human Language Technologies (NAACL-HLT), 2021. Electr Network
2. Ji, Z., et al.: Survey of hallucination in natural language generation. ACM Comput. Surv. **55**(12), 1–38 (2023)
3. Lee, M.: A mathematical investigation of hallucination and creativity in GPT models. Mathematics **11**(10), 2320 (2023)

4. Zeng, A., et al.: Glm-130b: An open bilingual pre-trained model. arXiv preprint arXiv:2210.02414, 2022
5. Hu, E., et al.: LORA: low-rank adaptation of large language models. In: 10th International Conference on Learning Representations, ICLR 2022, April 25, 2022–29 April 2022, 2022. Virtual, Online: International Conference on Learning Representations, ICLR
6. Li, X.L.S., Liang, P., Assoc Computat., L.: Prefix-tuning: optimizing continuous prompts for generation. In: Joint Conference of 59th Annual Meeting of the Association-for-Computational-Linguistics (ACL)/11th International Joint Conference on Natural Language Processing (IJCNLP)/6th Workshop on Representation Learning for NLP (RepL4NLP), 2021. Electr Network. L. Assoc Computat.
7. Lester, B., et al.: The power of scale for parameter-efficient prompt tuning. In: Conference on Empirical Methods in Natural Language Processing (EMNLP), 2021. Punta Cana, DOMINICAN REP
8. Liu, X., et al.: P-Tuning: prompt tuning can be comparable to fine-tuning across scales and tasks. In: 60th Annual Meeting of the Association-for-Computational-Linguistics (ACL), 2022, Dublin, IRELAND
9. Hasan, T., et al.: XL-sum: large-scale multilingual abstractive summarization for 44 languages. arXiv preprint arXiv:2106.13822, 2021
10. Papineni, K., et al.: A method for automatic evaluation of machine translation. In: The Proceedings of ACL-2002, ACL, Philadelphia, PA, July 2002, 2001
11. Rouge, L.C. A package for automatic evaluation of summaries. In: Proceedings of Workshop on Text Summarization of ACL, Spain, 2004

TUN-GCA: A Novel Approach for Organ Segmentation in Nasopharyngeal Carcinoma CT Images

Wenxin Che[1], Penghui Du[1], Rihan Huang[1], Quanying Liu[1], Youzhi Qu[1], and Ziyuan Ye[2(✉)]

[1] Department of Biomedical Engineering, Southern University of Science and Technology, Shenzhen 518055, China
[2] Department of Computing, The Hong Kong Polytechnic University, Hong Kong SAR 999077, China
ziyuanye9801@gmail.com

Abstract. Nasopharyngeal cancer (NPC) is a highly prevalent malignant tumor. Effective radiation therapy planning for NPC requires precise segmentation of nasopharyngeal structures and adjacent organs that may be affected by radiation, based on CT imaging. However, it is challenging to segment the nasopharynx and surrounding organs because they are relatively small and exhibit intricate patterns in CT images. In this work, we introduce a novel three-stage framework, termed TransUNet with Gradually Converging Attention (`TUN-GCA`), to address this problem. Our framework utilizes a TransUNet model pretrained on 3D surrogate labels, allowing the model to initially focus on a small region that encompasses all organs of interest, similar to how human doctors look for anatomical landmarks. Subsequently, the model is trained with actual labels to precisely segment these organs. Finally, DBSCAN clustering is employed to explore multiple organs and eliminate potential voxel outliers, enhancing overall robustness. We evaluated the proposed framework using a dataset of 99 subjects with CT images from nasopharyngeal cancer patients, comprising both left and right parotid and submandibular glands. The extensive experimental results demonstrate that our proposed model substantially outperforms other competitive baselines in terms of accuracy and efficiency.

Keywords: Nasopharyngeal Carcinoma · Computed Tomography · Medical Image Segmentation · Deep Learning · TransUNet

1 Introduction

Nasopharyngeal carcinoma (NPC) is a malignant tumor that occurs in the nasopharynx. Its treatment mainly relies on radiation therapy, which requires precisely targeting the tumor while protecting the surrounding organs, such as

W. Che, P. Du, and R. Huang—These authors contributed equally to this work.

the parotid glands and submandibular glands. Outlining the key organs in radiation therapy is mainly performed by experienced experts, which is subjective, time-consuming and labor-intensive. Therefore, the development of automated segmentation techniques is of great importance.

Deep learning approaches have shown extraordinary potential in medical image segmentation for various diseases [4,16,17,19]. These convolution-based segmentation methods often struggle to handle global and long-range semantic information. To overcome this limitation, some studies have explored and verified the effectiveness of combining convolution with Transformer blocks [6]. However, the Transformer-based models often require a substantial quantity of data to achieve optimal training outcomes, whereas medical image datasets tend to be limited in scale.

Unlike other medical image segmentation tasks with large segmentation targets, nasopharyngeal carcinoma often requires the segmentation of relatively small regions. These regions are structurally complex and exhibit significant inter-individual variability. The intricate anatomical structure of the nasopharyngeal region, coupled with the indistinct boundaries between neoplastic and healthy tissues, poses significant challenges for automated segmentation. Furthermore, to enhance the robustness and precision of segmentation tasks, it is crucial to address the challenge where small regions within an image are disproportionately affected by minor pixel shifts, leading to amplified noise and potentially compromising the accuracy of the segmentation results.

To tackle the above challenges, we propose a three-stage framework, called TransUNet with Gradually Converging Attention (TUN-GCA). Specifically, in the initial stage, we leverage existing gland labels to autonomously generate surrogate labels that encompass regions larger than the original targets. These surrogate labels are subsequently employed to train TransUNet, facilitating the rapid identification of approximate locations of key organs within large-scale images. In the subsequent stage, we fine-tune the pretrained TransUNet using the original segmentation labels. The Density-Based Spatial Clustering of Applications with Noise (DBSCAN) clustering algorithm [8] is further employed to eliminate potential noise points, thereby enhancing the robustness of the results. We verify this framework using a dataset of 99 subjects with CT images from nasopharyngeal cancer patients, comprising both left and right parotid and submandibular glands.

The main contribution of this paper can be summarized into the following three aspects:

(1) We proposed a novel three-stage framework for the segmentation of critical organs in nasopharyngeal carcinoma (Sect. 3). In this framework, an automated surrogate label generation is introduced in the pre-training stage for fast localizing the organs of interest. It is noteworthy that this surrogate-label based pre-training technique does not necessitate the acquisition of additional training data, thereby effectively alleviating the limited data problem prevalent in this field. Through further model finetuning and noise

reduction techniques, our framework can achieve precise and robust segmentation given the limited number of CT images.
(2) We conducted experiments on a nasopharyngeal carcinoma dataset. The extensive comparison in Sect. 4.6 demonstrates that the introduction of surrogate labels contributes to a significant improvement in convergence stability, especially in scenarios where the number of subjects is limited. Furthermore, the incorporation of surrogate labels renders the model more resilient to outliers present in the training data. By reducing the model's sensitivity to these anomalous data points, the overall robustness of the model is enhanced.
(3) The quantitative and qualitative results in Sect. 4.7 illustrate our model's capability in capturing fine-grained semantic information of key organs for nasopharyngeal carcinoma diagnosis, showing superior performance compared to other competitive baselines.

2 Related Work

In the field of automated medical imaging segmentation, convolutional neural networks (CNNs) based methods [2,12,20] are effective in feature extraction. However, these variants are limited in processing complex structures and medical images [22,24].

In contrast to the typical CNN-based approaches, UNet and its variants [7,23] employ a unique architecture that leverages downsampling to generate images of varying resolutions and performs convolution operations on these multi-resolution representations. This hierarchical processing enables the network to capture and integrate features at different scales, leading to a more comprehensive understanding of the image content. During the upsampling phase, where low-resolution feature maps are reconstructed to the original resolution, UNet extracts and synthesizes information from multiple levels of the network. By combining feature representations from different scales during the decoding stage, UNet is able to recover fine-grained details while maintaining a global context.

Since the Transformer model has been widely used in different fields and achieved incredible results, it was subsequently introduced to the image processing field [1,14,18,21]. In recent years, the variations of the combination of UNet and Transformer block have attracted much attention due to the excellent performance they show in different scenarios [9–11,13]. However, the high computational cost of Transformer in processing image data limits its application in real-time medical image processing to some extent. To alleviate the shortcomings of the separate Transformer and UNet models mentioned above, the TransUNet model combines the local detail-capturing ability of Unet and the global perception ability of the Transformer, which effectively improves the efficiency of fine structure segmentation in medical images [3].

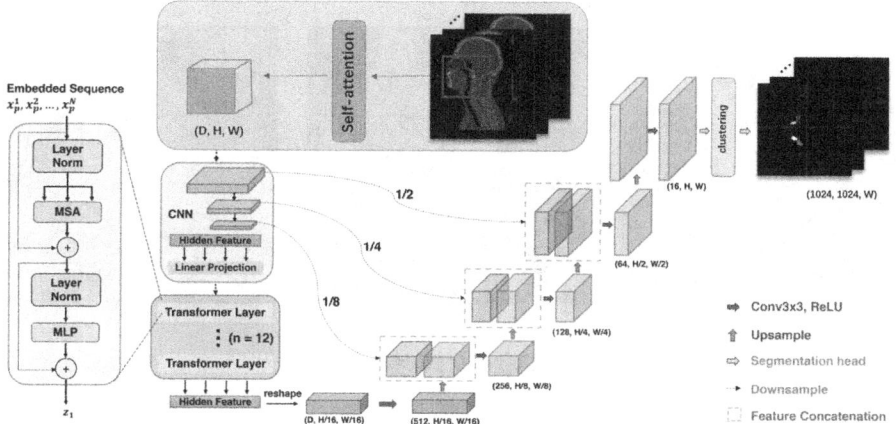

Fig. 1. Overview of the TUN-GCA. To capture global context and dependencies. The CNN backbone extracts hidden features, which are further refined through 12 Transformer layers. The output feature map is reshaped and downsampled to generate multiscale feature maps. The decoder path follows the U-Net architecture, progressively upsampling and concatenating feature maps from the encoder.

3 TUN-GCA

In this section, we introduce our proposed TransUNet with Gradually Converging Attention (TUN-GCA) framework for multi-organ segmentation. Figure 1 illustrates the backbone of TUN-GCA. Our training framework contains three stages. In the first stage (detailed in Sect. 3.1), we leverage an automated surrogate label generation technique for model pre-training. In the second stage (detailed in Sect. 3.2), the pre-trained model undergoes a fine-tuning process using the original label and later derives a binary segmentation mask. In the final stage, the segmentation results will be further refined by employing a clustering technique (detailed in Sect. 3.3), which results in a multi-organ segmentation mask.

3.1 Surrogate Label Generation for Model Pre-training

The parotid and submandibular glands are small in size, not obvious in the original image, and difficult to distinguish by the model. In order to reduce the number of training iterations and optimize the final performance of the model, we designed a pre-training module based on surrogate label generation on the basis of the original TransUNet model. Figure 2 visualize the surrogate label sample generated in this stage.

Specifically, our model is based on the TransUNet architecture [3], utilizing the ResNet-50 together with ViT-16 configuration pre-trained on the ImageNet21K dataset for weight initialization. To further pre-train the model for

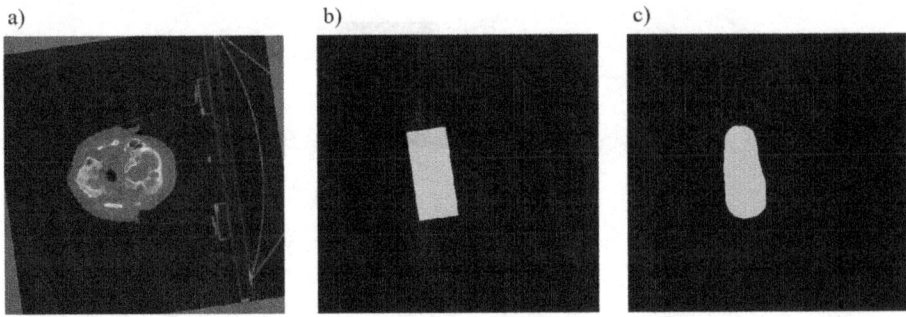

Fig. 2. Example of surrogate label generation and pre-training. (a) Input image; (b) Surrogate label generated at the pre-training stage; (c) Prediction results of surrogate labels.

NPC CT imaging segmentation, the original NPC CT imaging labels are automatically augmented with 3D cubes and designated as surrogate labels. These cube-like surrogate labels cover exactly the x, y, and z-axis range where the label exists, and are employed to pre-train the model, enabling it to focus its attention predominantly on the crucial regions encompassing the left and right parotid glands, as well as the left and right submandibular glands.

3.2 Model Fine-Tuning

In the second stage of our proposed TUN-GCA framework, we employ a fine-tuning process to further refine the pre-trained model using the original labels. During this stage, the model is presented with the ground truth segmentation masks corresponding to the input images. By minimizing the discrepancy between the model's predictions and the ground truth labels, the fine-tuning process enables the model to adapt its learned features and parameters to the multi-organ segmentation task.

This fine-tuning step allows the model to capture more nuanced and task-specific characteristics of the organs of interest, leading to improved segmentation accuracy. Once the fine-tuning process is completed, the model is utilized to generate binary segmentation masks for each input image. These binary masks provide a preliminary region of the organs, which will be further refined in the subsequent stage.

3.3 Multi-organ Exploration

In this stage, we employ the Density-Based Spatial Clustering of Applications with Noise (DBSCAN) algorithm [8] to refine the binary segmentation masks obtained from the previous stage and explore the multiple related organs in nasopharyngeal carcinoma. DBSCAN is particularly well-suited due to its ability to group together spatially connected and densely packed voxels that belong to the same organ, while simultaneously identifying and excluding isolated or spurious voxels that do not conform to the organ's spatial characteristics.

The DBSCAN algorithm analyzes the spatial proximity and density of the voxels and forms clusters based on these criteria. Voxels that are densely packed and spatially connected are grouped together into the same cluster, representing a specific organ. On the other hand, voxels that are isolated or do not exhibit sufficient spatial proximity to other voxels are considered noise points and are excluded from the clusters. This refinement process effectively explores multiple organs and ensures a more coherent and anatomically consistent outline.

4 Experiments

4.1 Dataset

We conducted experiments on a dataset comprising CT images from 99 subjects, each diagnosed and labeled into one of four categories: left parotid gland, right parotid gland, left submandibular gland, and right submandibular gland. All data were presented in three-dimensional NIfTI format (nii.gz).

4.2 Data Preprocessing

Given the variation in the dimensions of axial slices across the dataset, a preprocessing step was implemented to standardize the size of all slices to 512×512 pixels. This standardization ensures uniformity in input data, facilitating efficient training and testing of the model.

To maintain the integrity and clinical relevance of the original data, the predicted outputs were resized back to their original dimensions post-training and testing. This step ensures that the segmentation results are directly comparable to the original CT images, preserving the necessity for clinical interpretation.

4.3 Metrics and Loss Function

To assess the effectiveness of our model's segmentation capabilities, we employ two primary metrics: the Dice Similarity Coefficient (DSC) and the Average Surface Distance (ASD). These technical indicators are instrumental in quantitatively evaluating the accuracy and precision of the segmented outputs relative to the ground truth annotations. Specifically, DSC is defined as:

$$DSC = \frac{2\|Vol_{GT} \cap Vol_{SEG}\|}{\|Vol_{GT}\| + \|Vol_{SEG}\|} \qquad (1)$$

where Vol_{GT} denotes ground truth segmentation results, and Vol_{SEG} denotes the segmentation prediction by our framework.

In terms of ASD, it is defined as the average distance from the points on the segmented region boundary to the real labeled region boundary:

$$ASD = \frac{1}{2}(\frac{\Sigma_{z \in SEG} d(z, GT)}{|SEG|} + \frac{\sigma_{z \in GT} d(u, SEG)}{|GT|}) \qquad (2)$$

where $d(z, GT)$ refers to the shortest distance from the voxel to the ground truth surface; $d(u, SEG)$ denotes the shortest distance from the voxel u to the surface of the SEG which is the result of the automatic segmentation of the algorithm.

We adopt the value in Eq. 3 as the loss function, which effectively quantifies the discrepancy between the predicted segmentation and the ground truth.

$$Loss = 1 - DSC \qquad (3)$$

4.4 Experimental Settings

We performed the 10-fold cross-validation by randomly dividing the dataset into the training group (90%) and validation group (10%). The stochastic gradient descent (SGD) optimizer was utilized in the training stage. The hyperparameter configuration for the optimizer includes an initial learning rate of 0.005, a momentum value of 0.9, and a weight decay coefficient of 10^{-4}. The training was conducted with a batch size of 10, and the process was designed to run for a maximum of 300 iterations to optimize performance without overfitting. All training computations were performed on four NVIDIA V100 GPUs.

4.5 Alternative Baselines

We applied various clustering methods in the final multi-organs exploration stage for comparison, including Balanced iterative reducing and clustering using hierarchies (Birch) [25], Gaussian Mixture Model (GMM) with the number of steps for the Expectation-Maximization (EM) [5], and K-Means [15].

- **Birch**: The Birch method is a hierarchical clustering technique that incrementally constructs a tree structure for clustering large datasets by grouping data points into small clusters and then performing global clustering on these clusters.
- **GMM**: The GMM method is a clustering technique that models the data as a mixture of multiple Gaussian distributions, each representing a cluster, and uses the Expectation-Maximization algorithm to estimate the parameters of these distributions. Specifically, GM denotes a GMM with n_{init} set to 1, and GM10 denotes a GMM with n_{init} set to 10. Here, n_{init} represents the number of initializations. When $n_{init} = 10$, it means that 10 different initializations are performed, and the optimal one is selected.
- **K-means**: The K-means method is a clustering technique that partitions a dataset into K distinct clusters by iteratively assigning data points to the nearest cluster centroid and then updating the centroids based on the mean of the assigned points.

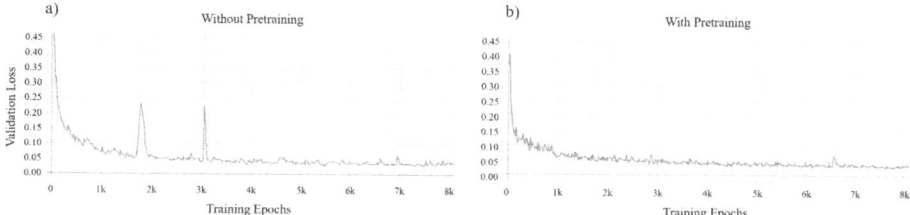

Fig. 3. (a) Change of Dice Loss with the number of iterations for a model trained with 1 epoch after pre-training process. (b) Variation of Dice Loss with the number of iterations when the model without pre-training is formally trained.

4.6 Qualitative Analysis

To verify the effectiveness of surrogate label generation, we compare the validation loss of the model with and without surrogate label pre-training. As Fig. 3 shows, with the surrogate label pre-training, the model required only one epoch (approximately 250 iterations) to accurately localize the target glands. Comparing the validation loss curves between the two scenarios, i.e., with pre-training and without pre-training, it is evident that the introduction of surrogate label pre-training leads to a more stable convergence of the model, particularly during the early stages of training (1k to 3k epochs) where the loss decreases significantly faster and more steadily. This observation suggests that pre-training contributes to enhancing the stability of model training, especially when the available training data is limited. Furthermore, compared to the model without pre-training, the loss curve of the pretrained model is generally smoother and exhibits notably fewer local fluctuations and spikes. This implies that pre-training renders the model more robust to outliers in the training data, preventing it from being biased by individual noisy data points, thereby improving the model's generalization and robustness.

Figures 4 and 5 show the comparison of segmentation results in the multi-organ exploration stage. It is evident that the proposed TUN-GCA method achieves the best performance in the segmentation tasks of both left and right parotid glands as well as left and right submandibular glands, significantly outperforming the other compared methods. In terms of the ASD metric, the curves corresponding to TUN-GCA consistently lie at the bottom for all four organs, indicating that its segmentation results have the smallest deviation from the gold standard, thus achieving the highest segmentation accuracy. In comparison, TUN-GCA with Birch and TUN-GCA with GMM-10 exhibit slightly inferior performance, while TUN-GCA with GMM and TUN-GCA with KMeans yield the largest errors. From the perspective of the DSC, the curves of TUN-GCA generally reside at the top, with the highest values approaching 0.9, demonstrating a high degree of overlap between its segmentation results and the gold standard.

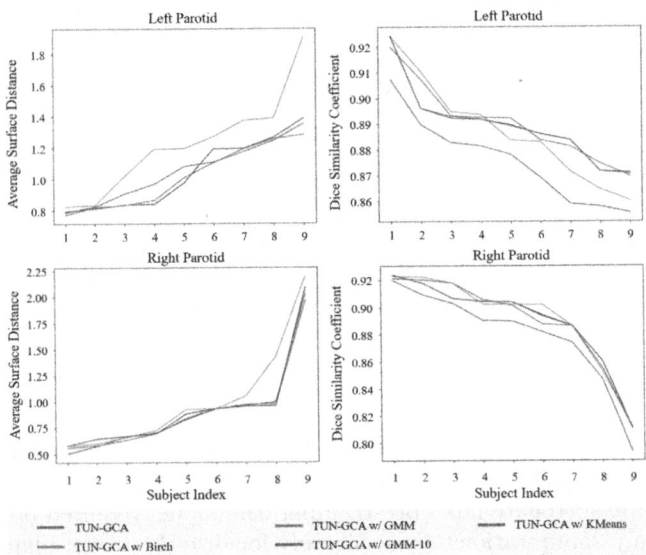

Fig. 4. Average Surface Distance (ASD ↓) and Dice Similarity Coefficient (DSC ↑) between the predicted and actual segmented areas of the parotid glands using 5 different clustering methods.

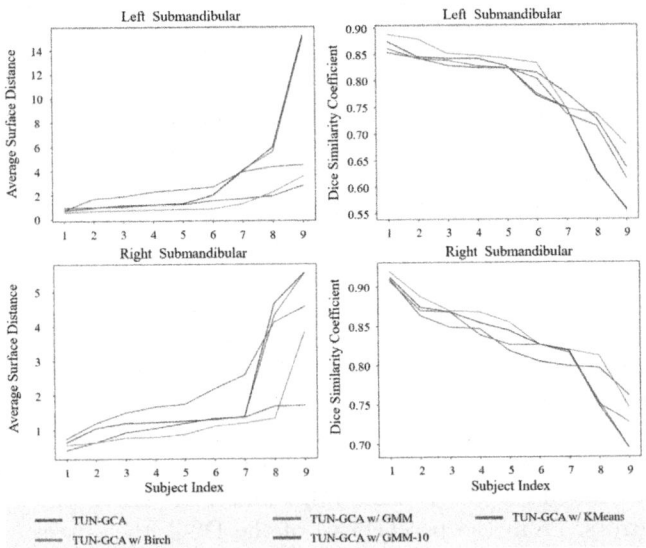

Fig. 5. Average Surface Distance (ASD ↓) and Dice Similarity Coefficient (DSC ↑) between the predicted and actual segmented areas of the submandibular glands using 5 different clustering methods.

The trends in the results under the two evaluation metrics are essentially consistent, indicating that the relative performance rankings of the different methods are stable. However, in the results for the submandibular glands, the advantage of TUN-GCA is more pronounced, possibly because the submandibular glands are smaller and more irregularly shaped, posing greater challenges for the segmentation algorithms, which DBSCAN is better equipped to handle. Apart from overall performance, TUN-GCA also exhibits better stability across different test subjects, as its curves are relatively smooth without significant fluctuations or sudden changes. This suggests that the method has stronger robustness and adaptability to individual anatomical variations.

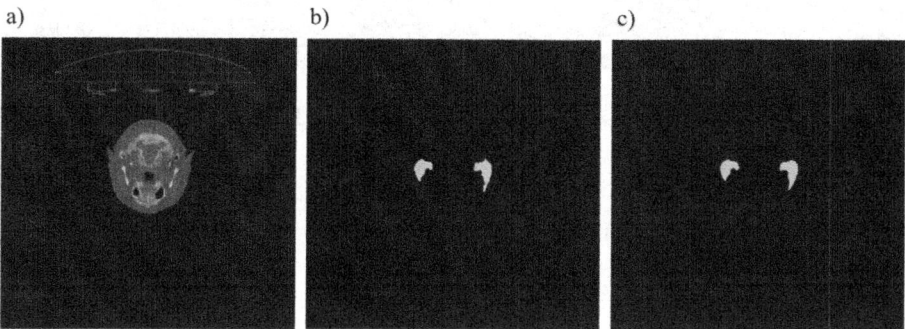

Fig. 6. Example of the segmentation results. (a) Original CT image. (b) Ground truth segmentation results that are marked by the expert. (c) Predicted results by TUN-GCA.

Figure 6 provides an example of the segmentation results obtained using the proposed TUN-GCA framework. The predicted segmentation masks exhibit a high degree of similarity and spatial correspondence with the ground truth annotations.

The predicted segmentations in the bottom row of Fig. 7 exhibit a high degree of spatial correspondence and similarity with their respective ground truth counterparts. The TUN-GCA model successfully captures the intricate morphologies and contours of the parotid glands and submandibular glands. The close agreement between the ground truth and predicted segmentations as shown in Figs. 6 and 7 underscore the effectiveness and accuracy of the TUN-GCA framework in automatically segmenting multiple organs in nasopharyngeal carcinoma CT images.

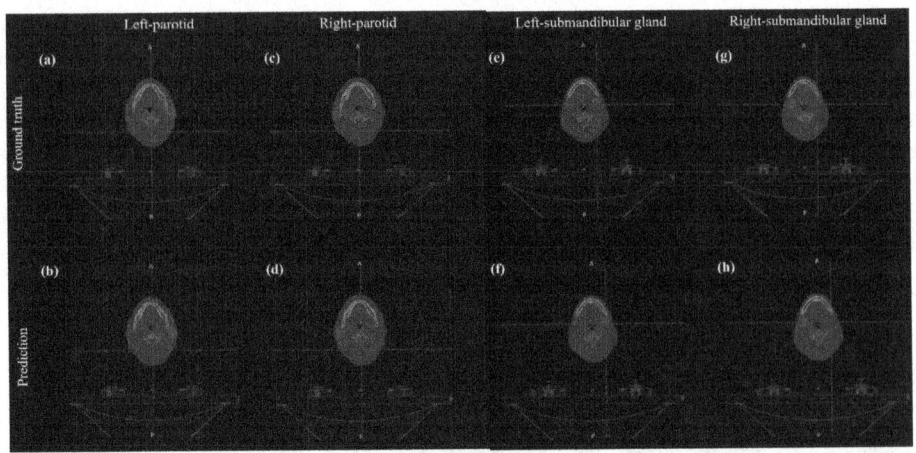

Fig. 7. Example of comparison between the ground truth and predicted glands using TUN-GCA.

4.7 Quantitative Analysis

Table 1. Comparison of different segmentation refinement methods using average ASD and DSC.

Metrics	NPC related organs	Birch	GM	GM10	K-Means	**DBSCAN**
Average ASD (↓)	Left Parotid	1.22	1.03	1.03	**1.02**	1.05
	Right Parotid	1.01	0.94	0.94	**0.91**	**0.91**
	Left Submandibular	**1.27**	3.54	3.59	2.72	1.47
	Right Submandibular	1.24	1.99	2.03	2.26	**1.15**
Average DSC (↑)	Left Parotid	**0.89**	**0.89**	**0.89**	**0.89**	0.88
	Right Parotid	0.89	0.89	**0.90**	**0.90**	0.88
	Left Submandibular	**0.81**	0.77	0.77	0.78	0.79
	Right Submandibular	**0.84**	0.83	0.83	0.83	0.83

Table 1 presents a comparative analysis of various segmentation refinement methods, namely Birch, GM, GM10, K-Means, and DBSCAN, in terms of their performance using average ASD and DSC. Across the four regions, DBSCAN consistently achieves competitive ASD scores, with the lowest values for the right parotid (0.91) and right submandibular (1.15) regions. K-Means also demonstrates strong performance, obtaining the best ASD for the left parotid (1.02) and right parotid (0.91) regions. In terms of DSC, Birch, GM, GM10, and K-Means exhibit comparable performance across the four regions, with values ranging from 0.77 to 0.90. DBSCAN, while still achieving high DSC scores, slightly underperforms in comparison to the other methods for the left and right parotid regions.

Table 2. Comparison of different model backbones using DSC (↑).

NPC related organs	AUnet	DDNN	Unet	**TransUNet**
Left Parotid Gland-average	0.83	0.77	0.55	**0.89**
Right Parotid Gland-average	0.88	0.82	0.58	**0.90**

The results in Table 2 demonstrate that TransUNet consistently outperforms the other models in segmenting both the left and right parotid glands, achieving average DSC scores of 0.89 and 0.90, respectively. These scores surpass those of AUnet (0.83 and 0.88), DDNN (0.77 and 0.82), and Unet (0.55 and 0.58) for the corresponding organs. The superior performance of TransUNet in segmenting the parotid glands is particularly noteworthy, as accurate delineation of these organs is crucial for precise radiation therapy planning and minimizing damage to surrounding tissues in nasopharyngeal carcinoma treatment.

5 Conclusion

In this paper, we introduced TransUNet with Gradually Converging Attention (TUN-GCA), a novel three-stage framework designed to enhance the segmentation of critical organs in nasopharyngeal carcinoma using CT images. Our approach addresses the challenges posed by the small and intricate anatomical structures of the nasopharyngeal region by employing surrogate label generation for model pre-training, and the DBSCAN clustering algorithm for multi-organ exploration. Extensive experiments on a dataset of 99 subjects demonstrated that TUN-GCA substantially outperforms competitive baselines in terms of accuracy and efficiency.

It is noteworthy that the potential applications of our TUN-GCA framework extend far beyond the realm of nasopharyngeal carcinoma. It also holds promise as a valuable tool for other multi-organ segmentation applications [24].

Acknowledgments. This work was funded in part by the National Key R&D Program of China (2021YFF1200804), National Natural Science Foundation of China (62001205), Shenzhen-Hong Kong-Macao Science and Technology Innovation Project (SGDX2020- 110309280100), Shenzhen Science and Technology Innovation Committee (2022410129, KCXFZ20201221173400001), Guangdong Provincial Key Laboratory of Advanced Biomaterials (2022B1212010003).

References

1. Achiam, J., et al.: GPT-4 technical report. arXiv preprint: arXiv:2303.08774 (2023)
2. Anwar, S.M., Majid, M., Qayyum, A., Awais, M., Alnowami, M., Khan, M.K.: Medical image analysis using convolutional neural networks: a review. J. Med. Syst. **42**, 1–13 (2018)
3. Chen, J., et al.: TransUNet: transformers make strong encoders for medical image segmentation. arXiv preprint: arXiv:2102.04306 (2021)
4. Çiçek, Ö., Abdulkadir, A., Lienkamp, S.S., Brox, T., Ronneberger, O.: 3D U-Net: learning dense volumetric segmentation from sparse annotation. In: Ourselin, S., Joskowicz, L., Sabuncu, M.R., Unal, G., Wells, W. (eds) MICCAI 2016. LNCS, vol. 9901, pp. 424–432. Springer, Cham (2016). https://doi.org/10.1007/978-3-319-46723-8_49
5. Dempster, A.P., Laird, N.M., Rubin, D.B.: Maximum likelihood from incomplete data via the em algorithm. J. Roy. Stat. Soc.: Ser. B (Methodol.) **39**(1), 1–22 (1977)
6. Dosovitskiy, A., et al.: An image is worth 16x16 words: transformers for image recognition at scale. arXiv preprint: arXiv:2010.11929 (2020)
7. Du, G., Cao, X., Liang, J., Chen, X., Zhan, Y.: Medical image segmentation based on U-Net: a review. J. Imag. Sci. Technol. **64**(2) (2020)
8. Ester, M., et al.: A density-based algorithm for discovering clusters in large spatial databases with noise. In: KDD, vol. 96, pp. 226–231 (1996)
9. Fan, L., Zhou, Y., Liu, H., Li, Y., Cao, D.: Combining Swin transformer with UNet for remote sensing image semantic segmentation. IEEE Trans. Geosci. Remote Sens. (2023)
10. He, S., Bao, R., Grant, P.E., Ou, Y.: U-netmer: U-Net meets transformer for medical image segmentation. arXiv preprint: arXiv:2304.01401 (2023)
11. Hou, R., Ye, Z., Yang, C., Fu, L., Liu, C., Liu, Q.: Immunofluorescence capillary imaging segmentation: cases study. In: Proceedings of the 30th ACM International Conference on Multimedia, pp. 4830–4838 (2022)
12. LeCun, Y., Bottou, L., Bengio, Y., Haffner, P.: Gradient-based learning applied to document recognition. Proc. IEEE **86**(11), 2278–2324 (1998)
13. Li, G., Jin, D., Yu, Q., Qi, M.: IB-TransUNet: combining information bottleneck and transformer for medical image segmentation. J. King Saud Univ.-Comput. Inf. Sci. **35**(3), 249–258 (2023)
14. Lin, T., Wang, Y., Liu, X., Qiu, X.: A survey of transformers. AI Open **3**, 111–132 (2022)
15. MacQueen, J., et al.: Some methods for classification and analysis of multivariate observations. In: Proceedings of the Fifth Berkeley Symposium on Mathematical Statistics and Probability, vol. 1, pp. 281–297. Oakland, CA, USA (1967)
16. Milletari, F., Navab, N., Ahmadi, S.A.: V-Net: fully convolutional neural networks for volumetric medical image segmentation. In: 2016 Fourth International Conference on 3D Vision (3DV), pp. 565–571. IEEE (2016)
17. Phan, N., Wu, X., Dou, D.: Preserving differential privacy in convolutional deep belief networks. Mach. Learn. **106**(9), 1681–1704 (2017)
18. Qu, Y., et al.: Integration of cognitive tasks into artificial general intelligence test for large models. iScience **27**(4) (2024)
19. Ronneberger, O., Fischer, P., Brox, T.: U-Net: convolutional networks for biomedical image segmentation. In: Medical Image Computing and Computer-assisted Intervention–MICCAI 2015: 18th International Conference, Munich, Germany, 5–9 October 2015, proceedings, part III 18, pp. 234–241. Springer (2015)

20. Sarvamangala, D., Kulkarni, R.V.: Convolutional neural networks in medical image understanding: a survey. Evol. Intel. **15**(1), 1–22 (2022)
21. Vaswani, A., et al.: Attention is all you need. In: Advances in Neural Information Processing Systems, vol. 30 (2017)
22. Wang, R., Lei, T., Cui, R., Zhang, B., Meng, H., Nandi, A.K.: Medical image segmentation using deep learning: A survey. IET Image Proc. **16**(5), 1243–1267 (2022)
23. Yin, X.X., Sun, L., Fu, Y., Lu, R., Zhang, Y.: U-Net-based medical image segmentation. J. Healthc. Eng. **2022** (2022)
24. Yuan, J., et al.: Machine learning applications on neuroimaging for diagnosis and prognosis of epilepsy: a review. J. Neurosci. Methods **368**, 109441 (2022)
25. Zhang, T., Ramakrishnan, R., Livny, M.: Birch: an efficient data clustering method for very large databases. ACM SIGMOD Rec. **25**(2), 103–114 (1996)

Convolutional Neural Networks Based on Axial Counting Attention for Deburring Cross-Sectional Images of Aluminum Profiles with Burrs

Weidong Huang[1], Dan Pan[2(✉)], An Zeng[1], and Baijing Liu[1]

[1] Guangdong University of Technology, Guangzhou, China
{2112305167,2112203066}@mail2.gdut.edu.cn, zengan@gdut.edu.cn
[2] Guangdong Polytechnic Normal University, Guangzhou, China
pandan@gpnu.edu.cn

Abstract. Burrs are thorny protrusions characterized by uneven edges or excess debris on the metal surface. In the aluminum profile extrusion industry, while a few burrs do not significantly affect the product's functionality, they can adversely impact the measurement of critical parameters such as inner and outer diameters. Traditionally, burrs on aluminum profiles have been manually removed before measurement. However, it is feasible and highly efficient to directly remove burrs from cross-sectional images of aluminum profiles with burrs and measure certain parameters using computer vision methods, considering the labor-intensive nature of manual removal and the sporadic occurrence of burrs. Therefore, designing an automatic and robust method to deburr cross-sectional images of aluminum profiles with burrs is imperative. This paper investigates the efficacy of multiple image restoration models for deburring cross-sectional images of aluminum profiles with burrs. Acquiring numerous pairs of cross-sectional images of aluminum profile from the real world is prohibitively expensive. Therefore, we propose a semi-automatic method to synthesize cross-sectional burrs and non-burrs images for training deburring models. Additionally, we design an explainable mechanism called Axial Counting Attention to enhance burr removal effectiveness. This mechanism is effective and capable to focus on the edge areas with burrs in convolution-based networks. Experimental results demonstrate that our simple deburring model effectively removes burrs comparable to manual methods or even surpasses human performance, while maintaining competitive computational costs and inference times.

Keywords: Image deburring · Smooth edge · Attention mechanism · Cross-sectional images · Aluminum profiles

1 Introduction

Due to mechanical processing limitations, excess protruding metal, known as burrs in the industry, occurs occasionally at the edges of metal workpieces. According to the Oxford English Dictionary, a burr is defined as an uneven ridge or edge formed on a material during machining processes. These materials can include metal, wood, plastic, or any other substance undergoing machining processes such as cutting, drilling, punching, etc. [1]. Burrs are undesirable obstructions commonly generated during machining processes, adversely affecting the quality of the workpiece in various aspects [1]. However, in the aluminum profile extrusion industry, the existence of a few burrs is allowed, which does not affect the normal use of metal parts but impacts the measurement of critical parameters including thickness, inner diameter, outer diameter, etc. Traditionally, burrs are manually removed before measuring these criteria on cross-sectional images using computer vision methods or by manual measurement on aluminum profiles. Although manual deburring removes burrs, this additional process is time-consuming, costly, and may compromise dimensional accuracy [1]. Therefore, to reduce labor costs and meet the demands of efficient assembly line production, it is crucial to establish an automatic, robust, and reliable deburring process using cross-sectional images of aluminum profiles. So far, there are a few methods employ machine vision or deep learning techniques for burr detection. [3] used edge detection, Hough line detection, parameter calibration and HSV color segmentation technology to obtain metal ingot contour information to form a mask image, in which the burr position can be accurately located. [4] used the Single Shot MultiBox Detector (SSD) network which was combined with the VGG16 convolutional neural network (CNN) to detect sharpness and burrs along its edges. At present, no one utilizes deep learning algorithms to directly remove burrs through cross-sectional images of aluminum profiles as we do.

We utilize image restoration models to remove burrs, considering that burrs occurring during aluminum profile cutting can be seen as a form of regular noise or corruption that these models can effectively learn to handle. However, there are no relevant datasets for this image deburring task. Capturing numerous cross-sectional images from the real world is time-consuming and expensive. Besides, interference factors like pollutants and obstructions may exist in those images captured from real-world. Therefore, based on the aforementioned reasons, it is necessary to establish a synthetic dataset. The details of how to create a synthetic dataset is describe in Sect. 2.1. We investigated the effectiveness of multiple models for this deburring task. These models are typically used in image restoration tasks such as denoising, deraining, and deblurring. To enhance burr removal effectiveness, we propose a explicable mechanism called Axial Counting Attention. This mechanism focuses on edge areas with burrs and facilitates edge smoothing, specifically tailored for convolution-based networks. We substitute simple channel attention with Axial Counting Attention in NAFNet [5] and CascadedGaze [6], denoting their variants as ACANet and CascadedGaze-ACA, respectively. Both qualitative and quantitative experimental results demonstrate the beneficial impact of Axial Counting Attention on edge smoothing.

The contributions of this paper are summarized as follows:

1. We propose a semi-automatic method to create a training dataset for this novel task of deburring cross-sectional images of aluminum profiles with burrs.
2. We evaluate the effectiveness of various image restoration models in this image deburring task.
3. We propose a mechanism called Axial Counting Attention to enhance the smoothness of edge areas.

2 Cross-Sectional Image Pairs Dataset

As mentioned above, capturing too many cross-sectional images of aluminum profile from the real world is time-consuming and expensive, and such interference factors as pollutants and obstructions may exist in those real cross-sectional images. In this section, we build synthetic image pairs as a training dataset for training deburring models. There are two types of image pairs in the training dataset. One type is constructed with pairs of images, one with burrs and one without, while the other type consists of pairs of the same non-burrs image. The code of synthesizing cross-sectional burrs and non-burrs image as well as the whole training dataset is available at https://github.com/Qing9/deburring_task.

2.1 Synthesize Cross-Sectional Burrs Image and Non-burrs Image

The real aluminum profile cross-section resembles a rectangular shape (see Fig. 1). The burrs on the aluminum profile typically appear as fine-toothed and uneven edges (see Fig. 1). Therefore, it is straightforward to conceive a simple approach: generating an cross-sectional image of aluminum profile by creating a rectangular contour and using mathematical formulas to simulate the burrs. Once we have captured the entire aluminum profile cross-section contour in an image, the next step is to fill in the pixels. To appropriately simulate the complex contours of burrs, we primarily employ trigonometric and power functions. The entire processing, consisting of 3 steps, and an example of synthesizing cross-sectional burrs and non-burrs images are shown in Fig. 2. Step 1 is to create a rectangular contour, Step 2 is to use trigonometric and power functions to simulate burrs on the edge area, and Step 3 is to fill pixels in both the burrs and non-burrs contours.

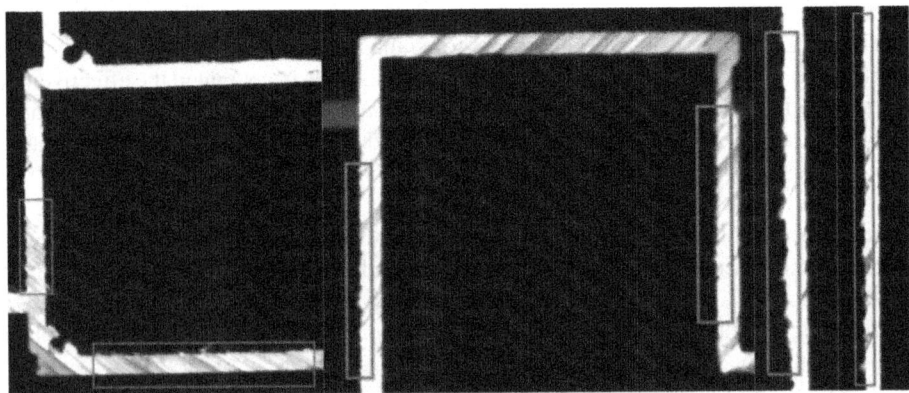

Fig. 1. Burrs on the aluminum profile cross-section

2.2 Image Pair Constructed with the Two Same Non-burrs Image

Models need to learn not only how to remove burrs but also how not to alter other parts without burrs in an aluminum profile cross-section. In other words, we do not want to

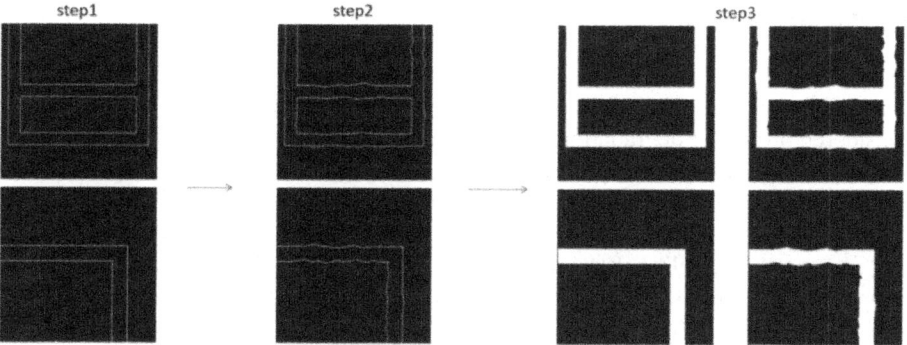

Fig. 2. Three steps to synthesize cross-sectional burrs image and non-burrs image

change the correctness of the original contour of the aluminum profile cross-section. Especially for those complex contours, we do not want a trained model to mistake these complex contours as burrs. For this reason, we add some image pairs constructed with the same non-burrs image. First, we crop some sub-images from real cross-sectional images of aluminum profile with complex contours and non-burrs. We use a precise manual intervention technique to smooth out those slightly uneven edges that are not burrs (see Fig. 3). After obtaining the sub-image with smooth edges, we duplicate it to create another identical image pair.

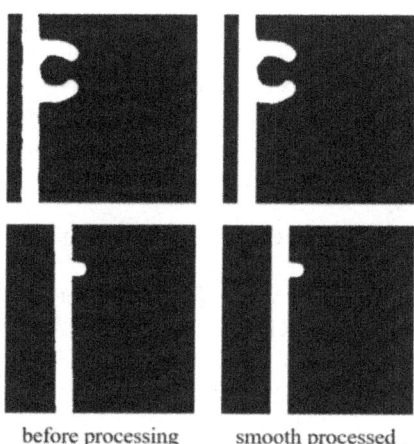

before processing   smooth processed

Fig. 3. Smooth processing

3 Deburring Model

In this section, we propose a mechanism named Axial Counting Attention and establish a simple baseline called ACANet. We illustrate the reasons for constructing such a simple network.

3.1 Model Selection

As mentioned above, we follow the intuition of utilizing image restoration models to remove burrs because we adopt the perspective that burrs on aluminum profiles can be considered a form of degradation. Considering the need for processing cross-sectional images of aluminum profile promptly during assembly line product testing, we do not choose diffusion-based models as our reference model. Although many excellent diffusion models [7–12] perform well in image restoration tasks, and techniques like DDIM [13] and parallel sampling [8] have been developed to address the common issue of slow sampling speeds in diffusion models, the time required to generate an image of normal size, such as 256 × 256, is still unacceptable, not to mention larger sizes. We begin with a simple baseline without too many complex mechanisms or components because we are unsure of what is important and necessary for this deburring task. We want to know if simply a simple network works in this task. We investigated several image restoration models. Significantly, we found that NAFNet [5] adheres to the principle of Occam's Razor, it is a convolution-based model. Owing to its low inter-block complexity and intra-block complexity, it is a good reference model to explore what components are necessary for this deburring task.

3.2 Network Architecture

We adopt the network architecture of NAFNet, including encoder blocks, middle blocks, and decoder blocks. But we replace simple channel attention, a simplified version of channel attention [14], with Axial Counting Attention proposed by us in the basic network block. We refer to it as ACANet in this paper. In addition, we utilize SimpleGate in our basic network block, proposed by [5] and adopted in NAFNet [5] and CascadedGaze [6]. SimpleGate directly divides the feature map into two parts in the channel dimension and performs multiplication. It aims to replace many nonlinear activation functions. It is formulated as:

$$\text{SimpleGate}(X, Y) = X \odot Y \qquad (1)$$

Thus, ACANet is a variant of NAFNet. The ACANet containing 8 encoder blocks, 2 middle blocks, and 8 decoder blocks is shown in Fig. 4, where ACA refers to Axial Counting Attention discussed in next section.

3.3 Axial Counting Attention

We propose a simple and explainable mechanism named Axial Counting Attention (shown in Fig. 5) for this deburring task. This mechanism effectively focuses on areas of the edge that contain burrs. Axial Counting Attention is a mechanism that re-weights the feature maps using two-channel descriptors, wherein the feature maps are compressed through axial counting operations (shown in Fig. 6) in the x-axis and y-axis directions. Each channel descriptor captures the long-range dependence of input feature maps along a spatial direction. The idea of re-weighting feature maps in the x-axis and y-axis directions is derived from coordination attention [15] and adopted in [16]. However, we

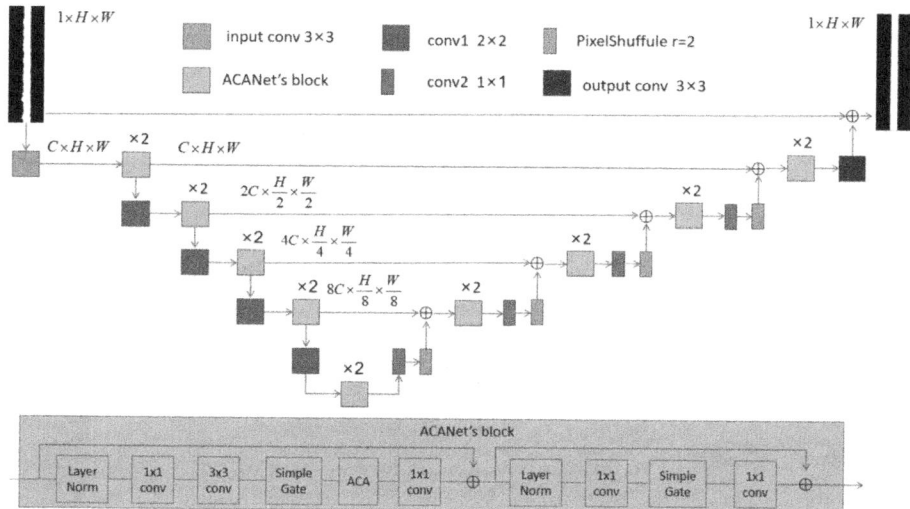

Fig. 4. The architecture of ACANet

simplified it to ensure the low intra-block complexity and made pertinent changes to adapt it for this deburring task better.

In this deburring task, we consider burrs occur on the aluminum profile when there are uneven protrusions or depressions in a local area, causing the contour of that area to deviate significantly from the surrounding contour. Therefore, it is critical to capture the feature of fluctuations of surrounding contour in a cross-sectional image of aluminum profile. The Axial Counting Attention aims to compute weights in the axial direction, which relates to the presence of burrs. These weights are obtained by computing the difference between adjacent pixels in the axial direction, a process referred to as axial counting operation in this paper. Moreover, the sum of the difference between adjacent pixels in the axial direction correlates with the presence of burrs, especially for aluminum profile cross-sections with simple contours. We can imagine this scenario. In the cross-sectional image of aluminum profile, white pixels represent the aluminum profile entity, while the black pixels represent the background. If two adjacent pixels have close values, we consider they belong to the same continuous part. Conversely, if their values differ significantly, they should be the boundary points between the background and the aluminum profile entity. Therefore, aside from extreme cases such as the aluminum profile itself containing parts with a serrated shape, detecting many boundary points in one axial direction implies the presence of burrs. The number of boundary points in one axial direction is linearly correlated with the sum of the differences between all adjacent pixels in that direction. The examples of our proposed Axial Counting Attention are illustrated in Fig. 7.

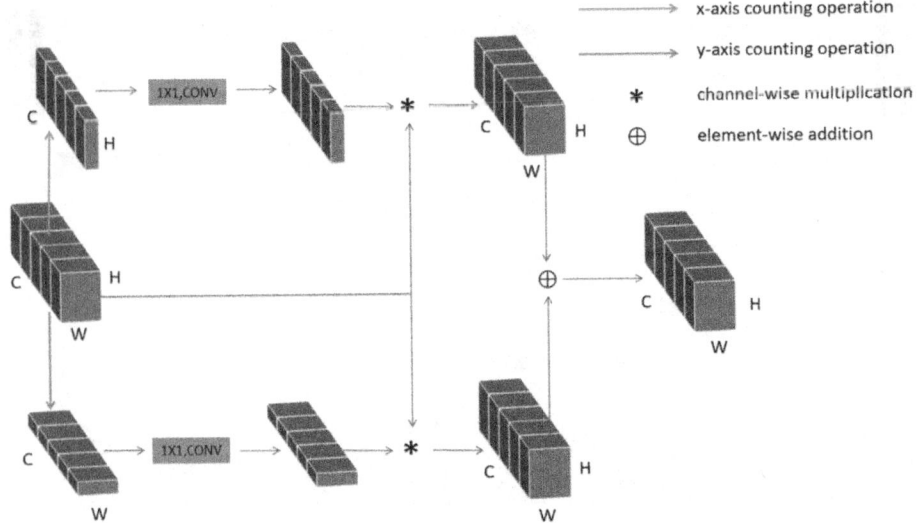

Fig. 5. Axial Counting Attention

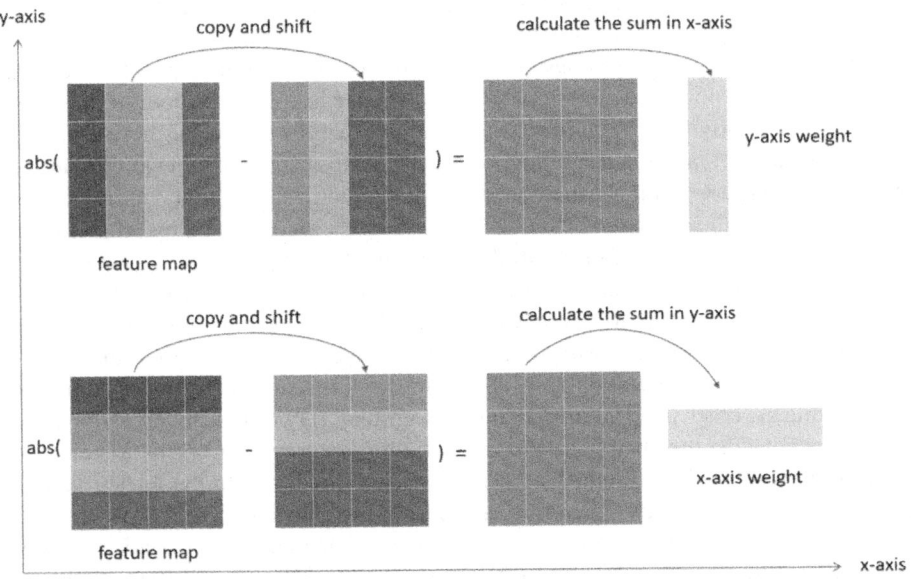

Fig. 6. Axial counting operation

4 Experiments

In this section, we verify the effectiveness of various models in the deburring task, including Restormer [17], NAFNet [5], SERT [18], KBNet [19], MambaIR [20], AST [21], MASH [22], and CascadedGaze [6]. As a result, we discovered that only Restormer,

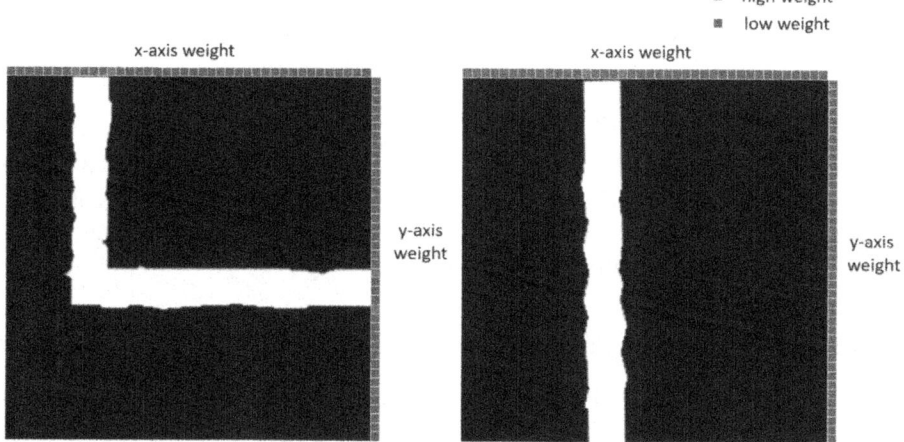

Fig. 7. Examples of Axial Counting Attention

NAFNet, SERT, and CascadedGaze work effectively in this deburring task. Furthermore, we try to combine Axial Counting Attention with Restormer, NAFNet, SERT, and CascadedGaze. We found that Axial Counting Attention does not work in transformer-based models such as Restormer and SERT, but it is effective in convolution-based models like NAFNet and CascadedGaze. We substitute Axial Counting Attention for simple channel attention in NAFNet and CascadedGaze. We denote these variants as ACANet and CascadedGaze-ACA. For ACANet, we build three variants of it containing 9 blocks, 18 blocks, and 36 blocks respectively.

We modify only some necessary parameters, such as the image input and output channels. We completely follow their default parameter settings for other parameters such as the optimizer and specific settings of the model. We stop training when the difference between minimum loss and maximum loss within 10 epochs is less than 1. The learning rate is divided by 10 when the model cannot keep training and we only do it once. For ACANet and CascadedGaze-ACA, we set the initial learning rate to 1e-3, with Adam as an optimizer, and the strategy of changing the learning rate and stopping training is the same as mentioned above. We do not verify the effectiveness of the diffusion-based model because, as mentioned above, the backward denoising process is too lengthy to generate an image. We train and test models on the machine with AMD EPYC 7302 16-Core Processor and GeForce RTX 3090.

4.1 Training Dataset

We employ the method described in Sect. 2.1 to create 500 pairs of cross-sectional images, each containing images with and without burrs. Furthermore, as outlined in Sect. 2.2, we extracted 20 sub-images from real cross-sectional images of aluminum profiles and applied data augmentations such as shifting and rotating, resulting in a total of 80 images. These 80 non-burr images and their duplicates constitute 80 image pairs. This approach ensures that the model does not alter parts without burrs, especially those with complex contours. In summary, our training dataset consists of 580 image pairs, each sized at 256 × 256 pixels. All models are trained on this dataset.

4.2 Experiments on Images Without Burrs

As previously discussed, it is crucial that deburring models do not alter the original contour of an aluminum profile when no burrs are present. By including pairs of images composed of the same non-burr image, we verified that all models maintain the original contour in the absence of burrs, as shown in Fig. 8.

4.3 Experiments on Real Cross-Sectional Images of Aluminum Profile with Burrs

We measured smoothness using curvature as a metric, where smaller values indicate smoother aluminum profile edges. Real cross-sectional images of aluminum profile with burrs were binarized and evaluated using a third-degree polynomial function to simulate aluminum profile edges. Curvature was computed for each side edge of the aluminum profile images, averaging the curvature values for both sides (top and bottom, or left and right). Three groups of images were tested: Group 1 (15 images), Group 2 (10 images), and Group 3 (20 images). Quantitative results are summarized in Tables 1, 2 and 3, while qualitative assessments are depicted in Figs. 9, 10 and 11.

In Group 1, focusing on the third image in Fig. 9 for qualitative analysis, Restormer, NAFNet, and SERT show less effective burr removal. Quantitatively, ACANet and CascadedGaze-ACA achieve superior deburring results, as indicated in Table 1. Restormer, NAFNet, SERT, and CascadedGaze perform comparably to human deburring capabilities. ACANet and CascadedGaze-ACA even outperform human performance significantly, achieving curvature values lower by an order of magnitude compared to other models.

In Group 2, Restormer, NAFNet, and SERT exhibit less effective burr removal in the first and second-to-last images in Fig. 10. CascadedGaze-ACA performs best in qualitative evaluation. Quantitatively, CascadedGaze-ACA and ACANet achieve deburring results superior to manual deburring in the majority, whereas Restormer, NAFNet, and SERT perform worse, as shown in Table 2.

In Group 3, NAFNet demonstrates the least effective deburring in the first image of Fig. 11. Overall, all models achieve comparable smoothness effects to manual deburring, as indicated in Table 3.

4.4 Model Complexity

Table 4 compares the average inference time and GFLOPs per single 256 × 256 image across various models. Based on the quantitative and qualitative results in Sect. 4.3, ACANet achieves deburring results comparable to or even superior to manual methods. Thus, we conclude that ACANet with 9 blocks achieves competitive computation costs and inference times while effectively removing burrs, all with an acceptable parameter count.

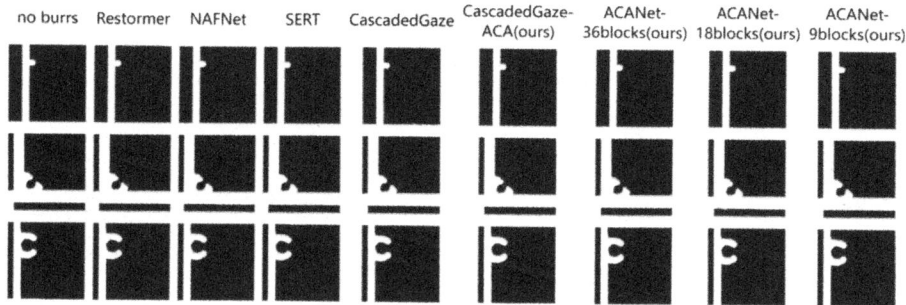

Fig. 8. Results on images without burrs

Table 1. Group 1 average curvature results.

	top-side average curvature	bottom-side average curvature
Manual deburring	1.75×10^{-4}	1.31×10^{-4}
Restormer	2.01×10^{-4}	1.27×10^{-4}
NAFNet	2.19×10^{-4}	1.16×10^{-4}
SERT	3.03×10^{-4}	1.54×10^{-4}
CascadedGaze	2.10×10^{-4}	1.28×10^{-4}
CascadedGaze-ACA	1.50×10^{-5}	9.64×10^{-5}
ACANet-36blocks	4.93×10^{-5}	1.35×10^{-5}
ACANet-18blocks	8.25×10^{-5}	7.29×10^{-5}
ACANet-9blocks	7.36×10^{-5}	3.50×10^{-5}

Table 2. Group 2 average curvature results.

	left-side average curvature	right-side average curvature
Manual deburring	3.91×10^{-5}	2.94×10^{-5}
Restormer	1.03×10^{-4}	6.04×10^{-5}
NAFNet	8.76×10^{-5}	4.94×10^{-5}
SERT	9.54×10^{-5}	5.57×10^{-5}
CascadedGaze	6.38×10^{-5}	4.10×10^{-5}
CascadedGaze-ACA	4.06×10^{-5}	2.22×10^{-5}
ACANet-36blocks	2.40×10^{-5}	1.63×10^{-5}
ACANet-18blocks	3.91×10^{-5}	2.79×10^{-5}
ACANet-9blocks	3.79×10^{-5}	3.19×10^{-5}

Table 3. Group 3 average curvature results.

	top-side average curvature	bottom-side average curvature
Manual deburring	2.12×10^{-4}	1.23×10^{-4}
Restormer	3.14×10^{-4}	1.83×10^{-4}
NAFNet	2.72×10^{-4}	1.83×10^{-4}
SERT	3.86×10^{-4}	1.76×10^{-4}
CascadedGaze	2.86×10^{-4}	2.37×10^{-4}
CascadedGaze-ACA	2.17×10^{-4}	2.13×10^{-4}
ACANet-36blocks	2.62×10^{-4}	1.88×10^{-4}
ACANet-18blocks	2.47×10^{-4}	1.85×10^{-4}
ACANet-9blocks	2.35×10^{-4}	1.65×10^{-4}

Fig. 9. Group 1 deburring results

Fig. 10. Group 2 deburring results

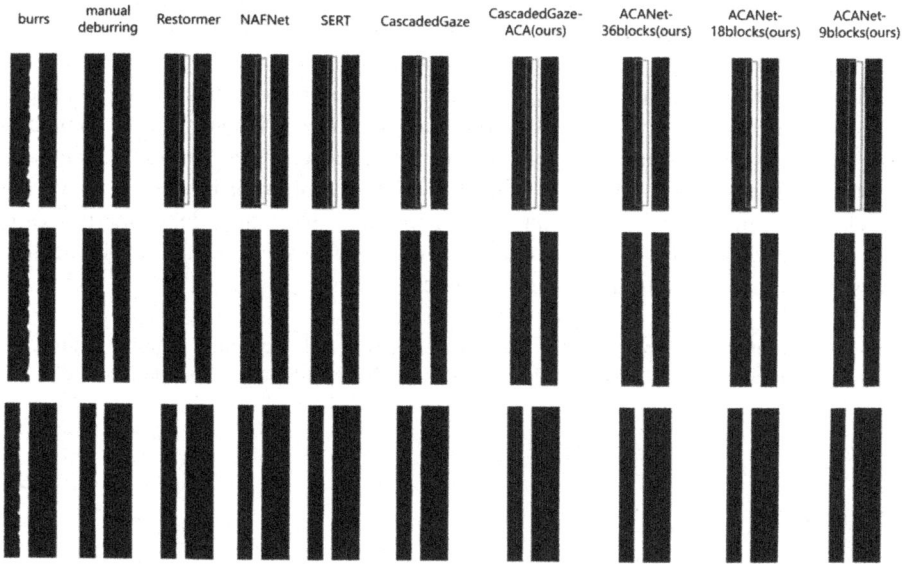

Fig. 11. Group 3 deburring results

Table 4. Comparisons of inference time, GFLOPs and Params of different models

	Time(s)	GFLOPs	Params(M)
Restormer	0.638	38.75	19.90
NAFNet	1.167	63.13	115.86
SERT	2.579	116.31	1.83
CascadedGaze	1.305	61.68	119.13
CascadedGaze-ACA	1.706	64.05	119.13
ACANet-36blocks	1.379	63.89	54.75
ACANet-18blocks	0.737	34.13	30.16
ACANet-9blocks	0.423	19.25	17.87

5 Conclusion

We have demonstrated the effectiveness of several image restoration models in the task of deburring cross-sectional images of aluminum profiles with burrs. By simulating complex burrs and uneven edges with synthetic images, we circumvent the costly and time-intensive process of capturing image pairs form real world. Experimental findings indicate that our deburring models can remove burrs as effectively as traditional manual methods, and in some cases, outperform human capability while operating at a faster speed. Importantly, our results suggest that complex network architectures and components are not necessary for effective burr removal. Simple architectures such as NAFNet and ACANet, leveraging basic components like convolution and Unet, perform well in this deburring task. We introduce Axial Counting Attention as a mechanism to enhance edge smoothness, effectively focusing on areas with burrs. However, it is compatible only with convolution-based models and not transformer-based models. Quantitative results presented in Sect. 4.3 demonstrate that ACANet outperforms other deep learning models, achieving comparable or superior performance to manual deburring processes. Moreover, ACANet offers practical feasibility due to its low computational cost, making it suitable for deployment in real-world applications.

Acknowledgments. This work was supported by Science and Technology Planning Project of Guangdong (Grant Nos. 2021B0101220006, National Natural Science Foundation under grant NO.92267107.

References

1. Jin, S.Y., Pramanik, A., Basak, A.K., et al.: Burr formation and its treatments—a review. Int. J. Adv. Manuf. Technol. **107**, 2189–2210 (2020). https://doi.org/10.1007/s00170-020-05203-2
2. Sung-Lim,K., Dornfeld, D.A.: Analysis of fracture in burr formation at the exit stage of metal cutting. J. Mater. Process. Technol. **58**(2–3), 189–200 (1996). https://doi.org/10.1016/0924-0136(95)02124-8. ISSN 0924–0136

3. Song, H., Yuan, T., Wang, Y., Zhang, D., Fan, R.: Non-ferrous metal defect recognition based on machine vision. In: 2022 IEEE International Conference on Manipulation, Manufacturing and Measurement on the Nanoscale (3M-NANO), Tianjin, China, pp. 60–64 (2022). https://doi.org/10.1109/3M-NANO56083.2022.9941598
4. Pranoto, K.A., Caesarendra, W., Petra, I., Królczyk, G., Surindra, M.D., Yoyo, P.W.: Burrs and sharp edge detection of metal workpiece using CNN image classification method for intelligent manufacturing application. In: 2023 IEEE 21st International Conference on Industrial Informatics (INDIN), Lemgo, Germany, pp. 1–7 (2023). https://doi.org/10.1109/INDIN51400.2023.10217849
5. Chen, L., et al.: Simple baselines for image restoration.In: European Conference on Computer Vision (2022)
6. Ghasemabadi, A., Janjua, M. K., Salameh, M., Zhou, C., Sun, F., Niu, D.: Cascadedgaze: Efficiency in global context extraction for image restoration. Trans. Mach. Learn. Res.(2024)
7. Bhunia, A.K., et al.: Person image synthesis via denoising diffusion model. In: Proceedings of the IEEE/CVF Conference on Computer Vision and Pattern Recognition (2023)
8. Cao, J., et al.: Deep equilibrium diffusion restoration with parallel sampling. In: Proceedings of the IEEE/CVF Conference on Computer Vision and Pattern Recognition (2024)
9. Choi, J., et al.: Conditioning method for denoising diffusion probabilistic models (2021). https://doi.org/10.1109/iccv48922
10. Fei, B., et al.: Generative diffusion prior for unified image restoration and enhancement. In: Proceedings of the IEEE/CVF Conference on Computer Vision and Pattern Recognition (2023)
11. Wang, Z., et al.: Dr2: diffusion-based robust degradation remover for blind face restoration. In: Proceedings of the IEEE/CVF Conference on Computer Vision and Pattern Recognition (2023)
12. Bin X., et al.: DiffIR: efficient diffusion model for image restoration. In: ICCV (2023)
13. Song, J., Meng, C., Ermon, S.: Denoising diffusion implicit models. In: ICLR (2021)
14. Hu, J., Li, S., Gang, S.: Squeeze-and-excitation networks. In: Proceedings of the IEEE Conference on Computer Vision and Pattern Recognition (2018)
15. Hou, Q., et al.: Coordinate attention for efficient mobile network design. In: 2021 IEEE/CVF Conference on Computer Vision and Pattern Recognition (CVPR), 13708–13717 (2021)
16. Wan, Q., Huang, Z., Lu, J., Gang, Y.U., Zhang, L.: Seaformer: squeeze-enhanced axial transformer for mobile semantic segmentation.In: The Eleventh International Conference on Learning Representations (2023)
17. Zamir, S.W., et al.: Restormer: efficient transformer for high-resolution image restoration. In: 2022 IEEE/CVF Conference on Computer Vision and Pattern Recognition (CVPR), pp. 5718–5729 (2021)
18. Li, M., et al.: Spectral enhanced rectangle transformer for hyperspectral image denoising. In: 2023 IEEE/CVF Conference on Computer Vision and Pattern Recognition (CVPR), pp. 5805–5814 (2023)
19. Zhang, Y., et al.: KBNet: kernel basis network for image restoration. ArXiv: abs/2303.02881 (2023)
20. Guo, H., et al.: MambaIR: a simple baseline for image restoration with state-space model. ArXiv: abs/2402.15648 (2024)
21. Zhou, S., Chen, D., Pan, J., Shi, J., Yang, J.: Adapt or perish: adaptive sparse transformer with attentive feature refinement for image restoration. In: Proceedings of the IEEE/CVF Conference on Computer Vision and Pattern Recognition (CVPR), pp. 2952–2963 (2024)
22. Chihaoui, H., Paolo, F.: Masked and shuffled blind spot denoising for real-world images. In: Proceedings of the IEEE/CVF Conference on Computer Vision and Pattern Recognition (2024)

Assessing the Feasibility of Using AI Models to Simplify Brain Imaging Reports for Patients: A Comparative Analysis of Four Large Language Models

Min Xu[1,2(✉)] and Yiwen Wang[1,2]

[1] School of Economics and Management, Fuzhou University, Fuzhou 350108, China
xuminhn@sina.com
[2] Institute of Psychological and Cognitive Sciences, Fuzhou University, Fuzhou 350108, China

Abstract. Brain imaging reports often contain complex medical jargon that is difficult for patients without a medical background to understand. In this study, we examined the ability of four popular large language models (LLMs; ChatGPT-3.5, ChatGPT-4, Google Bard, and Microsoft Bing) to simplify brain magnetic resonance imaging (MRI) reports for patients and compared physicians' and patients' satisfaction with the simplified reports generated by various types of LLMs. Specifically, 24 physicians and 40 participants without a medical background evaluated simplified reports generated from a full MRI report by each of the four LLMs, respectively. The results showed that physicians were satisfied with the comprehensibility, factual correctness, and completeness of the four versions of the simplified reports, but were neutral regarding the potential harm and overall quality of the simplified reports. They were also less likely to send the simplified reports to patients. Additionally, physicians were more satisfied with the reports generated by Microsoft Bing and ChatGPT-4 compared to those generated by the other two LLMs, in terms of potential harm, overall quality, and likelihood of sending the report to patients. Participants, on the other hand, reported that they easily understood the content of the simplified reports, drew correct conclusions, and expressed a high willingness to receive the simplified reports. There was no significant difference in participants' ratings of the four simplified reports. These findings suggest great potential for using AI models to improve patient-centered care in brain imaging and other medical fields.

Keywords: Artificial Intelligence · Brain Imaging · Large Language Models · Magnetic Resonance Imaging

1 Introduction

Brain imaging, such as magnetic resonance imaging (MRI), plays a critical role in providing physicians with important information about patients' brain-related conditions. It helps physicians identify abnormalities, make diagnoses, and monitor treatment effectiveness. However, interpreting MRI reports can be complex and time-consuming. These reports often contain a plethora of information, filled with complex medical jargon and technical terms, making them challenging for patients without a medical background to understand [1]. Providing patients with easily understandable imaging reports can empower them to play a more active role in their treatment process. Ideally, this communication should occur through active dialogue between physicians and patients. However, due to limitations in medical resources, such conversations are often delayed [2]. In this study, we explored whether utilizing artificial intelligence (AI) to generate simplified reports of brain imaging could serve as a coping strategy for this issue.

AI is a branch of computer science that strives to mimic human-like intelligence using computer software and algorithms [3]. One of the most exciting developments in the field of AI is the emergence of large language models (LLMs), including ChatGPT, Google Bard, and Microsoft Bing. These models, based on deep learning algorithms trained on massive datasets, possess the ability to recognize, summarize, translate, predict, and generate text and other forms of content. Their development has ushered in a new era of natural language processing, enabling machines to comprehend and generate human-like text at an unprecedented scale and quality [4, 5]. These LLMs can comprehend user queries and generate appropriate responses, providing assistance to users across various domains.

LLMs are widely utilized in the medical field, particularly in medical diagnosis [6–9]. For instance, Hirosawa et al. [10] found that Google Bard exhibited superior diagnostic performance in common cases. Levkovich and Elyoseph evaluated ChatGPT's ability to assess suicide risk and found that ChatGPT-4's assessment of the likelihood of suicide attempts was similar to that of mental health professionals [8]. LLMs can also provide valuable decision support for physicians [6, 11]. Ahmed et al. [12] compared ChatGPT's concordance with the recommendations set forth by The North American Spine Society (NASS) Clinical Guideline for the Diagnosis and Treatment of Degenerative Spondylolisthesis and found that ChatGPT was concordant for most clinical questions for which NASS offered recommendations. Another study found that although the treatment options suggested by LLMs in precision oncology did not reach the quality and credibility of those by human experts, they generated helpful ideas that might complement established procedures [13]. Elyoseph et al. [14] examined the role of LLMs in prognosis prediction and found that ChatGPT-4 and Google Bard were closely aligned with the perspectives of mental health professionals and the general public in predicting the prognosis of depression. Additionally, LLMs can be used to enhance medical education [15, 16]. For instance, Kung et al. [15] found that ChatGPT-4 has the ability to pass the United States Medical Licensing Exam (USMLE). Cheong et al. [17] found that ChatGPT-4 could accurately answer common questions about obstructive sleep apnea. Patil et al. [18] also found that ChatGPT-4 could accurately provide information on the risks, benefits, and alternatives of computed tomography (CT) and MRI scenarios.

In addition, one of the remarkable abilities showcased by LLMs is their capacity to simplify complex text, making it more accessible to a wider audience. This includes transforming intricate language used in various specialties, thus enhancing comprehension. Given the excellence of LLMs in simplifying text, researchers have also begun to explore the potential of LLMs in simplifying medical imaging reports. For instance,

Chung et al. [19] examined radiologists' perceptions of MRI reports for prostate cancer patients simplified by ChatGPT-3.5. Most radiologists found the simplified reports to be understandable, correct, and comprehensive. However, they expressed concerns about potential harm and the overall quality of the reports. In another study, Jeblick et al. [20] evaluated the ability of ChatGPT (version December 15th, 2022) to simplify full-body CT scans, as well as MRI reports for the knee and head. Professional physicians' assessments indicated that the simplified reports were generally accurate, complete, and unlikely to cause harm to patients. Nevertheless, there were occasional inaccuracies, omissions of relevant medical information, and potentially harmful passages. Similarly, Butler et al. [21] investigated the use of ChatGPT-3.5 in simplifying orthopedic radiology reports, including those for X-rays, CT scans, and MRIs of the foot and ankle. While the simplified reports generally contained accurate information, they sometimes included incorrect statements, as noted by two radiologists. Schmidt et al. [2] investigated physician and patient evaluations of simplified versions of knee MRI reports created by ChatGPT (version December 15th, 2022), varying in complexity from simple to moderate to complex. Physicians found that the simple version did not compromise the accuracy of the reports compared to the complex version. Patients also reported better understanding of the simple versions compared to the complex ones. In a recent study, Kuckelman et al. [22] explored the feasibility of using ChatGPT-4 to simplify musculoskeletal radiology reports. Musculoskeletal radiologists gave high overall ratings for the accuracy and completeness of the simplified reports generated by ChatGPT-4. Only a small portion of the content was deemed potentially confusing or inaccurate.

These studies provide insights into the feasibility of using LLMs to simplify brain imaging reports. However, most of these studies evaluated the quality of the simplified reports primarily from the perspective of professional physicians [19–21], with few focusing on the responses of patients without medical backgrounds [2], who are the primary recipients of these simplified reports. Understanding their reactions is crucial for optimizing the effectiveness of these simplified reports.

Furthermore, these studies primarily focused on using ChatGPT to simplify medical imaging reports, with fewer investigations into other LLMs. Although some studies found ChatGPT's performance to be superior to other LLMs [23, 24], others found no significant differences [25, 26] or even lower performance compared to alternative LLMs [27, 28]. For example, Ilgaz and Celik found no significant difference between ChatGPT-3.5 and Google Bard in answering anatomical questions [29]. The average overall scores for providing accurate responses to common myopia-related queries were also not significantly different between ChatGPT-4 and Google Bard, with Google Bard even slightly outperforming ChatGPT-4 (4.35 vs. 4.23) [25]. Makrygiannakis et al. [27] compared the performance differences of ChatGPT-3.5, ChatGPT-4, Google Bard, and Microsoft Bing in answering clinically relevant questions within the field of orthodontics and found that Microsoft Bing outperformed ChatGPT-3.5 and ChatGPT-4. Understanding the performance differences of these LLMs in simplifying brain imaging reports is essential for selecting the best tool.

To explore these issues, we conducted a study to evaluate the effectiveness of four popular LLMs - ChatGPT-3.5, ChatGPT-4, Google Bard, and Microsoft Bing - in simplifying brain MRI reports. We aimed to compare physician and patient satisfaction with the simplified reports generated by these four LLMs. Our goal is to explore the potential of LLMs in improving patient-centered care by providing easily understandable brain imaging reports.

2 Method

2.1 Participants

24 radiologists (10 females; 39.08 ± 4.34 years) who had been practicing for more than 3 years and 40 participants with no medical background (18 females; 43.05 ± 6.65 years) were recruited. They signed an informed consent form before participating in this survey. The study was performed in accordance with the Declaration of Helsinki and was reviewed and approved by the local Ethics Committee.

2.2 Materials and Procedures

This study utilized materials from Jeblick et al. [20], in which a fictitious brain MRI report was created by a radiologist with 10 years of experience. The report, of moderate complexity, mimicked common clinical cases, including previous medical information, findings on imaging, and conclusions.

On March 1st, 2024, researchers submitted the complete brain MRI report to each LLM, with the prompt to "explain this MRI report to a child using simple language." The LLMs used in this study were not permitted to alter any model settings, resulting in uncertain outputs. To minimize uncertainty, each LLM generated three different simplified reports. Additionally, to avoid response bias caused by cached information, after each simplification, the chat session was restarted and the cache was cleared for each LLM [2].

Table 1 shows the original report and an example of simplified reports generated by each LLM. Between March 2024 and April 2024, every week, a set of three simplified reports (generated by the same LLM) along with the original report were sent to all physicians and participants.

2.3 Questionnaires

To assess the quality of the simplified reports, we created two questionnaires: one for physicians and one for participants without medical backgrounds.

The questionnaire for physicians included questions regarding comprehensibility, factual correctness, completeness, potential harm, overall quality, and the likelihood of sending the report to a patient. The questionnaire utilized a 7-point Likert scale, ranging from 1 (strongly disagree) to 7 (strongly agree).

For participants without medical backgrounds, the questionnaire included questions such as whether they understood the content of the report, whether they could draw the

Table 1. The original report and an example of simplified reports generated by each LLM.

Condition	Reports
Original report [20]	• Several previous examinations are available for comparison The contrast-enhancing mass on the right parietooccipital side shows a progression in size in the short-term follow-up (currently max. 22 mm, in the previous examination max. 12 mm), with significantly increasing perifocal edema. The resection area on the right frontal side remains essentially unchanged, and there is still no definite evidence of a new contrast-enhancing mass. Otherwise, no evidence of recent or intermediate ischemia. No evidence of intracranial hemorrhage. No signs of cerebrospinal fluid circulation disturbance Conclusion: In the short-term follow-up, significantly size-progressive contrast-enhancing space-occupying lesion right parietooccipital with significantly increasing perifocal edema, DD distant GBM manifestation. No evidence of a contrast-enhancing mass in the right frontal resection area
ChatGPT-3.5	• This report is about one's brain. It shows a mass on the right side of the back of the head that has gotten bigger. There's also swelling around this mass. On the front right side, where something was removed before, there's no new lump. There's no evidence of recent stroke, bleeding in the brain, or problems with the fluid around the brain. The conclusion is that that the new mass on the back right side of the head could be a type of brain cancer called glioblastoma
ChatGPT-4	• This is a report of an MRI of the brain that can be compared to previous examinations. It shows a mass in the right side of the brain, near the back, has gotten bigger since your last scan. There's also swelling around this mass. The area of the right front side of the brain where the surgery was performed appears unchanged, with no signs of a new mass. The report says there is no bleeding in the brain or any problems with the fluid that surrounds it. The conclusion of the report says that the mass on the back right side of the head is likely a type of brain cancer called glioblastoma
Google Bard	• This MRI report is talking about a person's brain. The report shows a enhancing mass on the right side of the back the brain that has grown larger and is causing more swelling around it. The area where a previous surgery was done on the right side of the brain hasn't changed. There's no new mass found. There is no bleeding in the brain. The flow of cerebrospinal fluid is normal. The report suggests that the mass in the right back part of the brain may be a manifestation of a type of brain cancer called glioblastoma
Microsoft Bing	• This report is about a person's brain. Physicians can compare current examination with previous examinations. The report shows that there is a mass on the right side of the back of the head that has grown bigger in size from the last scan. The area around this mass has also shown more swelling. The area on the right side of the front of the head, which had been previously treated, is unchanged and shows no new mass. The report shows no ischemia or bleeding in the brain, as well as no problems with the fluid that surrounds the brain. The conclusion is that the mass on the right side of the back of the head may be a type of brain tumor called a glioblastoma

correct conclusions, whether they fully understood the injury or illness being discussed, and whether they would be willing to receive such a simplified report. The questionnaire also used a 7-point Likert scale.

3 Results

3.1 Physicians' Evaluation

The one-sample t-test revealed that physicians rated the comprehensibility of the simplified reports generated by the four LLMs higher than the median score (median = 4; $ps < 0.001$; See Fig. 1). Additionally, the results of the repeated-measures ANOVA indicated no significant difference in comprehensibility scores among the reports generated by the four different LLMs ($F(1, 23) = 0.42, p = 0.74, \eta2\,p = 0.018$).

Physicians also rated the factual correctness of the simplified reports generated by the four LLMs higher than the median score (median = 4; $ps \leq 0.023$). Furthermore, there was no significant difference in factual correctness scores among the reports generated by the four different LLMs ($F(1, 23) = 1.31, p = 0.28, \eta2\,p = 0.054$).

Scores for the completeness of the four versions of the simplified reports were all higher than the median (median = 4; $ps \leq 0.001$). Moreover, there were no significant differences in completeness scores among the four simplified reports ($F(1, 23) = 0.37, p = 0.77, \eta2\,p = 0.016$).

The simplified reports generated by ChatGPT-3.5 and Google Bard received potential harm scores below the median (median = 4), while the scores for ChatGPT-4 and Microsoft Bing were above the median. However, none of these differences reached statistical significance ($ps \geq 0.13$). Moreover, the repeated-measures ANOVA found a significant difference in potential harm scores among the simplified reports generated by the four different LLMs ($F(1, 23) = 3.06, p = 0.048, \eta2\,p = 0.12$). Post-hoc paired t-tests revealed that the potential harm score of Microsoft Bing was higher than those of ChatGPT-3.5 ($t(1, 23) = 2.88, p = 0.008$, Cohen's $d = 0.59$) and Google Bard ($t(1, 23) = 2.01, p = 0.057$, Cohen's $d = 0.41$). Additionally, the potential harm score of ChatGPT-4 was higher than that of ChatGPT-3.5 ($t(1, 23) = 2.14, p = 0.043$, Cohen's $d = 0.44$). No other significant differences were found ($ps \geq 0.20$).

Both ChatGPT-3.5 and Google Bard generated simplified reports that scored significantly lower than the median on the overall quality dimension (median = 4; $ps \leq 0.032$). Although the scores for ChatGPT-4 and Microsoft Bing were above the median, they did not reach statistical significance ($ps \geq 0.42$). There was a significant difference in the scores for overall quality among the simplified reports generated by the four different LLMs ($F(1, 23) = 3.65, p = 0.026, \eta2\,p = 0.14$). Microsoft Bing's score was higher than those of ChatGPT-3.5 ($t(1, 23) = 2.50, p = 0.020$, Cohen's $d = 0.51$) and Google Bard ($t(1, 23) = 2.54, p = 0.018$, Cohen's $d = 0.52$). Additionally, the score of ChatGPT-4 was higher than those of ChatGPT-3.5 ($t(1, 23) = 2.04, p = 0.053$, Cohen's $d = 0.42$) and Google Bard ($t(1, 23) = 3.19, p = 0.004$, Cohen's $d = 0.65$). No other significant differences were found ($ps \geq 0.69$).

Physicians were less likely to send the simplified reports generated by ChatGPT-3.5 and Google Bard to patients, with scores significantly lower than the median (median = 4; $ps \leq 0.022$). Similarly, ChatGPT-4 and Microsoft Bing also scored below the median, but these differences did not reach statistical significance ($ps \geq 0.41$). There was a significant difference in the scores of the four simplified reports ($F(1, 23) = 3.43, p = 0.031, \eta2\,p = 0.13$). Microsoft Bing's score was higher than those of ChatGPT-3.5 ($t(1, 23) = 2.22, p = 0.037$, Cohen's $d = 0.45$) and Google Bard ($t(1, 23) = 2.41, p = 0.025$,

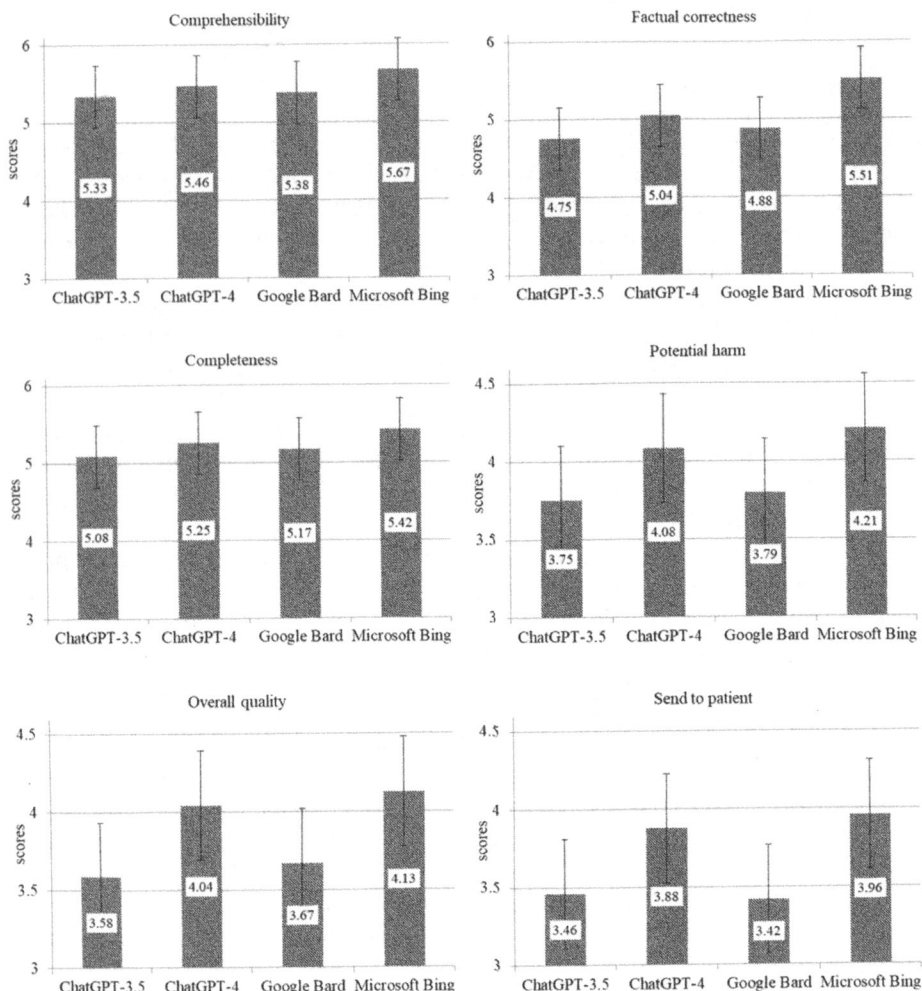

Fig. 1. Rating of statements in simplified reports generated by different LLMs by physicians.

Cohen's $d = 0.49$). The score of ChatGPT-4 was also higher than those of ChatGPT-3.5 ($t(1, 23) = 2.32, p = 0.030$, Cohen's $d = 0.47$) and Google Bard ($t(1, 23) = 2.11, p = 0.046$, Cohen's $d = 0.43$). No other significant differences were found ($ps \geq 0.69$).

3.2 Patient Evaluation

Participants without medical backgrounds reported that they could understand the content, draw correct conclusions, and comprehend the injury or illness discussed in the simplified versions of the MRI reports generated by all four LLMs. They also expressed willingness to receive such simplified reports. The one-sample t-test revealed that their ratings for these items were significantly higher than the median (median = 4; $ps < 0.001$; See Fig. 2). Additionally, the results of the repeated-measures ANOVA showed that the participants' ratings for the four versions of the simplified reports did not differ significantly on any of the items ($ps \geq 0.21$).

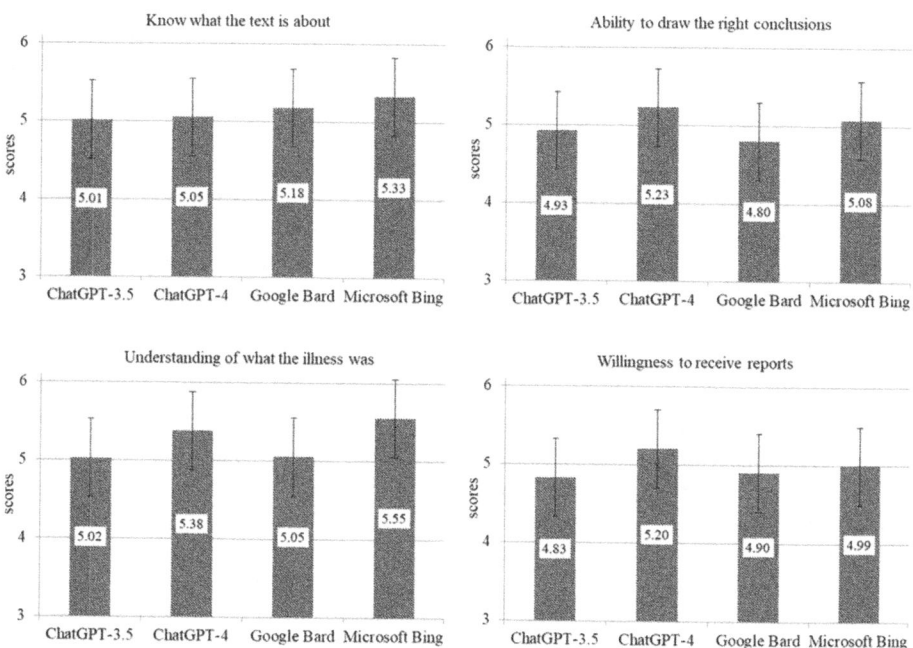

Fig. 2. Rating of statements in simplified reports generated by different LLMs by participants without medical backgrounds.

4 Discussion

The development of LLMs, such as ChatGPT, has emerged as a promising solution for generating simplified medical imaging reports for patients without medical backgrounds. In our study, we aimed to assess the feasibility of using LLMs to generate simplified brain imaging reports for patients. Specifically, we examined the ability of four popular LLMs (ChatGPT-3.5, ChatGPT-4, Google Bard, and Microsoft Bing) to simplify brain MRI reports for patients and compared physician and patient satisfaction with the simplified reports generated by these different LLMs.

Physicians' evaluations indicated satisfaction with the comprehensibility, factual correctness, and completeness of the simplified reports. However, they expressed neutrality or even negativity regarding the potential harm and overall quality of the reports, and were unlikely to send them to their patients. These findings are consistent with previous research. For instance, Jeblick et al. [20] found that physicians held a neutral stance on the potential harm of simplified head MRI reports generated by ChatGPT. Similarly, Chung et al. [19] discovered that radiologists were unlikely to be satisfied with the potential harm and overall quality of knee MRI reports generated by ChatGPT, and were also unlikely to send these reports to patients. These results suggest that physicians still hold a relatively cautious attitude towards using LLMs to generate simplified imaging reports.

In contrast, patients without medical backgrounds reported understanding the content of the simplified reports and drawing correct conclusions. They also expressed a strong willingness to receive these simplified reports. This could be attributed to the fact that patients without medical backgrounds often have a greater need for simplified reports to understand their own medical conditions. Previous research has also found that non-expert participants are more likely to use artificial intelligence models to assist in decision-making compared to industry experts [30]. These results underscore the necessity of bridging the gap in understanding between physicians and patients before using LLMs to generate medical imaging reports for patients without medical backgrounds.

Furthermore, we explored differences in physician and patient attitudes toward simplified reports generated by different LLMs. The results showed that there was no significant difference in participants' ratings of the four simplified reports. However, physicians had more positive evaluations of the simplified reports generated by ChatGPT-4 and Microsoft Bing, and they were more inclined to send them to their patients compared to those generated by ChatGPT-3.5 and Google Bard. As mentioned in the Introduction, different LLMs operate based on different underlying logics, resulting in variations in their performance across different tasks [23, 27]. The findings of this study suggest that ChatGPT-4 and Microsoft Bing are superior tools for simplifying brain imaging reporting for patients compared to ChatGPT-3.5 and Google Bard.

This study has several limitations that should be considered. The first limitation is that the evaluation of the simplified reports was conducted in a simulated setting, which may not fully reflect real-world scenarios. Future studies could involve real patients and clinical settings to validate the findings of this study. Secondly, this study exclusively focused on brain imaging reports. While this allowed us to delve deeply into the specific context of neuroimaging, the findings may not be applicable to other types of medical imaging reports. Future research should explore the use of LLMs in generating simplified reports for different types of medical imaging, such as musculoskeletal or abdominal imaging. Lastly, this study only evaluated four LLMs. There are many other LLMs available, and their performance in generating simplified imaging reports may vary. Future research could include a broader range of LLMs to provide a more comprehensive comparison.

In summary, the present study revealed differences in physician and patient attitudes toward simplified versions of brain imaging reports generated by LLMs. While patients showed positive evaluations of the simplified reports, physicians maintained a relatively cautious attitude towards using LLMs to generate such reports. Furthermore, our findings indicated differences in the performance of different LLMs in generating simplified brain imaging reports, with ChatGPT-4 and Microsoft Bing outperforming ChatGPT-3.5 and Google Bard. These findings offer valuable insights into the feasibility and acceptance of using LLMs to generate simplified brain imaging reports for patients without medical backgrounds.

Disclosure of Interests.. The authors have no competing interests to declare that are relevant to the content of this article.

References

1. Martin-Carreras, T., Cook, T.S., Kahn, C.E.: Readability of radiology reports: implications for patient-centered care. Clin. Imaging **54**, 116–120 (2019)
2. Schmidt, S., Zimmerer, A., Cucos, T., Feucht, M., Navas, L.: Simplifying radiologic reports with natural language processing: a novel approach using ChatGPT in enhancing patient understanding of MRI results. Arch. Orthop. Trauma. Surg. **144**(2), 611–618 (2024)
3. Shi, Z.Z., Zheng, N.N.: Progress and challenge of artificial intelligence. J. Comput. Sci. Technol. **21**(5), 810–822 (2006)
4. Qi, S.H., Cao, Z.Y., Rao, J., Wang, L., Xiao, J., Wang, X.: What is the limitation of multimodal LLMs? A deeper look into multimodal LLMs through prompt probing. Inf. Process. Manage. **60**(6), 103510 (2023)
5. Tsai, M.L., Ong, C.W., Chen, C.L.: Exploring the use of large language models (LLMs) in chemical engineering education: Building core course problem models with Chat-GPT. Educ. Chem. Eng. **44**, 71–95 (2023)
6. Sandmann, S., Riepenhausen, S., Plagwitz, L., Varghese, J.: Systematic analysis of ChatGPT, Google search and Llama 2 for clinical decision support tasks. Nat. Commun. **15**(1), 1–8 (2024)
7. Levkovich, I., Elyoseph, Z.: Identifying depression and its determinants upon initiating treatment: ChatGPT versus primary care physicians. Fam. Med. Community Health **11**(4), 1–12 (2023)
8. Levkovich, I., Elyoseph, Z.: Suicide risk assessments through the eyes of ChatGPT-3.5 Versus ChatGPT-4: vignette study. JMIR Ment. Health **10**, 1–11 (2023)
9. Caruccio, L., Cirillo, S., Polese, G., Solimando, G., Sundaramurthy, S., Tortora, G.: Can ChatGPT provide intelligent diagnoses? A comparative study between predictive models and ChatGPT to define a new medical diagnostic bot. Expert Syst. Appl. **235**, 121186 (2024)
10. Hirosawa, T., Mizuta, K., Harada, Y., Shimizu, T.: Comparative evaluation of diagnostic accuracy between Google bard and physicians. Am. J. Med. **136**(11), 1–23 (2023)
11. McLean, A.L., Wu, Y.H., McLean, A.C.L., Hristidis, V.: Large language models as decision aids in neuro-oncology: a review of shared decision-making applications. J. Cancer Res. Clin. Oncol. **150**(3) (2024)
12. Ahmed, W., et al.: ChatGPT versus NASS clinical guidelines for degenerative spondylolisthesis: a comparative analysis. Eur. Spine J. **1**, 1–22 (2024)
13. Benary, M., et al.: Leveraging large language models for decision support in personalized oncology. Jama Netw. Open **6**(11) (2023)
14. Elyoseph, Z., Levkovich, I., Shinan-Altman, S.: Assessing prognosis in depression: comparing perspectives of AI models, mental health professionals and the general public. Fam. Med. Community Health **12**(1), 1–9 (2024)
15. Kung, T.H., et al.: Performance of ChatGPT on USMLE: potential for AI-assisted medical education using large language models. PLOS Digit. Health **2**(2), e0000198 (2023)
16. Safranek, C.W., Sidamon-Eristoff, A.E., Gilson, A., Chartash, D.: The role of large language models in medical education: applications and implications. JMIR Med. Educ. **9**, e50945 (2023)
17. Cheong, R.C.T., et al.: Artificial intelligence chatbots as sources of patient education material for obstructive sleep apnoea: ChatGPT versus Google Bard. Eur. Arch. Otorhinolaryngol. **281**(2), 985–993 (2024)
18. Patil, N.S., et al.: Artificial intelligence Chatbots' understanding of the risks and benefits of computed tomography and magnetic resonance imaging scenarios. Can. Assoc. Radiol. J. **1**, 1–7 (2024)

19. Chung, E.M., Zhang, S.C., Nguyen, A.T., Atkins, K.M., Sandler, H.M., Kamrava, M.: Feasibility and acceptability of ChatGPT generated radiology report summaries for cancer patients. Digital Health **9**, 1–7 (2023)
20. Jeblick, K., et al.: ChatGPT makes medicine easy to swallow: an exploratory case study on simplified radiology reports. Eur. Radiol. **1**, 1–9 (2023)
21. Butler, J.J., et al.: From jargon to clarity: Improving the readability of foot and ankle radiology reports with an artificial intelligence large language model. Foot Ankle Surg., 1–7 (2024)
22. Kuckelman, I.J., Wetley, K., Yi, P.H., Ross, A.B.: Translating musculoskeletal radiology reports into patient-friendly summaries using ChatGPT-4. Skeletal Radiol., 1–4 (2024)
23. Koga, S., Martin, N.B., Dickson, D.W.: Evaluating the performance of large language models: ChatGPT and Google Bard in generating differential diagnoses in clinicopathological conferences of neurodegenerative disorders. Brain Pathol. **34**(3), 1–4 (2024)
24. Patil, N.S., Huang, R.S., van der Pol, C.B., Larocque, N.: Comparative performance of ChatGPT and bard in a text-based radiology knowledge assessment. Can. Assoc. Radiol. J.-J. De L Assoc. Canadienne Des Radiologistes. **75**(2), 344–350 (2024)
25. Lim, Z.W., et al.: Benchmarking large language models' performances for myopia care: a comparative analysis of ChatGPT-3.5, ChatGPT-4.0, and Google Bard. Ebiomedicine **95**, 1–11 (2023)
26. Thibaut, G., Dabbagh, A., Liverneaux, P.: Does Google's Bard Chatbot perform better than ChatGPT on the European hand surgery exam? Int. Orthop. **48**, 151–158 (2023)
27. Makrygiannakis, M.A., Giannakopoulos, K., Kaklamanos, E.G.: Evidence-based potential of generative artificial intelligence large language models in orthodontics: a comparative study of ChatGPT, Google Bard, and Microsoft Bing. Eur. J. Orthod., cjae017 (2024)
28. Gan, R.K., Ogbodo, J.C., Wee, Y.Z., Gan, A.Z., González, P.A.: Performance of Google bard and ChatGPT in mass casualty incidents triage. Am. J. Emerg. Med. **75**, 72–78 (2024)
29. Ilgaz, H.B., Çelik, Z.: The significance of artificial intelligence platforms in anatomy education: an experience with ChatGPT and Google Bard. Cureus J. Med. Sci. **15**(9), 1–11 (2023)
30. Logg, J.M., Minson, J.A., Moore, D.A.: Algorithm appreciation: people prefer algorithmic to human judgment. Organ. Behav. Hum. Decis. Process. **151**, 90–103 (2019)

How Do Transformers Integrate Meanings? An Investigation Using Interpretable Brain-Based Componential Semantics in Two-Word Phrases

Shaonan Wang[1,2(✉)]

[1] State Key Laboratory of Multimodal Artificial Intelligence Systems, Institute of Automation, CAS, Beijing, China
shaonan.wang@nlpr.ia.ac.cn
[2] School of Artificial Intelligence, University of Chinese Academy of Sciences, Beijing, China

Abstract. Transformer models have proven to be an efficient architecture for large language models, enabling them to exhibit human-like behaviors. However, it remains unclear how they combine the properties of individual words. To investigate this, we focus on the simplest compositional unit—two-word phrases—and employ four simple, parameter-free composition operations: addition, multiplication, first word, and second word. This approach serves as the first step toward understanding the composition mechanisms in transformer modules. We propose straightforward interpretation methods based on brain-based componential semantics. Specifically, we map the distributed vector space in transformers to an interpretable brain-based componential space to explore the intrinsic properties of representation and their semantic compositionality. Our findings show that, like humans, phrase types and semantic features influence the combination process in transformers. However, unlike humans, most phrase types paired with semantic features require more complex combination operations than simple addition or multiplication.

Keywords: Transformer · Word representations · Composition operations · Brain-based componential semantics

1 Introduction

For much of history, humans were the only beings capable of creating and understanding complex natural language. However, with the advent of large language models like ChatGPT, machines now exhibit language abilities comparable to those of humans. These models have even passed the Turing Test, making it increasingly difficult to distinguish between human and machine-generated responses [1,16]. Despite this achievement, how these models work remains underexplored.

At the heart of large language models is the transformer architecture, which leverages multiple self-attention mechanisms to integrate meanings [10], a process central to language comprehension (Li et al., 2023). To further the development of advanced language processing models, it is crucial to examine and compare the mechanisms of composition within transformer-based models. This paper investigates how composition functions in transformers using simple two-word phrases. We employ an interpretable, brain-based componential semantic representation to demystify the inner workings of these 'black boxes.'

The brain-based componential semantic representation was developed to study how the brain represents word meanings. This question has been a longstanding focus of cognitive psychology. Currently, most researchers agree that meanings are represented across multiple brain regions and are at least partly embodied in perception, action, and other neural systems related to individual experiences [3,11,18]. Summarizing previous work, Binder et al. [2] proposed "brain-based componential semantic representations," which are based entirely on functional divisions in the human brain. They represent concepts using properties like vision, somatic, audition, spatial, and emotion. This semantic space is interpretable and offers the most comprehensive explanation of human concept representation to date, making it an ideal tool for interpreting machine learning models.

The two-word phrase paradigm was proposed to study combinatory processing in the brain using basic composition units. Regarding conceptual combination in the brain, substantial evidence indicates that the left anterior temporal lobe (LATL) activates during conceptual composition within 200–250 milliseconds [9,13]. To understand this process, Wang et al., [17] used both human annotations of semantic features and Magnetoencephalography (MEG) data for phrases and their component words. They found that conceptual combination is not a one-size-fits-all process; phrase types and semantic features influence the choice of composition operations. Specifically, sensory-motor features favored simple first and second-word models as composition operations, relying solely on the semantic ratings of one component word to determine the rating at the phrasal level. In contrast, non-sensory feature ratings largely depended on the combination of both words, favoring multiplication as the composition operation. Additionally, VerbNoun phrases showed a distinct pattern compared to other phrase types. For perceptual features, VerbNoun constructions blended component words through multiplication rather than relying on individual word models.

This paper builds on prior research into human compositional abilities [17] to examine how meanings are integrated in transformer-based models. We extract representations of phrases and their component words from GPT and BERT architectures, project them into an interpretable semantic space, identify encoded properties across layers, and determine the composition operations used. Results show that, like humans, phrase types (e.g., adjective-noun, verb-noun) and semantic features (e.g., vision, audition, motor) significantly influence the choice of composition function. However, unlike humans, most phrase types paired with most semantic features require more complex operations than simple addition or multiplication. Similar trends are observed across BERT models, while GPT and BERT exhibit distinct characteristics, with different layers showing unique patterns.

2 Brain-Based Componential Semantic Representations

The brain-based componential semantic dataset, proposed by Binder et al. [2], contains 535 different types of concepts, comprising 122 abstract words and 413 concrete words, including nouns, verbs, and adjectives[1]. Each concept is evaluated across 65 properties including vision, somatic, audition, gustation, olfaction, motor and so on. Crowd-sourced rating experiments provide saliency scores (ranging from 0 to 6) for each attribute of all 535 concepts. Through extensive experiments, Binder et al. [2] showed that the brain-based semantic vectors effectively capture semantic similarities and align well with prior conceptual categories, supporting the validity of the dataset.

Wang et al. [17] extend this dataset to the phrase level, rating 216 phrases and 107 component words across the 65 semantic features defined by Binder et al. [2]. Figure 1 demonstrates an example of the brain-based semantic vectors for the phrase 'green cake' and its two component words. As expected, the abstract adjective 'green' carries more weight on abstract properties, while the concrete noun 'cake' emphasizes sensory and motor properties. For the feature "color," 'green cake' receives the same rating as 'green,' whereas for other features, 'green cake' follows the ratings of 'cake.'

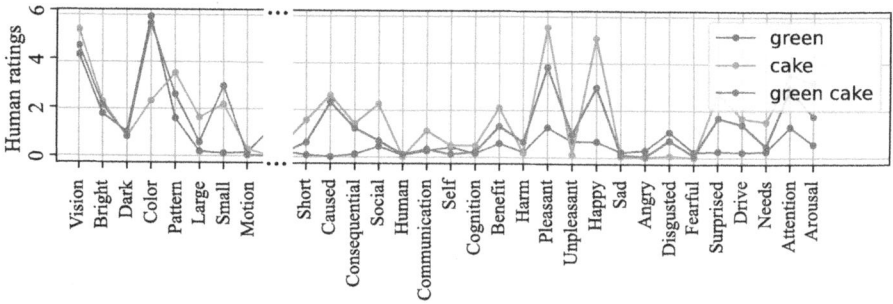

Fig. 1. An example of brain-based semantic vectors, each dimension rated and averaged by 30 individuals.

---

[1] The dataset can be found at: http://www.neuro.mcw.edu/resources.html.

3 Methodology

3.1 From Transformer Semantic Space to Experiential Semantic Space

The core algorithm of large language models uses a transformer structure with self-attention for composition. As illustrated in Fig. 2, its composition process functions by first learning word embeddings, and then generating three vectors per word: query, key, and value. The query vector of one word is dot-multiplied with the key vectors of others to produce scores, which are processed through division and softmax to derive an importance score. This score is multiplied with the value vector, and the results are averaged to create the phrase representation.

To investigate how meanings are combined in transformer-based models, we project their representations from transformer semantic space into experiential semantic space. We also examine the composition operations employed by different types of phrases, referred to as linguistic relations, when associated with various semantic features (relation-feature pairs).

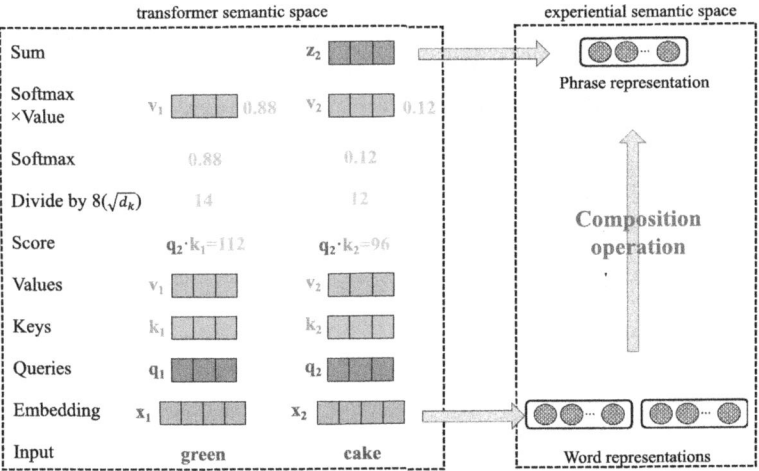

Fig. 2. Framework of our method by projecting the semantic space of transformers into an experiential semantic space and exploring the compositional operations used by different types of phrases.

3.2 Composition Operations

For composition operations, we employed Addition and Multiplication to examine the composition process between the component words, denoted as w1 (the first word) and w2 (the second word) in a two-word phrase. Additionally, we utilized two simpler models for comparison: Word1 (corresponding to the first word) and Word2 (corresponding to the second word).

- Addition: (w1 + w2)/2
- Multiplication: (w1 × w2)/6
- Word1: w1
- Word2: w2

For instance, given w1 = [1, 2, 3] and w2 = [4, 5, 6], predicted ratings using Addition, Multiplication, Word1, and Word2 methods would be [5/2, 7/2, 9/2], [2/3, 5/3, 3], [1, 2, 3], and [4, 5, 6], respectively. The four operations mentioned above represent distinct methods of semantic composition. Addition involves directly adding corresponding word features, reflecting the superposition effect of brain activation. Multiplication, on the other hand, multiplies word features to form phrase meaning, representing the interactive effect of componential words. The predicted phrasal ratings by Addition or Multiplication are normalized to 0–6 range by dividing 2 and 6, respectively. The Word1 and Word2 methods solely consider the features of the first and second words to determine phrase meaning.

3.3 Composition Error

To quantify the composition process, we developed a metric known as 'composition error'. This metric quantifies the challenge of conceptual composition by measuring the discrepancy between predicted and observed phrase ratings. The difference is computed for each feature independently, achieved by employing the composition operation on the constituent words to derive the predicted feature ratings and then subtracting them from the observed phrase ratings. The resulting error score indicates how effectively the operation captures the relationship between observed word and phrase features, serving as an indicator of the suitability of employing the composition operation to combine words into phrases. For instance in Fig. 3, we first independently apply different composition operations to the word vectors for 'green' and 'cake.' This process yields predicted phrase vectors for 'green cake.' We then subtract these from the human-annotated phrase vector for 'green cake,' resulting in the composition error. This value measures the discrepancy between the predicted and observed phrase ratings, thus gauging the accuracy of our compositional predictions.

3.4 Dominant Composition Operation

To further explore the preference for composition operations, we identified dominant composition operations for each relation-feature pair. If one composition operation outperformed the other three in more than half of the cases, we deemed this operation the dominant composition operation. For instance, when using scalar noun phrases for vision features, observing a significantly lower composition error with addition, compared to other composition functions, suggests that scalar-noun phrases predominantly utilize addition as their composition operation when addressing vision features.

This method enhances our understanding of how phrasal meanings are composed within the experiential semantic space.

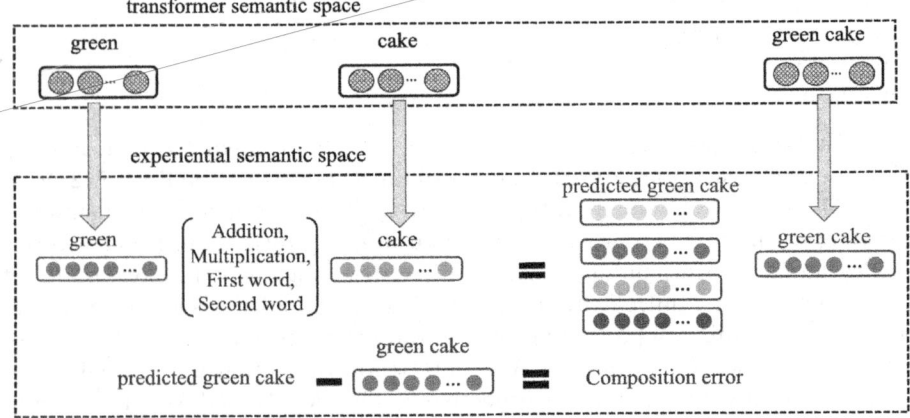

Fig. 3. Conceptual composition quantification method: composition error that assesses the difficulty of conceptual composition by measuring the disparity between predicted and actual phrase ratings.

4 Experimental Settings

4.1 Dataset

We used the 36 unique phrases for each of the 6 phrasal types created in [17]: ScalarNoun (e.g., 'small train'), IntersectiveNoun (e.g., 'dusty train'), VerbNoun (e.g., 'stop trains'), HasNounNoun (e.g., 'power train'), ForNounNoun (e.g., 'tourist train'), and MadeOfNounNoun (e.g., 'iron train'). This resulted in a total of 216 phrases. Each phrase type contains 13 different first words, appearing 1 to 5 times. The second noun in these phrases belonged to one of 6 categories: food-plant, furniture, human, place, vehicle, and tool. Each phrase and it component words are rated by the 65-dimensional componential semantic vectors.

4.2 Models

We evaluated four popular models from two distinct families: 1) The BERT Family, which includes BERT, BERT-large, and distillBERT, all utilizing bidirectional learning techniques. 2) The GPT Family, featuring models like GPT-2 that employ autoregressive training, ideal for sequential text generation tasks that require creativity.

All models used in this study are sourced from OpenAI's pretrained models in the Transformers library[2].

---

[2] https://huggingface.co/transformers/v3.3.1/pretrained_models.html.

4.3 Data Preprocessing

There is a total of 390 relation-feature pairs by combining 6 relations and 65 features (e.g., Scalar-Vision, Scalar-Bright, Intersective-Vision), with each relation-feature pair encompassing 36 phrases. We filtered out pairs if more than half of the phrases exhibited scores lower than 1 on the given feature for either word (W1 or W2), since overall low scores indicates irrelevance of that feature to the concepts in question. For instance, the relation-feature pair ScalarNoun-Color was excluded as more than half of the ScalarNoun phrases received ratings below 1 on the Color feature, reflecting the fact that color is not a relevant dimension to many of the scalar adjectives such as big, hot, soft, long, new, heavy. Instead, we focused on understanding how words combine, concentrating on relation-feature pairs where both words share the relevant feature above a certain degree. Consequently, pairs like the one mentioned are filtered out in our analysis. This process resulted in 132, 137, 135, and 134 relation-feature pairs for GPT-2, BERT, BERT-large, and distillBERT, respectively. Subsequently, we assessed their compositional suitability. If one composition operation outperformed the other three in more than half of the cases, we deemed this operation the dominant composition operation (as defined in Sect. 3.4), i.e., an operation that produces the closest match between predicted and observed feature values.

5 Results

5.1 Properties Encoded in Representations

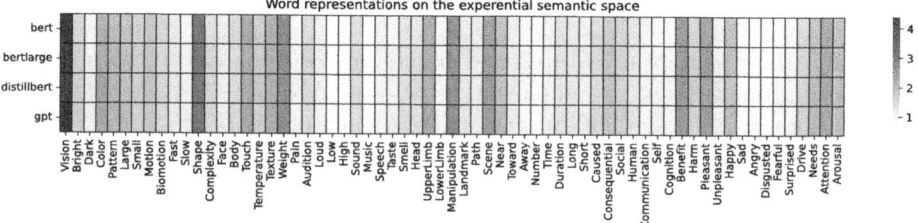

Fig. 4. Semantic properties encoded in word representations across different models.

We first examined the average ratings of the word representations mapped in the experiential semantic space, analyzing 107 words from various models. As depicted in Fig. 4, all models scored similarly. Previous experiments revealed distinctions in the original transformer space [8]; however, these differences diminished once mapped to the experiential semantic space. This observation aligns with the findings reported in [4]. In this space, the vision feature scored the highest, while semantic features such as Low and High scored the lowest. This result is logical given that the majority of the words evaluated are either color descriptors or concrete nouns.

We analyzed the average ratings of two-word phrase representations in the experiential semantic space, evaluating 216 phrase stimuli from various model layers. As depicted in Fig. 5, the ratings across different layers are notably similar, mirroring the trend observed with word representations where the vision model scored highest. These findings suggest that the representations generated by the transformer-based model encode similar information, indicating that they are indistinguishable from one another within the experiential semantic space.

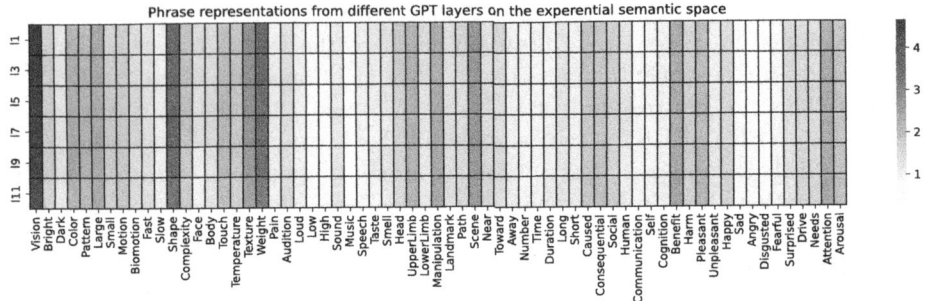

Fig. 5. Semantic properties encoded in two-word phrase representations in GPT2 model across different layers.

5.2 Significant Interaction Effects on Conceptual Composition

To explore how semantic features and linguistic relations affect the choice of composition operations, we employed a two-way ANOVA. We calculated the composition error (as detailed in Sect. 3.3) for each of the 65 semantic features across six types of linguistic relation phrases, using this error as the dependent variable. Both semantic features and linguistic relations were treated as independent variables.

The results, summarized in Table 1 and derived from the GPT model (similar outcomes were observed with other models, hence they are not included), demonstrated highly significant interaction effects ($p < 0.001$) between these variables across all four composition operations. These findings highlight the substantial role of both linguistic relations and semantic features in how transformers process and combine meanings (Fig. 6).

To further investigate the interactions, we computed the composition errors for linguistic relations, averaging across the four composition operations. As shown in Fig. 4, unlike the human results where VerbNoun phrases demonstrated significantly larger errors compared to other phrase types-indicating a poor fit of

Table 1. Results of the two-way ANOVA using the GPT model in which F is F statistic and p denotes p-value.

	Addition		Multiplication		Word1		Word2	
	F	p	F	p	F	p	F	p
Feature x Relation	2.636	< 0.001	4.946	< 0.001	5.315	< 0.001	1.977	< 0.001

the composition operations to these phrases, likely due to their increased complexity [17]-we observed no significant differences between the various linguistic relations in transformer-based models[3].

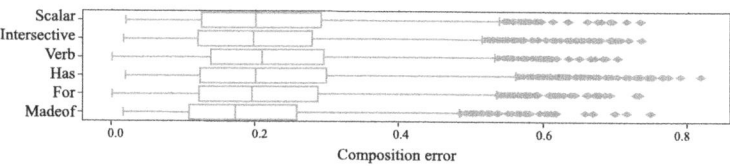

Fig. 6. Average Composition Errors in Linguistic Relations using the GPT model.

5.3 Preferred Composition Operations for Individual Features and Relations

The previous analysis has demonstrated significant interactions between semantic features and linguistic relationships, influencing the effectiveness of various composition operations on our dataset. However, it remains unclear which specific composition operations are preferred by different semantic features and linguistic relationships, and what factors drive these preferences. To explore these questions, we identified the primary composition operations for each relation-feature pair, as detailed in Sect. 3.4. Figure 7 illustrates that, among the 130+ relation-feature pairs we examined in GPT and BERT-based models, only a minority displayed a dominant composition operation. This indicates that most phrase pairs utilize multiple operations to integrate meanings effectively. Specifically, for the GPT model, most pairs predominantly employ either 'Word1' or 'Word2' as the composition method. Meanwhile, the BERT model frequently uses 'Word1', 'Word2', and 'multiplication' to merge word meanings across various BERT model types, suggesting a consistent trend.

To assess generalizability, we used the t-SNE method to visualize relation-feature pairs in two dimensions, focusing on composition errors. This visualization helped us intuitively see the variations across different layers. Figure 8

---

[3] Results from BERT, BERT-large, and DistilBERT were consistent and are therefore not included here.

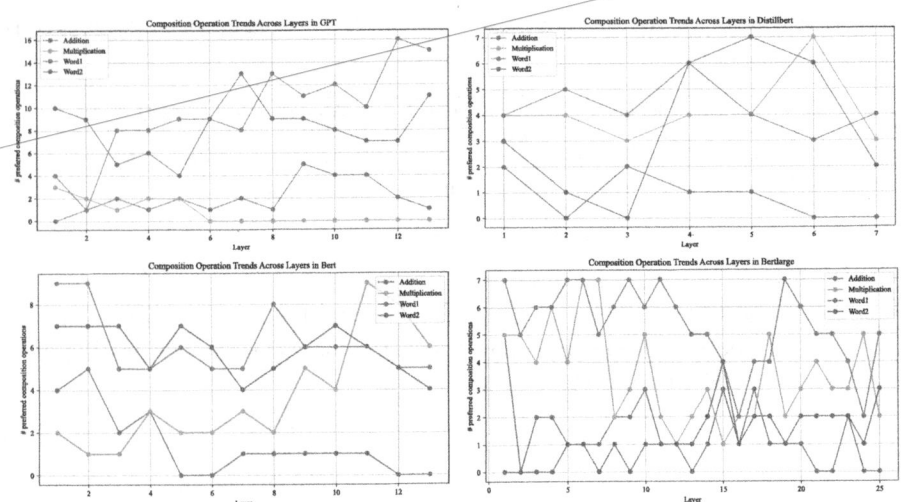

Fig. 7. Dominant composition operation trends across layers in transformer-based models.

illustrates that as the number of layers increases, the emphasis on Multiplication decreases, while the presence of 'Word1' and 'Word2' becomes more pronounced. Initially, no clear patterns emerge in Layer 1, but by Layer 13, 'Word1' is predominantly found in verb-noun and noun-noun combinations, and 'Word2' is frequently seen in adjective-noun pairings. Notably, these layers show far fewer prevalent composition operations than those typically found in human data observed in [17].

In summary, the clustering results highlight the prominent divisions of linguistic relation (i.e., AdjNoun/NounNoun versus VerbNoun) and composition operation (i.e., Word1/Word2 versus Multiplication). Like humans, phrase types and semantic features influence the combination process in transformers. However, unlike humans, most phrase types paired with semantic features require more complex combination operations than simple addition or multiplication.

6 Related Work

6.1 Investigation of Word Representations

Numerous studies have focused on interpreting word representations, with most research examining the internal characteristics of semantic representations by correlating them with linguistic features [6,14,15]. Additionally, Wang et al. [19] and Chersoni et al. [4] have investigated the semantic properties encoded in word embeddings by mapping them onto interpretable vectors. These vectors comprise explicit and neurobiologically motivated semantic features [2], with the former study evaluating multimodal word representations and the latter analyzing differences between various types of word embeddings. Building on

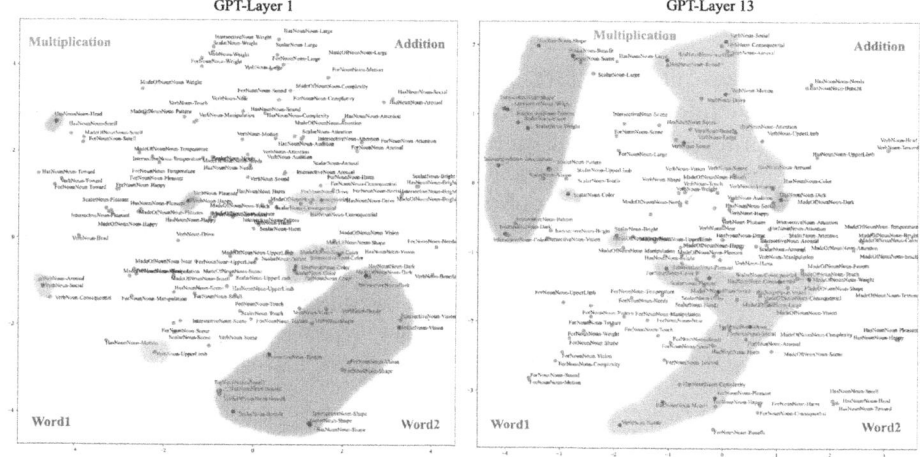

Fig. 8. t-SNE visualization of relation-feature pairs based on composition errors. The green, red, blue, and purple dots represent relation-feature pairs where Multiplication, Word1 Addition, Word2 Addition, and Multiplication are the dominant composition operations, respectively. The gray color dots represent relation-feature pairs without dominant composition operations. (Color figure online)

these foundations, our work aims to explore how different properties are encoded in word representations and, more importantly, how compositions are processed in this interpretable semantic space.

6.2 Investigation of Semantic Compositionality

Semantic compositionality has been addressed by various composition models [5,12,20,21]. However, the dimensions in many semantic vector spaces lack clear meanings, making it challenging to interpret how different composition models function. Fyshe et al. [7] addressed this issue by employing sparse vector spaces, using the intruder task to measure the interpretability of semantic dimensions. This approach requires manual labeling, and the outcomes are often not intuitive. Furthermore, Wang et al. [19] explored the interpretable experiential semantic space and visualized differences between word and phrase representations. Recent work by Yu and Ettinger [22] analyzes phrase representations in advanced pre-trained transformers. They used tests based on human evaluations of phrase similarity and meaning shifts to distinguish between lexical and compositional effects, adjusting for word overlap. Their findings reveal that these models' phrase representations rely primarily on individual words, with limited evidence of complex compositional understanding. Despite these advances, there remains a lack of quantitative results that illustrate the composition process in two-word combinations, which our study seeks to address by introducing a new methodological framework that combines quantitative analysis with cognitive linguistic theory.

7 Conclusion and Future Work

This paper utilizes interpretable, brain-based componential semantics in two-word phrases to explore how transformer models integrate meaning, in comparison to human processes. Drawing on findings from [17], we observe that, for most phrases, machines employ complex compositional operations that vary across different models. Additionally, similar information encoded across different layers indicates information loss when projecting transformer representations into experiential semantic space. Similar to humans, the type of phrase and semantic features influence how transformers combine meanings. However, in contrast to human methods, most combinations of phrase types and semantic features in transformers require more intricate operations than mere addition or multiplication.

References

1. Biever, C.: ChatGPT broke the Turing test-the race is on for new ways to assess AI. Nature **619**(7971), 686–689 (2023)
2. Binder, J.R., et al.: Toward a brain-based componential semantic representation. Cogn. Neuropsychol. **33**(3–4), 130–174 (2016)
3. Binder, J.R., Desai, R.H.: The neurobiology of semantic memory. Trends Cogn. Sci. **15**(11), 527–536 (2011)
4. Chersoni, E., Santus, E., Huang, C.R., Lenci, A., et al.: Decoding word embeddings with brain-based semantic features. Comput. Linguist. **47**(3), 663–698 (2021)
5. Dinu, G., et al.: General estimation and evaluation of compositional distributional semantic models. In: Proceedings of the Workshop on Continuous Vector Space Models and their Compositionality, pp. 50–58 (2013)
6. Eichler, M., Şahin, G.G., Gurevych, I.: LINSPECTOR WEB: a multilingual probing suite for word representations. arXiv preprint: arXiv:1907.11438 (2019)
7. Fyshe, A., Wehbe, L., Talukdar, P., Murphy, B., Mitchell, T.: A compositional and interpretable semantic space. In: Proceedings of the 2015 Conference of the North American Chapter of the Association for Computational Linguistics: Human Language Technologies, pp. 32–41 (2015)
8. Jawahar, G., Sagot, B., Seddah, D.: What does BERT learn about the structure of language? In: ACL 2019-57th Annual Meeting of the Association for Computational Linguistics (2019)
9. Kim, S., Pylkkänen, L.: Composition of event concepts: evidence for distinct roles for the left and right anterior temporal lobes. Brain Lang. **188**, 18–27 (2019)
10. Li, C., Wang, S., Zhang, Y., Zhang, J., Zong, C.: Interpreting and exploiting functional specialization in multi-head attention under multi-task learning. arXiv preprint: arXiv:2310.10318 (2023)
11. Lin, N., Zhang, X., Wang, X., Wang, S.: The organization of the semantic network as reflected by the neural correlates of six semantic dimensions. Brain Lang. **250**, 105388 (2024)
12. Mitchell, J., Lapata, M.: Composition in distributional models of semantics. Cogn. Sci. **34**(8), 1388–1429 (2010)
13. Pylkkänen, L.: The neural basis of combinatory syntax and semantics. Science **366**(6461), 62–66 (2019)

14. Qian, P., Qiu, X., Huang, X.J.: Investigating language universal and specific properties in word embeddings. In: Proceedings of the 54th Annual Meeting of the Association for Computational Linguistics (Volume 1: Long Papers), pp. 1478–1488 (2016)
15. Tsvetkov, Y., Faruqui, M., Ling, W., Lample, G., Dyer, C.: Evaluation of word vector representations by subspace alignment. In: Proceedings of the 2015 Conference on Empirical Methods in Natural Language Processing, pp. 2049–2054 (2015)
16. Vaswani, A., et al.: Attention is all you need. In: Advances in Neural Information Processing Systems, vol. 30 (2017)
17. Wang, S., Kim, S., Binder, J.R., Pylkkänen, L.: Unlocking the complexity of phrasal composition: an interplay between semantic features and linguistic relations
18. Wang, S., Sun, J., Zhang, Y., Lin, N., Moens, M.F., Zong, C.: Computational models to study language processing in the human brain: a survey. arXiv preprint: arXiv:2403.13368 (2024)
19. Wang, S., Zhang, J., Lin, N., Zong, C.: Investigating inner properties of multimodal representation and semantic compositionality with brain-based componential semantics. In: Proceedings of the AAAI Conference on Artificial Intelligence, vol. 32 (2018)
20. Wang, S., Zhang, J., Zong, C.: Exploiting word internal structures for generic Chinese sentence representation. In: Proceedings of the 2017 Conference on Empirical Methods in Natural Language Processing, pp. 298–303 (2017)
21. Wang, S., Zong, C.: Comparison study on critical components in composition model for phrase representation. ACM Trans. Asian Low-Res. Lang. Inf. Process. (TALLIP) **16**(3), 1–25 (2017)
22. Yu, L., Ettinger, A.: Assessing phrasal representation and composition in transformers. In: Proceedings of the 2020 Conference on Empirical Methods in Natural Language Processing (EMNLP), pp. 4896–4907 (2020)

Author Index

C
Calhoun, Vince D. 120, 265, 287
Cao, Siyuan 120, 265
Chang, Liang 240
Che, Wenxin 368
Chen, Xuyang 105
Cheng, Ran 192
Chu, Congying 3

D
Du, Penghui 368
Du, Xinxin 51
Du, Zonghan 312

F
Fan, Lingzhong 3
Feng, Tao 130
Fu, Zening 120, 287

G
Gan, Anan 240
Gao, Ci-Jun 67
Gao, Shuzhan 120
Guo, Qinghai 192
Guo, Xinyue 51

H
Han, Yaning 181
Han, Yuyang 331
He, Guohang 192
He, Xingye 240
Hei, Xinhong 222
Heinke, Dietmar 18
Huang, Liwei 161
Huang, Rihan 368
Huang, Ruey-Song 67
Huang, Weidong 382
Huang, Yongjian 359

I
Ieong, Chio-In 207

J
Ji, Wenqi 51
Jia, Gufeng 67
Jia, Tianyuan 331
Jian, Xinyao 35
Jiang, Rongtao 120, 265, 287
Jiang, Zhiwei 181
Ju, Furong 181

K
Kang, Dong Woo 297
Ke, Huihang 130
Kim, Mansu 297

L
Lai, Zhongyuan 312
Leng, Luziwei 192
Li, Deying 3
Li, Dongyang 250
Li, Dongzhe 275
Li, Wendu 35
Li, Xiuxing 331
Liang, Chuang 120, 265, 287
Liang, Zhichao 35, 144
Liao, Jianxing 192
Lim, Hyun Kook 297
Lin, Jingsong 275
Lin, Jingzhe 144
Liu, Baijing 382
Liu, Fang 51
Liu, Longzhao 89
Liu, Niqi 51
Liu, Quanying 18, 35, 105, 144, 181, 250, 344, 368
Liu, Xucheng 67
Liu, Yong-Jin 51
Lu, Xiaofeng 222
Luo, Jing 222

© The Editor(s) (if applicable) and The Author(s), under exclusive license to Springer Nature Singapore Pte Ltd. 2025
Q. Liu (Ed.): HBAI 2024, CCIS 2438, pp. 421–422, 2025.
https://doi.org/10.1007/978-981-96-4001-0

M
Ma, Zhengyu 161
Mao, Qi 222

P
Pan, Dan 382
Pan, Guandong 89
Peng, Kaining 35

Q
Qi, Shile 120, 265, 287
Qian, Fuyuan 344
Qu, Youzhi 35, 368

R
Rong, Fenqi 240

S
Shao, Wei 287
Shen, Xinke 105
Shi, Weiwei 222
Shi, Zhenghao 222
Song, Xuan 192
Sun, Fangling 287
Sun, Weiting 144

T
Tang, Shaoting 89
Tao, Wei 67, 207
Tian, Yonghong 161

W
Wan, Feng 67, 207
Wang, Jialin 144
Wang, Liping 181
Wang, Qixin 331
Wang, Shaonan 407
Wang, Song 250
Wang, Xiaofan 222
Wang, Xin 89
Wang, Yiwen 396
Wei, Chen 18
Wei, Pengfei 181
Wen, Xuyun 120
Wu, Chao 130
Wu, Haiyan 35, 67, 207

Wu, Lei 120
Wu, Simin 359
Wu, Xia 331
Wu, Zhaobang 89

X
Xia, Junfeng 35
Xie, Yuebin 275
Xu, Min 396
Xu, Xijia 120
Xu, Xin 105

Y
Yang, Aolei 240
Yang, Banghua 240
Yang, Baoyao 275
Yang, Hongzhe 67
Yang, Yaqian 89
Yao, Hui 130
Ye, Jiawo 207
Ye, Ziyuan 368
Youn, Jiwon 297
Yu, Liutao 161
Yu, Minjing 51

Z
Zeng, An 382
Zhan, Ruichao 250
Zhan, Weide 275
Zhang, Dan 105
Zhang, Daoqiang 120, 265, 287
Zhang, Han 161
Zhang, Menglong 344
Zhang, Yinuo 144
Zhang, Yonghuai 240
Zhang, Zheng 192
Zhao, Guanyi 144
Zhao, Guozhen 51
Zhao, Qiande 3
Zheng, Zhiming 89
Zhou, Chao 51
Zhou, Chenlin 161
Zhou, Huihui 161
Zhou, Zhou 192
Zhu, Qi 265
Zou, Jiachen 18

Made in the USA
Monee, IL
03 May 2026

49438529R00240